W9-BFE-036

3 1611 00200 1995

Ecological Studies, Vol. 82

Analysis and Synthesis

Edited by

M.M. Caldwell, Logan, USA
G. Heldmaier, Marburg, Germany
O.L. Lange, Würzburg, Germany
H.A. Mooney, Stanford, USA
E.-D. Schulze, Bayreuth, Germany
U. Sommer, Kiel, Germany

Ecological Studies

Volumes published since 1995 are listed at the end of this book.

Springer
New York
Berlin
Heidelberg
Barcelona
Hong Kong
London
Milan
Paris
Singapore
Tokyo

Monica G. Turner
Robert H. Gardner
Editors

Quantitative Methods in Landscape Ecology
The Analysis and Interpretation of Landscape Heterogeneity

With 108 Illustrations

 Springer

GOVERNORS STATE UNIVERSITY
UNIVERSITY PARK
IL 60466

C. Bergstrom, Walt Conley, Valerie Cullinan, Virginia H. Dale, Janet Franklin, James R. Gosz, Robin L. Graham, Richard A. Houghton, Alan R. Johnson, Carol Johnston, Donald W. Kaufman, William K. Lauenroth, Simon A. Levin, Thomas Lillesand, Kamlesh Lulla, Mark MacKenzie, Vernon Meentemeyer, William K. Michener, Redwood Nero, Richard Park, John Pastor, W. Mac Post, Marguerite M. Remillard, Steven W. Running, George Sugihara, John Thomas, Donald E. Weller, and John A. Wiens.

Finally, special thanks are due to Michael Turner and Sandi Gardner, whose patience, encouragement, and assistance were invaluable to us.

Monica G. Turner
Robert H. Gardner

Contents

Contributors

Bartell, Steven M.

Environmental Sciences Division, Oak Ridge National Laboratory, Oak Ridge, Tennessee 37831 USA

Brenkert, Antoinette L.

Science Applications International Corporation, Oak Ridge, Tennessee 37830 USA

Conley, Marsha R.

Department of Mathematical Sciences, New Mexico State University, Las Cruces, New Mexico 88003 USA

Conley, Walt

Department of Biology, New Mexico State University, Las Cruces, New Mexico 88003 USA

Costanza, Robert

Chesapeake Biological Laboratory, University of Maryland, Solomons, Maryland 20688-0038 USA

Coulson, Robert N.

Department of Entomology, Texas A&M University, College Station, Texas 77843 USA

Dale, Virginia H.   Environmental Sciences Division, Oak
                                 Ridge National Laboratory, Oak Ridge,
                                 Tennessee 37831 USA

DeAngelis, Donald L.   Environmental Sciences Division, Oak
                                 Ridge National Laboratory, Oak Ridge,
                                 Tennessee 37831 USA

Dunn, Christopher P.   Department of Biology, Ball State Uni-
                                 versity, Muncie, Indiana 43706-0440
                                 USA

Fahrig, Lenore   Northwest Atlantic Fisheries Centre, Sci-
                                 ence Branch, P.O. Box 5667, St. John's,
                                 Newfoundland, Canada A1C 5X1

Flamm, Richard O.   Department of Entomology, Texas A&M
                                 University, College Station, Texas 77843
                                 USA

Gardner, Robert H.   Environmental Sciences Division, Oak
                                 Ridge National Laboratory, Oak Ridge,
                                 Tennessee 37831 USA

Giblin, Anne E.   The Ecosystems Center, Marine Biolog-
                                 ical Laboratory, Woods Hole, Massachu-
                                 setts 02543 USA

Grover, Herbert D.   Department of Biology, University of
                                 New Mexico, Albuquerque, New Mexico
                                 87131 USA

Guntenspergen, Glenn R.   Department of Botany, Louisiana State
                                 University, Baton Rouge, Louisiana
                                 70803-1705 USA

Henein, Kringen   Ottawa-Carleton Institute of Biology,
                                 Department of Biology, Carleton Uni-
                                 versity, Ottawa, Canada K1S 5B6

Holbo, H. Richard   Department of Forest Products, Forest
                                 Resources Laboratory, Oregon State
                                 University, Corvallis, Oregon 97331
                                 USA

Humphries, Hope C.

Range Science Department, Colorado State University, Fort Collins, Colorado 80521 USA

Hyman, Jeffrey B.

Graduate Program in Ecology, University of Tennessee, Knoxville, Tennessee 37996 USA

King, Anthony W.

Environmental Sciences Division. Oak Ridge National Laboratory, Oak Ridge, Tennessee 37831 USA

Lovelady, Clark N.

Department of Entomology, Texas A&M University, College Station, Texas 77843 USA

Luvall, Jeffrey C.

National Aeronautics and Space Administration, John C. Stennis Space Center, Mississippi 39529 USA

McAninch, Jay B.

Minnesota Department of Natural Resources, Farmland Wildlife Research Group, Route 1, Box 181, Madelia, Minnesota 56062 USA

Merriam, Gray

Ottawa-Carleton Institute of Biology, Department of Biology, Carleton University, Ottawa, Canada K1S 5B6

Milne, Bruce T.

Department of Biology, University of New Mexico, Albuquerque, New Mexico 87131 USA

Musick, H. Brad

Department of Biology, University of New Mexico, Albuquerque, New Mexico 87131 USA

Nadelhoffer, Knute J.

The Ecosystems Center, Marine Biological Laboratory, Woods Hole, Massachusetts 02543 USA

O'Neill, Robert V.

Environmental Sciences Division, Oak Ridge National Laboratory, Oak Ridge, Tennessee 37831 USA

Parks, Peter J.

School of Forestry and Environmental
Studies, Duke University, Durham,
North Carolina 27706 USA

Pelletier, Ramona E.

National Aeronautics and Space Administration, John C. Stennis Space Center,
Mississippi 39539 USA

Quattrochi, Dale A.

National Aeronautics and Space Administration, John C. Stennis Space Center,
Mississippi 39529 USA

Rastetter, Edward B.

The Ecosystems Center, Marine Biological Laboratory, Woods Hole, Massachusetts 02543 USA

Saunders, Michael C.

Department of Entomology, Texas A&M
University, College Station, Texas 77843
USA

Sharpe, David M.

Department of Geography, Southern Illinois University, Carbondale, Illinois
62901 USA

Shaver, Gaius R.

The Ecosystems Center, Maine Biological Laboratory, Woods Hole, Massachusetts 02543 USA

Sklar, Fred H.

Baruch Marine Laboratory, P.O. Box
1630, University of South Carolina,
Georgetown, South Carolina 29442 USA

Spradling, Sharon L.

Department of Entomology, Texas A&M
University, College Station, Texas 77843
USA

Stearns, Forrest

Forestry Sciences Laboratory, North
Central Forest Experiment Station, U.S.
Forest Service, P.O. Box 898, Rhine-
lander, Wisconsin 54501 USA

Stuart-Smith, Kari

Ottawa-Carleton Institute of Biology,
Department of Biology, Carleton University, Ottawa, Canada K1S 5B6

Turner, Monica G.   Environmental Sciences Division, Oak
                                     Ridge National Laboratory, Oak Ridge,
                                     Tennessee 37831 USA

Turner, Sandra J.   Division of Wildlife and Ecology,
                                     CSIRO, P.O. Box 84, Lyneham, ACT
                                     2606, Australia

Yang, Zhao   Department of Geography, Southern Illi-
                                     nois University, Carbondale, Illinois
                                     62901 USA

1. Introduction and Concepts

1. Quantitative Methods in Landscape Ecology: An Introduction

Monica G. Turner and Robert H. Gardner

1.1 Introduction

Quantitative methods that link spatial patterns and ecological processes at broad spatial and temporal scales are needed both in basic ecological research and in applied environmental problems. Ecological processes such as plant succession, biodiversity, foraging patterns, predator-prey interactions, dispersal, nutrient dynamics, and the spread of disturbance all have important spatial components (Huffaker 1958; Holling 1966; May 1975; Peterjohn and Correll 1984; McNaughton 1985; Turner 1987; Senft et al. 1987; Burke 1989; Hardt and Forman 1989). However, the difficulty in analyzing these processes has often caused the spatial dynamics to be ignored. Recent developments in landscape ecology have reemphasized the important relationships between spatial patterns and many ecological processes (Turner 1989). In turn, this increased attention on spatial dynamics has highlighted the need for new quantitative methods that can analyze patterns, determine the importance of spatially explicit processes, and develop reliable landscape models.

The dramatic expansion of the spatial and temporal scales at which ecological problems must be considered has presented another difficult quantitative challenge. Ecologists now face a series of problems (e.g., acid precipitation, global carbon cycling, and climatic change) that requires landscape, continental, and even global levels of information and an understanding of both short- and long-term consequences. The nature of these problems has created new demands to understand the effects of spatial heterogeneity and to make broad-scale predictions by extrapolating from fine-scale measurements. Ecologists must also be able to predict responses that may not be manifested in the time frame of typical experimental studies. These challenges underscore the need for quantitative methods that can be applied to ecological data at multiple spatial-temporal scales.

New methods to analyze and interpret landscape heterogeneity and to simulate spatially explicit ecological processes have developed rapidly during the past decade. However, this development occurred so quickly that the array of quantitative methods is fragmented across a diverse literature and is not easily accessible. This volume is designed to synthesize and present much of this recent development, reflecting a diversity of approaches to the study of pattern and process. In this chapter, we define some key terms and concepts and introduce the remaining sections of the book.

1.2 Landscape Ecology

Landscape ecology emphasizes large areas and the ecological effects of the spatial patterning of ecosystems. Specifically, it considers (1) the development and dynamics of spatial heterogeneity, (2) interactions and exchanges across heterogeneous landscapes, (3) the influences of spatial heterogeneity on biotic and abiotic processes, and (4) the management of spatial heterogeneity (Risser et al. 1984). The consideration of spatial patterns distinguishes landscape ecology from traditional ecological studies, which frequently assume that systems are spatially homogeneous.

The term *landscape ecology* was first used by Troll (1939), arising from European traditions of regional geography and vegetation science (the historical development is reviewed in Naveh 1982 and Naveh and Lieberman 1984). Landscape ecology is well integrated into land-use planning and decision-making in Europe (e.g., Buchwald and Engelhart 1968; Ruzicka 1987; Ruzicka et al. 1988; Schreiber 1977; Van der Maarel 1978; Vink 1983). In Czechoslovakia, for example, landscape-level studies serve as a basis for determining the optimal uses of land across whole regions (Ruzicka et al. 1988). Landscape ecology is also developing along more theoretical avenues of research with an emphasis on ecological processes (Risser et al. 1984; Forman and Godron 1986; Turner 1987, 1989; Urban et al. 1987), and a variety of practical applications are being developed concurrently (e.g., Noss 1983; Forman 1986; Hayes et al. 1987; Joyce et al. 1987; Baker 1989).

Landscapes can be observed from many points of view, and ecological processes in landscapes can be studied at different spatial and temporal scales (Risser 1987). *Landscape* commonly refers to the landforms of a region in the aggregate (Webster's New Collegiate Dictionary 1980) or to the land surface and its associated habitats at scales of hectares to many square kilometers. Most simply, a landscape can be considered to be a spatially heterogeneous area. Three landscape characteristics that are especially useful to consider are structure, function, and change (Forman and Godron 1986). *Structure* refers to the spatial relationships between distinctive ecosystems, that is, the distribution of energy, materials, and species in relation to the sizes, shapes, numbers, kinds, and configurations of components. *Function* refers to the interactions between the spatial elements, that is, the flow of energy, materials, and organisms among the component ecosystems. *Change* refers to alteration in the structure and function of the ecological mosaic through time.

Recent theoretical developments in landscape ecology have emphasized the relationship between pattern and process and the effect that changes in spatial scale have on our ability to extrapolate information across scales. New insights into ecological dynamics have emerged from landscape studies and have led to hypotheses that can be tested in a diversity of systems and at many scales. Several studies have suggested that the landscape has critical thresholds at which ecological processes will show dramatic qualitative changes (e.g., Gardner et al. 1987; Krummel et al. 1987; O'Neill et al. 1988a; Turner et al. 1989a; Gosz and Sharpe 1989; Rosen 1989). For instance, the number or length of edges in a landscape changes rapidly near the critical threshold (Gardner et al. 1987), and this change may have important implications for species persistence. Habitat fragmentation may progress with little effect on a population until the critical pathways of connectivity are disrupted; then, a slight change near a critical threshold can have dramatic consequences for the persistence of the population. Similarly, the spread of disturbance across a landscape may be controlled by disturbance frequency when the habitat is below the critical threshold, but it may be controlled by disturbance intensity when the habitat is above the critical threshold (Turner et al. 1989a). Hypotheses regarding the existence and effects of critical thresholds in spatial patterns can now be tested through the use of a diversity of landscapes, processes, and scales.

Current theoretical research also suggests that different landscape indices may reflect processes operating at different scales (e.g., Krummel et al. 1987). The relationships among indices, processes, and scale need more study to provide understanding of the factors that create pattern and the ecological effects of changing patterns on processes. Broad-scale indices of landscape structure (e.g., O'Neill et al. 1988b) may provide an appropriate metric for monitoring regional ecological changes. Such applications are of particular importance because changes in broad-scale patterns (e.g., in response to global change) can be measured with remote-sensing technology, and an understanding of the pattern-process relationship will allow functional changes to be inferred.

A few variables may be adequate to predict many landscape patterns. The relative importance of parameters controlling ecological processes appears to vary with spatial scale (Meentemeyer and Box 1987; Meentemeyer 1989). Several studies suggest that, at the landscape level, only a few variables may be required to predict landscape patterns, the spread of disturbances, or ecosystem processes such as NPP or the distribution of soil organic matter. These observations could simplify the prediction of landscape dynamics if a significant amount of fine-scale variation can be incorporated into a few parameters. A better understanding of the parameters essential for predicting patterns at different scales is necessary (Turner et al. 1989b).

1.3 Definitions and Concepts of Space, Time, and Scale

The effects of spatial and temporal scale must be considered in landscape ecology (e.g., Meentemeyer and Box 1987; Urban et al. 1987; Turner et al. 1989b; Wiens 1989). Because landscapes are spatially heterogeneous areas (i.e., environmental mosaics), the structure, function, and change of landscapes are themselves scale dependent. The measurement of spatial pattern and heterogeneity is dependent upon the scale at which the measurements are made. For example, Gardner et al. (1987) demonstrated that the number, sizes, and shapes of patches in a landscape were dependent upon the linear dimension of the map. Observations of landscape function, such as the flow of organisms, also depend on scale. The scale at which humans perceive boundaries and patches in the landscape may have little relevance for numerous flows or fluxes. For example, if we are interested in a particular organism, we are unlikely to discern the elements of patch structure or dynamics that are important unless we adopt an organism-centered view of the environment (Wiens 1985). Similarly, abiotic processes such as gas fluxes may be controlled by spatial heterogeneity that is not intuitively obvious or visually apparent to a human observer (Gosz and Sharpe 1989). Finally, changes in landscape structure or function are scale dependent. For example, a dynamic landscape may exhibit a stable mosaic at one spatial scale but not at another. Thus, conclusions or inferences regarding landscape patterns and processes must be drawn with an acute awareness of scale.

The word *scale* is used in many contexts and often connotes different aspects of space and time. A common vocabulary and set of working definitions of scale-related concepts are necessary for discussion (Turner et al. 1989b). We follow the definitions listed in Table 1.1 and distinguish between scale and level of biotic organization. *Scale* refers to the spatial or temporal dimension (e.g., size of area or length of time), whereas *level of organization* refers to the place within some biotic hierarchy (e.g., organism, deme, population). We categorize scale as is commonly done in ecology (i.e., *fine scale* refers to minute resolution or small study area, and *broad scale* refers to coarse resolution or large study area) rather than use the cartographic scale of geography (i.e., *large scale* refers to small resolution). We also differentiate between grain and extent, both spatially and temporally (see also Turner et al. 1989b). *Grain* refers to the finest level of spatial or temporal resolution available within a given data set. *Extent* refers to the size of the study area or the duration of the study.

Discussing the extrapolation of information from one scale to another requires the use of several other terms. The term *extrapolation* refers to the process of estimating unknown values from a known set of conditions. Extrapolation can be used to transfer information from a landscape with one set of dimensions to a larger set of dimensions or from a reference landscape to another area of equal dimensions. However, the transformation is difficult if it exceeds some limit, or "critical threshold," at which there is an abrupt change in some quality, property, or phenomenon of the system. Scale transformations may also be performed by

Table 1.1 Definitions of Scale-Related Terminology and Concepts[a]

Term	Definition
Scale	The spatial or temporal dimension of an object or process, characterized by both grain and extent.
Level of organization	The place within a biotic hierarchy (e.g., organism, deme, population).
Cartographic scale	The degree of spatial reduction indicating the length used to represent a larger unit of measure; ratio of distance on a map to distance on the earth surface represented by the map, usually expressed in terms such as 1:10,000.
Resolution	Precision of measurement: grain size, if spatial.
Grain	The finest level of spatial resolution possible with a given data set; e.g., pixel size for raster data.
Extent	The size of the study area or the duration of time under consideration.
Extrapolate	To infer from known values; to estimate a value from conditions of the argument not used in the process of estimation; to transfer information (a) from one scale to another (either grain size or extent) or (b) from one system (or data set) to another system at the same scale.
Critical threshold	The point at which there is an abrupt change in a quality, property, or phenomenon.
Absolute scale	The actual distance, direction, shape, and geometry.
Relative scale	A transformation of absolute scale to a scale that describes the relative distance, direction, or geometry based on some functional relationship (e.g., the relative distance between two locations based on the effort required by an organism to move between them).

[a] Turner et al. (1989b).

using measures of "relative scale," in which units of absolute scale (e.g., distance or geometry) are changed to units based on some functional relationship (e.g., units representing the energy to move from one location to another) (see also Meentemeyer 1989).

1.4 Quantitative Methods in Landscape Ecology

Recent works have provided a good conceptual basis for landscape studies (Naveh and Lieberman 1984; Risser et al. 1984; Forman and Godron 1986; Urban et al. 1987; Turner 1989), addressing particular topics such as disturbance (Turner 1987) and presenting an international perspective on landscape ecology (e.g., Naveh and Lieberman 1984; Zonneveld and Forman 1990). However, many ecologists remain limited in their access to the recently developed suite of quantitative techniques that are geared toward landscape analyses. Landscape-level research requires new methods to quantify spatial patterns, compare landscapes, identify significant differences, and determine relationships of functional processes to landscape patterns. Models are necessary because of (1) the large spatial and temporal scales integral to landscape studies, (2) the many components and processes, and (3) the difficulties, particularly the time and expense, associated with large-scale experiments. However, a number of new developments have profoundly changed our approach to these problems. There have been significant advances in computer hardware, data handling and analysis, geographic information systems (GIS), remote sensing, and model development. The array of quantitative approaches to landscape ecology has not previously been assembled in a reference volume.

The objectives of this book are to describe and evaluate current quantitative methods in landscape ecology, providing ideas and sufficient detail for other scientists to apply the methods in their own research. This book does not attempt to provide a comprehensive treatment of remote sensing or GISs, although several chapters address these topics. Our emphasis (hence the structure of the book) is on the analysis and interpretation of landscape heterogeneity (Section 2) and model development and simulation (Section 3). However, these two sections are not mutually exclusive, and several chapters would be appropriately placed in either section. The final two chapters (Section 4) present methods of extrapolation across scales and synthesize the volume.

1.4.1 Analysis and Interpretation of Patterns in the Landscape

The identification of scales is an important component of pattern analysis. The application of spatial statistics to patterns and scales in landscape data is reviewed by S. Turner et al. (Chapter 2). S. Turner et al. compare and contrast a variety of techniques designed to identify the intrinsic spatial scales within a spatial data set. These approaches may be important for many ecological processes, particularly when the scale of heterogeneity differs from that which is easily perceived by humans.

Capabilities in remote sensing and GISs have been expanding rapidly and clearly have much to offer to landscape ecology. Remote sensing opportunities are described by Quattrochi and Pelletier (Chapter 3). This chapter introduces landscape ecologists to the kinds of data and analyses that are available and directs them toward more detailed treatments of specific topics. Remote sensing also offers new ways of conceptualizing pattern in the landscape. In Chapter 4, Musick and Grover examine the applications of texture measures that do not require an a priori definition of landscape elements (e.g., patches) but rather rely on attributes of the remote imagery. The effects of scale on the texture measures are also explored.

Landscape patterns can be quantified in relation to particular processes. For example, consideration of horizontal nutrient transport across a landscape requires an understanding of biogeochemical diversity. Shaver et al. (Chapter 5) describe an approach to developing a spatially explicit nutrient budget for a heterogeneous landscape in the arctic. Their approach views heterogeneity from the process level and allows the importance of spatial pattern for nutrient transport to be estimated. The heterogeneity of processes across large areas can also be measured by using remote sensing, as demonstrated by Luvall and Holbo (Chapter 6) for thermal characteristics of forested landscapes. Thermal characteristics are important in estimating water balance at the landscape scale and in extrapolating site-specific measures across heterogeneous landscapes. It is important to remember that the pattern or heterogeneity of processes may or may not correspond to the heterogeneity of the patches observed by a human.

GISs are receiving increasing use in landscape studies (e.g., Iverson 1988; Milne et al. 1989; Johnston and Bonde 1989; Burke et al. 1990). We have not attempted to provide a comprehensive introduction or description of GIS methods (but see Burrough 1986). Rather, some illustrative applications are included. In Chapter 7, Coulson et al. combine GISs with another new quantitative approach, artificial intelligence, to develop an intelligent geographic information system (IGIS). The IGIS approach offers great potential for land and resource management applications, such as the pest and grazing issues discussed in their chapter. In Chapter 8, Dunn et al. analyze the dynamics of an agricultural landscape by using remote sensing and a GIS. More generally, Dunn et al. consider the appropriate methods for studying temporal changes in landscapes.

The complexity of landscape patterns can also be studied by using fractals (Milne, Chapter 9). Fractals can be used to measure the spatial patterns of a diversity of quantities (e.g., patch mosaics, movement patterns, and density of organisms) that contribute to heterogeneity within a landscape. In addition, fractals offer a means of examining landscapes and ecological processes at a multitude of scales. Milne begins by presenting some exciting theoretical ideas regarding fractal applications, then proceeds to review the variety of applications of fractal geometry to landscape studies.

1.4.2 Model Development and Simulation

Prediction of landscape-level phenomena requires model development. Models are necessary for landscape studies because experiments frequently cannot be performed at the ideal spatial or temporal scale. The linking of models with GISs and remote sensing technologies has begun (e.g., Hall et al. 1988; Burke et al. 1989; Kesner and Meetemeyer 1989; Running et al. 1989), and functional models are being constructed.

The section begins with a review of spatial modeling approaches that are applicable to landscape studies (Sklar and Costanza, Chapter 10). These approaches include geographic models, fluid dynamics, population and ecosystem models, and spatial models that are stochastic and process oriented. The application of neutral models (*sensu* Caswell 1976) to landscape studies is then described by Gardner and O'Neill (Chapter 11). Use of a neutral model allows the development of a quantitative expectation for a particular phenomenon (e.g., the effect of landscape pattern on species abundance). This expectation can then be tested with actual landscapes and processes.

Landscape changes often result from socioeconomic factors, such as prices of agricultural or forestry products. These causal factors frequently are not incorporated into spatial models of changing landscape patterns. In Chapter 12, Parks examines ways in which economic considerations may be incorporated into landscape models, emphasizing descriptive, optimization, and econometric approaches to modeling. Parks then explores implementation of this linkage with a spatially explicit data set.

A variety of research questions and modeling approaches are described in the remaining chapters in this section. Turner and Dale (Chapter 13) examine the simulation of landscape disturbances and compare models that are spatially aggregated and spatially explicit. Rastetter (Chapter 14) presents a transect model in which interactive geomorphologic and vegetation processes on a barrier island are simulated. The dynamic spatial interactions are explored by simulating annual changes in vegetation, geomorphology, water table depth, and average groundwater salinity on a cross-sectional transect of a barrier island.

Nutrient cycling frequently has a spatial component, but the effect of spatial pattern on the horizontal flow of nutrients has only been considered recently (e.g., Peterjohn and Correll 1984; Kesner and Meentemeyer 1989). In Chapter 15, Bartell and Brenkert present a pixel-based spatial model of nutrient dynamics in a forested watershed. The model is hierarchically structured, allowing simulations to be conducted at different spatial scales and the importance of parameters at each scale to be evaluated.

Several chapters then address the interaction of organisms and landscape pattern. Merriam et al. (Chapter 16) examine models that simulate systems in which landscape elements interact in space and time with demographically and spatially dynamic populations. The metapopulation (Levins 1970) is emphasized, referring to a population of several subpopulations in scattered patches that are separated by unsuitable habitat. Subpopulations are potentially connected by the dispersal of individuals, which is simulated by using corridor, noncorridor, and pool models. In Chapter 17, Fahrig argues for a modeling approach in which the simplest possible simulation model is developed to address a specific question and a rigorously designed simulation experiment is conducted to identify hypotheses. General models of population dynamics in a patchy landscape and population response to disturbance are used to illustrate the approach. Hyman et al. (Chapter 18) develop an individual-based landscape model of herbivory at different spatial scales. The model simulates grazing by mammalian herbivores (voles and deer) in a landscape with patchy resources. Movement patterns, foraging, and social behaviors of individual herbivores are included, as are recruitment, growth, and damage of individual tree seedlings.

1.4.3 Synthesis

The final section of the book contains two chapters. Because scaling remains such an important research topic in landscape ecology, a chapter is devoted to methods to extrapolate results across scales (King, Chapter 19). A variety of scaling methods are presented and applied to several simple models, and their relative success is evaluated. The last chapter (Gardner and Turner, Chapter 20) synthesizes the approaches, methods, and results from the previous chapters and discusses future research directions in quantitative landscape ecology.

1.5 Summary

All ecological processes occur in a spatial context. The development of landscape ecology has reemphasized the importance of spatial pattern in constraining many ecological processes. In addition, the broad-scale environmental problems that presently challenge ecologists have created new demands to understand the effects of spatial heterogeneity and to extrapolate predictions across spatial and temporal scales. In this chapter, the development of landscape ecological theory is reviewed, and a set of definitions for space, time, and scale is presented. Landscape-level research requires new methods to quantify spatial patterns, compare landscapes, identify significant differences, and determine relationships of functional processes to landscape patterns. Recent studies have provided a sound conceptual basis for landscape studies, but many ecologists remain limited in their access to the recently developed suite of quantitative techniques that are geared toward landscape analyses. The objectives of this book are to describe and evaluate current quantitative methods in landscape ecology, providing ideas and sufficient detail for other scientists to apply the methods in their own research.

Acknowledgments

Preparation of this chapter was supported by the Ecological Research Division, Office of Health and Environmental Research, U.S. Department of Energy, under contract DE-AC05-84OR21400 with Martin Marietta Energy Systems, Inc. Publication No. 3539 of the Environmental Sciences Division, Oak Ridge National Laboratory.

References

Baker, W.L. 1989. Landscape ecology and nature reserve design in the Boundary Waters Canoe Area, Minnesota. *Ecology* 70:23–35.

Buchwald, K. and Engelhart, W., eds. 1968. *Handbuch fur Landschaftpflege und Naturschutz*. Bd. 1. Grundlagen. Munich: BLV Verlagsgesellschaft.

Burke, I.C. 1989. Control of nitrogen mineralization in a sagebrush steppe landscape. *Ecology* 70:1115–26.

Burke, I.C.; Schimel, D.S.; Yonker, C.M.; Parton, W.J.; Joyce, L.A.; and Lauenroth, W.K. 1990. Regional modeling of grassland biogeochemistry using GIS. *Landscape Ecology* 4:45–54.

Burrough, P.A. 1986. *Principles of Geographical Information Systems for Land Resources Assessment*. Oxford: Clarendon Press.

Caswell, H. 1976. Community structure: a neutral model analysis. *Ecological Monographs* 46:327–54.

Forman, R.T.T. 1986. Emerging directions in landscape ecology and applications in natural resource management. In *Proceedings of the Conference on Science in the National Parks*, eds. R. Herrmann and T. Bostedt-Craig, pp. 59–88. Washington, D.C.: George Wright Society.

Forman, R.T.T., and Godron, M. 1986. *Landscape Ecology*. New York: John Wiley and Sons.

Gardner, R.H.; Milne, B.T.; Turner, M.G.; and O'Neill, R.V. 1987. Neutral models for the analysis of broad-scale landscape pattern. *Landscape Ecology* 1:19–28.

Gosz, J.R., and Sharpe, P.J.H. 1989. Broad-scale concepts for interactions of climate, topography, and biota at biome transitions. *Landscape Ecology* 3:229–43.

Hall, F.G.; Strebel, D.E.; and Sellers, P.J. 1988. Linking knowledge among spatial and temporal scales: vegetation, atmosphere, climate and remote sensing. *Landscape Ecology* 2:3–22.

Hardt, R.A., and Forman, R.T.T. 1989. Boundary form effects on woody colonization of reclaimed surface mines. *Ecology* 70:1252–1260.

Hayes, T.D.; Riskind, D.H.; and Pace, W.L., III. 1987. Patch-within-patch restoration of man-modified landscapes within Texas state parks. In *Landscape Heterogeneity and Disturbance*, ed. M.G. Turner, pp. 173–98. New York: Springer-Verlag.

Holling, C.S. 1966. The functional response of invertebrate predators to prey density. *Memoirs of the Entomological Society of Canada* 48:1–85.

Huffaker, C.B. 1958. Experimental studies on predation: dispersion factors and predator-prey oscillations. *Hilgardia* 27:343–83.

Iverson, L.R. 1988. Land-use changes in Illinois, USA: the influence of landscape attributes on current and historic land use. *Landscape Ecology* 2:45–62.

Johnston, C.A., and Bonde, J. 1989. Quantitative analysis of ecotones using a geographic information system. *Photogrammetric Engineering and Remote Sensing* 55:1643–47.

Joyce, L.A.; Hoekstra, T.W.; and Alig, R.J. 1987. Regional multiresource models in a national framework. *Environment Management* 10:761–71.

Kesner, B.T., and Meentemeyer, B. 1989. A regional analysis of total nitrogen in an agricultural landscape. *Landscape Ecology* 2:151–64.

Krummel, J.R.; Gardner, R.H.; Sugihara, G.; O'Neill, R.V.; and Coleman, P.R. 1987. Landscape patterns in a disturbed environment. *Oikos* 48:321–24.

Levins, R. 1970. Extinction. In *Some Mathematical Questions in Biology. Lectures on Mathematics in the Life Sciences*, vol. 2, ed. M. Gerstenhaber, pp. 77–107. Providence, R.I.: American Mathematical Society.

McNaughton, S.J. 1985. Ecology of a grazing ecosystem: the Serengeti. *Ecological Monographs* 55:259–94.

May, R.M. 1975. Patterns of species abundance and diversity. In *Ecology and the Evolution of Communities*, eds. M.L. Cody and J.M. Diamond, pp. 81–120. Cambridge, Mass.: Belknap Press.

Meentemeyer, V. 1989. Geographical perspectives of space, time, and scale. *Landscape Ecology* 3:163–73.

Meentemeyer, V., and Box, E.O. 1987. Scale effects in landscape studies. In *Landscape Heterogeneity and Disturbance*, ed. M.G. Turner, pp. 15–36. New York: Springer-Verlag.

Milne, B.T.; Johnston, K.M.; Forman, R.T.T. 1989. Scale-dependent proximity of wildlife habitat in a spatially-neutral Bayesian model. *Landscape Ecology* 2:101–10.

Naveh, Z. 1982. Landscape ecology as an emerging branch of human ecosystem science. *Advances in Ecological Research* 12:189–237.

Naveh, Z., and Lieberman, A.S. 1984. *Landscape Ecology, Theory and Application*. New York: Springer-Verlag.

Noss, R.F. 1983. A regional landscape approach to maintain diversity. *BioScience* 33:700–6.

O'Neill, R.V.; Milne, B.T.; Turner, M.G.; and Gardner, R.H. 1988. Resource utilization scales and landscape pattern. *Landscape Ecology* 2:63–9.

O'Neill, R.V.; Krummel, J.R.; Gardner, R.H.; Sugihara, G.; Jackson, B.; DeAngelis, D.L.; Milne, B.T.; Turner, M.G.; Zygmunt, B.; Christensen, S.W.; Dale, V.H.; and Graham, R.L. 1988b. Indices of landscape pattern. *Landscape Ecology* 1:153–62.

Peterjohn, W.T., and Correll, D.L. 1984. Nutrient dynamics in an agricultural watershed: observations on the role of a riparian forest. *Ecology* 65:1466–75.

Risser, P.G. 1987. Landscape ecology: state-of-the-art. *In Landscape Heterogeneity and Disturbance*, ed. M.G. Turner, pp. 3–14. New York: Springer-Verlag.

Risser, P.G.; Karr, J.R.; and Forman, R.T.T. 1984. *Landscape ecology: directions and approaches*. Special Pub. No. 2. Illinois Natural History Survey, Champaign.

Rosen, R. 1989. Similitude, similarity, and scaling. *Landscape Ecology* 3:207–16.

Running, S.W.; Nemani, R.R.; Peterson, D.L.; Band, L.E.; Potts, D.F.; Pierce, L.L.; and Spanner, M.A. 1989. Mapping regional forest evapotranspiration and photosynthesis by coupling satellite data with ecosystem simulation. *Ecology* 70:1090–1101.

Ruzicka, M. 1987. Topical problems of landscape ecological research and planning. *Ekologia-CSSR* 5:233–38.

Ruzicka, M.; Hrnciarova, T.; and Miklos, L., eds. 1988. *Proceedings of the VIIIth International Symposium on Problems in Landscape Ecological Research*, Vol. 1. Institute of Experimental Biology and Ecology, CBES SAS, Bratislava, Czechoslovakia.

Schreiber, K.F. 1977. Landscape planning and protection of the environment: the contribution of landscape ecology. *Appl. Sci. Dev.* 9:128–39

Senft, R.L.; Coughenour, M.B.; Bailey, D.W.; Rittenhouse, L.R.; Sala, O.E.; and Swift, D.M. 1987. Large herbivore foraging and ecological hierarchies. *BioScience* 37:789–99.

Troll, C. 1939. Luftbildplan and ökologische Bodenforschung. *Z. Ges. Erdkunde*: 241–98, Berlin.

Turner, M.G. 1989. Landscape ecology: the effect of pattern on process. *Annual Review of Ecology and Systematics* 20:171–97.

Turner, M.G., ed. 1987. *Landscape Heterogeneity and Disturbance*. New York: Springer-Verlag.

Turner, M.G.; Gardner, R.H.; Dale, V.H.; and O'Neill, R.V. 1989a. Predicting the spread of disturbance across a heterogeneous landscape. *Oikos* 55:121–29.

Turner, M.G.; Dale, V.H.; and Gardner, R.H. 1989b. Predicting across scales: theory development and testing. *Landscape Ecology* 3:245–52.

Urban, D.L.; O'Neill, R.V.; and Shugart, H.H. 1987. Landscape ecology. *BioScience* 37:119–27.

Van der Maarel, E. 1978. Ecological principles for physical planning. In *The Breakdown and Restoration of Ecosystems*, eds. M.W. Holgate and M.J. Woodman, pp. 413–50. NATO Conference Series 1 (Ecology), Vol. 3. New York: Plenum Press.

Vink, A.P.A. 1983. *Landscape Ecology and Land Use*. London: Longman.

Wiens, J.A. 1985. Vertebrate responses to environmental patchiness in arid and semiarid ecosystems. In *The Ecology of Natural Disturbance and Patch Dynamics*, eds. S.T.A. Pickett and P.S. White, pp. 169–93. New York: Academic Press.

Wiens, J.A. 1989. Spatial scaling in ecology. *Functional Ecology* 3:385–97.

Zonneveld, I.S., and Forman, R.T.T., eds. 1990. *Changing Landscapes: An Ecological Perspective*. New York: Springer-Verlag.

2. Analysis and Interpretation of Patterns in the Landscape

2. Pattern and Scale: Statistics for Landscape Ecology

Sandra J. Turner, Robert V. O'Neill, Walt Conley,
Marsha R. Conley, and Hope C. Humphries

2.1 Introduction

The current interest in landscape ecology (Risser et al. 1984; Forman and Godron 1986; Naveh and Lieberman 1984) represents a renewed awareness of the effects of spatial pattern on ecological processes. This awareness reinforces decades of research in the spatial patterning of plant communities (McIntosh 1985). For example, the insightful analysis of A. S. Watt (1947) clearly indicates the important relationships that exist between pattern and process.

Fully acknowledging its roots in the past, the burgeoning field of landscape ecology differs greatly from vegetation analysis. Physical, biological, and anthropogenic elements interact, forming heterogeneous mosaics of patches. Landscapes are often thought of in terms of patches and their characteristics. Patch descriptors (Forman and Godron 1986), such as shape, degree of isolation, accessibility, interaction, and dispersion, refer to spatial relationships defined by location. These describe the structure of the landscape. Landscape structure, rather than the functional interactions or the change in the structure through time, is the focus of this chapter.

The landscape view also emphasizes the importance of scale (e.g., O'Neill 1989; Wiens, in press). Developments in hierarchy theory (Allen and Starr 1982; O'Neill et al. 1986) demonstrate how processes and constraints change across scales. Thus, the scale at which the landscape exhibits patchiness is important for understanding ecological processes (Wiens 1986; Maurer 1985; Steele 1985).

The emphasis on scale poses statistical problems for landscape ecology. Patch definition is often accomplished through an arbitrary selection process. As the concept of landscape heterogeneity is extended beyond geography, geomorphology, and plant communities to levels where what constitutes a "patch" becomes less obvious visually, quantitative methods assume prime importance. It is not a trivial task to extract from the literature those techniques appropriate for detecting the scale of landscape pattern. The purpose of this chapter, therefore, is to present a review of statistical methods available for the detection of scale in landscape data.

This chapter is intended for ecologists. The intent is to provide a road map of available scale-detection methods, permitting landscape ecologists to determine the techniques appropriate for their problem. To this end, we have minimized mathematical detail and attempted to communicate the advantages and disadvantages of each technique with intuitive arguments and illustrative examples. For each technique, we reference the statistical literature where mathematical detail can be found. This literature also provides the reader with detail related to characteristics of spatial data sets that are beyond the scope of this chapter.

To simplify presentation of the diverse methods, we have divided the chapter into two major sections. The first section treats methods appropriate for landscapes that exhibit regularly repeating pattern. These techniques are based on variance measures and are generally univariate in nature. The second section deals with landscape pattern that varies in an irregular manner. Patterns may not repeat on a landscape. These techniques focus on the detection of edges and are generally multivariate. A third section provides the reader with a brief summary and a table of attributes for the methods discussed in the first two sections.

2.2 Landscapes that Exhibit Repeated Pattern

The first set of techniques shares a common assumption: the landscape being sampled exhibits regular patterning: that is, regardless of whether the structure of the pattern is homogenous or heterogenous, there is some pattern on the landscape which is repeated over and over again in a regular way. For instance, in a grassland where several species of grasses grow in ring form, the rings may be of similar size (possibly under abiotic control), but the species may be different (data from S. J. Turner). The pattern is regular. The species composition of the pattern is heterogenous.

The techniques in this first section are not useful for landscapes that exhibit patterns that are not regular. For instance, patches may well be distinct, but their size or distribution in space may be quite irregular. Or conversely the pattern may be continually changing as it does along an altitude gradient.

Some of the techniques in this first section are useful for detection of pattern that changes in space as long as the change occurs in a regular manner. Some approaches, such as spectral analysis, can remove the influence of simple changes, e.g., linear trends, in the data. Nevertheless, all of the methods detect the distance or area at which the pattern tends to repeat itself. The dependence on repetition requires that the sampled landscape be sufficiently large that a given pattern can be repeated several times.

The following discussions are developed from vegetation analysis. They are, however, generally applicable to all types of landscapes, including those structured by anthropogenic influences.

2.2.1 Analysis of Data by Blocking

The search for a method to detect the scale of landscape patterns can begin with the blocking technique developed by Greig-Smith (1952). Greig-Smith's procedure requires systematic sampling of a grid of contiguous quadrats. Kershaw (1957) expanded the method to the use of line transects. The dimensions of the transect or each side of a grid must be a power of two, e.g., 2, 4, or 8 for a line and 16 by 16 or 32 by 64 for a grid. The number of individuals per quadrat may be counted in the traditional manner, but numerous abundance measures are appropriate, including dimensions of individuals, percent cover, frequency, or biomass.

The field data are blocked by combining adjacent pairs of quadrats to give two-quadrat blocks. Adjacent two-quadrat blocks are then combined into four-quadrat blocks. The blocking is iterated until there are two blocks, each containing one half of the total data set (Pielou 1977). Analysis of the data involves nested analysis of variance at different block sizes. The variance is partitioned between and within block size. Mean square variance is then plotted against block size. Different scales of pattern in the data appear as peaks or troughs at a block size corresponding to the mean area of a clump or unit of pattern. A peak represents a block size where adjacent blocks are dissimilar, while troughs represent sizes where adjacent blocks tend to be particularly similar (Goodall 1974).

Severe criticisms have been leveled at this technique. The result has been the development of a number of allied techniques that, at least in part, mitigate the faults. Together these techniques have a long history of application. They continue to be very useful. The following commentary is a compilation of critiques based on both field data and artificial data sets with known pattern (Pielou 1977; Ludwig 1979; Ripley 1978, 1981; Greig-Smith 1983; Kershaw and Looney 1985).

1. No statistical technique is available to test the significance of the peaks found in the mean square by block size plot. The variance estimate for a peak at a given block size is not independent of the variance calculated for the remainder of the blocks, because all data are used for every blocking. Ripley (1978) suggests that this difficulty is partially overcome by comparing plots from a set of parallel transects.

2. A set of problems arises from restrictions on the blocking strategy and secondarily from the choice of quadrat size. Blocks are constructed in a power-of-two series (2, 4, 8, 16), and pattern related to lengths not equal to a power of two would not be seen. Quadrat size is influential because of a power-of-two series would not detect a pattern of 60 cm if the sample unit were 10 cm (20 cm, 40 cm, 80 cm). But the pattern might be found if the sample unit were 8

cm (16 cm, 32 cm, 64 cm) (Ripley 1978). Block size is confounded with spacing. And because the center-to-center distance between adjacent data pairs is doubled at each second blocking, there is an increase in the variance/mean ratio with block size even in the absence of heterogeneity.

3. The degrees of freedom decline with increasing block size. For a transect 16,000 m long and a sample interval of 10 cm, the first blocked unit has 7,999 degrees of freedom, while the last has only 1. Obviously an increase in block size leads to a decrease in precision. Pattern suggested by peaks at the largest block sizes should be questioned.

4. The method is sensitive to the starting position of the analysis. For example, if the pattern has a period of about $2y$ times the quadrat size, then for some starting positions a large mean square will appear at block size y and for some other starting position the peak will be at $y/2$ (Ripley 1981). Sensitivity to starting position suggests that simple blocking is most useful for patterns that are not very complex. This particular problem is mitigated with a stepped blocking technique suggested by Usher (1975). The procedure averages a number of blocked-quadrat analyses with different starting positions that are obtained by stepping progressively along the transect. However, this technique still has the other problems described above.

2.2.1.1 Improved Blocking Techniques

The above criticisms represent serious restrictions on the utility of quadrat blocking despite its widespread use. Further development of the technique has led to a considerable improvement in precision. Hill (1973) proposed a modification that solved many problems of quadrat blocking for exploration of scale, and Goodall (1974) suggested a version that allows the statistical comparison of variance at different block sizes.

Hill (1973) proposed a technique called a "two-term local variance" that reduces the problems associated with starting position and can be done for any block size up to one-half the total number of quadrats. Calculation of the average variance over all block sizes up to the maximum possible eliminates problems with power-of-two sampling. Computation time is increased over the original blocking design. Quadrat size and spacing are still confounded in the variance estimate (Ludwig 1979).

Ludwig (1979), in a review based on artificial data with known scale, suggested that the two-term local variance method provided the most accurate evaluation of the existing pattern. He also noted that, when two scales of pattern exist, the technique emphasizes the pattern at the larger scale.

Two-term local variance is a useful technique for exploration of pattern in data because the estimates of variance are calculated for all possible block sizes up to one-half the total number of quadrats. The method will find scale in both artificial and field data although (1) quadrat and block size remain confounded; (2) the lack of independent variance estimates preclude hypothesis testing; and (3) a loss of degrees of freedom reduces confidence in the detection of scale at larger block sizes (Hill 1973; Ludwig 1979).

In considering Hill's method for identifying scale, remember that the systematic sampling method allows no missing data. A missed quadrat negates this method of analysis. Further, the relationship between peaks and troughs is complex and may be difficult to interpret (Ripley 1978). Finally, it is important to note that contiguous blocking samples the difference between the sums of observations over the two halves of the block divided by its length. The actual peak may represent the scale of a patch, the scale of the space between patches, or one-half the total scale of the pattern, that is, the patch plus the interpatch space (Errington 1973).

Goodall (1974) suggested selecting random pairs of quadrats for variance estimates. Paired quadrats are chosen at random from the set of all those in the data with a selected distance separation or spacing. Variance calculated from pairs of quadrats is plotted against separation distance, and the peaks have the traditional interpretation. Quadrat pairs are taken without replacement, which means independent estimates of variance for different spacings. Hypotheses about the difference in variance at two spacings may be tested. This technique overcomes the first criticism of the blocking methods.

The degrees of freedom for the variance estimates at all spacings cannot exceed one-half the total number of quadrats. Degrees of freedom are not partitioned into a decreasing geometric series but are more evenly divided among spacings. As a consequence, no single spacing will be left with only one degree of freedom, but all spacings may have a low number. Thus, information is lost because of the inability to pair all quadrats at all spacings (Goodall 1974).

Because quadrats are not combined into blocks, there is no confounding of the block center-to-center distance and block size (Pielou 1977). The investigator is able to examine a variety of different quadrat spacings. Therefore, there is no restriction on what scale can be examined in the data. This suggests that it may also be possible to identify important spacings by scrutiny and then to test that spacing against other spacings.

Paired quadrats allow for missing data. The data set need not be complete. It is necessary only that there be sufficient pairs of quadrats to provide the necessary degrees of freedom. Further, the investigator is not restricted to sampling from a line or grid. A landscape point could be chosen, and quadrats along various radii of various lengths could be used.

Ludwig and Goodall (1978) and Ludwig (1979) conclude that the paired quadrat method (Goodall 1974) is not particularly good for scale exploration. The loss of information which results from the distribution of the degrees of freedom results in a weak and variable signal when a large number of spacings is compared. They suggest that the power in the method lies in the independence of its variance estimates. It is well suited to testing scale when only a few spacings are examined (degrees of freedom will be greater). They suggest further that the data set could be divided in half, with one-half of the data used for exploration and the other half used for testing the pattern found by exploration.

Blocking and related methods require a complete sampling of the population. This can involve very large data sets, and procedures may be practically limited to detecting scale on fairly small landscapes or remotely sensed data for large areas. Remotely sensed data is particularly amenable to blocking techniques because the landscape is completely sampled at fixed equal intervals. Larger real areas might be sampled if the pattern being investigated is sufficiently large grained (*sensu* Allen and Starr 1982). The term *grain*, as used in this chapter, refers to the finest resolution in time or space that is possible with a given data set. *Large grain* implies that only large patterns may be identified.

2.2.1.2 Applications to Field Data

We have applied these blocking techniques to transect data collected at the Sevilleta National Wildlife Refuge Long Term Ecological Research Site, New Mexico (data from S. J. Turner). The Sevilleta transect was a 1600-m-long line in a grassland with a slope $\leq=1\%$. The area was a mixture of C_4 perennial and annual species. Several of the grass species encountered along the transect grow in rings. The area was chosen to represent regularly repeating pattern in the vegetation. *Bouteloua gracilis* (BOGR) and *Bouteloua eriopoda* (BOER) were the dominant species on the transect and in the area in general. Data were read consecutively from west (0 m) to east (1,600 m). Presence or absence of species was counted in decimeter sampling units. The smallest pattern that we might expect to find in the data with this sampling regime would be 30 cm to 40 cm (3 to 4 dm).

Figures 2.1a, 1b, and 1c present the blocked variance of the Sevilleta transect data for a single species, *Bouteloua gracilis*, following the methods of Greig-Smith (1952), Hill (1973), and Goodall (1974), respectively. In each case a doubling of the block-size protocol was used so that the methods could be directly compared. The power-of-two sampling of the data is not required by the Hill (1973) or the Goodall (1974) methods.

Note that Figs. 2.1a, 1b, and 1c are not the same. Other investigators have found better correspondence between the methods (Ludwig 1979; Goodall 1974). Peaks or changes of the inflection of the line are present in or near some of the same areas on each plot. The difference between the Hill plot and the Greig-Smith plot may be related to any of the problems previously discussed. However the random pairs method of Goodall (1974) produced a plot that is similar to the Greig-Smith plot.

Peaks at the large end of these plots (Fig. 2.1) are supported by few degrees of freedom. At 400 m there are two for the Greig-Smith method, but for Goodall's method the degrees of freedom were evenly distributed and each peak has 550 df. Information at the low end of the scale, where there are many data points, is congested because of the large number of degrees of freedom assigned to smaller scales by the Greig-Smith and Hill techniques. Thus, peaks related to fine-grained pattern are not evident.

These methods were not designed for such long data sets. In Fig. 2.2 we present an enlargement of the low end of the graphs presented in Fig. 2.1. This expansion of both axes of the plot suggests that the information is present at the small spacings. Goodall's method (1974) appears to give the clearest information about fine-grain detail.

Hill's (1973) and Goodall's (1974) methods do not require doubling of the blocks, and this is part of their strength. However, the mass of data produced from

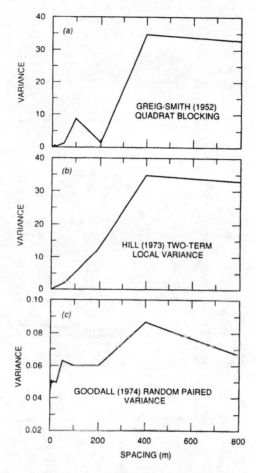

Figure 2.1. Blocked-quadrat variance. Block size is doubled at each spacing for all three methods: Greig-Smith (1952), Hill (1973), Goodall (1974). Data are from the Sevilleta Long-Term Ecological Reserve, New Mexico. There are 16,000 data points from a 1,600-m grassland transect with sampling intervals of 10 cm. The species is *Bouteloua gracilis*.

the Sevilleta transect presents a computational time problem when either method is used as an unrestricted tool in the search for scale.

Figure 2.3 is a Hill (1973) two-term local variance plot of *Bouteloua gracilis*. These are the same data presented in Fig. 1b. However, in Fig. 2.3 the restriction of quadrat doubling was removed and all possible pairs were considered. The result is a smoothing of the information so that no distinct peaks appear.

Figure 2.4 is a plot of Goodall's (1974) random-paired variance of *Bouteloua gracilis*. The data are the same as those presented in Fig. 2.1c. For Fig. 2.4 all evenly spaced comparisons were analyzed at 4-m increments. This protocol was chosen to reduce the computational time required for the analysis. Degrees of

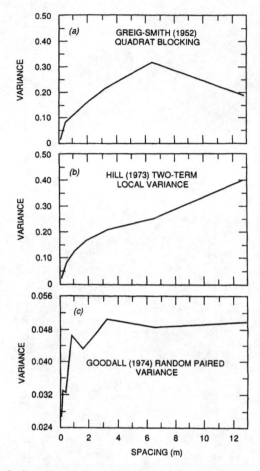

Figure 2.2. Blocked-quadrat variance. Block size is doubled at each spacing for all three methods, and axes are expanded with no other changes from Fig. 1.

freedom were 45 per spacing for this protocol. The plot is difficult to interpret because of the large number of peaks, which suggests a large random component to the data. The investigator might choose to look more closely at information at about 350 m and 550 m.

With both the Hill (1973) and the Goodall (1974) methods, excessive amounts of computer time were required and the results were less than definitive. However, from our own exploration of the data, we note, as expected, that replication serves to make the peaks increasingly distinct from the background variance. For smaller scale patterns, a sampling protocol of several transects, each three to four times the size of the expected pattern, would provide better information than a single longer transect. A review of Figs. 2.1 through 2.4 reveals a series of peaks at various scales of resolution. There are peaks at small scales in the vicinity of 1 and 3 m

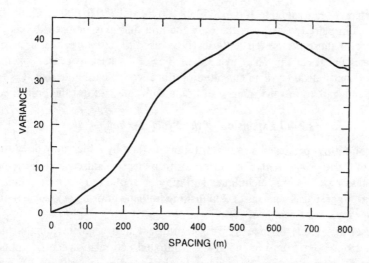

Figure 2.3. Hill two-term local variance method (1973) of partitioning variance between spacings. All possible pairs of plot spacings were considered. Data are the same as Figs. 1 and 2.

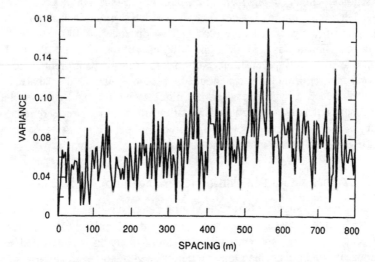

Figure 2.4. Goodall (1974) random-paired variance method (1974) of partitioning variance between spacings. All evenly spaced comparisons were analyzed at 4-m increments. Data are the same as Figs. 1–3.

and at the larger scales at 60 to 100 m, 350 to 400 m, and 550 to 600 m. Goodall's (1974) random-paired variance is the only method that may be used to test the significance of these peaks. Bartlett's test was used to determine the statistical significance of peaks at 1 m, 60 m, and 350 m. The peaks at each of these spacings were significantly different than surrounding distances. Please note that it is not statistically "legal" to use the same set of data to locate and test the points.

2.2.1.3 Expansion of the Block Technique

Carlile et al. (1989) presented a rather different method by which they determine the scale of *Agropyron spicatum* cover on a shrub-steppe landscape. The method is an expansion of blocking techniques and may be applied to the same kinds of data. It was suggested as a means of establishing both the grain and extent at which a landscape should be sampled.

Carlile et al. (1989) divided a 2050-m transect into segment lengths in a geometric progression from 1 to 265 m and computed a correlation coefficient (see below) for adjacent segments. Each estimate of correlation is based on 100 pairs of observations so that the degrees of freedom remain constant for all segment lengths. Correlation was shown to increase as the segment length increased to 64 m and then to monotonically decline. They interpret this result to mean that 64 m is the basic patch size or grain of the pattern.

They then computed the correlation between randomly selected pairs with various segment lengths (or lags) within the pair. For instance, the segment lengths may have been 2 m, and the lag between the pair, 100 m. At inclusive distances (i.e., the sum of the transect length plus the lag distance) of 400 to 700 m, the correlation coefficient was zero for segment lengths of 64, 128, and 256 m. They conclude that a 64-m transect is the appropriate scale of measurement for this species and that replicate transects should be placed at least 200 m apart.

This technique is a hybrid between the block techniques and autocorrelation analysis described in the next section. The statistical rigor of this procedure has not been subjected to the same review as the blocking techniques. As a result, it is difficult to evaluate its advantages and limitations.

2.2.2 Techniques that Partition Variance into Spatial Lags

Time-series analysis suggests that scale can be detected by locating the repetition of a pattern in time or space. A common example of truly periodic time series data is tidal behavior. Landscape ecologists are more likely to be interested in the distribution of windmills, which may be determined by the average farm size, or in mammal mound occurrence, which could be related to territorial behavior or habitat constraints. It is the relative locations of data points that provide information about the scale of pattern. Data arranged along a continuum in time or space are appropriate for time-series analysis. We take the view that time and space relationships can be considered interchangeably. Time-series analysis considers long sequences of equally spaced data (as did the blocking techniques previously discussed). Lattice data (i.e., grid cell data), as well as quadrat and transect data, are appropriate for these analyses. A lattice may be considered to be a two-dimensional analogue of a time series (Ripley 1981). It is assumed that the data have statistical properties that are stationary over space (that is, there are no significant trends in the data) so that the mean, variance, and the autocorrelation or its Fourier transform are good summary descriptors of the series (Cliff et al. 1975).

The nature of the data and the data sequence determine the questions that can be considered. For the purpose of discovery of important scales of pattern on landscapes, we describe several subtypes of the general time-series analysis that have been found to be useful in ecological analysis.

2.2.2.1 Autocorrelation

Autocorrelation tests whether the observed value of a variable at one locality is significantly dependent on values of the variable at other localities (Sokal and Oden 1978a, b). It assumes a normal distribution in the data. The data may be any variable measured at equally distant successive points. Sampling from the data may be done with or without replacement.

Spatial autocorrelation is a method where repetitions of the sequence are found by computing a measure of self-similarity in the data. The sequence is compared to itself at successive positions. The degree of similarity between adjacent intervals is computed. If every point is compared with every other point, the positions of strong similarity will be found (Cliff et al. 1975). The positions of strong dissimilarity will also be evident.

The distance between any two points is referred to as a lag. For a lag-one analysis, the sequence of data is compared to an identical sequence offset from the original by one sample unit. It is an analysis of the displacement between the series and itself at a previous location (Davis 1986). For a lag-two analysis, the cross-comparison is made at a displacement length of two sample units.

Comparisons are usually calculated for lags from zero to one-fourth the total number of samples (Davis 1986). Results may be expressed as the autocovariance function, but they are usually standardized to autocorrelation. The autocovariance is divided by the variance of the entire sequence, and the resultant "autocorrelation" is then plotted against lag length in a correlogram.

Correlograms show the autocorrelation coefficients as a function of distance between pairs of localities and summarize the patterns of geographic variability (Sokal and Oden 1978a,b). Peaks above zero covariance suggest the scale of the pattern. Intermediate points on the sides of the peak suggest the increasing correlation of pattern nearing the significant scale. The negative troughs are the offset mirror image of the positive peaks, the result of the relative movement of the two sequences of identical data in the lagging procedure. In practice only the peaks are considered. Many computer software programs for autocorrelation print only the peaks. Thus the autocorrelation function allows a search for possible periodic variations or repetitions of pattern through the covariance structure of the data (Ord 1979).

Comparison of the correlogram estimated from data to correlograms generated by an idealized model helps to reveal characteristics of the data. The simplest model is that of spatial independence and normal distribution. A simple test of the hypothesis of no relationship between the observations at a particular lag may be carried out by using a normal approximation, provided the sample size is adequate (Ord 1979). If there is no relationship between observations at one location and observations at any other, the correlogram is a flat line with minor fluctuations and no significant trend. Other models assume dependence between successive observations. A wide variety of increasingly complex models can be proposed. However, the investigator should be aware that, while identical patterns give identical correlograms, different patterns may or may not yield different correlograms (Sokal and Oden 1978a, b). Sokol and Oden (1978b) suggest that similarity in correlograms of different patterns indicates similarity in the mechanisms that are responsible for the pattern.

Autocorrelation is a useful technique for exploration of data to determine the grain and extent of repeated spatial pattern and for understanding special structure. It has also been suggested as a method for exploration in model building, estimation, and forecasting (Cliff et al. 1975). Autocorrelation can be defined in sufficient generality to be applied to observations in an irregularly spaced lattice. Such data require multivariate time-series analyses (Bennett 1979).

2.2.2.2 Cross-Correlation

Cross-correlation is the comparison of two sequences to determine positions of correspondence. It is appropriate for comparison of sequences that may have a dependency between them. In comparison with autocorrelation, in which the denominator of the coefficient is based on the variance of the entire sequence and is considered to be constant for all lags, the denominator for cross-correlation is based only on those segments of the two sequences that overlap (Davis 1986); thus the technique is less stable.

Cross-correlation is a test of the significance of correlation between two samples drawn from populations that are normally distributed. There are a small number of observations at long lags, and, therefore, peaks at those distances must be considered nonsignificant.

2.2.2.3 Spectral Analysis

As with blocking methods, spectral analysis identifies scales of repeated pattern in a sequence of contiguous, equally spaced samples. But unlike blocking techniques, spectral analysis identifies pattern in a data sequence by comparing successive values to known wave-form patterns. This is also quite different from comparing data points with each other as in autocorrelation. A wide range of wave-form patterns is known and the distribution of pattern between wave shapes is simplified (Ripley 1978). The result is the partitioning of the sequence variation into components according to the length of the intervals within which the variation occurs (Davis 1986).

The observations in the sequence are expressed as a linear combination of a set of sinusoidal waves (Ripley 1978). The data series is considered to be the sum of many simpler series that have the form of regular sine waves of different amplitudes, wave lengths, and starting points. The sum of all the waves is equal to the original series, and the sum of variation in all the waves must be equal to the total variation in the series. Thus the data are represented as the sum of a series of wave forms that add increasing complexity to the pattern with each successive addition of information. The process is called Fourier analysis after Jean Baptiste Fourier (1768–1830), who proved that a continuous function wave could be represented by a series of sinusoids (Davis 1986). The data are transformed trigonometrically to sines and cosines. This Fourier relationship is used to decompose the additive or multiplicative patterns into successive multiples or harmonics of the fundamental frequency of the series (Jenkins and Watts 1968; Cliff et al. 1975).

Spectral analysis can handle a complexity of information that is insurmountable by other methods. Because it is a sine and cosine transformation, it is not subject to error caused by the location of the starting point in the data (Ripley 1978). However, there are constraints upon the method. The data must be continuous and without missing values. Further, there can be only a single variable for each equally spaced data point. Equally spaced data may be obtained from unequal spacings by interpolation (Platt and Denman 1975). The data for spectral analysis are assumed to be normally distributed, but lack of normality may not present a great problem in practice (Platt and Denman 1975).

It is possible to test for the existence of the dominant periodic component in the data based on departure from a series composed of random points. The choice of the length of the series and the sampling interval excludes the possibility of obtaining information on all frequencies (Platt and Denman 1975).

As in autocorrelation, spectral analysis assumes that the data are statistically stationary (there are no spatial trends). In practice, elements of nonstationarity, such as a trend in the mean, may be present and distort the spectral estimates, particularly at the low-frequency end of the spectrum (Cliff et al. 1975), i.e., those that have few replicates in the transect and are large scale. Data that are not stationary can be made stationary by subtracting trends in the data from the observations. Large-scale trends can be separated by filtering or linear or polynomial regression. Small-scale trends are smoothed with moving averages (Davis 1986). These techniques should leave large- or small-scale systematic periodic components untouched.

The least-squares fit of a number of sine and cosine functions of different frequencies is known as a periodogram (Platt and Denman 1975). The periodogram is often called a *line power spectrum* or *discrete spectrum* (Davis 1986). It is simply a plot of variance (also called power, intensity, or spectral density) versus the frequency or period of the harmonic (or harmonic number). The periodogram indicates periodic variation as a strong peak and the random components of the data as lesser variation in the plot. The area under a peak is the variance attributable to pattern at that frequency.

This "raw spectrum" periodogram is subject to interpretational problems. There may be large parametric standard errors that cannot be reduced by increasing the number of observations. Pattern discovered at a frequency of five, for instance, may be repeated as peaks at multiples of five. Also, aliasing, the use of a sampling interval that is too large to resolve the smallest periodicities in the data, may confuse interpretation. Aliasing results in a high-frequency pattern (one with wavelength less than two times the sampling interval) being erroneously incorporated into lower frequencies (Davis 1986). Platt and Denman (1975) recommend placing at least four sample points within the smallest-scale periodicity expected to be present. Patterns less than this are not found because sampling is of lower frequency than the frequency of the pattern. There are two main techniques within spectral analysis that address the problems of the raw spectrum and the stochastic continuous nature of plot data. These techniques both estimate the *continuous spectral analysis* or spectral density functions or *power spectrums*. The continuous spectrum has the form of a continuous plot of variance versus frequency and is analogous to a continuous probability distribution (Davis 1986).

The older technique produces the power spectrum with calculations performed on the positive autocorrelation functions. Weighting and windowing techniques are used to partition the variance over adjacent frequencies. This approach achieves the same results as the newer Fast Fourier Transformation (FFT) method and is easier to understand mathematically; however, it is computationally inefficient (Davis 1986).

The second and newer technique is FFT. The FFT computer algorithm produces many values of the line spectrum and averages the spectral values across frequencies to produce a smoothed estimate of the continuous spectrum (Ripley 1978). Development of the algorithm is mathematically complex and beyond the scope of this discussion. Currently FFT is the method of choice for continuous spectrum analysis because it is a direct partition of the variance. Computer packages are available.

Fourier transformation of autocorrelation function and FFT are both good descriptions of how variance of a series is distributed over frequency. Hill (1973) cautions that, while spectral analysis is a natural way of dealing with linear and quasi-linear trends, organism responses are not linear. Therefore, spectral analysis bears no important relation to any presumed underlying structure in the data. Note that this criticism is equally applicable to all techniques we have described so far.

Spectral analysis is the most statistically sophisticated of the methods yet discussed and computationally quite a different technique than the various blocking methods. It identifies cyclic phenomena in the data that may be extracted as patch size or scale (Ripley 1978) and gives a complete description of variation in the data series. It is sensitive to small-scale patterns, though somewhat less sensitive to large-scale pattern. Spectral analysis is appropriate for both exploration of a data set and hypothesis testing.

Two-dimensional spectral analysis (Ripley 1981, Rayner 1971, Davis 1986) is an important technique, which to date has not been applied widely in landscape ecology. It should be. *Two-dimensional* means that the dependent variable is a function of two coordinates—for instance, two dimensions in space or time and space. An important advantage is that two dimensional time-series analysis allows for an assessment of the effect of orientation on the pattern and scale (Getis and Franklin 1987). Cross-spectral analysis enables spectral methods to be extended to measure directly the relationships between different regional data series in much the same way that cross-correlation is used (Cliff et al. 1975).

A periodogram of data for *Bouteloua gracilis* and *Bouteloua eriopoda* from the Sevilleta transect is presented as Fig. 2.5. Period in the graph is analogous to spatial extent. A period of 40 suggests that there is a wave form that corresponds to a data length of 40 m. Figure 2.5 is a portion of the whole periodogram that has

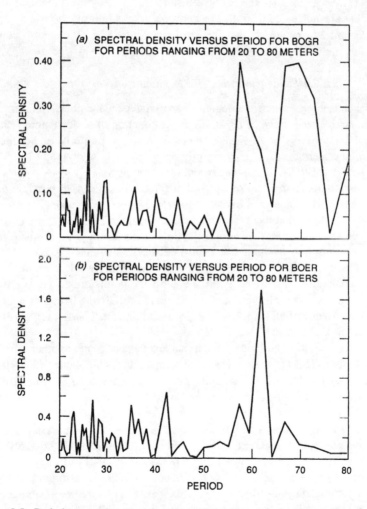

Figure 2.5. Periodogram of spectral density for *Bouteloua gracilis* (BOGR) and *Bouteloua eriopoda* (BOER). The plot is an expansion of a section of the periodogram in the range 20–80 m taken from a 1,600-m data set. Data are from the Sevilleta Long-Term Ecological Reserve, New Mexico.

been expanded so that the peak structure is clear. As in the other figures, all of the 16,000 data points were analyzed. The axes were expanded for clarity. The Kolmogorov-Smirnov procedure (Davis 1986) for testing significance was used on the two periodograms. In this procedure the sample distribution is compared to a hypothetical model of the population.

Bouteloua gracilis (Fig. 2.5a) has statistically significant points at 58 m and 26 m. The points at 61 m, 42 m, and 26 m are statistically significant for *Bouteloua eriopoda* (Fig. 2.5b). The numerous smaller peaks on the graphs represent noise in the data. Note that spectral analysis suggests that there is a pattern that has a scale of about 60 m and that the pattern occurs in both dominant species. This pattern is evident in the blocked analysis also (Fig. 2.1); however, there are other patterns that are not evident in all the analyses. The similarity between graphs (Figs. 2.5a and 2.5b) suggests that the process that underlies the pattern evident in these data may be the same.

2.2.2.4 Semivariograms and Regionalized Variables

Semivariance is a measure of the degree of spatial dependence between samples, which summarizes the variance as a continuous function of scale (Palmer 1988). The samples may be a point measurement of a single property or a suite of properties. The measurements may be uniformly spaced along straight lines, but this is not necessary. The quadrats need not be contiguous.

The semivariance is the sum of squared differences between all possible pairs of points separated by a chosen distance. Estimates of semivariance are procedurally similar to those of autocorrelation. A semivariogram, similar to the correlogram, is a plot of the semivariances at different distances or lags. At distance zero, the value of a point is compared to itself and the semivariance is zero. As the distance is increased, if the compared points are increasingly different, the semivariance increases. This increase continues until the points are so far apart that they are not related to each other and their squared difference becomes equal to the average variance of all samples. The line becomes flat. The flat region of the graph is called a sill.

The area or neighborhood where locations are related to one another is called the range or span. This is the area where the semivariance approaches the variance at the sill. Beyond the sill the samples can be considered to be spatially independent. For instance, at the sill there may be a transition from forest to grassland.

There may be more than one sill and range in a plot. The semivariogram can suggest the scale(s) of patchiness of a vegetation mosaic (Palmer 1988). Independent replicates of a system would be placed at distances further apart than the occurrence of the sill at the greatest distance on the semivariogram. A sill at a very small spatial distance might suggest the scale of a quadrat for a homogenous vegetation sample. Thus the semivariogram can determine both the grain and extent of sampling needed to characterize landscape pattern.

Often data are not stationary and/or the data set represents samples that are not random but instead have continuity from place to place. The concept of the regionalized variable was introduced to facilitate analysis of data that are not statistically stationary. Regionalized variables have properties that are intermediate between truly random variables and completely deterministic ones. They are functions that describe natural phenomena that have geographic distributions and spatial continuity from point to point (Davis 1986). For instance, a community might be thought of in these complex terms because it cannot be described by a tractable deterministic function and is spatially correlated over short distances. Regionalized variables are useful descriptors of natural phenomena, such as the occurrence of a pest, that have geographic distribution.

The characteristics of regionalized variables reflect the size, shape, orientation, and spatial arrangement of the samples collected and change as these parameters are changed. The rate of change of a regionalized variable along a specific orientation is expressed in terms of its semivariance. But the regionalized variable may have properties not present in the similar autocorrelation function. For example, it is not necessary that the variable remain stationary; it may exhibit changes in average value from place to place.

Nonstationary data are analyzed by calculating the residual portion of the regionalized variable and its "drift" or spatial trend. The drift is subtracted from the regionalized variable, and the residual is then stationary so that the semivariogram can be calculated (Davis 1986). This is not quite as simple as it might seem because the drift is seldom known and is difficult to estimate.

A process of estimation, experimentation, calculation, and comparison is employed to determine the appropriate semivariogram. This is a trial-and-error process that does not yield a unique result. Fitting a curve to the semivariogram can be very subjective. Semivariogram model, drift, and neighborhood may combine in different ways to give similar results. Despite this difficulty, regionalized variables provide a means of describing the autocorrelation in data and a means to use the knowledge about the autocorrelation to derive precise, unbiased estimates of the sample values within the sampling unit and thereby to resolve detailed spatial pattern with known variance for each interpolated point (Robertson 1987). This technique is the focus of the chapter on fractal analysis (Milne, Chapter 9) where the topic is covered in greater detail.

If the regionalized variable is random and has no spatial correlation, the line of the semivariogram will be straight and horizontal. A linear form suggests moderate continuity in the data, while a parabolic form suggests excellent continuity of the regionalized variable (Davis 1986). Davis (1986) suggests that, if the line of the graph does not pass through zero, the data are highly variable at intervals shorter than the sampling distance.

2.2.2.5 Nearest-Neighbor Analysis

Ecologists are often interested in how points are distributed on a two-dimensional surface. Mapped point pattern analysis is a group of techniques that relates the spatial dispersion of points on a plane. The points may represent trees on a landscape, for instance. Mapped points may be random, regular, or show aspects of both random and regular distributions (Davis 1986). This group of analyses is generally referred to as locational analysis by geographers. Bartlett (1975) provides a discussion of the relationship between these methods and time-series analysis.

The point pattern technique most familiar to ecologists is nearest-neighbor analysis. The data used are the distances between pairs of closest points. No quadrat size is selected. Therefore, all scales of pattern may be distinguished. The major drawback of the method is the assumption that edge or boundary effects are not present (Ripley 1981; Davis 1986) and that the pattern extends to infinity (Davis 1986). As expected from this problem alone, nearest-neighbor methods can reveal only the smallest spatial patterns (Cormack 1979). Nearest-neighbor distances have been used successfully to investigate community processes such as competition for space. Nearest-neighbor distances have not traditionally been applied to landscape problems.

Recently Getis and Franklin (1987) developed a second-order neighbor analysis of mapped point patterns. As with traditional nearest-neighbor methods, this analysis examines pairs within a specified distance of each other. Rather than studying the mean distances between the pairs, Getis and Franklin (1987) investigate the variation. Their model includes a boundary correction, and the data are compared to a random process.

Getis and Franklin (1987) demonstrated the method by using aerial photography to locate 5,000 ponderosa pine trees in the Kalamath National Forest, northern California. In plots of observed and expected values, they were able to demonstrate relationships between the trees at various scales. They concluded that the method identified different dominant patterns at different scales for mapped point data. They suggest that the technique identifies several important scales of pattern: (1) the distance to the nearest neighbor, (2) the distance at which heterogeneity begins, (3) the distance at which clustering becomes significant, and (4) the distance at which maximum clustering is observed.

Nearest-neighbor methods are well known, but, as with several of the techniques in this chapter, Getis and Franklin's (1987) modification has not been subjected to the statistical scrutiny and modification that tends to enhance its applicability. The method holds promise for landscape analysis and should be investigated further.

2.2.2.6 Trend Surface Analysis

Trend surface analysis is a second technique that may be applied to points mapped onto a coordinate system. Its utility is dependent upon the hypothesis that the spatial pattern results from a large-scale regional trend, upon which smaller-scale local effects are superimposed (Gittins 1968). For example, a species' regional distribution may be controlled by abiotic factors and also influenced locally by the presence of other species. What we consider "regional" and "local" is largely subjective. The designation depends upon the size of the region being examined (Davis 1986) and varies with the experiment. Trend surface analysis is a technique that can identify a broad regional pattern or trend and separate it from smaller-scale nonsystematic local variation (Chorley and Haggett 1965). It may be used to identify local structures (nesting territories) or to define regional trends in, for instance, geochemical data, or it may be used to remove trends from spatial data and accurately represent trends within the boundaries of the study area (Cormack and Ord 1979).

As with all mapped point patterns, the observation is considered to be in part a function of location (Davis 1986). The spatial location of the data has a very real influence on the outcome of the analysis. It is impossible to identify features whose size approaches the spacing between sample points. Spacing of locations can be even or irregular. Irregular spacings lead to further problems with identification of the scale of detectable processes.

This technique uses multiple-regression methods to separate trend from residual variation (Cliff et al. 1975). The trend is represented by a linear or higher polynomial function (Gittins 1968). The regression proceeds through linear, quadratic, cubic or higher-degree polynomials that describe the trend in the data until the polynomial is identified that has the smallest residual variance, i.e., minimizes the squared deviations from the trend (Davis 1986). Trend-surface analysis provides contour maps showing the trend surfaces obtained with successively higher-degree polynomials as well as maps of trend deviations or local variation. A statistical model describing the trend surface is also provided in computer software packages.

The goodness of fit of the trend surface polynomial to the actual observations is expressed as a percentage of the total sums of squares (Chorley and Haggett 1965). The trend surface sum of squares can be partitioned among successive trend components to determine when a satisfactory fit to the observations has been achieved. The purpose of the investigation will influence the interpretation of the goodness of fit. Goodness of fit may be tested by comparison of variance due to the trend with the variance due to deviations from the trend. The results of analysis of variance cannot be rigorously interpreted because the deviations from the trends may themselves contain trend components in addition to true residuals. The analysis can provide a conservative estimate of significance (Gittins 1968). When a high percentage of variance is explained by a trend surface, the surface and its interpretation are more likely to have ecological meaning. When the percentage variance explained by a trend surface is low, the deviations from the surface may be of greater interest (Gittins 1968).

Statistical tests of trends are also restricted by the assumptions incumbent upon regression. The dependent variable is assumed to be normally distributed, and its variance is assumed to be stationary with regard to its geographic location. It is also assumed that the samples are chosen at random. A planned experimental design increases the likelihood of upholding these assumptions (Davis 1986).

Maps of trend surfaces and of deviations from the trend may be useful descriptive tools for examining the spatial distribution of a variable. Comparisons among variables or among regions of patterns produced by trend surface analysis have been accomplished by inspection (Chorley and Haggett 1965; Gittins 1968). Problems involved in visual comparisons include lack of rigor and, for comparisons among regions, the need to use a similar sampling grid. Cliff and Kelly (1977) suggest a method for nonstatistical map comparisons that assumes a common underlying process.

Cliff et al. (1975) regard the fitting of trend polynomials as a search process. The results of trend surface analysis may point to alternative models or hypotheses regarding processes that may be refined in subsequent analyses. They also note that in systems where processes operate at many scales, a distinction between regional and local processes may be arbitrary and therefore not appropriately analyzed using a trend surface model.

Trend surface analysis has been extended to include a third geographic coordinate as an independent variable. This is four-dimensional trend surface analysis. Problems noted in the original analysis tend to be exaggerated with the addition of another dimension. The procedure has proved useful in geology (Davis 1986).

2.2.2.7 Variance Staircase

Patterns in the heterogeneity of the data can reveal much about the structure of the system. The variance staircase method begins with the observation that the sample variance, s^2, associated with a mean is inversely proportional to sample size or area, n. If $\ln s^2$ is plotted as a function of $\ln n$, the result is a straight line with a slope of -1 as long as the sampled population is randomly distributed in space so that each new sample is independent of previous samples. However, if significant spatial correlation exists with correlation coefficient, r, then the next sample will not be independent and the straight line will deviate from a slope of -1. The slope will lie between -1 and 0 ($0.0 <= r < 1$) (Smith 1938).

Weigert (1962) used this approach with nested quadrats to determine the appropriate quadrat size to sample vegetation. Recently, Levin and Buttel (1986) proposed that the deviation of the slope from -1 is a measure of the spatial scale or patchiness of the landscape.

A specific application of this method for detecting scale has been proposed by O'Neill et al. (in press). Though new, the method may prove to be very useful in landscape ecology because it is applicable to numerous remotely sensed data and the analysis is relatively straightforward. They assume that specific processes will operate at some scales corresponding to hierarchical levels of the system to structure the landscape and produce pattern. At such scales, some level of spatial correlation is expected, with a concomitant decrease in variance and a slope that approaches 0. At scales between hierarchical levels, processes that may be operating do not form a coherent pattern and therefore cannot be detected. For these scales the slope of $\ln s^2$ plotted against $\ln n$ equals -1.0, indicating random distribution.

The expectation is that the log-log plot would appear like a staircase having discrete scales with shallow slopes (steps) alternating with slopes of -1.0 (risers).

The scales with shallow slopes would correspond to significant scales at which important processes operate to structure the landscape.

O'Neill et al. (in press) tested the method on land use data (LUDA tapes) available from the U.S. Geological Survey (USGS). The LUDA tapes (Fegeas et al. 1983) are a digitized analog of aerial photographs. The sampling design involved thirty two radiating transects. At each of a number of transect lengths corresponding to increasing sample area there were 32 samples from which variance was calculated. The area of the circle increased as the length of the radii increased, but the number of sample radii (32) remained the same. The measured parameter was the percentage of cover by specific land use category found along the length of each transect.

A piecewise linear regression method was used to detect changes in the slopes of the variance at several scales. The procedure takes the first two points (two smallest sample sizes) and fits a straight line. This regression is used to predict the next point to an error tolerance specified by the investigator. If the next point is predicted well, the regression is repeated using all three points. The procedure continues until the next point cannot be predicted. At this point, a break in the line is assumed to occur and the procedure begins again, taking the next two points as a starting point. In this manner the log-log regression of variance against scale is broken into a series of linear regressions.

Figure 2.6 is the graph of the variance in the number of land use categories for West Palm Beach, Florida. Analysis shows two discrete scales at which spatial

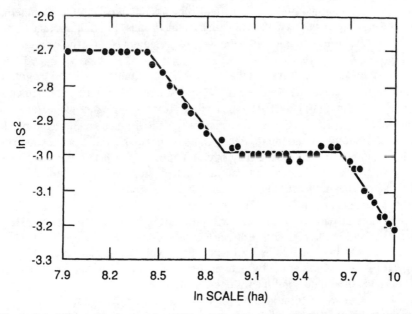

Figure 2.6. Variance staircase plot of West Palm Beach, FL. Analysis shows two discrete scales at which spatial pattern appears to exist. The two steep portions of the curve have slopes of −0.52 and −0.65, and the two shallow slopes are −0.003 and −0.018.

pattern appears to exist. The two steep portions of the curve have slopes of -0.52 and -0.65, and the two shallow slopes are -0.003 and -0.018.

This approach is specifically designed to detect multiple scales. In many landscape problems there may be several different scales at which different processes dominate pattern (Krummel et al. 1987). The variance staircase approach identifies not only the scale at which pattern exists but also the width or span of the pattern. However, pattern over a range of scales tends to produce noisy or ambiguous results.

This approach is data intensive. Multiple transects must be available, and gaps in the data are not allowed. In addition, this method suffers many of the disadvantages of other techniques reviewed above. The assumption is made that the data are stationary over the scales being sampled. If sampling is done over a gradient, the results may be ambiguous or misleading. This approach is exploratory. The regression method introduces a subjective element and does not permit hypothesis testing of the precise points where the line changes slope. O'Neill et al. (in press) also discuss several additional biases that must be considered in applying this approach. Like the Carlile et al. (in press) technique, this new methodology has not been subjected to close scrutiny.

2.3 Landscapes that Exhibit Irregular Pattern

The methods covered in earlier sections of this chapter assume that a pattern repeats itself over the length of a transect. However, landscapes are heterogeneous and may show sharp transitions at irregular intervals. Pattern and scale may change rapidly (Krummel et al. 1987), and methods that assume repeatability of pattern, even in detrended data, may not detect the scale changes. In addition, landscape patterns are often multivariate, while autoregressive models are primarily univariate.

Carlile et al. (in press) point out that even a shallow gradient over the length of a sampling transect can (1) obscure a repeatable pattern and (2) introduce elements of change that cannot be detected by the methods that assume pattern repeatability. Thus methods that rely on the decay of similarity with distance are limited if patch size varies randomly along the transect or monotonically becomes larger or smaller. For example, the landscape may have a characteristic patch size randomly interspersed with other patches of various smaller sizes. As a result, the characteristic patch does not recur at a detectable lag or periodicity.

For these reasons an approach that emphasizes the identification of edges between units of pattern of any size will prove useful for scale detection (Marr 1980; Marr and Hildreth 1980). Edge detection methods may be applied to data sets with repeatable patterns as well as to those with the greater complexity of irregular pattern. Multivariate by nature, they may be used to analyze suites of human and natural data combined. Edge detection has a history of application in analysis of species distributions and abundances (Leopold 1933; Wertz 1965; Schuerholz 1974; Patton 1975; Thomas et al. 1977; Ranney et al. 1981), population cycles and periodicity (Cole 1954), edges along altitude gradients (Beals

1969) and composition gradients (Wilson and Mohler 1983). Patton (1975) proposed an index of patch diversity as a function of perimeter and area. This approach has been extended with the methods of fractal analysis of shape complexity (Burrough 1981; Loehle 1983). Krummel et al. (1987) applied fractal analysis to contrast forest patches caused by human influences to patches shaped by natural factors. Milne (Chapter 9) discusses fractal analysis in greater detail.

Edge detection has been applied in the field of image analysis for computer vision (Canny 1983). The basic analyses use images expressed as gray-color-scale values. Edges are identified by the local difference in gray values of adjacent cells by using spatial derivatives (Overington and Greenway 1987). This is a univariate application of edge analysis. Image textural analysis applied to landscape pattern is the subject of the chapter by Musick and Grover (Chapter 4).

The methods for edge detection are free from many of the assumptions that cause problems in other approaches. The methods may utilize point or transect data, but the samples need not be continuous or equally spaced. Missing samples do not preclude analysis, but the precision of placing an edge along the transect is affected. If a data gap is sufficiently large, an entire patch (i.e., a pair of edges) may remain undetected. As is true of all techniques in this chapter, the greatest precision in detecting scale is achieved if the sampling interval is small relative to the size of the smallest scale of interest.

2.3.1 Global Zonation

Global zonation searches for edges by breaking the total transect into segments that are as internally homogeneous as possible and as distinct as possible from adjacent segments (Davis 1986). The technique has been developed in geology and hydrology but has real potential for applicability in landscape ecology. Paleontologists, for example, may want to zone a stratigraphic sequence on the basis of consistent abundances of fossils (Davis 1986).

The procedure developed by Gill (1970) divides the total transect into two segments by placing a boundary somewhere along the transect length. The within-segment sum of squares (SS_w) is calculated for each segment, as is the between-segment sum of squares (SS_b) and the ratio between the two [$R = (SS_b - SS_w)/SS_b$]. The boundary is then moved to all possible positions along the length of the transect, and the procedure is repeated. The position at which R is maximum is chosen as the first significant edge. In essence, this point represents the boundary that divides the transect into segments that are most internally homogeneous and most distinct from each other.

The procedure is then repeated for each of the two segments identified in the first step. The transect is subdivided into segments until the ratio, R, no longer increases with the addition of more boundaries. The result is a series of subdivisions of the original transect representing distinct, internally homogeneous zones or patches.

A related technique by Hawkins and Merriam (1973, 1974) uses optimality principles to ensure that the final set of zone boundaries is the best of all possible partitioning of the data. This technique is not only iterative (all possible locations of the initial boundary) but also recursive (all possible locations of the second boundary for all possible locations of the first).

Global zonation makes no assumptions about the repetitiveness of a pattern and so is applicable to landscapes with abrupt transitions. The sums of squares can be calculated for multiple attributes at each location. The approach should be particularly applicable to situations in which the landscape is composed of discrete patches of homogeneous composition, whether embedded in a matrix or contiguous.

One important drawback of the approach may be the computational burden for a long transect with many attributes. This would be a particular problem with the recursive and iterative approach of Hawkins and Merriam (1973, 1974). Global zonation is a multivariate technique and thus exhibits the attendant problems of independence and multivariate normality. Any correlation between attributes (e.g., species abundances or relative frequency) would tend to distort the ratio R, so that the edges would be overly influenced by a few attributes.

Global zonation techniques assume that the landscape is composed of discrete, homogeneous patches of any size. When this assumption is reasonable for a study, consideration should be given to the technique. However, global zonation may be less useful for problems dealing with a transect along a gradient. In this case, there may be considerable variability within a patch, and this method, which relies upon minimizing within-segment variability to identify edges, may experience difficulties.

2.3.2 Moving Window Analysis

Another set of techniques involves placing a window containing data from several adjacent samples on a transect and dividing the window in two halves. The window is an arbitrarily chosen width through which the data is viewed during the analysis. For each half window the mean for each variable is calculated. Then the "distance" or "dissimilarity" between the halves is calculated. The window is then moved along the transect by one sample station, and the process is repeated until the entire transect is covered. This technique is univariate or multivariate and may consider an array of variable species or attributes. The analysis produces a plot of distances that represent differences between adjacent window halves along the transect. In all cases a strong peak on the plot indicates the scale of the edge. Smaller peaks on the plot represent small-scale background variance (or noise) in the data.

Consider what would happen as the window was moved along a transect from one homogeneous patch, across an edge, to a different homogeneous patch. Within the first patch, adjacent window halves are nearly identical, and the distance measure is small. At the point that the window spans the edge between the dissimilar patches, the two window halves are most different and the distance · measure is large. Then, once the window is within the second homogeneous patch, the distance measure is, once again, small.

This example demonstrates that peaks in a plot of the distance measure against transect position indicate the location of discontinuities. Choice of window width and distance between sampling points determines resolution of the edge detection algorithm. Smallest-scale resolution occurs when the window contains only two adjacent samples, and large-scale resolution occurs when the window width is equal to the entire transect. Window sizes of interest can be determined by trial and error and vary with the purposes of the study, the resolution of the data used, and the characteristics of the attributes involved.

Windowing techniques have been applied in landscape ecology (Forman and Godron 1986; Delcourt and Delcourt 1988), statistical genetics (Manly 1985), statistical ecology (Legendre and Legendre 1983 and Orloci 1978), numerical taxonomy (Sneath and Sokal 1973), and geostatistics (Webster 1977; Warrick et al. 1986). Many types of data used to measure landscape elements are appropriate for moving window analysis: for instance, presence-absence data (Forman and Godron 1986; Delcourt and Delcourt 1988), percent dominance (Delcourt and Delcourt 1988), soil data (Webster 1973), and contiguous comparisons for similarity (Whittaker 1956, 1960). Within the general framework of sliding a window along a transect, many variations are possible. We consider some window techniques that are likely to be of value in ecological pattern recognition and landscape ecology. Others may be useful as well.

2.3.2.1 Euclidean Distance Metrics

Euclidean distance may be used to calculate the linear "distance" or difference between two points that are characterized by many variables. Squared Euclidean distance (SED) is the square of the difference between the means of each variable in adjacent windows, summed across all variables measured. Standardized Euclidean distance divides the measure of each variable in each sample by the largest value observed for each variable in the entire data set.

Metrics based on Euclidean distance have been effectively applied to a wide array of ecological data and are mathematically simple. However, because SED places no upper bound of the value of the metric (Legendre and Legendre 1983), an area composed of attributes (for instance, species) with low overall abundance and no attributes in common can exhibit a smaller distance measure than an area with identical attribute compositions, where abundances are large and areas differ only in relative abundance among attributes (Orloci 1978). Standardization provides an upper bound to the distance and overcomes this problem.

Euclidean distance metrics assume independence in the behavior of variables. If there is a significant correlation among variables, the distance measure will underestimate the actual dissimilarity among samples.

In a computational simulation study, Brunt and Conley (in press) found that, at a given window width, the effectiveness of SED in detecting edges is a function of the level of heterogeneity within landscape units. Edges are readily detectable when background variance (small-scale heterogeneity) is low relative to the peak produced by the patch edge (large-scale transition). When background variance produces high peaks, detection of the large-scale edge fails. By systematically varying window size, the investigator can effectively identify different levels of spatial heterogeneity within a single data set.

2.3.2.2 Mahalanobis Distance and Other Multivariate Approaches

A number of other multivariate approaches have been applied to the problem of detecting a significant difference between window halves. The Mahalanobis distance between windows defined through discriminate analysis of covariance avoids an assumption implicit in SED, that attributes are independent (Webster 1973). However, accounting for this lack of independence by covariance can also create problems. Covariance is an efficient descriptor only for linear relationships. Experience with field data often reveals that strong nonlinear relationships, in turn, generate strong nonlinear correlations among the responses. As a result, the theoretical advantages of using the Mahalanobis distance measure may be outweighed by the problems of the linearity assumption.

Discriminant analysis independent of the Mahalanobis distance is widely used for detecting significant differences between window halves. However, this method also assumes that the attributes are linearly related to each other, and this assumption may be questionable for real landscapes. The same problem exists for canonical correlation analysis, which may be used to describe the correlation between linear compounds of variables. The double linear constraint and the assumption of multivariate normality weakens the potential of all these measures.

2.3.2.3 Applications to Field Data

We have applied moving window analysis to data for animal populations collected along a pair of 2700-m transects (data of W. Conley and M. Conley) on the Jornada LTER site, New Mexico. The transects were on a northern Chihuahuan Desert watershed that is characterized by an elevation gradient, extending from the edge of a rocky mountain slope to a low-basin ephemeral (playa) lake. The area features a complex mixture of grasslands and shrublands that have developed in association with soil and elevation gradients and land use. The direction of the transects was chosen to maximize change along the gradient. Gillison and Brewer (1985) demonstrate that a gradient-directed transect is the most efficient approach to detecting pattern in such an area. Ecological edges in geological and soil series intrusions (Wierenga et al. 1987) and in annual and perennial vegetation (Lugwig and Cornelius 1987; Wierenga et al. 1987) have previously been explored along these transects. Data on species representations were collected at 30-m intervals.

Squared Euclidean distance was used in the window analysis. Window widths of 2- to 20-sample intervals were chosen to examine variation in edge patterns at different scales within a given group and difference in edge patterns between groups at the same scale.

Locations of peaks for window width six (Fig. 2.7) can be visually compared for the consumer groups. It is immediately evident that, at this window width, many more discontinuities (edges) appear in the organization of the ant species

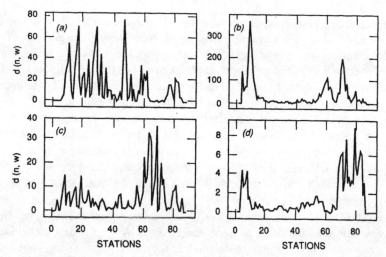

Figure 2.7. Hierarchy of zonation on a desert transect. Moving-window edge analysis of abundance data, deriving squared Euclidean distamce (d), with window width 6 for ants, mammals, lizards, and birds. Data are from a 2700-m transect with sampling stations at 30-m intervals. Stations were numbered consecutively beginning at lower elevations (north end of the transect). Ticks at the top of each plot show the location of zone boundaries of perennial vegetation as identified by Ludwig and Cornelius (1987).

(Fig. 2.7a) than in the mammals (Fig. 2.7b). There appear to be four distinct edges for mammals, corresponding roughly in location with some (but notably not all) of the perennial plant zones identified by Wierenga et al. (1987). The position of Wierenga's (1987) perennial plant zones is noted at the top of each graph in Fig. 2.7. Most of the heterogeneity demonstrated (at this scale) for ants does not correspond to any of the previously demonstrated vegetation zones. Subsequent analyses indicate that some of the "edges" in the ant community correspond to shifts in distributions of annual plants, and some of the background variation may result from a small-scale patchiness resulting from interspecific interactions.

Similarly, edge patterns differ between lizards (Fig. 2.7c) and birds (Fig. 2.7d). Lizards in this study were predatory, while birds included both granivorous and predatory species. Bird species observed across this watershed appear to exhibit two community boundaries that include combinations of two to three of the perennial plant zones reported by Wierenga et al. (1987). Thus, the distribution pattern for birds does appear more spatially integrative than those for the other consumers. Birds are a mixed consumer group and are also highly mobile.

For these examples, the magnitude of peaks is relevant only within a group. Potential maximum distance SED values are a function of the number of variables (species) sampled and maximum abundances per species for the sample unit employed and thus vary between groups (note the different scales for ordinate axes in Figs. 2.7a through 2.7d).

2.4 Summary

This chapter focuses on statistical methods for determining the spatial structure of a landscape. Most of our examples have come from vegetation analysis. The examples chosen represent the bias of the authors, not the bias of the techniques. The attributes for analysis depend upon the background of the investigator and the questions to be answered. Most of the methods presented will have broad applicability to spatial analysis of landscapes. Nonetheless, some methods will serve specific types of analysis better than other methods. With this in mind, we provide a summary table of the techniques discussed in this chapter (Table 2.1). The table may be considered a preliminary guide to the contents of this chapter or a review for those readers who managed to wade through this admittedly dense material.

We have described thirteen methods that fall into several categories of analysis useful for spatially explicit data. This should in no way be considered an exhaustive list. Most of these methods have withstood sufficient statistical scrutiny to ensure continued application to ecological analysis. A few are new techniques. All are useful to landscape ecologists whether they investigate urban, suburban, agricultural, or natural landscapes and whether they are interested in the present state of the system, study paleo systems, manage future land use, or attempt to decipher the effects of man's activity on global processes.

Research into the statistical analysis of spacial pattern is synthesized in a series of excellent textbooks that are recommended for those interested in greater detail than that presented here. Kershaw and Looney (1985) provide a good introduction to vegetation analysis, and Greig-Smith (1983) provides an extensive discussion of many of the techniques. Pielou (1977) is particularly useful for readers seeking more mathematical detail. Cormack and Ord (1979) provide a variety of detailed examples. Ripley (1981) and Diggle (1983) present the techniques from the viewpoint of statisticians. Ecological interest in spatial pattern has a parallel development in geology. The developments in geology are also summarized in texts such as Webster (1977) and Journel and Huijbregts (1978). In many instances, the textbook by Davis (1986) provides the clearest explanations of techniques.

Acknowledgments

Research supported in part by the Ecological Research Division, Office of Health and Environmental Research, U.S. Department of Energy under contract No. DE–AC05–84OR21400 with Martin Marietta Energy Systems, Inc.; in part by the Oak Ridge National Environmental Research Park; in part by NSF: BSR–8419790 to W. Conley and NSF: BSR–8612106, the Jornada LTER Project; and in part by NSF: BSR–8811906 to the Biology Department, University of New Mexico, the Sevilleta LTER Project. Sevilleta LTER Publication No. 4 ORNL-NERP publication No. 4 and Environmental Sciences Division Publication No. 3586. The first author would like to acknowledge the assistance of computer wizard Greg Shore, Biology Department, UNM.

Table 2.1 Summary of Data Requirements and Characteristics of Methods

	Pattern Must Repeat?	Transect (T) Quadrat (Q) Point (P)	Contiguous? Equal Spacing?	Multi-Variate?	Exploring (E) Testing (T)	Assumes Stationary Data?
Blocking[a]	Yes	T,Q	Yes	No	E,T	Yes
Auto-correlation	Yes	T,Q,P	No	Yes	E,T	Yes
Spectral analysis[b]	Yes	T,Q	Yes	No	E,T	Yes
Semi-variance	No	T,Q,P	No	Yes	E	No
Nearest neighbor[c]	Yes	P	No	No	E,T	Yes
Trend surface	Yes	P	No	Yes	E	Yes
Variance staircase[d]	Yes	T	Yes	No	E	Yes
Global zonation[e]	No	T,P	No	Yes	E,T	Yes
Moving window[f]	No	T,P	No	Yes	E,T	Yes

[a] Blocking may confound quadrat size and spacing. Sensitive to starting position. Degrees of freedom may decline with increased block size. Missing data are not permitted for some analyses. Computation time is long for large data sets.
[b] No starting point sensitivity. No missing data are allowed.
[c] Sensitive to small-scale pattern. Edge or boundary definition problems.
[d] Sensitive to starting point and requires multiple transects.
[e] Analysis may be computationally burdensome.
[f] Computationally simple. Choice of a window size is subjective.

References

Allen, T.F.H.; Starr, T.B. 1982. *Hierarchy: Perspectives for Ecological Complexity*. Chicago: Univ. of Chicago Press.

Bartlett, M.S. 1975. *The Statistical Analysis of Spatial Pattern*. London: Chapman and Hall.

Beals, E.W. 1969. Vegetation change along altitudinal gradients. *Science* 165:981–85.

Bennett, R.J. 1979. *Spatial Time Series*. London: Pion.

Brunt, J., and Conley, W. 1990. Behavior of a multivariate algorithm for ecological edge detection. *Ecological Modelling*, 49:179–203.

Burrough, P.A. 1981. Fractal dimension of landscapes and other environmental data. *Nature* 294:240–42.

Canny, J.F. 1983. Findings edges and lines in images. MIT Technical Report No. 720. Cambridge, Mass.

Carlile, D.W.; Skalski, J.R.; Batker, J.; and Thomas, J.M. 1989. Determination of ecological scale. *Landscape Ecology* 2:203–13.

Chorley, R.J., and Haggett, P. 1965. Trend-surface mapping in geographical research. In *Spatial Analysis: A Reader in Statistical Geography*, eds. B.J.L. Berry and D.F. Marble, pp. 195–217. Englewood Cliffs, N.J.: Prentice-Hall.

Clampitt, C. 1985. DECORANA for IBM-PC's. *Bioscience* 35:738.

Cliff, A.D.; Haggett, P.; Ord, J.K.; Bassett, K.A.; Davies, R.B. 1975. *Elements of Spatial Structure: A Quantitative Approach*. Cambridge, Mass.: Cambridge Univ. Press.

Cliff, A.D. and Kelly, F.P. 1977. Regional taxonomy using trend-surface coefficients and invariants. *Environment and Planning: A* 9:945–55.

Cole, L.C. 1954. Some features of random population cycles. *Journal of Wildlife Management* 18:2–24.

Cormack, R.M. 1979. Spatial aspects of competition between individuals. In *Spatial and Temporal Analysis in Ecology*, eds. R.M. Cormack and J.K. Ord, pp. 151–212. Fairland, Md.: International Cooperative Publishing House.

Davis, J.C. 1986. *Statistics and Data Analysis in Geology*. 2nd ed. New York: John Wiley and Sons.

Delcourt, H.R. and Delcourt, P.A. 1988. Quaternary landscape ecology: Relevant scales in space and time. *Landscape Ecology* 2:45–61.

Diggle, P.J. 1983. *Statistical Analysis of Spatial Point Patterns*. London: Academic Press.

Errington, J.G. 1973. The effect of regular and random distributions on the analysis of pattern. *Journal of Ecology* 61:99–105.

Fegeas, R.G.; Claire, R.W.; Guptill, S.C.; Anderson, K.E.; Hallam, C.A. 1983. Land use and land cover digital data. U.S. Geological Survey Circular 895–E, Alexandria. Va.

Forman, R.T.T. and Godron, M. 1986. *Landscape Ecology*. New York: John Wiley and Sons.

Getis, A. and Franklin, J. 1987. Second-order neighborhood analysis of mapped Point Patterns. *Ecology* 69:473–77.

Gill, D. 1970. Application of a statistical zonation method to reservoir evaluation and digitized log analysis. *Bulletin of the American Association of Petroleum Geologists* 54:719–29.

Gillison, A.N. and Brewer, K.R.W. 1985. The use of gradient directed transects or gradsects on natural resource surveys. *Journal of Environmental Management* 20:103–27.

Gittins, R. 1968. Trend-surface Analysis of Ecological Data. *Journal of Ecology* 56:845–29.

Goodall, D.W. 1974. A new method for analysis of spatial pattern by random pairing of quadrats. *Vegetatio* 29:135–46.

Greig-Smith, P. 1952. The use of random and contiguous quadrats in the study of the structure of plant communities. *Annals of Botany, New Series* 16:293–316.

Greig-Smith, P. 1983. *Quantitative Plant Ecology*. 3rd ed. Berkeley: Univ. of California Press.

Hawkins, D.M. and Merriam, D.F. 1973. Optimal zonation of digitized sequential data. *Journal of the International Association for Mathematical Geology* 5:389–95.

Hawkins, D.M. and Merriam, D.F. 1974. Zonation of multivariate sequences of digitized geologic data. *Journal of the International Association of Mathematical Geology* 6:263–69.

Hill, M.O. 1973. The Intensity of Spatial Pattern in Plant Communities. *Journal of Ecology* 61:225–35.

Jenkins, G.M. and Watts, D.G. 1968. *Spectral analysis and its applications*. San Francisco: Holden Day.

Journel, A.G. and Huigbregts, C.J. 1978. *Mining Geostatistics*. London: Academic Press.

Kershaw, K.A. 1957. The use of cover and frequency in the detection of pattern in plant communities. *Ecology* 38:291–99.

Kershaw, K.A. and Looney, J.H.H. 1985. *Quantitative and Dynamic Plant Ecology*. 3rd ed. Victoria, B.C. Canada: Edward Arnold.

Krummel, J.R.; Gardner, R.H.; Sugihara, G.; O'Neill, R.V.; and Coleman, P.R. 1987. Landscape patterns in a disturbed environment. *Oikos* 48:321–24.

Legendre, L. and Legendre, P. 1983. *Numerical Ecology*. New York: Elsevier.

Leopold, A. 1933. *Game Management*. New York: Scribner.

Levin, S.A. and Buttel, L. 1986. Measures of patchiness in ecological systems. Publication No. ERC–130, Ecosystem Research Center, Cornell University, Ithica, N.Y.

Loehle, C. 1983. The fractal dimension and ecology. *Speculations in Science and Technology* 6:131–42.

Ludwig, J.A. 1979. A test of different quadrat variance methods for the analysis of spatial pattern. In *Spatial and Temporal Analysis in Ecology*, Statistical Ecology Series Vol. 8, eds. R.M. Cormack and J.K. Ord, pp. 298–304. Fairland, Md.: International Cooperative Publishing House.

Ludwig, J.A. and Cornelius, J.M. 1987. Locating discontinuities along ecological gradients. *Ecology* 68:448–50.

Ludwig, J.A. and Goodall, D.W. 1978. A comparison of paired- with blocked-quadrat variance methods for the analysis of spectral pattern. *Vegetation* 38:49–59.

McIntosh, R.P. 1985. *The Background of Ecology*. Cambridge, England: Cambridge Univ. Press.

Manly, B.F.J. 1985. *The Statistics of Natural Selection*. New York: Chapman and Hall.

Marr, D. 1980. *Vision*. San Francisco: W.H. Freeman.

Marr, D. and Hildreth, E. 1980. Theory of edge detection. *Proceedings of the Royal Society of London Series B* 207:187–212.

Maurer, B.A. 1985. Avian community dynamics in desert grassland: Observational scale and hierarchical structure. *Ecological Monographs* 55:295–312.

Naveh, Z. and Lieberman, A.S. 1984. *Landscape Ecology: Theory and Application*. New York: Springer-Verlag.

O'Neill, R.V. 1989. Perspectives in hierarchy and scale. In *Perspectives in Ecological Theory*, ed. R.M. May, and S.A. Levin. pp. 140–156. Princeton Univ. Press.

O'Neill, R.V.; DeAngelis, D.L., Waide, J.B.; and Allen, T.F.H. 1986. *A Hierarchical Concept of Ecosystems*. Princeton, N.J.: Princeton Univ. Press.

O'Neill, R.V.; Gardner, R.H.; Milne, B.T.; Turner, M.G.; and Jackson, B. Heterogeneity and spatial hierarchies. In *Heterogeneity in Ecological Systems*, eds. J. Kolasa and S.T.A. Pickett. New York: Springer-Verlag, in press.

Ord, J.K. 1979. Time-series and spatial patterns in ecology. In *Spatial and Temporal Analysis in Ecology*, Statistical Ecology Series Vol. 8, eds. R.M. Cormack and J.K. Ord, pp. 1–94. Fairland, Md.: International Cooperative Publishing House.

Orloci, L. 1978. *Multivariate Analysis in Vegetation Research.* 2nd ed. The Hague: Dr. W. Junk.

Overington, I. and Greenway, P. 1987. Practical first-difference edge detection with sub-pixel accuracy. *Image Vision Computing* 5:217–24.

Palmer, M.W. 1988. Fractal geometry: A tool for describing spatial patterns of plant communities. *Vegetatio* 75:91–102.

Patton, D.R. 1975. A diversity index for quantifying habitat edge. *Wildlife Society Bulletin* 3:171–73.

Pielou, E.C. 1977. *Mathematical Ecology.* New York: John Wiley and Sons.

Platt, T. and Denman, K.L. 1975. Spectral Analysis in Ecology. *Annual Review of Ecology and Systematics* 6:189–210.

Ranney, J.W.; McBruner, J.W.; and Levenson, J.B. 1981. The importance of edges in the structure and dynamics of forest islands. In *Forest Island Dynamics in Man-Dominated Landscapes,* Ecological Studies No. 41, eds. R.L. Burgess and D.M. Sharpe, pp. 67–95. New York: Springer-Verlag.

Rayner, J.N. 1971. *An Introduction to Spectral Analysis.* London: Pion Limited.

Ripley, B.D. 1978. Spectral analysis and the analysis of pattern in plant communities. *Journal of Ecology* 66:965–81.

Ripley, B.D. 1981. *Spatial Statistics.* New York: John Wiley and Sons.

Risser, P.G.; Karr, J.R.; and Forman, R.T.T. 1984. Landscape ecology: directions and approaches. Special Pub. No. 2, Illinois Natural History Survey, Champaign.

Robertson, G.P. 1987. Geostatistics in ecology: Interpolating with known variance. *Ecology* 68:744–48.

Schuerholz, G. 1974. Quantitative evaluation of edge from aerial photographs. *Journal of Wildlife Management* 38:913–20.

Smith, H.F. 1938. An empirical law describing heterogeneity in the yields of agricultural crops. *Journal of Agricultural Science* 28:1–23.

Sneath, P.H.A. and Sokal, R.R. 1973. *Numerical Taxonomy.* San Francisco: W.H. Freeman.

Sokal, R. and Oden, N.L. 1978a. Spatial autocorrelation in biology 1. Methodology. *Biological Journal of the Linnean Society* 10:199–228.

Sokal, R. and Oden, N.L. 1978b. Spatial autocorrelation in biology 2. Some biological implications and four applications of evolutionary and ecological interest. *Biological Journal of the Linnean Society* 10:229–49.

Steele, J.H., ed. 1985. *Spatial Pattern in Plankton Communities.* New York: Plenum Press.

Thomas, J.W.; DeGraaf, R.M.; and Mawson, J.C. 1977. Determination of habitat requirements for birds in suburban areas. USDA Forest Service Research Paper, NE 357, Upper Darby, Pa.

Usher, M.B. 1975. Analysis of pattern in real and artificial plant populations. *Journal of Ecology,* 63:569–86.

Warrick, A.; Myers, D.; and Nielsen, D. 1986. Geostatistical methods applied to soil science. In *Methods of Soil Analysis,* Part 1, *Physical and Mineralogical Methods,* 2nd ed., pp. 53–82. American Society for Agronomy and Soil Science Society of America, Agronomy Monograph No. 9.

Watt, A.S. 1947. Pattern and process in the plant community. *Journal of Ecology* 13:27–73.

Webster, R. 1973. Automatic soil-boundary location from transect data. *Mathematical Geology* 5:27–37.

Webster, R. 1977. *Quantitative and Numerical Methods in Soil Classification and Survey.* New York: Clarenndon Press.

Weigert, R.G. 1962. The selection of an optimum quadrat size for sampling the standing crop of grasses and forbs. *Ecology* 43:125–29.

Wertz, W.A. 1965. Interpretation of soil surveys for wildlife management. *American Midland Naturalist* 75:221–31.

Whittaker, R.H. 1956. Vegetation of the Great Smoky Mountains. *Ecological Monographs* 26:1–80.

Whittaker, R.H. 1960. Vegetation of the Siskiyou Mountains, Oregon and California. *Ecological Monographs* 30:279–338.

Wiens, J.A. 1986. Spatial scale and temporal variation in studies of shrubsteppe birds. In *Community Ecology*, eds. J. Diamond and T.J. Case, pp. 154–72. New York: Harper and Row.

Wiens, J.A. Spatial scaling in ecology. *Functional Ecology*, in press.

Wierenga, P.J.; Hendrickx, J.M.H.; Nash, M.H.; Ludwig, J.; and Daugherty, L.A. 1987. Variation of soil and vegetation with distance along a transect in the Chihuahuan Desert. *Journal of Arid Environments* 13:53–63.

Wilson, M.V. and Mohler, C.L. 1983. Measuring compositional change along gradients. *Vegetation* 54:129–41.

3. Remote Sensing for Analysis of Landscapes: An Introduction

Dale A. Quattrochi and Ramona E. Pelletier

3.1 Introduction

The use of remote sensing as a tool for analyses of environmental, cultural, and natural resource management characteristics is well documented (see, for example, Richason 1983; Holz 1985; Colwell 1983; Lo 1986; Campbell 1987; Lillesand and Kiefer 1987; Jensen 1989). Because of its multispectral capabilities, remote sensing presents a unique perspective for observation and measurement of biophysical characteristics (Reeves 1975; Colwell 1983). Remotely sensed data can be collected at multiple scales and at multiple times, thereby offering the opportunity for analyses of various phenomena synoptically from local to global scales through time. These attributes make remote sensing appealing for application to landscape ecology. Moreover, remote sensing can be used as an important data source for the development and refinement of models and can be used to validate models that predict landscape change.

To illustrate the benefits of remote sensing to landscape studies, we first describe how remote sensing can be applied within a landscape ecological framework. We then describe the concept of a "remotely sensed landscape" as a means for observing and measuring landscape components from a multidimensional perspective. Such a discussion is useful for elucidating the spatial, scaling, and temporal attributes of remote sensing. A review of some commonly used image processing techniques is then presented to illustrate how remotely sensed data can be processed and enhanced to extract information through digital techniques.

Analytical approaches that use image analysis are important in remote sensing because they can be applied to large areas with relative efficiency. The remainder of the chapter gives examples of how remote sensing has been used within a landscape framework. This review focuses on the cross-disciplinary nature of remote sensing as a way of illustrating the applicability of this technology to the interdisciplinary science of landscape ecology. Foremost in this section are discussions of space, time, and scale as they relate to remote sensing of landscape characteristics.

3.2 Remote Sensing and Landscape Ecology

Remote sensing promises to bridge the gap between intensive ecological research and the better planning and management of landscapes (Johnson 1969). Three distinct advantages of remote sensing for environmental studies are that (1) observer interference is prevented because of the large distance between the sensor and the object; (2) regional or even global measurements can be done repeatedly; and (3) there is a wide variety of spectral ranges and sensors available to provide remotely sensed data (Lulla and Mausel 1983). Examples of the types of natural resource and planning/environmental management applications where remote sensing can be used are presented in Table 3.1. This list is not all-inclusive, and in many cases the examples overlap between the three general categories. Common to almost all the examples is that each is either a cause or a result of a landscape process. It is evident that remote sensing can provide an added dimension to landscape studies and, therefore, can greatly benefit quantitative model development and refinement.

Determining the meaning and validity of remote sensing data requires familiarity with the ecosystem that is being measured, along with a basic understanding of the matter-energy relationships manifested in the remotely sensed data (Johnson 1969). Four kinds of ecological inquiry are amenable to remote sensing techniques (Johnson 1969): (1) inventory and mapping of resources, (2) quantification of environmental characteristics, (3) describing the flow of matter and energy in the ecosystem, and (4) evaluating change and alternative solutions for ecosystems management. A series of ecological problems that can be studied by using remote sensing are shown in Table 3.2. The table illustrates that remote sensing can be employed in a variety of ways to acquire important ecological information at different levels of analysis. Community level information is essential to ecosystem studies, and it is apparent from Table 3.2 that remote sensing offers a capability for identifying community characteristics (e.g., pattern, extent, vigor, and disturbance) from a synoptic perspective. One of the most dynamic attributes of remote sensing, which may be inferred from the table, is the ability to detect and observe temporal changes in habitat conditions.

Table 3.3 offers a number of research questions that are germane to the application of remote sensing to landscape ecology. These questions are not all-encompassing and should be addressed as perhaps more of a starting point for determining how remote sensing can be used in landscape studies. The individual questions

Table 3.1. General Remote Sensing Application Types[a]

Renewable Resources	Nonrenewable Resources	Planning/Environmental Management
Crop inventory	Geologic structure	Land use/cover classification
Crop yield	Landforms	Land use/cover change
Crop condition	Lithology	Urban ecosystem analysis
Crop irrigation		Environmental impact
Agricultural episodal event	Thermal anomalies	Coastal zone monitoring
Soil classification	Geobotanical anomalies	Currents (near shore)
Soil erosion	Topography	Tides
Soil moisture	Gravity fields	Bathymetry
Forest inventory	Magnetic fields	Ocean pollution
Forest stand		Drainage patterns
Forest stand evaluation		Inland water inventory
Forest condition		Snow pack parameters
Forest episodal event		Ice (inland and near shore)
Range and natural vegetation inventory		Water quality (inland and near shore)
Range forage condition		Wetland episodal event
Range episodal event		Wetland estuaries inventory
		Hydrologic episodal event
		Wildlife habitat inventory
		Wildlife habitat evaluation

[a] Source: Adapted from Lulla and Mausel (1983). Ecological Applications of Remotely Sensed Multispectral Data. In *Introduction to Remote Sensing of the Environment.* 2nd ed. Benjamin F. Richason, Jr., ed. Copyright 1983. Reprinted by permission of Kendall/Hunt Publishing Company.

associated with space, time, and dynamics listed in Table 3.3 are not discrete but are interrelated. Hence, the spatial framework for remote sensing analysis must be considered in reference to both the temporal conditions and dynamics associated with the landscape processes to be observed and measured. It is also evident from Table 3.3 that a number of different types of remotely sensed data can be employed within the scope of an investigation to provide answers to the specific landscape questions under consideration. An appropriate data collection design, which uses data from multiple remote sensing sources, can then be implemented. For example, as a first-level evaluation, color infrared aerial photos can be used to observe the general condition of the landscape. At a second level, analyses may require satellite data (e.g., Landsat or Systeme Pour l'observation de la Terre [SPOT]) to provide information at a broader spatial scale on the patterns or distribution of landscape phenomena. To adequately address temporal dynamics, a third level of analyses may be required. This evaluation would involve the acquisition of remotely sensed data from satellites over time as a means for detecting and measuring change. The questions posed in Table 3.3, therefore, can be approached from a hierarchical perspective by using remote sensing as a means for resolving complex landscape interrelationships.

Table 3.2. Ecological Concepts and Implications for Remote Sensing Ecological Concepts/Framework[a]

Ecological Concepts/Framework	Implications for Remote Sensing
1. Species, populations, and communities are uniquely organized and ordered as determined by their niche requirements (with the exception of people-manipulated agroecosystems).	1. Spectral response patterns represent this community/population/species organization and thus provide information for identification, mapping, and areal differentiation.
2. Ecological viability of the habitat determines the requirements of seedlings and vegetative offsprings and mature individuals in vegetation.	2. Variation in spectral response patterns at various phenological stages in the same test site is useful for remote phenological observations.
3. Biomass and appearance of plants in any habitat are altered by several factors classified under "stress" and "disturbance."	3. Variation in spectral response patterns due to stress or disturbance is of value to measure stress or disturbance remotely.
4. The effects of both stress and disturbance change according to their severity.	4. Remotely sensed data can be used to estimate severity of stress vital for economic-management decisions.
5. Organismic ecological amplitude is determined by the law of tolerance. This has practical value. Species capable of tolerating the habitats containing mineral deposits become "indicator species" or "indicator vegetation."	5. Distinct spectral responses from indicator vegetation are valuable for remotely locating mineral deposits.
6. Disturbances such as clearance cause successional or cyclical changes in the habitat.	6. Comparative analysis of multi-temporal spectral responses is useful in remote change detection-monitoring activities.
7. Dominance in vegetation is the process by which the resources in a certain habitat are monopolized by one species or by a group of (usually small) species.	7. The dominant species contribute to the major portion of spectral response, thus making it possible to estimate such important ecological parameters as biomass and primary production.
8. The vegetation of any area is the function of its physical environment, edaphic environment, and its genetic potential, thus growing in a specific habitat, and each habitat has specific carrying capacity. Wildlife species deriving the energy from vegetation or biomass are thus tied to the specific habitat.	8. Remote sensing of vegetation gives a precise estimate of habitat conditions and biomass-energy and is, thus, useful for wildlife studies.

[a] Source: Adapted from Lulla and Mausel (1983). Ecological Applications of Remotely Sensed Multispectral Data. In *Introduction to Remote Sensing of the Environment*. 2nd ed. Benjamin F. Richason, Jr., ed. Copyright 1983. Reprinted by permission of Kendall/Hunt Publishing Company.

Table 3.3. Pertinent Questions Regarding the Use of Remote Sensing in Landscape Ecology

---

Questions of Space

What is there?
  Land cover attributes (e.g., forest, field, urban)
  Type of landscape elements present (e.g., vegetation and animal species and surface materials)
  Terrain Attributes (e.g., topography, aspect, and slope)
  Land cover texture (e.g., rough or smooth)
What is the arrangement, distribution, and pattern of the landscape elements?
  Heterogeneous
  Continuous or dispersed
  Linear
  Vertical (e.g., tree canopy, buildings)
  Horizontal
  Matrix
What is the spatial scale required for analysis?
  Micro
  Meso
  Regional
  Global
  Multiple

Questions of Time

What are the temporal dynamics of the landscape?
  Discrete
  Continuous
  Random
  Chaotic
What are the time frames required for analysis?
  Short (e.g., minutes, hours)
  Medium (e.g., days, weeks)
  Long (e.g., years)
  Multitemporal
  Indeterminate

Questions of Dynamics

What kinds of processes are involved that shape the landscape (why does the landscape exist in its present form)?
  Explicit (i.e., easily observable)
  Implicit (i.e., not easily observable)
  Natural
  Anthropogenic
  Stochastic
  Heuristic

(*continued*)

Table 3.3. *continued*

What is the nature of these processes?
    Static (i.e., stable)
    Entropic (i.e., continuous change)
    Chaotic
    Disturbed
    Undisturbed
    Constant (or consistent)
What types of models can be used to define, measure, or simulate these processes?
    Descriptive
    Quantitative
    Bio-physical
    Individual and population level
    Interaction-redistribution
    Optimization
    Integrative
    Nested

3.3 A Conceptual Model of the Remotely Sensed Landscape

In addressing the questions presented in Table 3.3, it is useful to consider a conceptual model of a remotely sensed landscape. Landscape attributes are time and space dependent and are a function of the processes that shape the landscape (Forman and Godron 1986). The remotely sensed landscape is multidimensional: horizontal, vertical, and multispectral. All of these dimensions (or axes) exist within and transcend the space, time, and dynamics (i.e., process) domains illustrated in Table 3.3.

The horizontal axis is comprised of an array of spatial attributes that define or describe the relative composition and position of land cover elements within the landscape. Patterns of land covers and the adjacency or dispersion (i.e., distributions) of land units, for example, are fundamental attributes of this horizontal dimension.

The vertical domain is predominated by factors related to the height above the horizontal axis and extends into the atmosphere. Important vertical characteristics are topography and the height of trees, buildings and other features that protrude upward from the horizontal surface, and atmospheric elements (e.g., precipitation, wind, clouds, air mass, etc.). The distinction between purely horizontal or vertical factors is in many cases unclear, depending upon the scale of observation. Shrubs, forbs, and grasses, for example, are horizontal parameters in their distributions across a landscape, and they also have a vertical aspect (e.g., the height of blades of grass above the surface of the soil or the extent of the leaf mass above the ground). Similarly, trees may be perceived in the vertical domain (i.e., the height of the tree), but they have a horizontal axis relative to their spatial distributions (i.e., aggregation and dispersion) in comparison to other trees within the forest community and in the horizontal width of their canopies. We see, then, where the identification, categorization, and measurement of landscape attributes in the horizontal and vertical dimensions are a function of the spatial scale of observation (Meentemeyer and Box 1987; Meentemeyer 1989). The effects of spatial scale must be of paramount consideration since they permeate landscape studies of all types and origins. This is particularly important in studies that employ remote sensors for observation or measurement of land cover types and related landscape characteristics (White and MacKenzie 1986; Woodcock and Strahler 1987).

Remote sensing exists as a way to integrate spatially heterogeneous responses into a more easily measurable format by quantifying spectral responses at a specific scale (e.g., 10, 20, or 30 m). The picture element or "pixel" represents the smallest unit of spatial area on the ground for which data are collected through the use of digital remote sensing systems. A pixel, therefore, can be of various sizes depending upon the spatial resolution of the sensing system. For example, the Landsat Thematic Mapper (TM) sensor has a spatial resolution or pixel size of 30 m, which represents a 30- by 30-m area on the ground. In an area of heterogeneous land covers, spectral responses for different objects within a 30-m pixel will be averaged or aggregated into a composite spectral response for any particular pixel that falls over a specific area on the ground. Thus the multispectral domain exists as both an additional axis of information available for analysis and as an integrating factor of scale-related phenomena.

As every object on this theoretical landscape has a vertical and horizontal spatial structure, so does it possess a characteristic multispectral response. This response, referred to as a "signature," is a function of the way objects or phenomena reflect, emit, or transmit electromagnetic energy. Identification and separation of objects can be accomplished in part through analysis of spectral signatures (Jensen 1986). The discrimination of a signature is not limited to only one portion of the spectrum at any one time, because remote sensors can record simultaneous responses across a wide range of the electromagnetic spectrum (EMS) (Colwell 1983). Some disciplines have a narrow perspective about which region of the EMS to use, whereas the actual signature of any object or phenomenon covers the entire spectrum. This view may result from (1) the varying characteristics or information content of the phenomena under observation which depend upon the region of the EMS used or (2) researchers becoming accustomed to or familiar with using certain sensors that, due to engineering considerations, generally deal with relatively small and spectrally similar portions of the spectrum. While these perspectives may be sufficient for many applications (e.g., simple land cover classifications), an evaluation of a wider range of the spectrum may be necessary to gather information for more complicated ecological analysis (e.g., vegetation composition structure).

Most sensors in the reflective portion of the spectrum provide a two-dimensional horizontal perspective of the landscape. Other portions of the spectrum, especially the longer wavelengths, provide a means to characterize a third, vertical dimension. Microwave data (e.g., Synthetic Aperture Radar data) have been used to determine prostrate and standing crop residue and directional orientation of rows in cropland (Brunfeldt and Ulaby 1984). These data have also been used to determine landscape macrotopography and microtopography (Wu 1984) and vegetation canopy density and structure (Wu and Sader 1987). An instrument that has unique operational characteristics is the LASER Profiler, an active sensing system that emits a LASER beam to the earth's surface. This beam, in turn, is reflected back to the sensor by objects on the ground. The differential in time between beam emittance and reflective beam return is used to determine the height of an object. The LASER Profiler has been used to determine forest biomass and canopy height (Nelson et al. 1988). Other instruments, such as airborne electromagnetic (AEM) profilers, utilize an induced three-dimensional magnetic field to measure a profile of earth material electrical conductivity. Because of their ability to penetrate water and soil, AEM have been useful in mapping shallow water bathymetry and many subsurface sediment physical and chemical characteristics (Pelletier and Wu 1989). Characterization of this vertical dimension of landscapes, either above ground or below ground, provides a means to develop and test three-dimensional ecological models.

Aside from the information available on spectral signatures of objects within the multispectral domain, the electro-optical engineering capabilities of remote sensing systems allow the quantification of these responses into measurement data. For example, estimations of biomass and primary productivity have been developed by using Leaf Area Indices (LAI) and remotely sensed data for a variety of vegetative land covers (Colwell 1983; Goetz et al. 1983; Lulla and Mausel 1983; Rock et al. 1986; Waring et al. 1986; Peterson et al. 1988). The LAI is based on a ratio of spectral responses in the red wavelengths (in the visible portion of the EMS from 0.6 to 0.7 μm) to the reflective near-infrared wavelengths (from 0.7 to 1.0 μm). Green LAI response is negatively related to red reflectance but is positively related to near-infrared reflectance (Lo 1986). The ratio of infrared/red (IR/R) expresses the increasing difference between red and near-infrared reflectance with increasing green LAI (Lo 1986). Another useful indicator of biomass vigor and productivity is the vegetation index (VI), which represents a ratio of subtractive and additive spectral responses (IR−R)/(IR+R). Vegetation indices have been used in a number of studies with good results (Jensen 1983; Tucker et al. 1985; Goward et al. 1985; Jensen 1986; Lillesand and Kiefer 1987; Running and Nemani 1988; Nemani and Running 1989).

Spectral responses can also be categorized into spectral classes for use in mapping land covers and related ecological phenomena (Colwell 1983; Lo 1986). The inventory and mapping of land cover phenomena are crucial to understanding why landscapes appear as they do and are a requisite for understanding how they came to exist. It is useful to employ multiple regions of the EMS for analysis by using remotely sensed data collected in two or more portions of the spectrum (e.g., visible, reflective infrared, or thermal). Multispectral remote sensing, therefore, offers the capability for identification and measurement of biophysical characteristics that otherwise could not be made if only a single region of the EMS were used. Table 3.4 provides insight into how multispectral remote sensing can collectively provide more information than is available in one portion of the EMS alone.

Table 3.4. Thematic Mapper Spectral Bands and a General Description of Uses for Each Spectral Region[a]

Band 1 (0.45–0.52 µm) (Visible - Blue Region):

Designed for penetration of water, making it useful for coastal water mapping. Also useful for discrimination of soil from vegetation and deciduous from coniferous vegetation. Useful for cultural feature identification.

Band 2 (0.52–0.60 µm) (Visible - Green Region):

Designed to measure green reflectance peak of vegetation. Useful for vegetation discrimination and vigor assessment. Also useful for cultural feature identification.

Band 3 (0.63–0.69 µm) (Visible - Red Region):

Designed for chlorophyll absorption detection important for vegetation discrimination. Also useful for cultural feature identification.

Band 4 (0.76–0.90 µm) (Reflective Near-Infrared):

Useful for determining vegetation types, vigor, and biomass content. Also useful for delineating water bodies and for soil moisture discrimination.

Band 5 (1.55–1.75 µ) (Reflective Mid-Infrared):

Indicative of soil and vegetation moisture content. Also useful for differentiating snow from clouds.

Band 6 (10.4–12.5 µm) (Thermal Infrared):

Useful for vegetation stress analysis, soil moisture discrimination, and thermal mapping applications.

Band 7 (2.08–2.35 µ) (Reflective Mid-Infrared):

Useful for discriminating mineral and rock types. Also sensitive to vegetation moisture content.

[a]Adapted from Lillesand and Kiefer (1987) p. 567 and Lo (1986) p. 31.

The table illustrates some general uses for each of the spectral regions or "bands" available on the Landsat TM as a means for showing the utility of multispectral remote sensing. (See also Colwell 1983; Jensen 1983, 1986; Lo 1986, Campbell 1987; and Lillesand and Kiefer 1987 for other examples of the attributes of multispectral remote sensing for landscape analysis.)

The multispectral and multiple spatial domain (i.e., vertical and horizontal) data provided by remote sensors are ideally suited for integration into a geographic information system (GIS). The combination of remote sensing and GIS technology has become important for the analysis and manipulation of data for landscape studies and ecological assessment (Marble and Peuquet 1983; Krummel 1986; Ripple 1987; Parker 1988; Ehlers 1989; Gessler et al. 1989; Jakubauskas 1989), where GIS methods offer the capability for readily combining, integrating, and analyzing disparate data types within a spatial framework.

As we have seen from this discussion, the remotely sensed landscape is multi-dimensional, multitemporal, and multiscaled. The sum of the horizontal, vertical, and multispectral dimensions is synergistic, where the integration of the three axes yields much more information than do the individual domains. The application of remote sensing techniques, though, cannot be done without due consideration of the parameters to be measured, the scale of measurement for analysis, the time frame for observation, and an awareness or understanding of the processes that affect the phenomena under study. A summary of the interrelationships between remote sensing and landscape ecology is presented in Table 3.5. The presentation of these relationships is useful in linking the conceptual ideas discussed above with the practical applications that will be described in the following section.

Table 3.5. Summary of Interrelationships Between Remote Sensing and Landscape Ecology[a]

Landscape Ecological Characteristics

Study of pattern, structure, and morphology of phenomena:
Measured attributes may generally be classed into:
  Physical properties
  Spatial properties (e.g., shape, size, pattern, arrangement, and texture)
  Scale properties (micro, meso, regional, or global)
  Temporal properties

Remote Sensing Attributes

Measurement of physical properties:
An object's signature or spectral response can provide data on:
  Spectral color or visual appearance
  Temperature
  Moisture content
  Organic and inorganic composition
Measurement of spatial properties:
  Provides information on geometry and position (e.g., size, shape, arrangement, and texture)
  Provides both point (per pixel) and areal information (i.e., integration of spatial properties)
  Contributes to better preparation of thematic maps for use in spatial analysis
  Provides for multiple scales of analysis (micro, meso, regional, or global)
Measurement of temporal properties:
  Multitemporal analysis capability (i.e., time-dependent repeatability of data collection)
  Provides capability for analysis of landscape change through time

[a] Some information adapted from Estes et al. (1980).

3.4 Image Processing Techniques

After making the decision to use remotely sensed data in one's investigation, it may be necessary to consider the variety of types of computer processing techniques available for data analysis and information extraction. Unlike the manual interpretation of aerial photography, the digital nature of satellite data permits the employment of multiple automated processing techniques, often referred to as digital image processing. These techniques allow analytical approaches (e.g., land cover classifications and a variety of qualitative measures) to be applied quickly and objectively to large areas.

The central idea behind digital image processing is rather simple, but the processing techniques employed are diverse and may be mathematically complex. Digital data are entered as an array of pixels into the computer and stored for subsequent processing by one or more algorithms. As each pixel is modified by these algorithms, a new file of digital data is developed. These new files may be displayed or further manipulated by additional algorithms until the analysis process is completed.

Virtually all digital image processing falls into one or more of the following categories: (1) image rectification and restoration, (2) image enhancement, (3) image classification and modeling, (or 4) data merging. Often processing stages follow a sequence from category 1 through 4, but not all categories need be addressed for each application. Before processing can take place, however, it is wise to clean up the data, if necessary, through image rectification and restoration techniques. Such activities involve removing noise in the signal recorded by the sensor, replacing missing data due to sensor recording problems, and correcting for radiometric problems (i.e., problems with the inherent spectral signal from the target) perhaps caused by sun angle or atmospheric haze. A common practice is to geometrically rectify the data to a base map early in the processing cycle, but such rectification procedures can alter the appearance of linear features, which could later impact spatial filtering techniques. Spatial filtering is used to emphasize or de-emphasize the spatial heterogeneity of tonal variations (often caused by land cover variations) in an image. Filtering techniques are often used to identify linear patterns in the data. In such situations it may be more desirable to geometrically rectify the data after analysis or bypass this step if rectification is not a requirement for the application.

Image enhancement is usually the next phase in data processing. The purpose of image enhancement is to increase the visual distinction between features in a scene in order to increase the amount of information that can be visually interpreted from the data. Enhancement procedures may be applied to a single image or may involve combinations of multiple images to produce the enhanced product. Enhancement techniques generally fall into one of three categories: (1) contrast manipulation, which operates in the spectral domain of the data; (2) spatial feature manipulation, which operates within the spatial domain; (or 3) multi-image manipulation, which principally deals with spectral differences between images.

In contrast manipulation, spectral responses are modified to either highlight or subdue a certain range of values in the data. Typically, spectral values are very different between water and terrestrial surfaces or between vegetation and bare soil or man-made materials (i.e., there is a contrast between spectral values). In cases such as this, techniques can be used to segment the data into two parts, thereby masking out water surfaces to focus on the vegetated terrestrial surface. The dynamic range (or spread of class-specific data values) of the focal vegetated area could be increased through contrast stretching techniques, such as a linear stretch, to highlight differences between vegetation types. Density or level slicing techniques can also be used to regroup relatively similar spectral ranges, thus reducing or collapsing the number of individual spectral values for analysis.

Digital data may also be enhanced according to spatial characteristics. Spatial feature manipulation generally involves some type of convolution technique that evaluates pixels based on the spectral value of neighboring pixels (Moik 1980). Spatial filtering can be used to highlight or subdue edges and spatial relationships among neighboring pixels. High-pass filters enhance differences, while low-pass filters smooth differences between adjacent land cover types. Edge enhancement techniques are especially useful in determining abrupt changes between adjacent pixels due to significant differences in their spectral values. Such differences can come about as boundaries between vegetation types, between water bodies and inert soils or vegetation, or as a function of man-made features that have linear or regular patterns such as roads, fence lines, field boundaries, and center pivot irrigation. Edge enhancement techniques can be useful in better understanding the spatial relationships of differing landscape types (e.g., adjacency, dispersion, roughness, or texture). Other spatial feature manipulation techniques can be applied to digital data for landscape fractal dimension determination (De Cola 1989; Lam 1990). The fractal dimension of the boundaries of landscape ecological units (e.g., species habitat) can determine the complexity of these boundaries and infer spatial relationships between units (Mandelbrot 1977; Burrough 1981; Goodchild and Mark 1987).

Multi-image manipulation is often the most appropriate image enhancement method for many ecological applications. Some techniques under this approach include the Kauth-Thomas or Tasseled Cap Transform (Kauth and Thomas 1976; Crist and Cicone 1984; Crist and Kauth 1986), the Normalized Difference Vegetation Index (Rouse et al. 1973; Jensen 1986; Lillesand and Kiefer 1987), Principal Component Analysis (Byrne et al. 1980; Lillesand and Kiefer 1987; Conese et al. 1988), Canonical Correlation (Jensen and Waltz 1979; Lillesand and Kiefer 1987), and the Intensity-Hue-Saturation Color Space Transformation (Schowengerdt 1983). These methods typically tend to reduce data correlation between two or more images and result in one or more output products that accentuate differences in the data. Additional techniques that can be employed to enhance remotely sensed data for landscape analysis include the ratioing of selected spectral bands, the use of differencing indices to emphasize the contrast between two images, and other more complex algorithms that can be applied to multiple data sets over a wide range of spectral bands and temporal periods. (Numerous references exist that describe image enhancement techniques. See, for example, Moik 1980; Colwell 1983; Schowengerdt 1983; Jensen 1986; and Lillesand and Kiefer 1987.)

The output from a variety of image enhancement techniques may be the end products for a number of applications. Other approaches follow the enhancement procedures with image classification techniques or bypass the need for enhancement procedures before employing pattern recognition and classification procedures. Hence, there is no single way to approach digital image analysis. The methodology chosen depends on the type and scale of the data, the intended application, and the computer resources available.

The overall objective of pattern recognition or image classification procedures is to automatically categorize all pixels in an image into land cover classes or themes. Three types of classification approaches include spectral, spatial, and temporal pattern recognition schemes. Spectral classification has been the most widely used approach and resides at the core of most classification activities. Classification typically involves a series of mathematical and statistical algorithms to cluster pixels of similar spectral value by using techniques such as the minimum-distance to means classifier (Hixson et al. 1980; Colwell 1983), a parallelepiped classifier (Addington 1975; Colwell 1983; Jensen 1986), and the Gaussian maximum-likelihood classifier (Anuta 1977; Colwell 1983; Jensen 1986). Spatial classification procedures typically involve more artificial intelligence capability than spectral classifiers by attempting to replicate the spatial-spectral synthesis that a human analyst may employ in visual interpretation. Consequently, spatial classifiers tend to be more complex and more computationally demanding than spectral classifiers. Temporal classifiers generally employ some type of spectral classifier that is modified to account for changes between multidate data.

3.5 Some Examples of Space, Time, and Scale Interrelationships in Remote Sensing for Landscape Ecology

Although detailed examples of how remote sensing can be used for measurement of landscape ecological properties are provided in Chapters 4 and 6, an overview of studies that have utilized remote sensing technology within a landscape framework will be useful to illustrate the interrelationships among space, time, and scale. Our presentation will be illustrative rather than all-inclusive. We give these examples to elucidate the wide application that remote sensing has to landscape ecology and to expand upon the questions that may arise in using this technology.

3.5.1 Spatial Analysis and Remote Sensing

An important application of remote sensing has been the spatial assessment of landscape characteristics (Reeves 1975; Williams and Carter 1976; Short 1982; Short and Stuart 1982; Colwell 1983; Short and Blair 1986). For example, forest silviculture and management has made use of the multispectral, multitemporal, and synoptic attributes of remote sensing techniques (Cibula and Ovrebo 1988;

Fryar and Clerke 1988; Sader 1988; Teuber 1988; Vogelmann 1988; Brockhaus 1989). When coupled with adequate ground-based studies, satellite data have potential for determining the stress agents responsible for forest decline, where degeneration or changes in key trace metals (Cu, Zn, and Pb) may be manifested as spectral signatures (Rock et al. 1986). With its synoptic spatial coverage, remote sensing offers the capability to delineate, map, and measure forest damage (Dull 1988; Jones et al. 1988). Satellite data have also been shown to be useful for monitoring forest vegetation vigor and species distributions (Walsh 1980; Benson and DeGloria 1985; Franklin et al. 1986). In a study by Franklin (1986), Landsat Thematic Mapper Simulator data (30-m spatial resolution) were employed for analysis of specific forest structure and composition components. These data show potential for discriminating spectral response characteristics, such as basal area and leaf biomass, within stands dominated by red fir (*Abies magnifica* var. *shastensis*) and white fir (*Abies concolor*) in a portion of northern California.

Disturbance or disruption has been recognized as a significant influence in landscape patterns (Burgess and Sharpe 1981; Mooney and Godron 1983; Pickett and White 1985; Waring and Schlesinger 1985; Parker 1987; Turner 1987). The spread of the pine bark beetle (e.g., southern pine bark beetle, *Dendroctonus fontalis* Zimmerman), for example, can be viewed as an episodic landscape phenomenon (Rykiel et al. 1988). Because of the widespread occurrence of this insect, studies of the areal extent affected by pine bark beetle infestation and its severity are of interest for determining the effects of large-scale landscape disturbances. Past attempts at identifying areas of pine beetle disturbance with remote sensing have relied principally on interpretation of color infrared aerial photography (Reeves 1975; Colwell 1983). Digital remotely sensed data from satellites offer the ability to survey large areas at one time to establish the pattern of insect disturbance across the landscape (Rencz and Nemeth 1985; Mukai and Sugimura 1987). Other factors in landscape disturbance such as fire or tree species defoliation can be readily observed with remote sensing (Hall et al. 1983; Morin et al. 1988; Vogelmann and Rock 1988).

The analysis of ecosystem pattern is of important concern to landscape ecology (White and MacKenzie 1986; O'Neill et al. 1988). The patchy distribution of habitats across the landscape results from both natural or human (i.e., fire or settlement history) and environmental gradients (i.e., elevational, topographic, and climatic factors) (White and MacKenzie 1986). Ecological surveys and evaluation of patch dynamics can be accomplished with satellite data as a means for analysis of broad-scale landscape patterns (Rubec 1983; Muchoki 1988; Ringrose et al. 1988; Wood and Foody 1989). Walsh (1987) investigated the impact of forest stand and site characteristics on spectral responses from Landsat Multispectral Scanner (MSS) data (80-m spatial resolution). He found that topography (slope and aspect) was a significant factor that affects spectral signatures within a forested environment and suggested that ancillary topographic data be used to reduce this variability, particularly in mountainous areas. The integrated use of remotely sensed and terrain data has been illustrated by Cibula and Nyquist (1987) and Frank (1988). In both of these studies, satellite remotely sensed data were combined with digital elevation data to obtain detailed bioclimatic vegetation distribution assessments over large geographic areas in mountainous terrain.

Although the broad-scale characteristics of satellite and other remotely sensed data are important in providing a qualitative, synoptic view of the land, even more significant is the quantitative analysis of landscapes from these data. Remote sensing is a viable method for measuring and modeling the cause and effect processes that influence the structure, function, and change in landscape evolution or development (e.g., Ecology from Space 1986; Dyer and Crossley 1986). Figure 3.1 illustrates an example of this capability in a model estimating the productive capacity of wetland ecosystems in southern Louisiana (Butera et al. 1984; Dow et al. 1987). Quadrant D shows a digital classification of wetland and nonwetland environments derived from Landsat MSS data. Based on the land cover classification and in situ measurements, net community production in terms of exportable organic matter is determined and illustrated in Quadrant B. Quadrant A indicates the distance in meters for every pixel in the image to the nearest water body. Proximity to water is very important to the immature life stages of many estuarine and marine faunal species. Quadrant C factors both the net community production and distance to water to determine a generalized productive capacity for each pixel in the image. Modeling techniques, such as those illustrated in Fig. 1, can be applied to large geographic areas easily and quickly once the algorithms have been developed.

3.5.2 Analyzing Temporal Change with Remote Sensing

A second important aspect of remotely sensed data is its capacity to record temporal landscape change. United States commercial and operational satellite programs have well-identified sampling intervals for data acquisition for any spot on the ground, ranging from four times daily from the NOAA-6 and NOAA-7 Advanced Very High Resolution Radiometer (AVHRR) to once every sixteen days with the MSS and TM. Other airborne and ground-based systems can be employed as frequently as necessary (within economic feasibility). Satellite-acquired data are often sufficient for relatively infrequent analyses (seasonally or longer) or for baseline ecological characterization.

To support high frequency analysis, remotely sensed data can be acquired on a regular basis (e.g., by using satellite data). For example, data can be collected during the day or night to obtain the diurnal thermal variation of landscapes, at high and low tides to determine tidal inundation, or at high and low sun angle to exaggerate landscape slope and canopy height differences. Data for less frequent periodic analyses can be acquired weekly, monthly or seasonally. This is useful to monitor seasonal changes in crop growth and natural vegetation, to evaluate snowpack and surface water changes and model hydrological consequences, or to trace the spread of drought or disease and monitor their effects on vegetation. For landscape conditions that require more limited temporal analyses, data can be acquired every year or every few years. Examples are the phenological inventory of crop types and planting trends, the identification of vegetational succession, and

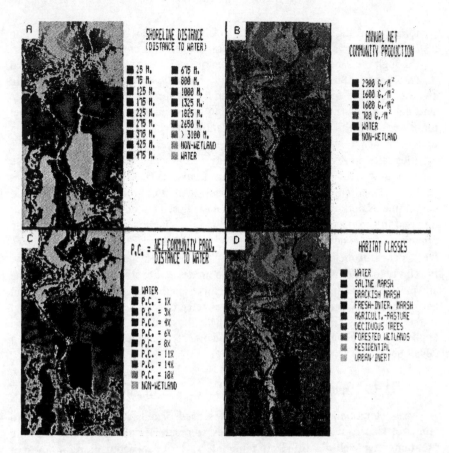

Figure 3.1. Components of a wetlands productive capacity model utilizing remotely sensed (Landsat MSS) data. Quadrant A illustrates shoreline density by measuring distance to water for each land pixel. Quadrant B illustrates net community production of exportable organic matter for each wetland type. Quadrant C depicts the productive capacity for each pixel based on the interactions of shoreline density and net community productivity. Quadrant D is a digital classification of habitat type. (D.D. Dow, NASA, Science and Technology Laboratory, John C. Stennis Space Center.)

the monitoring of urban growth. Remotely sensed data can also be collected for both man-induced and natural episodic events such as pollutant discharge and its effects on the environment, algal blooms, and flood impacts.

Because satellite and airborne systems can typically acquire voluminous amounts of data that can be spatially rectified and processed, characterizing the temporal dynamics of large areas may be easier with remote sensing than through traditional field approaches. Field work is often necessary, however, especially when new or different relationships must be established between the remotely sensed data and field conditions. After such relationships have been established, much field work can be eliminated and restricted to a spot checking of selected sites. Thus, most types of remotely sensed data lend themselves to broad-scale analysis on temporal frequencies not feasible in the past.

Airborne sensing systems have much greater flexibility for temporally dependent data acquisition than do satellite platforms, but are generally more expensive
and sometimes less reliable to operate. The greatest benefit of airborne sensors is
that data can be acquired to best fit the temporal needs of a particular investigation.
These data provide greater flexibility for frequent analyses (more often than
seasonally) of episodic events, or when spectral or spatial resolution requirements
cannot be met by a specific satellite system.

The temporal nature of the data is sometimes limited by the sensing system. For
example, Landsat maintains a sun-synchronous pattern that generally acquires
data during the late morning hours. This time was selected to optimize conditions
for relatively full sun exposure with minimal shadowing and cloud conditions. The
sixteen-day acquisition period of Landsat theoretically permits at least monthly
data collection for the same site, but for regions of the world where even moderate
cloud coverage is common, four to five data collections per year may be the best
that can be obtained. The AVHRR offers up to four acquisitions daily but sacri-
fices spatial resolution (1-km resolution) to attain this frequency.

Data from the AVHRR polar orbiting satellite, with its broad-scale, repetitive
coverage, are emerging as an important tool for modeling vegetation dynamics
over large geographic regions. The AVHRR provides high temporal resolution
with 1-km or 4-km spatial resolution that can offer weekly analyses of earth
surface conditions across the globe. The temporal and spatial capabilities of these
data for mapping environmental conditions and vegetative patterns have been
demonstrated by Tucker and his colleagues (Tucker et al. 1983; Tucker et al. 1984;
Gatlin et al. 1984; Tucker et al. 1985; Goward et al. 1985; Justice et al. 1985;
Tucker et al. 1986). These studies have shown the usefulness of AVHRR data for
developing vegetation indices that can measure primary productivity and for
relating the variations in atmospheric CO_2 to vegetation spectral response. The
application of multitemporal satellite data for measurement of broad-scale vegetation spatial variability and land use has also been demonstrated by Matthews
(1983) and Ormsby et al. (1987).

3.5.3 Landscape Scale and Remote Sensing

The structure and function of a landscape can be perceived differently at different
scales, and it is important for the observer to decide upon appropriate scales for
a study (Allen and Hoekstra 1988; Turner et al. 1989). It is also important to
recognize that the concepts of heterogeneity and homogeneity are scale-dependent
because they describe how individual land cover components or processes are
interrelated across a landscape (O'Neill et al. 1988). The influence of scale on
landscape studies is discussed in-depth by Meentemeyer and Box (1987), and
Meentemeyer (1989) and we direct readers to these citations for a more complete
understanding of scale, the effects of changing scale, and the impact of spatial
scales on analyses.

Scale terminology used in our discussion relates "fine scale" to minute resolution or small study area and "broad scale" to coarse resolution or large study area

(Turner et al. 1989). From a remote sensing context, therefore, fine scale is synonymous with a pixel that encompasses a small amount of area on the ground (e.g., 5 m); a pixel of broad scale would include a large area on the ground (e.g., 1 km). Another differentiation used in relating landscape scales are the terms *grain* and *extent*. *Grain* implies the finest level of spatial resolution available (i.e., pixel size) within a specific data set (Turner et al. 1989). *Extent* refers to the areal breadth or size of a study area (e.g., local, regional, and global) (Turner et al. 1989). For efficiency in cost, data processing time, and analysis, it is necessary to choose the broadest scale data available for identifying the landscape characteristics under consideration. It is important to note, however, that although data can always be resampled to a broader grain size using averaging or percent dominance techniques to recalculate a pixel's value, data cannot be resampled to a finer grain size to improve spatial and spectral resolvability should greater detail be necessary. Fine-scale data, however, are typically more expensive to purchase and process because of the quantity of information involved. Monitoring programs (e.g., environmental or natural resource surveys) usually require spatially broader-scale data (i.e., regional extent) for ease in processing large amounts of temporal data over broad geographic areas. Experimental programs, on the other hand, may require finer-scale data over more frequent time periods for detailed landscape or ecological analysis, or for determining the minimum grain-size requisite for eventual use within an operational framework.

In remote sensing, spatial resolution (or the ability of a sensing system to resolve objects on the ground) is analogous to grain size. The selection of the resolution of the remotely sensed data to be used is a question of importance concomitant with that of scale. Of concern is the problem of mixed pixels, where the spatial resolving power of the sensor is too coarse to capture the intrinsic spatial heterogeneity of a landscape. If the spatial resolution of a sensor is too coarse (i.e., coarse grain size), it will combine signatures from independent objects on the ground into an aggregated spectral response for a particular pixel. Thus, a heterogeneous landscape can become more homogeneous by virtue of a sensor's pixel ground resolution. More spatial resolution (i.e., finer grain size), however, is not necessarily better. If spatial resolution is too fine, objects may appear more heterogeneous than they really are; this misrepresentation masks their inherent homogeneity. The challenge of selecting scales in remote sensing may be illustrated by an example taken from White and MacKenzie (1986), who used remotely sensed data with different spatial resolutions to define vegetation patterns within the Great Smoky Mountains in Tennessee and North Carolina. This area is vegetationally complex, with existing ecosystems often patchy in both time and space. The objective of this study was to determine the optimum scale, or resolution, for differentiating these diverse vegetation patterns. As noted by these investigators:

> Key areas of interest in determining optimum scale of resolution include the size of tree crowns, canopy roughness (i.e., smooth canopies will be sensed as more uniform than canopies with irregular shadowing due to the presence of emergent tree crowns), number of species present within the vegetation type to be mapped (species may have very different spectral properties), the

spectral contrast with the matrix around the forest type, and heterogeneity produced by patchiness within the type (e.g., treefall gaps). Since all of these parameters vary between different vegetation types and, within one vegetation type, between different successional states or sites, no one scale of resolution will be perfect for even a single vegetation mapping goal (White and MacKenzie 1986, p. 63).

The optimal scale of remote sensing will vary with the objectives for analysis and the inherent characteristics of the landscape in question. An 80-m pixel resolution may only be appropriate in many mountain forested regions if the forest types are homogeneous and exist in nonlinear patches over 100 ha. For the Great Smoky Mountains, some vegetation types may be mapped at this scale, but most will only be discernible at a 30-m pixel resolution (White and MacKenzie 1986). A few vegetation types in this area may only be mapped from pixels with resolutions of 13 m, while remote sensing of within-crown stress characteristics will require data at a pixel size of 1 to 2 m. Hence, landscape processes appear to be hierarchical in pattern and structure. Measurement of these processes is a function of scale (Urban et al. 1987) and a function of the resolving capabilities of the sensors used to observe landscape phenomena (Nellis and Briggs 1989).

The effects of spatial resolution can be illustrated by comparing the images obtained from different sensors for the same geographic area (Fig. 3.2). Image I depicts the southern portion of the Florida peninsula with AVHRR data at a 1-km pixel resolution. Image II (representing the area enclosed in the larger box in Image I) depicts much of the Florida Everglades and surrounding area by using Landsat MSS data with 80-m pixel resolution. Image III (representing the region enclosed in the larger box of Image II) depicts an area along the Shark River Slough in the Everglades National Park from Landsat TM data with 30-m pixel resolution. The small boxes in each of the images on the left represent the same 9-km^2 area and are enlarged to illustrate the inherent detail due to sensor resolution in the three images on the right (i.e., images A, B, and C). As seen in Fig. 3.2, the necessity of matching sensor resolution and scene area coverage with the appropriate scale of the investigation becomes evident. AVHRR data appear useful for broad regional to global scales, MSS data may be best utilized at smaller regional scales, and TM data are best suited for local-scale investigations.

3.6 Summary

As we have described, remote sensing exists as an important observation and measurement tool for analysis of landscape ecological relationships. Remote sensing offers the capability to collect data on landscape characteristics and interactions without disturbing the surrounding environment. We have also seen where the multispectral, multitemporal, and multiscaled attributes of the remotely sensed landscape can provide a look at biophysical and land cover components that previously was unachievable. The identification and measurement of landscape

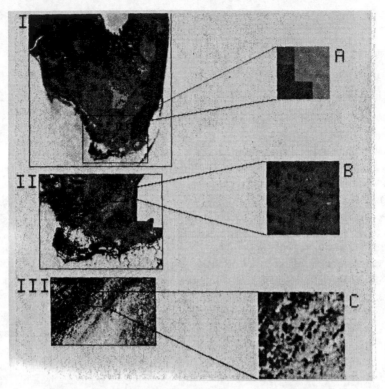

Figure 3.2. An illustration of multiple spatial scale data for the Everglades area in south Florida. Image I depicts 1-km pixel resolution data from the AVHRR; Image II depicts 80-m pixel resolution data from the Landsat MSS; and Image III shows 30-m pixel resolution data from the Landsat TM. Images A, B, and C to the right show resolution effects from each sensor data type for the same 9 km² geographic region.

variables from remotely sensed data include a range of interpretation techniques, from photointerpretation to advanced image-processing algorithms. We have shown that in all of these techniques, however, the factors of space, time, and scale are critical in selecting both the appropriate data and analysis methods.

In summary, we have provided an overview of the many types of applications that remote sensing has to landscape studies. Further applications of remote sensing technology to landscape ecology are presented in other chapters of this book. One of the greatest potential benefits of applying remote sensing to landscape ecology, however, is the ability to measure the state and dynamics of ecological variables and the processes that drive these variables (Botkin et al. 1984; Dyer and Crossley 1986; Hall et al. 1988). Within a GIS framework, the extrapolation of remotely sensed information from point sources to spatially large scales will enhance the capabilities for modeling landscape properties and processes (Krummel 1985, 1986). The testing of quantitative models of important land-atmosphere processes (e.g., gas exchanges) and understanding the role of vegetation (i.e., estimating biomass and net primary productivity) at a global scale, for example, can only be addressed by using remote sensing techniques in conjunction with GIS technology. In fact, the integration of multiple data sources over many different time periods and at different scales is a cornerstone for analyses of very broad-scale land-atmosphere studies, such as the International Geosphere/Biosphere Program (National Research Council 1986) and the NASA Earth Observing System (NASA 1988) initiatives, which will address questions of global ecosystem dynamics and functioning in the 1990s and beyond.

References

Addington, J.D. 1975. A hybrid maximum likelihood classifier using the parallelepiped and Bayesian techniques. *Proceedings: Technical Papers of the 50th Annual Meeting of the American Society of Photogrammetry*, pp. 772–84.

Allen, T.F.H., and Hockstra, T.W. 1988. The critical role of scaling in land modelling. In *Perspectives on Land Modelling*, ed. R. Gelinas, D. Bond, and B. Smit, pp. 9–13. Workshop Proceedings, Nov. 17–20, 1986. Montreal: Polyscience Publications.

Anuta, P.E. 1977. Computer-assisted analysis techniques for remote sensing data interpretation. *Geophysics* 42.468–81.

Benson, A.S., and DeGloria, S.D. 1985. Interpretation of Landsat-4 Thematic Mapper and Multispectral Scanner Data for forest surveys. *Photogrammetric Engineering and Remote Sensing* 51:1281–89.

Botkin, D.B.; Estes, J.E.; MacDonald, R.M.; and Wilson, M.V. 1984. Studying the earth's vegetation from space. *Bioscience* 34:508–34.

Brockhaus, J.A. 1989. An assessment of remotely sensed imagery for use in hardwood stand density distribution mapping in central California. In *Technical Papers*, vol. 4, GIS/LIS, pp. 109–17. American Society for Photogrammetry and Remote Sensing/American Congress on Surveying and Mapping Annual Convention, April 2–7, 1989.

Brunfeldt, D.R., and Ulaby, F.W. 1984. Measured microwave emission and scattering in vegetation canopies. *IEEE Transactions, Geoscience and Remote Sensing* GE–22: 520–24.

Burgess, R.L., and Sharpe, D.M. 1981. *Forest Island Dynamics in Man-Dominated Landscapes*. New York: Springer-Verlag.

Burrough, P.A. 1981. Fractal dimensions of landscapes and other environmental data. *Nature* 294:240–42.

Butera, M.K.; Browder, J.A.; and Frick, A.L. 1984. A preliminary report on the assessment of wetland productive capacity from a remote-sensing-based Model-A NASA/NMFS joint research project. *IEEE Transactions, Geoscience and Remote Sensing* GE–22:502–11.

Byrne, G.F.; Crapper, P.F.; and Mayo, K.K. 1980. Change by principal component analysis of multispectral Landsat data. *Remote Sensing of Environment* 10:175–84.

Campbell, J.B. 1987. *Introduction to Remote Sensing*. New York: Guilford Press.

Cibula, W.G., and Nyquist, M.O. 1987. Use of topographic and climatological models in a geographical data base to improve Landsat MSS classification for Olympic National Park. *Photogrammetric Engineering and Remote Sensing* 53:67–75.

Cibula, W.G., and Ovrebo, C.L. 1988. Mycosociological studies of mycorrhizal fungi in two loblolly pine plots in Mississippi and some relationships with remote sensing. In *Remote Sensing for Resource Inventory, Planning, and Monitoring*, ed. J.D. Greer, pp. 268–307. Proceedings of the Second Forest Service Remote Sensing Applications Conference, April 11–15, 1988. Falls Church, Va.: American Society for Photogrammetry and Remote Sensing.

Colwell, R.N., ed. 1983. *Manual of Remote Sensing*. 2nd ed. Falls Church, Va.: American Society for Photogrammetry and Remote Sensing.

Conese, C.; Maracchi, G.; Miglietta, F.; Maselli, F.; and Sacco, V.M. 1988. Forest classification by principal component analyses of TM data. *International Journal of Remote Sensing* 9:1597–1612.

Crist, E.P., and Cicone, R.C. 1984. Application of the tasseled cap concept to simulated thematic mapper data. *Photogrammetric Engineering and Remote Sensing* 50:343–52.

Crist, E.P., and Kauth, R.J. 1986. The tassled cap de-mystified. *Photogrammetric Engineering and Remote Sensing* 52:81–86.

De Cola, L.D. 1989. Fractal analysis of a classified Landsat scene. *Photogrammetric Engineering and Remote Sensing* 55:601–10.

Dow, D.D.; Browder, J.A.; and Frick, A.L. 1987. Modelling the effects of coastal wetland change on marine resources. In *Proceedings, 8th Annual Meeting of the Society of Wetland Scientists: Wetland and Riparian Ecosystems of the American West*, eds. K. Mutz and L. Lee, pp. 221–27.

Dull, C.W. 1988. Forest insect and disease activity assessment utilizing geographic information systems and remote sensing technology. In *Remote Sensing for Resource Inventory, Planning, and Monitoring*, ed. J.D. Greer, pp. 355–59. Proceedings of the Second Forest Service Remote Sensing Applications Conference, April 11–15, 1988. Falls Church, Va.: American Society for Photogrammetry and Remote Sensing.

Dyer, M.I., and Crossley, D.A., Jr. 1986. Coupling of ecological studies with remote sensing: potentials at four biosphere reserves in the United States. U.S. Dept. of State Publ. 9504.

Ecology from space. 1986. *Bioscience*, vol. 36, no. 7.

Ehlers, M. 1989. The potential of multisensor satellite remote sensing for geographic information systems. In *Technical Papers*, vol. 4, *GIS/LIS*, pp. 40–45. American Society for Photogrammetry and Remote Sensing/American Congress on Surveying and Mapping Annual Convention, April 2–7, 1989.

Estes, J.E.; Jensen, J.R.; and Simonett, D.S. 1980. Impacts of remote sensing on U.S. geography. *Remote Sensing of Environment* 10:43–80.

Forman, R.T.T., and Godron, M. 1986. *Landscape Ecology*. New York: John Wiley and Sons.

Frank, T.D. 1988. Mapping dominant vegetation communities in the Colorado Rocky Mountain front range with Landsat Thematic Mapper and digital terrain data. *Photogrammetric Engineering and Remote Sensing* 54:1727–34.

Franklin, J. 1986. Thematic mapper analysis of coniferous forest structure and composition. *International Journal of Remote Sensing* 7:1287–1301.

Franklin, J.; Logan, T.L.; Woodcock, C.E.; and Strahler, A.H. 1986. Coniferous forest classification and inventory using Landsat digital terrain data. *IEEE Transactions, Geoscience Remote and Sensing* GE–24 24:139–49.

Fryar, R., and Clerke, W.H. 1988. An evaluation of remote sensing technology for assessing the vegetative diversity of pine plantations. In *Remote Sensing for Resource Inventory, Planning, and Monitoring*, ed. J.D. Greer, pp. 99–105. Proceedings of the Second Forest Service Remote Sensing Applications Conference, April 11–15, 1988. Falls Church, Va.: American Society for Photogrammetry and Remote Sensing.

Gatlin, J.A.; Sullivan, R.J.; and Tucker, C.J. 1984. Considerations of and improvements to large-scale vegetation monitoring. *IEEE Transactions, Geoscience and Remote Sensing* GE–22:496–502.

Gessler, P.; McSweeney, K.; Kiefer, R.; and Morrison, L. 1989. Analysis of contemporary and historical soil/vegetation/landuse patterns in southwest Wisconsin utilizing GIS and remote sensing technologies. In *Technical Papers*, vol. 4, *GIS/LIS*, pp. 85–92. American Society for Photogrammetry and Remote Sensing/American Congress on Surveying and Mapping Annual Convention, April 2–7, 1989.

Goetz, A.F.H.; Rock, B.N.; and Rowan, L.C. 1983. Remote sensing for exploration: an overview. *Economic Geology* 78:573–90.

Goodchild, M.F. and Mark, D.M. 1987. The fractal nature of geographic phenomena. *Annals of the Association of American Geographers* 77:265–278.

Goward, S.N.; Tucker, C.J.; and Dye, D. 1985. North American vegetation patterns observed with the NOAA-7 advanced very high resolution radiometer. *Vegetatio* 64:3–14.

Hall, F.G.; Strebel, D.E.; and Sellers, P.J. 1988. Linking knowledge among spatial and temporal scales: vegetation, atmosphere, climate and remote sensing. *Landscape Ecology* 2:3–22.

Hall, R.J.; Still, G.N.; and Crown, P.H. 1983. Mapping the distribution of Aspen defoliation using LANDSAT colour components. *Canadian Journal of Remote Sensing* 9:86–91.

Hixson, M.; Scholz, D.; and Akiyama, T. 1980. Evaluation of several schemes for classification of remotely sensed data. *Photogrammetric Engineering and Remote Sensing* 46:1547–53.

Holz, R.K. 1985. *The Surveillant Science: Remote Sensing of the Environment*. 2nd ed. New York: John Wiley and Sons.

Jakubauskas, M.E. 1989. Utilizing a geographic information system for vegetation change detection. In *Technical Papers*, vol. 4., *GIS/LIS*, pp. 56–64. American Society for Photogrammetry and Remote Sensing/American Congress on Surveying and Mapping Annual Convention, April 2–7, 1989.

Jensen, J.R. 1983. Biophysical remote sensing. *Annals of the Association of American Geographers* 73:111–32.

Jensen, J.R. 1986. *Introductory Digital Image Processing*. Englewood Cliffs, N.J.: Prentice-Hall.

Jensen, J.R., ed. 1989. Remote sensing. In *Geography in American*, eds. G.L. Gaile and C.L. Willmott. Columbus, Ohio: Merrill Publishing.

Jensen, S.K., and Waltz, F.A. 1979. Principal component analysis and canonical analysis in remote sensing. *Proceedings, 5th Annual Meeting of the American Society of Photogrammetry* 1:337–48.

Johnson, P.L. 1969. *Remote Sensing in Ecology*. Athens: University of Georgia Press.

Jones, W.C.; Hill, C.; Thetford, W.; and Sellers, R. 1988. Identifying and mapping forest clearcut areas from Landsat Thematic Mapper transparencies. In *Remote Sensing for Resource Inventory, Planning, and Monitoring*, ed. J.D. Greer, pp. 186–88. Proceedings of the Second Forest Service Remote Sensing Applications Conference, April 11–15, 1988. Falls Church, Va.: American Society for Photogrammetry and Remote Sensing.

Justice, C.O.; Townshend, J.R.G.; Holben, B.N.; and Tucker, C.J. 1985. Analysis of the phenology of global vegetation using meteorological satellite data. *International Journal of Remote Sensing* 6:1271–1318.

Kauth, R.J., and Thomas, G.S. 1976. The tassled cap—A graphic description of spectral-temporal development of agricultural crops as seen by Landsat. *Proceedings, Symposium on Machine Processing of Remotely Sensed Data*, pp. 4B41–51. West Lafayette, Ind.: Purdue University.

Krummel, J.R. 1985. Landscape ecology: Spatial data and analytical approaches. Report presented at the Workshop on Remote Sensing, Athens, Ga., April 22. Nat. Tech. Information Ser. Report N88–70076.

Krummel, J.R. 1986. Landscape ecology: Spatial data and analytical approaches. In *Coupling of Ecological Studies with Remote Sensing*, eds. M.I. Dyer and D.A. Crossley, Jr., pp. 125–32. Washington, D.C.: U.S. Dept. of State.

Lam, N.S. 1989. Description and measurement of Landsat TM images using fractals. *Photogrammetric Engineering and Remote Sensing*, 56:187–195.

Lillesand, T.M., and Kiefer, R.W. 1987. *Remote Sensing and Image Interpretation*. 2nd ed. New York: John Wiley and Sons.

Lo, C.P. 1986. *Applied Remote Sensing*. New York: Longman.

Lulla, K., and Mausel, P. 1983. Ecological applications of remotely sensed multispectral data. In *Introduction to Remote Sensing of the Environment*, 2nd ed., ed. B.F. Richason, Jr., pp. 354–77. Dubuque, Iowa: Kendall/Hunt.

Mandelbrot, B. 1977. *Fractals, Form, Chance and Dimension*. San Francisco: Freeman.

Marble, D.F., and Peuquet, D.J. 1983. Geographic information systems and remote sensing. In *Manual of Remote Sensing*, 2nd ed., ed. R.N. Colwell, pp. 923–58. Falls Church, Va.: American Society for Photogrammetry and Remote Sensing.

Matthews, E. 1983. Global vegetation and land use: New high-resolution data bases for climate studies. *Journal of Climate and Applied Meteorology* 22:474–87.

Meentemeyer, V. 1989. Geographical perspectives of space, time, and scale. *Landscape Ecology* 3:163–73.

Meentemeyer, V., and Box, E.O. 1987. Scale effects in landscape studies. In *Landscape Heterogeneity and Disturbance*, ed. Monica Goigel Turner, pp. 15–34. New York: Springer-Verlag.

Moik, J. 1980. Digital processing of remotely sensed images. NASA SP–431. Washington, D.C.: NASA Scientific and Technical Information Branch.

Mooney, H.A., and Godron, M. 1983. *Disturbance and Ecosystems: Components of Response*. New York: Springer-Verlag.

Morin, R.L.; Derenyi, E.E.; Wein, R.W.; and Yazdani, R. 1988. Up-dating landscape data for natural resource management through Landsat imagery. In *Perspectives on Land Modelling*, eds. R. Gelinas, D. Bond, and B. Smit, pp. 167–71. Workshop Proceedings, November 17–20, 1986. Montreal: Polyscience Publications.

Muchoki, C.H.K. 1988. Remotely sensed relationships between wooded patch habitats and agricultural landscape type: A basis for ecological planning. In *Landscape Ecology and Management*, ed. M.R. Moss, pp. 85–94. Proceedings of the First Symposium of the Canadian Society for Landscape Ecology and Management, May 1987. Montreal: Polyscience Publications.

Mukai, Y., Sugimura, T. 1987. Extraction of areas infested by pine park beetle using Landsat MSS data. *Photogrammetric Engineering and Remote Sensing*. 53:77–81.

NASA 1988. From pattern to process: The strategy of the earth observing system. Eos Science Steering Committee Report, vol. 2. Washington, D.C.

National Research Council 1986. *Global Change in the Geosphere-Biosphere: Initial Priorities for an IGBP*. Washington, D.C.: National Academy Press.

Nellis, M.D., and Briggs, J.M. 1989. The effect of spatial scale on konza landscape classification using textural analysis. *Landscape Ecology* 2:93–100.

Nelson, R.F.; Krabill, W.B.; and Tonelli, J. 1988. Estimating forest biomass and volume using airborne laser data. *Remote Sensing of Environment* 15:201–12.

Nemani, R.R., and Running, S.W. 1989. Estimation of regional surface resistance to evapotranspiration from NDVI and thermal-IR AVHRR data. *Journal Applied Meteorology* 28:276–84.

O'Neill, R.V.; B.T. Milne; M.G. Turner; and R.H. Gardner. 1988. Resource utilization scales and landscape pattern. *Landscape Ecology* 2:63–69.

Ormsby, J.P.; Choudhury, B.J.; and Owe, M. 1987. Vegetation spatial variability and its effects on vegetation indices. *International Journal of Remote Sensing* 8:1301–06.

Parker, A.J. 1987. Structural and functional features of mesophytic forests in the east-central United States. *Annals of the Association of American Geographers* 77:423–35.

Parker, H.D. 1988. The unique qualities of a geographic information system: a commentary. *Photogrammetric Engineering and Remote Sensing* 54:1547–49.

Pelletier, R.E., and Wu, S.T. 1989. A preliminary evaluation of the airborne electromagnetic bathymetry system for characterization of coastal sediments and marsh soils. In *Technical Papers*, vol. 3., *Remote Sensing*, pp. 366–75. American Society for Photogrammetry and Remote Sensing/American Congress on Surveying and Mapping Annual Convention. April 2–7, 1989.

Peterson, D.L.; Aber, J.D.; Matson, P.A.; Card, D.H.; Swanberg, N.; Wessman, C.; and Spanner, M. 1988. Remote sensing of forest canopy and leaf biochemical contents. *Remote Sensing of Environment* 24:85–108.

Pickett, S.T.A., and White, P.S. 1985. *The Ecology of Natural Disturbance and Patch Dynamics*. Orlando, Fla.: Academic Press.

Reeves, R.G., ed. 1975. *Manual of Remote Sensing*. Falls Church, Va.: American Society for Photogrammetry and Remote Sensing.

Rencz, A.N., and Nemeth, J. 1985. Detection of mountain pine beetle infestation using Landsat MSS and simulated Thematic Mapper data. *Canadian Journal of Remote Sensing* 11:50–58.

Richason, B.F., Jr., ed. 1983. *Introduction to Remote Sensing of the Environment*. 2nd ed. Dubuque, Iowa: Kendall/Hunt.

Ringrose, S.; Matheson, W.; and Boyle, T. 1988. Differentiation of ecological zones in the Okavango Delta, Botswana by classification and contextural analyses of Landsat MSS data. *Photogrammetric Engineering and Remote Sensing* 54:601–8.

Ripple, W.J., ed. 1987. *GIS For Resource Management: A Compendium*. Falls Church, Va.: American Society for Photogrammetry and Remote Sensing.

Rock, B.N.; Vogelmann, J.E.; Williams, D.L.; Vogelmann, A.F.; and Hoshizaki, T. 1986. Remote detection of forest damage. *Bioscience* 36:439–45.

Rouse, J.W.; Haas, R.H.; Schell, J.A.; and Deering, D.W. 1973. Monitoring vegetation systems in the Great Plains with ERTS. *Proceedings, Third ERTS Symposium* 1:48–62.

Rubec, C.D.A. 1983. Application of remote sensing in ecological land survey in Canada. *Canadian Journal of Remote Sensing* 9:19–30.

Running, S.W., and Nemani, R.R. 1988. Relating seasonal patterns of the AVHRR vegetation index to simulated photosynthesis and transpiration of forests in different climates. *Remote Sensing of Environment* 24:347–67.

Rykiel, E.J., Jr.; Coulson, R.N.; Sharpe, P.J.H.; Allen, T.F.H.; and Flamm, R.O. 1988. Disturbance propagation by bark beetles as an episodic landscape phenomenon. *Landscape Ecology* 1:129–39.

Sader, S.A. 1988. Satellite digital image classification of forest change using three Landsat data sets. In *Remote Sensing for Resource Inventory, Planning, and Monitoring*, ed. J.D. Greer, pp. 189–201. Proceedings of the Second Forest Service Remote Sensing Applications Conference, April 11–15, 1988. Falls Church, Va.: American Society for Photogrammetry and Remote Sensing.

Schowengerdt, R.A. 1983. *Techniques for Image Processing and Classification in Remote Sensing*. New York: Academic Press.

Short, N.M. 1982. The Landsat tutorial workbook. NASA Reference Publication 1078. Washington, D.C.: NASA Scientific and Technical Information Branch.

Short, N.M., and Blair, R.W., Jr. 1986. Geomorphology from space: A global overview of regional landforms. NASA SP-486. Washington, D.C.: NASA Scientific and Technical Information Branch.

Short, N.M., and Stuart, Jr., L.M. 1982. The heat capacity mapping mission (HCMM) anthology. NASA SP-465. Washington, D.C.: NASA Scientific and Technical Information Branch.

Teuber, K.B. 1988. Large-scale forest area estimation using advanced very high resolution radiometer (AVHRR) data. In *Remote Sensing for Resource Inventory, Planning, and Monitoring*, ed. J.D. Greer, pp. 145–52. Proceedings of the Second Forest Service Remote Sensing Applications Conference, April 11–15, 1988. Falls Church, Va.: American Society for Photogrammetry and Remote Sensing.

Tucker, C.J.; Fung, I.Y.; Keeling, C.D.; and Gammon, R.H. 1986. Relationship between atmospheric CO_2 variations and a satellite-derived vegetation index. *Nature* 319:195–99.

Tucker, C.J.; Gatlin, J.A.; and Schneider, S.R. 1984. Monitoring vegetation in the Nile delta with NOAA-6 and NOAA-7 AVHRR imagery. *Photogrammetric Engineering and Remote Sensing* 50:53–61.

Tucker, C.J.; Townshend, J.R.G.; and Goff, T.E. 1985. African land-cover classification using satellite data. *Science* 227:369–75.

Tucker, C.J.; Vanpraet, C.; Boerwinkel, E.; and Gaston, A. 1983. Satellite remote sensing of total dry matter production in the Senegalese Sahel. *Remote Sensing of Environment*

13:461–74.

Turner, M.G., ed. 1987. *Landscape Heterogeneity and Disturbance*. New York: Springer-Verlag.

Turner, M.G.; Dale, V.H.; and Gardner, R.H. 1989. Predicting across scales: theory development and testing. *Landscape Ecology*, 3:245–52.

Urban, D.L.; O'Neill, R.V.; and Shugart, H.H., Jr. 1987. Landscape ecology. *Bioscience* 37:119–27.

Vogelmann, J.E. 1988. Detection of forest change in the Green Mountains of Vermont using multispectral scanner data. *International Journal of Remote Sensing* 9:1187–1200.

Vogelmann, J.E., and Rock, B.N. 1988. Assessing forest damage in high-elevation coniferous forests in Vermont and New Hampshire using thematic mapper data. *Remote Sensing of Environment* 24:227–46.

Walsh, S.J. 1980. Coniferous tree species mapping using Landsat data. *Remote Sensing of Environment* 9:11–26.

Walsh, S.J. 1987. Variability of Landsat MSS spectral responses of forests in relation to stand and site characteristics. *International Journal of Remote Sensing* 8:1289–99.

Waring, R.H.; Aber, J.D.; Melillo, J.M.; and Moore III, B. 1986. Precursors of change in terrestrial ecosystems. *Bioscience* 36:433–38.

Waring, R.H., and Schlesinger, W.H. 1985. *Forest Ecosystems: Concepts and Management*. Orlando, Fla.: Academic Press.

White, P.S., and MacKenzie, M.D. 1986. Remote sensing and landscape pattern in Great Smoky Mountains National Park Biosphere Reserve, North Carolina and Tennessee. In *Coupling of Ecological Studies with Remote Sensing: Potentials at Four Biosphere Reserves in the United States*, eds. M.I. Dyer and D.A. Crossley, Jr., pp. 52–70. U.S. Dept. of State Publ. 9504.

Williams, R.S., and Carter, W.D., eds. 1976. ERTS-1: A new window on our planet. Geological Survey Professional Paper 929. Washington, D.C.: U.S.G.P.O.

Wood, T.F., and Foody, G.M. 1989. Analysis and representation of vegetation continua from Landsat Thematic Mapper data for lowland heaths. *International Journal of Remote Sensing* 10:181–91.

Woodcock, C.E., and Strahler, A.H. 1987. The factor of scale in remote sensing. *Remote Sensing of Environment* 21:311–32.

Wu, S.T. 1984. Analysis of synthetic aperture radar data acquired over a variety of land covers. *IEEE Transactions, Geoscience and Remote Sensing* GE–22:550–57.

Wu, S.T., and Sader, S.A. 1987. Multipolarization SAR data for surface feature delineation and forest vegetation characterization. *IEEE Transactions, Geoscience and Remote Sensing* GE–25:67–76.

4. Image Textural Measures as Indices of Landscape Pattern

H. Brad Musick and Herbert D. Grover

4.1 Introduction

Landscape ecology is concerned with relationships between ecological processes and spatial pattern, particularly at large spatial scales. Quantitative analysis of these relationships requires the development and use of quantitative measures of landscape pattern. A wide variety of pattern indices are necessary to accommodate the many different data types and formats used by landscape ecologists and to measure different aspects of pattern. This chapter describes methods developed by Haralick (Haralick et al. 1973; Haralick and Shanmugam 1974) to quantify aspects of texture (i.e., heterogeneity) in pictorial image data and describes how these methods may be used to measure some aspects of landscape pattern. In the example presented here, we examined the scale-dependence of these measures. The multiplicity of spatial variables involved in applying these measures led us to consider questions concerning concepts of scale and methods of changing scale.

4.2 Background

The texture measures described here were originally developed for engineering applications in the field of automated pattern recognition and analysis. The literature of this field provides a diverse array of techniques that may be potentially useful for landscape ecology (Risser et al. 1984). In attempting to adapt techniques developed in one discipline for use in another, it is important to recognize interdisciplinary differences and similarities in fundamental objectives and in the nature of the data to be analyzed. Differences in terminology sometimes complicate the process of technique transfer, and we will therefore attempt to relate the terms used in image processing to terms more familiar to ecologists.

Texture measures belong to a class of techniques developed for the analysis of digital pictorial data represented as a regularly spaced two-dimensional array of data values. Each cell in the array is a "picture element," usually referred to as a "pixel." The data values are usually measures of brightness or light intensity made by an imaging device but sometimes may be code numbers for classes that differ qualitatively (e.g., colors). Many analysis methods are appropriate for only one of these two data types. Computer hardware and software limitations often require that the data values be integers and have a finite range (often 0–63, 0–127, or 0–255). Pictorial data formatted as a two-dimensional array of data values may be produced from photographs by using optical devices that scan across and down the photograph, recording brightness at regular intervals. Other devices convert analog video images to digital data in this format, and images of Earth's surface are collected by satellite-borne sensors (e.g., Landsat Thematic Mapper) directly into this digital format.

A biological data set having the same format as digital imagery could be collected by sampling a geographical area with a regularly spaced rectangular array of quadrats or points; the measured variable could be biotic or abiotic, structural or functional. Until recently, such data sets have rarely been collected because sampling all points of the array by conventional methods often requires excessive time and labor or entails complete destruction of the study area. To overcome these problems, ecologists have relied upon statistical theory to provide methods for estimating parameters of the population of data values and for describing some aspects of spatial pattern from partial samples such as transects or arrays of randomly located plots. The remotely sensed data (including aerial photographs and maps derived therefrom) used by landscape ecologists often have the format of digital pictorial imagery or can readily be converted to this format. For example, a map of forested versus nonforested areas generated from remotely sensed data could be sampled with a regular array of quadrants and percent of forest cover determined for each quadrant. Because these landscape-scale ecological data sets have the same format as pictorial imagery, they can be analyzed using the methods for engineering applications in pattern analysis, such as the image texture measures described in this chapter.

In engineering, image analysis methods are usually designed for utilitarian objectives, such as distinguishing between images of unlike materials, identifying the imaged materials by comparison with images of known materials, or segmenting a complex image into relatively homogeneous regions corresponding to discrete objects or types of materials (Nevatia 1982). Because the human visual sensory system is highly adapted to these tasks, automated image analysis methods are sometimes based on an understanding of human visual perception, derived either by intuition or by experimental studies (Rosenfeld and Kak 1976). Success of an automated method is sometimes judged by comparison with the capabilities of human visual perception (e.g., does the method distinguish between visually unlike images?) and in other cases by comparison with objective reality (e.g., does the method distinguish between images of unlike materials?).

Landscape ecologists may also use pattern analysis methods for strictly utilitarian purposes such as conversion of raw data to a more usable form. For example, a variety of techniques may be used for classification of remotely sensed digital images to derive a map of land cover or vegetation type (e.g., Schowengerdt 1983; Quattrochi and Pelletier, Chapter 3). In this chapter we are concerned with a different use of pattern analysis methods—the quantitative measurement of aspects of pattern for descriptive and interpretive purposes. Quantitative indices of pattern are necessary to relate pattern to process and to measure changes in landscape structure. A wide variety of pattern indices have been applied to landscapes for this purpose (Burrough 1981; Romme and Knight 1982; O'Neill et al. 1988; S. Turner et al., Chapter 2; Milne, Chapter 9). Some pattern indices might be used both for utilitarian objectives and for descriptive and interpretive purposes, but the results would be judged by different criteria. For example, a pattern index that failed to distinguish between two visually unlike images might be judged unsatisfactory for utilitarian objectives, but from a descriptive and interpretive viewpoint, the finding that the two patterns were structurally equivalent in some aspect might be valuable information.

4.2.1 Texture

In the image processing literature, texture refers in the broadest sense to the pattern of brightness variations within an image or a region within an image. The breadth and complexity of the concept of texture has been difficult to capture in precise verbal or mathematical definitions; thus, most authors have relied on definition by example (e.g., Rosenfeld and Kak 1976). Texture includes such concepts as uniformity, coarseness, regularity, frequency, and linearity (Nevatia 1982). Differences in texture may be illustrated by images of different materials, such as straw, sand, cloth, and wood (Brodatz 1956).

Haralick (1979) and Nevatia (1982) contrast statistical and structural approaches to describing texture. The structural approach assumes that images are composed of primitive elements (e.g., patches of a given shape and size) repeated in a certain pattern, and differences in texture result from differences in the primitive elements, the pattern of repetition, or both. Using this structural approach, texture would be characterized by explicitly describing the shape, size, and other properties of the primitive elements and their pattern of repetition. Application of the structural approach to landscape pattern is exemplified by Forman and Godron's (1981,1986) description of landscape structure in terms of patch size, shape, and configuration within a matrix.

Image analysts have found the structural approach difficult to use in characterizing natural patterns because these patterns often have neither an obvious fixed primitive element nor a fixed pattern of repetition (Nevatia 1982). These difficulties have led to the characterization of texture by the statistical properties of a neighborhood of pixels. Local (neighborhood) intensity variance and other statistics derived from frequency distributions of individual pixel attributes are sometimes used as statistical texture measures (Woodcock and Strahler 1987), but these statistics have been criticized as lacking information on the spatial distribution of intensity values within the neighborhood (Nevatia 1982; Schowengerdt 1983). Haralick (1979) reviewed eight basis for statistical texture measures:

1. autocorrelation functions
2. optical transforms (e.g., Fraunhofer diffraction patterns)
3. digital transforms (e.g., Fourier and Hadamard transforms)
4. textural edgeness (e.g., local edge density)
5. structural element filtering
6. spatial intensity-value cooccurrence probabilities
7. intensity-value run lengths
8. autoregressive models

This list illustrates the breadth of the texture concept as used in image analysis. Some of these approaches to pattern analysis have been widely used by ecologists (see S. Turner et al., Chapter 2; Pielou 1969); others have been used rarely or perhaps never.

4.3 Haralick Texture Measures

4.3.1 Calculation

Haralick et al. (1973) proposed fourteen texture measures based on the spatial cooccurrence probabilities of intensity values. The spatial cooccurrence probability $p(i,j,d,\theta)$ is the probability of a pair of pixels separated by distance $|d|$ in direction θ having the intensity values i and j. If distance $|d| = 1$, then nearest-neighbor pixels are compared; if $|d| = 2$, then neighbors once removed, and so forth. The angle θ may be 0° (horizontal), 45° (right diagonal), 90° (vertical), or 135° (left diagonal). An image or portion of an image yields a matrix of $p(i,j)$ values when all pixels separated by given distance $|d|$ along a given direction θ are compared. Because distance is here defined as an absolute value, comparison of two pixels' values involves two reciprocal cooccurrences, and therefore the matrix of $p(i,j)$ values is symmetric. For example, a pair of pixels with intensity values 3 and 4 is counted both as a cooccurrence of 4 with 3 ($i = 3$, $j = 4$) and as a cooccurrence of 3 with 4 ($i = 4$, $j = 3$). Fig. 4.1A shows a 3 by 3 pixel block of data values, with arrows indicating all the cooccurrences for $|d| = 1$ and $\theta = 0°$. Fig. 4.1B is the resulting matrix of spatial cooccurrence frequencies, which are divided by the total number of cooccurrences (twelve in this example) to yield the matrix of cooccurrence probabilities in Fig. 1C.

The texture measures are functions of the $p(i,j)$ values. The measures we used in our investigations, Angular Second Moment (ASM) and Inverse Difference Moment (IDM), are both indices of homogeneity. These were selected because of their ease of computation with the software available to us (see following section).

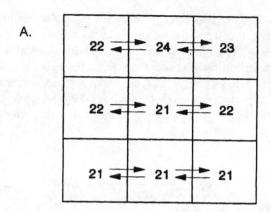

Figure 4.1. Example of derivation of co-occurrence probability $[p(i,j,d,\theta)]$ matrix for $d = 1$ (specifying that the values i and j to be compared shall be immediately adjacent) and $\theta = 0°$ (specifying co-occurrences along the horizontal): (A) Block of input data values, limited to 3 by 3 size to simplify this example; arrows, pointing from value i to value j, indicate all specified co-ocurrences. (B) Co-occurence frequency matrix for the block of data and the specified co-occurrences shown in (A). (C) Matrix of co-occurrence probabilities $[p(i,j,d,\theta)$ values] derived from (B) by dividing frequencies by the total number of co-occurrences. The texture measures are calculated from the matrix of $p(i,j,d,\theta)$ values (see Eq. 1 and EQ. 2).

ASM is the sum of the squared co-occurrence probabilities:

$$ASM = \sum_i \sum_j [p(i,j)]^2 \tag{4.1}$$

ASM increases with homogeneity because it is sensitive to the co-occurrence of identical values. A co-occurrence of two unlike values adds 1 to each of two off-diagonal cells in the co-occurrence frequency matrix, but a co-occurrence of two identical values adds 2 to the frequency of a cell along the matrix diagonal (Fig. 4.1). In relatively homogeneous blocks, most co-occurrences will be of identical values, and the squared probabilities will be higher. ASM attains a maximum value of 1 when all values in a block are identical. ASM is insensitive to the magnitude of difference between nonidentical values. Therefore, this index could be used for nominal data, where values are code numbers for qualitatively different categories (e.g., vegetation type), as well as for interval data, where values are measurements of intensity or amount (e.g., Biomass).

The other texture measure we tested, IDM, does take into account the difference in magnitude of spatially cooccurring pixel values. IDM is a weighted sum of the co-occurrence probabilities, with the weighting factor decreasing with the difference between values:

$$\text{IDM} = \sum_i \sum_j \left[\frac{1}{1 + (i - j)^2} \right] p(i, j) \qquad (4.2)$$

As with ASM, IDM attains a maximum value of 1 when all pixel values are identical. Because the magnitude of the pixel-value differences is meaningless when the values are simply arbitrary code numbers for qualitatively different categories (i.e., data are nominal measurements), this measure requires that the input data be measurements of intensity or amount (i.e., interval type data).

The procedures described above will yield (for a given comparison distance d) four direction-specific values of the texture measure for a block of data, and additional calculations or alternative procedures are necessary to determine the degree of directionality in texture or to obtain an overall measure of texture that is not specific to a particular direction. For some purposes, texture might be considered a property of pattern that is invariant with rotation of the data set; for example, a pattern of horizontal stripes might be considered equivalent to one of identical stripes oriented vertically. Haralick et al. (1973) suggest that texture measure values independent of image rotation be obtained by using the range and mean of the four angular measures. The range is a measure of the degree of directionality in texture, and the mean represents an overall or direction-independent measure. An alternative method for obtaining a direction-independent measure is to use all four directions in determining a single co-occurrence probability matrix and then to calculate a texture value from this matrix (Fig. 4.2).

4.3.1.1 Directionality

These two methods for obtaining direction-independent measures of texture will usually yield different values, and both methods have potential shortcomings. A rectangular block of regularly spaced data values has fewer total cooccurrences along the diagonals than along the major axes (horizontal and vertical). Therefore, a texture value derived from a tally of cooccurrences along all four directions will be biased in favor of the horizontal and vertical directions. This bias is greater for small blocks of data; for a 3 by 3 block, the ratio of diagonal to horizontal plus vertical cooccurrences is 2:3. This ratio approaches unity as the size of the data block increases (e.g., ratio is 0.9 for a 10 by 10 block, 0.95 for a 20 by 20 block). Averaging the four direction-specific texture values, as suggested by Haralick et

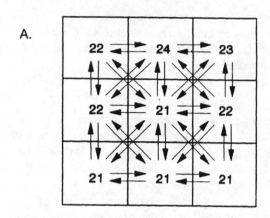

A.

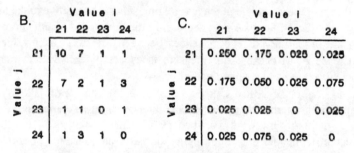

Figure 4.2. Example of derivation of an all-directional co-occurrence probability $[p(i,j,d)]$ matrix for the same block of input data values given in Fig. 1A; $d = 1$ in this example: (A) Block of input data values; arrows pointing from value i to value j, indicate all specified co-occurrences. (B) Co-occurrence frequency matrix for the block of data and specified co-occurrences shown in (A). (C) Co-occurence probability $[p(i,j,d)]$ matrix derived from (B).

al. (1973), avoids bias in favor of the major-axis directions but may be inappropriate for texture measures that are nonlinear functions of the $p(i,j)$ values (e.g., ASM). Perhaps the best procedure for obtaining an unbiased all-directional texture value would be to tally the cooccurrences separately for each direction, convert each frequency matrix to a probability matrix, average the probability matrices, and calculate the texture value from the average $p(i,j)$ values.

Another problem arising when texture measures for different directions are compared or combined is how to make the measurements of comparison distance d equivalent for the diagonal and major-axis directions. If d is in units of degree of adjacency (e.g., $d = 2$ for neighbors once removed), then the true Euclidean distance between a pair of pixels at distance $d_{45°,135°}$ along the diagonal will be $(2)^{\frac{1}{2}}$ times the Euclidean distance between a pair of pixels at distance $d_{0°,90°}$ along the horizontal or vertical. Thus the set of four directional texture values would not be invariant with image rotation. The best solution might be to use for the diagonal directions the value of $d_{45°,135°}$ that most closely approximated the true Euclidean distance used for the horizontal and vertical directions. For example, the nearest diagonal equivalent to $d_{0°,90°} = 3$ for the major-axis directions would be $d_{45°,135°} = 2$ (Euclidean distance = $(2)^{1/2} d = 2.83$).

4.3.1.2 Input Data Properties

The numerical properties and the aggregate or nonspatial statistical properties of the data are important considerations in calculating the texture measures and evaluating the results. An important numerical property is the number of quantization levels or possible values the data are allowed to take. The computer's data storage format may impose an absolute upper limit to the number of quantization levels, but this limit may be further reduced by transformation of the original data. Reducing the number of quantization levels reduces the computational effort but eliminates faint patterns. Some of the texture measures are sensitive to the number of quantization levels. For example, decreasing the number of quantization levels increases the probability of cooccurrences between identical pixel values; this would increase ASM because this measure is especially sensitive to identical-value cooccurrences, as described earlier.

The nonspatial or aggregate statistical properties of a block of data values are those statistical properties that may be calculated without reference to the spatial positions of the data values. As described earlier, image analysts have sometimes been reluctant to include these nonspatial statistics as measures of texture. An additional question is whether texture measures that are based on spatial arrangement should be independent of the nonspatial statistical properties of the data. The arguments for independence from nonspatial statistical properties have been based on both conceptual and practical grounds. An example of the conceptual argument is the statement that stretching or shrinking the range of values in an image does not change the degree of "busyness" of the image; that is, a pattern may be faint (i.e., low contrast) or strong (high contrast), yet have the same degree of busyness (Rosenfeld and Kak 1976). The major practical consideration for image analysts is that contrast and other aggregate statistical properties of images are subject to variations in illumination intensity and sensor response characteristics, and these variations are sometimes difficult to eliminate from the process of image acquisition.

Variations in illumination intensity and sensor response (assuming sensor response is a linear function of brightness) are equivalent to transformation of the image intensity values by linear functions. Haralick and Shanmugam (1974) therefore recommend that the texture measures be made invariant under linear intensity-value transformations. Some of the Haralick texture measures, including ASM, are inherently insensitive to linear data transformations. For others, including IDM, invariance under linear data transformations must be achieved by means of some data transformation prior to calculation of the texture measure. For this purpose Haralick and Shanmugam (1974) suggest equal-probability quantizing, in which the original values are assigned to new quantization levels so that all levels are equally frequent. Another possible method would be to transform all images by linear functions so that the resultant images had the same minimum and maximum values. Both of these procedures eliminate linear differences between images, but equal-probability quantizing also removes any differences in skewness or kurtosis of the intensity-value distributions. While there may sometimes be good practical reasons (e.g., nonlinear sensor response) for eliminating any effects of variation in skewness or kurtosis from measures of texture, there appears to have been little discussion of this issue from a conceptual viewpoint.

The problem of data sets containing an unknown but significant amount of linear measurement error is probably encountered less often in landscape ecology than in optical image analysis. Landscape ecologists may therefore have no compelling practical reasons for ensuring that their pattern indices are independent of the nonspatial statistical properties of their data, but the conceptual issues should nevertheless be considered. For example, Pielou (1969) describes two kinds of measures of spatial aggregation in single populations. Some measures would be affected if the mean population density were reduced by the death of randomly selected individuals, while other measures would not. Which of these two kinds of aggregation measures should be used might depend on whether we are interested in the effects of crowding, which are dependent on mean density as well as spatial pattern, or in the causes of spatial pattern. The sensitivity of landscape pattern indices to nonspatial statistical properties should be considered in comparing the results from different indices and in generalizing from the results obtained with a particular index.

4.3.1.3 Piecewise Analysis

There are two ways of calculating texture measure values for an image or a similar array of ecological data: (1) a single $p(i,j)$ matrix and resultant texture value may be calculated for the entire image or (2) the image may be analyzed piecewise, with a $p(i,j)$ matrix and texture value calculated for each of the subimage blocks of data comprising the image. In image analysis, the first approach is often used when each image consists entirely of a single type of material and the objective is to identify each image by comparison with known standards or to classify the set of images. Piecewise analysis is commonly used when a single image contains a variety of materials, and the objective is to determine which areas correspond to different materials. The initial product of piecewise analysis may itself be represented as an image (i.e., spatial array of data values) wherein each data value is the value of the texture measure for a defined neighborhood or subimage block.

Piecewise analysis may be accomplished by extracting the subimage blocks and creating a number of new and smaller data sets, but the more widely used procedure is sliding window analysis. Computer programs utilizing the latter approach slide a window of defined size through the entire image by defined increments, calculating a value of the texture measure for each window position. When the window position is incremented by a distance equal to window size, then texture values are obtained for contiguous nonoverlapping blocks of data, and the procedure is equivalent to subdividing the image into blocks before analysis. When the desired final product is a pixel-by-pixel classification of the original image, the window's position may be moved by single-pixel increments and the texture value for a given window position is then written to the spatial coordinates of the window's central pixel. This requires that the window have an odd number of pixels per side so that there will be a central pixel. Note that the sliding window technique described here differs from other "moving window" techniques for pattern analysis (S. Turner et al., Chapter 2) that involve comparisons between halves of the window.

In the preceding description of the Haralick texture measures we have followed the original assumption (Haralick et al. 1973; Haralick and Shanmugam 1974) that they would be applied to a two-dimensional spatial array of data values. However, these measures, unlike many other image analysis methods, could also be applied to one-dimensional linear arrays as might be obtained by transect sampling. Because collection of large, regular two-dimensional arrays of ecological data is often impractical, this capability for transect analysis is of greater value to ecologists than to analysts of optical images.

4.3.2 Interpretation

What is measured by the Haralick texture indices? A simple answer is difficult to formulate because what is measured depends on the numerical and statistical properties of the data, how these properties may have been modified by some transformation of the data prior to analysis, and which options are employed in calculating the measures. The method, in general, is a multipurpose approach, capable of being modified in its operation in order to extract information on different aspects of pattern. For example, the change in texture value with inter-pixel comparison distance could be used to detect spacing in repetitive patterns in a manner similar to autocorrelation analysis (S. Turner et al., Chapter 2). Directionality in repetitive patterns could also be detected by examining the change in texture value with sampling angle.

Another option is the choice of index to be calculated from a cooccurrence matrix. We have followed Haralick et al. (1973) in describing ASM and IDM as indices of homogeneity, but what is the meaning of homogeneity in the context of these methods? ASM is largely a measure of how often neighboring samples have identical values, but ASM will also take high values if repetitive patterns result in high frequencies of a restricted number of neighboring-value combinations. ASM is thus an inverse measure of interspersion. IDM gives an overall measure of the similarity in value of neighboring samples. The homogeneity measured by these methods has a spatial component lacking in measures of the homogeneity of a population of values considered without regard to spatial location. For example, an array of samples taken from a smooth gradient could be randomly scrambled with no effect on the population variance, but ASM and IDM would decrease.

IDM might best be described as an overall measure of the shallowness of local gradients, where "local" is explicitly defined as the intersample comparison distance d, and gradient is measured by the difference in value ($\equiv i - j$). The calculation of IDM by the Haralick procedures is in fact mathematically equivalent to an evaluation of the shallowness of local gradients by a direct approach. In a direct approach, the first step would be calculation of the gradient, as difference in intensity or amount, between each pair of neighboring samples. The gradient values would then be accumulated and the relative frequency of each ($\equiv p(i - j)$) determined. The $p(i - j)$ distribution would likely be truncated at zero and skewed toward higher values (i.e., small gradients more frequent than large). Standard statistics such as the mean and variance could be calculated from this distribution. A less widely known statistic, the inverse difference moment, could be used to measure how strongly the distribution is dominated by values near zero (that is, the inverse of how strongly the distribution differs from 100% zero values):

$$\text{inverse difference moment} = \sum_{i-j} \left[\frac{1}{1 + (i - j)^2} \right] p(i - j) \qquad (4.3)$$

Equation 4.3 is mathematically equivalent to Eq. 4.2, the formula previously given for IDM. In Eq. 4.2 the $p(i,j)$ values are weighted according to the value of $i - j$ and then summed, whereas Eq. 4.3 is equivalent to first summing along the diagonals of the $p(i,j)$ matrix to yield $p(i - j)$ values, then applying the appropriate weighting factor, and finally summing all the weighted $p(i - j)$ values.

4.3.3 Application to Landscape Ecology

How is an overall measure of the weakness of local gradients, as given by IDM, potentially useful in landscape ecology? Edges and boundaries are ecologically important features of the landscape because they may act as selective filters or amplifiers for the transfer of energy, matter, organisms, and disturbances between adjacent patches (Forman and Godron 1981, 1986; Wiens et al. 1985; Turner and Bratton 1987; Schonewald-Cox 1988). Thus the "edgeness" of a landscape may be an important ecological parameter. Edge density (amount of edge per unit area) is a useful measure of edgeness but requires that edges be defined as discrete entities. When the landscape is represented as a mosaic of qualitatively different types (e.g., forest vs. grassland), edges are defined simply as the locations where two types adjoin. However, when the landscapes is represented as a continuum in amount or intensity of some variable (e.g., tree density), then defining edges as discrete entities becomes a problem. One solution is to define edges as the points where the local gradient exceeds some threshold. This thresholding approach requires that some criterion be selected for determining a meaningful threshold, and information about the degree of contrast across the edges may be lost.

If IDM is used as an inverse measure of edgeness, these problems are avoided because patches and edges need not be defined as discrete entities. Because edge is an undefined term in the context of IDM calculation, the aspect of pattern measured by IDM cannot be precisely described in terms of edges. Nevertheless, IDM could be called an inverse measure of contrast-weighted edge density.

The practical problems in defining edges and patches as discrete entities and extracting these from continuous data are not insurmountable and may even be trivial for some landscapes. Even so, it may be desirable on theoretical grounds to analyze continuous data in terms of gradients rather than force the landscape into a structural model based on discrete entities. Statistical tests can be used to determine which model, continuum or discrete, best fits the data, but the results may be misleading. The processes of sampling and measurement that produced the data may have already imposed a model or at least restricted the types of models that the data might fit (Allen et al. 1984). Risser et al. (1984) note that each species views the landscape differently, and an area that is homogeneous for one species might be heterogeneous for another. Similarly, we suggest that some organisms or processes may respond to the landscape as a continuum, while others respond as if the landscape were a mosaic of discrete entities. The choice of model should be based, if possible, on the behavior of the organism or process.

The causes and consequences of landscape heterogeneity are questions of major importance in landscape ecology. Interest has focused especially on relationships between landscape heterogeneity and disturbance (Forman and Godron 1986), as evidenced by the recent publication of an entire volume on this topic (Turner 1987). Does disturbance increase or decrease landscape heterogeneity? Does landscape heterogeneity enhance or retard the spread of disturbances? The limited evidence available is inconsistent (Risser et al. 1984; Wiens et al. 1985; Risser 1987; Turner and Bratton 1987; Remillard et al. 1987; Forman 1987). This inconsistency arises largely from differences in the nature of the various disturbances and in properties of the various landscape elements, but the answers to these questions may also depend on whether the aspect of heterogeneity examined is spatial or nonspatial.

Nonspatial aspects of heterogeneity include both the richness and evenness components of landscape element diversity and the overall variance of continuous variables. Among the spatial aspects of heterogeneity are interspersion (of types) and the frequency of steep local gradients (in continuous measures of amount or intensity), as measured inversely by ASM and IDM. Whether spatial or nonspatial aspects of heterogeneity are more important may depend on the nature of the process or organism in question. For example, disturbances that lack means of long-distance spread probably respond more strongly and directly to the spatial than to the nonspatial aspects of heterogeneity (Forman 1987).

4.4 Spatial Scale

4.4.1 Landscape Heterogeneity and Scale

A landscape may appear homogeneous at one scale but heterogeneous at another; thus, spatial scale is inherent in definitions of landscape heterogeneity (Meentemeyer and Box 1987). Measurements of the change in landscape heterogeneity with spatial scale may be useful in several ways. Measurements of heterogeneity at multiple scales might be used to determine the scale at which heterogeneity controls a process or influences the behavior of an organism (Milne et al. 1989). For example, suppose an investigator wishes to test the hypothesis that the spread of a given disturbance is related to landscape heterogeneity. A reasonable approach would be to measure disturbance spread and heterogeneity for a variety of landscapes and compare these variables. At what special scale should heterogeneity be measured? The investigator might select a scale based on some knowledge of the disturbance-spreading process, but if no relation is found between disturbance spread and heterogeneity at this one scale, which hypothesis should be rejected—that landscape heterogeneity controls disturbance spread, or that the spatial scale selected is the appropriate scale to examine? If heterogeneity were measured at a wide range of spatial scales, the first hypothesis could be tested independently of assumptions about the appropriate scale. Furthermore, variations in the strength of the relationship with scale might reveal previously unknown characteristics of the disturbance-spreading process.

Often the landscape ecologist is faced with the problem of integrating and relating measurements of landscape properties (e.g., heterogeneity) made at disparate spatial scales. The measurements may be transformed to a common scale if some means are available for estimating values at the common scale from measurements made at another scale. This is possible if the landscape attribute has been shown not to change or to change predictably with spatial scale (Milne 1988; Turner et al. 1989). Thus, a knowledge of the scale-dependency of a landscape pattern index may be useful in linking observations made at different scales.

Some investigators have suggested that the pattern of change in heterogeneity and other landscapes attributes with spatial scale may provide important information about the fundamental organization of landscapes. For example, Meentemeyer and Box (1987) hypothesize that the "scale of structural heterogeneity dictates the scale of the main functional linkages." Observations of the changes in heterogeneity with scale may be useful in determining whether landscapes are hierarchically structured and in identifying the levels of the spatial hierarchy (O'Neill et al. 1986; O'Neill personal communication).

4.4.2 Response of the Haralick Texture Measures to Scale

Because heterogeneity is scale dependent, a measurement of landscape heterogeneity is clearly not very useful unless the spatial scale of the measurement can be specified. Also, scale-dependent change in a measure of heterogeneity cannot be examined without specifying the manner in which scale is varied. Therefore, we explored how spatial scale may be specified and varied in the application of the Haralick texture measures ASM and IDM. We first examined the response of ASM and IDM to spatial scale (or to spatial variables that might be considered aspects of scale) using a collection of landscape images as "natural test patterns."

How pattern changes with spatial scale depends on how scale is defined (Turner et al. 1989). We found that there are several spatial variables involved in the application of the Haralick texture measures, and this multiplicity of spatial variables raised questions concerning concepts of scale and methods of changing scale. These questions are discussed because they may also pertain to other methods of measuring landscape pattern.

4.4.2.1 Data Sets

The data sets used were derived from Landsat Thematic Mapper (TM) images from five current or former National Science Foundation Long-Term Ecological Research sites: Jornada (New Mexico), Okefenokee (Georgia), North Temperate Lakes (Wisconsin), Niwot Ridge/Green Lakes Valley (Colorado), and Konza Prairie (Kansas). These sites were chosen in the expectation that their variety in climate, landforms, and vegetation (see Halfpenny and Ingraham 1984; Brenneman 1989) would result in a variety of landscape patterns.

As part of the preprocessing of these images for an earlier study, raw digital count values had been converted to units of radiance using equations derived from calibration of the TM sensor response (Markham and Barker 1986). The images had also been spatially resampled (i.e., geometrically corrected) to conform to the Universal Transverse Mercator coordinate system, with each data value nominally representing 28.5 by 28.5 m on the ground. A block of 247 by 282 pixels (approximately 7 by 8 km, corresponding to the size of the smallest original image) was extracted from each image for analysis.

Because the Haralick texture measures are based on comparisons of neighboring pixel values, it is important to recognize that the relationship between adjacent pixel values has been influenced by atmospheric scattering of the reflected radiation, by scanner optical geometry and signal processing, and by postacquisition spatial resampling of the raw data (see Townshend 1981; Billingsley et al. 1983). The net results of these influences is that the ground area contributing radiance to a given pixel in the final data set is larger than the nominal pixel size (as defined by the distance between pixel centers), is not square, and does not contribute radiance uniformly over the area (Townshend 1981, Curran and Hay 1986). The resulting overlap in the ground areas contributing radiance to adjacent pixels increases the correlation between adjacent pixel values by some degree. To precisely determine or remove this increase in correlation is very difficult, and we chose instead to take its possible effects into account in choosing our analysis procedures and in interpreting the results.

The TM sensor measures radiance in seven wavebands, but the texture measures require univariate data as input. The variable we chose to extract was the ratio of near infrared (TM Band 4) to red (TM Band 3) radiance, an index of the amount of green vegetation (Curran 1980, 1983). The infrared/red ratio numbers were real numbers with a limited range, rarely exceeding 10. To convert the data to integer values within the range 0–255 (as required by our image processing system) yet preserve fine distinctions, it was necessary to rescale the raw ratio values before rounding to the nearest integer.

One alternative would have been to rescale all images by the same linear function (e.g., based on the range of values in the five images collectively). We rejected this approach because it would have eliminated most evidence of spatial pattern in images having a small range of raw ratio values, such as Jornada. We would then have been unable to assess how heterogeneity varied with scale if most of the faint, yet detectable, heterogeneity had thus been removed. We chose another approach, rescaling the ratio values for each image independently to a range of 0–255. This allowed us to examine changes in pattern with scale even in images where patterns were faint, but the use of a different rescaling function for each data set made between-image comparisons of absolute values of the texture measures difficult to interpret. For example, we earlier described IDM as a measure of the shallowness of local gradients. What is being measured when IDM is calculated from independently rescaled data sets might be described as the shallowness of local gradients relative to the overall range of values.

The choice of approaches we faced here, to normalize data sets or to preserve absolute-value distinctions, is not unique to the Haralick texture measures. As noted earlier, measures of pattern based on spatial arrangement may be formulated to be either independent of or partially sensitive to the nonspatial statistical properties of the data (Haralick et al. 1973; Rosenfeld and Kak 1976). Neither alternative is inherently superior; each measures a different aspect of pattern. The problem is somewhat similar to the problem of evaluating noise in a temporal signal: both the absolute variation and the variation normalized for the differences in signal strength (i.e., noise/signal ratio) may be important, depending on the application.

The ratio images exhibit a variety of patterns (Fig. 4.3). The pattern of the North Temperate Lakes and Okefenokee images is strongly influenced by the distribution of surface water bodies, which have very low ratio values in the absence of emergent green vegetation. In the North Temperate Lakes image, large and mostly open water bodies are seen as large, sharp-edged patches of uniformly low ratio values. These uniform areas contrast with the small-scale heterogeneity of the mostly forested uplands. In the Okefenokee image, there is a gradation from open water bodies (mostly small) through wetlands with various amounts of emergent vegetation to vegetated uplands, and these types and their corresponding ratio values are highly interspersed. In addition to lakes, the Niwot Ridge/Green Lakes Valley image also has large areas of snow cover with low ratio values. The highly irregular topography of this site results in a high degree of interspersion of lakes, snow cover, alpine tundra, and forest, but some of the patches, although irregular, are large. Pattern in the Konza Prairie image is related to topography and to disturbances. The upland prairie is penetrated by more heavily vegetated lowlands forming a dendritic drainage pattern, and there are patches of prairie varying in vegetation amount as a result of differences in grazing and burning history (Nellis and Briggs 1989). Much of the variation in the Jornada image is small-scale; there is little contrast between most large areas, perhaps because topographic gradients are very gentle over all but the southwest (lower left) quadrant of the image.

Thus, in these five images we encountered a set of natural test patterns composed of landscape elements with widely different scale properties as a result of different physical and biological processes contributing to their formation. In the context of this chapter, our central objective was to evaluate the sensitivity and behavior of texture measures to these patterns, and thereby gain insight into how these indices might be more broadly applied.

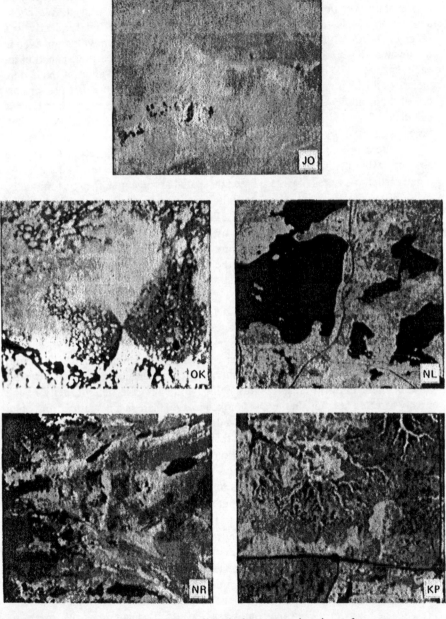

Figure 4.3. (TM Band 4)/(TM Band 3) ratio images used as input for texture measure calculation: Jornada (JO), Okefenokee (OK), North Temperate Lakes (NL), Niwot Ridge/Green Lakes Valley (NR), and Konza Prairie (KP). Images represent approximately 7 by 8 km.

4.4.2.2 Methods

To calculate the Haralick texture measures, we used TEXT, a program in the Earth Resources Laboratory Applications Software (ELAS) package (Earth Resources Laboratory 1987). TEXT permits many, but not all, of the previously described options for calculating these measures. The program was designed to produce a texture-value image that could easily be merged with the original data to produce a multilayer data set suitable for input to pixel-by-pixel clustering and classification algorithms. Therefore, the output data set has the same pixel size and number of pixels as the input image. TEXT uses the previously described sliding window procedure, writing the texture value of each window to the window's central pixel.

Windows must be square and have an odd number of pixels per side. Window position may be incremented by more than one pixel, in which case the texture values for pixel positions not at the center of any window are obtained by linear interpolation of neighboring window-center values. To avoid using linearly interpolated values of nonlinear texture measures (e.g., ASM), we extracted only the window-center values from the output image whenever we used a window movement increment greater than one.

TEXT calculates any of the four direction-specific texture measures or an all-directional value derived from a tally of cooccurrences along all four angles, as in Fig. 4.2. We chose the latter procedure to reduce computational time in obtaining an all-directional measure.

TEXT requires that the input data be quantized to no more than forty levels, and requantizing (if necessary) is accomplished internally to the program by a linear function. Our range-normalized ratio images were each independently requantized by this means to forty levels. As a result, patterns characterized by pixel-to-neighbor differences of less than one-fortieth of the image range were rendered partially invisible to the texture measures.

For our objectives we required not a texture-value image comprised of many pixels but a single texture value characterizing the whole areas of the input image. Because maximum window size in TEXT (77 by 77 pixels) was smaller than our images (247 by 282 pixels), we could not set window size equal to image size as a means of obtaining whole-image texture values. It was necessary to analyze each image piecewise by using the sliding window procedure, collect the texture values of all windows, and use the mean of these values to characterize the whole image. This procedure is essentially the same as subsampling an area by means of quadrats except that we could obtain values for the whole population of quadrats (or a large proportion thereof) rather than for a small sample of this population. Although we were forced by software limitations to use piecewise analysis methods, this circumstance gave us an opportunity to explore the consequences of varying quadrat size for the texture measures. Thus our first experiment was an examination of the response of ASM and IDM to varying window size.

4.4.2.3 Varying Window Size

In this experiment, windows were nonoverlapping and contiguous. We chose not to sample all possible overlapping windows in order to reduce computational time.

Not all window sizes had multiples exactly fitting the image, but window sizes were selected to minimize the amount of unsampled edge. Cooccurrences were examined over a distance of two pixels (neighbors once removed). Comparing immediately adjacent pixels was avoided because the partial overlap in ground area sampled by adjacent pixels was expected to result in spuriously high correlations, as previously described. This analysis was performed on three images selected to provide a wide range of patterns: Jornada, Okefenokee, and North Temperate Lakes.

The responses of ASM to window area were approximately linear on a log-log plot (Fig. 4.4). ASM decreased with window size for all three sites. In contrast, IDM was nearly constant with window size (Fig. 4.5). This contrast in the response of the two measures was unexpected, because they both measure neighbor-to-neighbor similarity, and whole-image values of the two indices are highly correlated (Haralick et al. 1973). This issue will be discussed before we consider the details of ASM response.

A consideration of the way IDM is calculated (Eq. 4.2) reveals that image mean values of this index are inherently insensitive to window size. IDM is a simple sum of the weighted cooccurrence frequencies, and averaging the window IDM values is also an additive process. Therefore, the same IDM value is obtained whether the image is analyzed piecewise or as a whole. The slight variation in IDM that we obtained resulted from edge effects (i.e., failure to count cooccurrences across the boundaries of the contiguous windows and failure of some window arrays to sample the entire image). If we had used a window-movement increment of one pixel and thereby sampled all possible overlapping windows, these edge effects

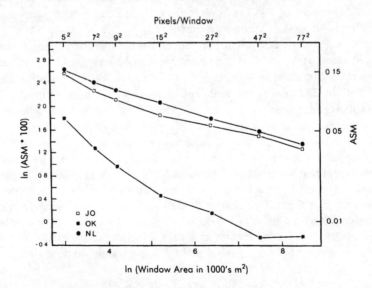

Figure 4.4. Response of image mean ASM to window area. Site abbreviations as given in Fig. 4.3.

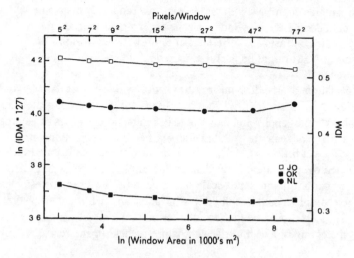

Figure 4.5. Response of image mean IDM to window area. Site abbreviations as given in Fig. 4.3.

would have been eliminated, and IDM would not have varied at all with window size.

These and other measures based on comparisons of adjacent values are more easily interpreted if we recognize that their calculation is, in effect, a stepwise operation. The stepwise nature of the process is also important in determining the scale properties of the final values, as will be explained later. The first step is, in effect, the derivation of a transformed image. The values in this image are a function of the pair of input data values being compared, and the pixels of this transformed image are located at spatial coordinates offset $d/2$ from pixels of the original image. This transformed image is then further transformed, aggregated, or otherwise processed according to the index formula. In the case of IDM, the operations performed on this transformed image are purely additive processes. ASM is sensitive to window size because the cooccurrence probabilities are squared before summation for each window (Eq. 4.1).

ASM decreases with window size (Fig. 4.4) because it is a nonlinear measure and therefore responds positively to between-window variance in the property that it measures (i.e., aggregation of like values). If windows vary in their degree of aggregation, then some windows have higher and some have lower $p(i,j)$ values than the whole image; the window-mean ASM will then exceed the whole-image value because ASM is a function of the square of the $p(i,j)$ values. The observed decreases in ASM with window size indicate that larger windows are less variable than smaller windows in their degree of aggregation.

The interpretation of window-size effects on ASM could be aided by procedural modifications and further studies beyond the scope of our preliminary investigations. For example, the response of ASM to window size could be compared with the response obtained if the data values were randomly scrambled; since variance between samples decreases with sample size in a randomly distributed population, ASM should decrease with window size even in a randomized image. Also, some measure of the contrast (i.e., difference or ratio) between window-mean ASM and whole-image ASM might be useful as an index of between-window variability in aggregation.

The deviation of window-mean ASM from whole-image ASM would be a measure of large-scale (i.e., window size) variability in small-scale (i.e., pixel-to-neighbor) homogeneity and thus would be related to two types of heterogeneity, each on a different scale. Images or landscapes could be equally heterogeneous on a small scale, yet differ markedly in how uniformly the small-scale heterogeneity is distributed over the area. How shall such complex measures of pattern as "the large-scale variability in small-scale heterogeneity" be used to test hypotheses about the ecological causes and effects of landscape heterogeneity? We suggest that the first step should be the formulation of new hypotheses that will take into account the complex and multifaceted nature of heterogeneity.

4.4.2.4 Varying Pixel Size

Ground area represented by each pixel was increased by averaging blocks of original pixels. To simplify the analysis, we used a window size of 3 by 3 pixels for every measurement. Because the ground area represented by a window thus increased along with pixel size, the procedure could be viewed as changing two scale-related aspects simultaneously. To ensure that the effects of varying pixel size would be independent of varying window area, we used only IDM for these analyses; as shown in the previous experiment, the window-mean value of IDM is unaffected by window size and equals the whole-image IDM value. As described earlier, IDM is a measure of the shallowness of local gradients. For this experiment we used a comparison distance of 1 pixel, so that immediately adjacent pixels were compared.

Partial overlap in the ground areas contributing radiance to adjacent pixels was expected to result in high IDM values for the original images, followed by a sharp drop as the overlap was reduced by aggregating blocks of pixels. Four of the five images did indeed exhibit an initial decrease in IDM that might be partially the result of overlap in pixel ground areas (Fig. 4.6). Pixel overlap may have made the local minimum in Jornada IDM at the 1-pixel scale less pronounced than it would have been otherwise.

Averaging is usually thought of as a strictly homogenizing process, but the results show that increasing pixel size may either increase or decrease local gradients, depending on the image characteristics and the range of pixel sizes examined (Fig. 4.6). Averaging blocks of samples does reduce overall population range and variance and is therefore a strictly homogenizing process for the non-spatial or aggregate statistical properties of the population. However, the block-averaging process does not always result in greater homogeneity in spatial relationships and the statistics that measure these relationships.

Woodcock and Strahler (1987) examined the response of mean local variance in 3 by 3 pixel windows to pixel size in a variety of remotely sensed images and

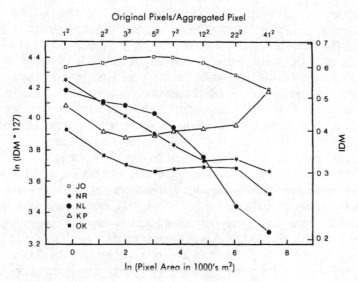

Figure 4.6. Response of image mean IDM to pixel area. Site abbreviations as given in Fig. 4.3.

simulated images with simple patterns. Their findings are useful in interpreting the IDM responses in our experiment because local variance and pixel-to-neighbor gradients should be highly correlated when windows are as small as 3 by 3 pixels. Responses are most easily described for a simple image comprised of uniformly black, approximately round, and uniformly sized patches randomly arranged on a uniformly white background. Initially, when pixel size and window size are less than patch size, most windows fall mostly or entirely inside either the black patches or the white background, and local variance is low. As pixel size approaches patch size, the probability that a neighboring pixel will have a similar value decreases, and local variance increases. As pixel size increases beyond patch size, local variance decreases as larger pixels incorporate an increasingly even mixture of patch and background. Local variance peaks at a pixel size about one-half to three-fourths the size of patches. Some patterns may exhibit only a portion of this curve. For example, if patches are near the size of the original pixels, only the decrease in local variance resulting from mixing of patches and background will be observed as pixel size is increased (Woodcock and Strahler 1987).

The response of local variance to pixel size for more complex patterns follows the principles described above. For example, the peak in local variance should be broad if patch size is evenly distributed over a broad range. Peak height should be related to the contrast between patch and background. If small patches are aggregated in clusters, there should be two peaks in local variance: one related to patch size and one related to cluster size.

IDM responses can be interpreted in the same way as responses in local variance, except that the inverse relationship between IDM and heterogeneity will result in local minima in IDM corresponding to peaks in local variance. For all sites except Konza Prairie, IDM decreased at the largest pixel size (Fig. 4.6), indicating the presence of patches too large to be completely incorporated in the largest pixels. In the North Temperate Lakes image (Fig. 3), this large dominant pattern element can easily be identified as the large-scale pattern of lakes versus uplands. The Konza Prairie and Okefenokee curves exhibit broad, shallow local minima at intermediate pixel sizes, indicating a broad range of patch sizes. The Jornada curve approaches local minima at both the smallest and largest pixel sizes, indicating the presence of both very small and very large patches.

A detailed explanation of the ecological causes and consequences of the observed patch-size distributions in each image is beyond the scope of this chapter, but some general conclusions may be drawn from the results. Because IDM responses to pixel size are related to patch size and contrast, the results might be used to indicate the most appropriate resolutions to use when converting these continuous data to a model of patches as discrete entities. If an image is dominated by patches of uniform size, then the maximum appropriate resolution should be significantly less than patch size if patches are to be clearly delineated as discrete entities with minimal loss of information. It might be useful to go beyond the analyses we performed and examine the frequency distribution of window IDM values at different pixel sizes. If local gradients were uniformly shallow at all pixel sizes, we might conclude that a model of discrete patches does not fit the data well at any scale. If a hierarchically organized landscape would be expected to exhibit a pattern of small patches aggregated into clusters, then we could conclude that there is little evidence of strong hierarchical organization in these landscape data sets at the scales examined. Another conclusion is that average local gradient in green vegetation amount is not constant with scale and does not change predictably with scale when scale is defined as pixel area. Therefore, values of this landscape property cannot be easily predicted at one scale from measurements at another scale.

Detecting patch size and contrast in continuous data can be accomplished not only by examining IDM response to pixel size, but also by a variety of other methods such as local variance versus pixel size. The IDM method would be especially appropriate if the interest was not in patch characteristics specifically but rather in the resultant effects of patch characteristics on the strength of local gradients, the property directly measured by IDM.

4.4.3 Aspects of Scale

Ecologists are becoming increasingly aware of the effects of scale on ecological phenomena and of the ways that scale of observation may limit observable relationships and possible conclusions (Wiens et al. 1986; Meentemeyer and Box 1987). As more investigators seek to explicitly account for the effects of scale in their studies, clear definitions and concepts of scale are needed. As described earlier, the use of TEXT to calculate texture measures requires that five spatial variables be specified: pixel area, image area, window area, window movement increment, and distance between pixels to be compared. Because these or similar variables are likely to be encountered in many other sampling and analysis procedures, their relationships to current concepts of scale should be examined.

Allen et al. (1984) described grain and extent as two aspects of the scale of observation sets. Grain determines the fineness of distinctions that can be made from an observation set, and extent determines the largest observable distinctions. These limits are initially set by the measurement and sampling procedures employed, but further limitations may be set by subsequent transformations and analysis. Although practical difficulties in collecting and analyzing large amounts of data often dictate a reduction in grain with an increase in extent (Meentemeyer and Box 1987), these two aspects of scale can often be varied independently. Turner et al. (1989) found that landscape pattern indices responded to scale differently depending on whether grain or extent was varied. In their analyses of digital land use maps, Turner et al. (1989) equated grain with pixel area and varied this aspect of scale by aggregating blocks of contiguous pixels. Extent was varied by extracting for analysis successively larger concentrically nested quadrats. For our texture measurements we may similarly identify grain as pixel area and extent as image area, but what aspects of scale do the remaining spatial variables represent?

First, it is useful to recognize that there are similarities between the sampling procedures used in the texture measurements and those used in collecting the original radiance data. That is, the sliding window technique employed by TEXT is analogous to operation of the orbiting optical scanner on the Landsat satellite. The sliding window scans across and periodically samples a digital image, just as the orbiting scanner samples the upward stream of radiation; window size is analogous to the instantaneous field of view (IFOV; the area from which an individual measurement of radiance is taken) of the optical device. The increment by which a window moves between texture measurements is analogous to the sampling interval of the scanner. The scanner and the texture analysis perform different operations on the within-IFOV or within-window signal; the scanner sums or integrates the signal, whereas the texture analysis first derives a frequency distribution of the cooccurrences and then performs additional operations on those frequencies as specified by the particular texture measure used. In the resultant image consisting of texture values, the area represented by each texture value is the window area used in deriving the texture image; window area has thus become the pixel area of the texture image. Perhaps it is useful to think of the block size used in a blockwise analysis as becoming the grain size of the derived observation set.

Specifying a value greater than 1 for the interpixel comparison distance is equivalent to creating a new input data set by deleting the pixels not used in the comparison. This undersampled image is equivalent to one that could theoretically be created by modifying the process of data collection by the scanner so that the interval between recorded IFOVs was increased while keeping IFOV area constant.

The relationship of window or IFOV movement distance to aspects of scale is unclear. The absolute value of this parameter may be of less significance than the relationship of movement distance to window or IFOV dimensions, because it is this relationship that determines whether the data will be oversampled or under-sampled. If window movement distance is much greater than window dimensions and the data are thus undersampled, information on fine-scale variation within windows and large-scale variation between windows could be extracted, but information on intermediate-scale patterns would appear to be lost. For example, suppose mean solar radiation was recorded each minute (= pixel size) of 1 h (= window size) at 24-h intervals (= window movement distance) for 1 year (= extent). Information on short-term variability and seasonal trends would be potentially extractable, but evidence of a diurnal cycle would be lost. Even complete sampling (window movement distance = window dimensions) could result in loss of information on pattern at some scales, particularly if windows were out of phase with a periodic pattern of variation. Complete oversampling (window movement distance = pixel size) samples the entire population of possible win-dows and captures information that might otherwise be lost. Perhaps window movement distance may be viewed as a filter to observable phenomena because it may eliminate information on pattern at some scales within the range of scales determined by grain and extent.

Allen et al. (1984) noted that a set of observations is more than simply a collection of raw measurements; the information captured is determined by the data collection and analysis procedures. Because scale cannot be attributed to a simple collection of numbers without knowledge of the sampling and analysis procedures that produced the numbers, it must be these procedures that determine scale. When applied to a data set, some analysis procedures (e.g., variance calcula-tion) will preserve some of the information on fine-grain pattern, but others (e.g., summation) will result in loss of this information and coarsening of the grain. Our method of calculating the texture measures is, in effect, a stepwise process produc-ing a series of observation sets, each derived from the preceding by a process of sampling and analysis: the input image, an implied image of spatially referenced cooccurrences, the output image consisting of a texture value for each window, and the mean of the window texture values. Because all of these observation sets pertain to the area of the input image, extent for all sets is readily defined as image area, a constant. Grain of the input image is clearly equivalent to pixel area, but assignment of a single value for grain to the derived observation sets is difficult. Analyses of landscape data often involve stepwise procedures, and a concept-ualization of scale applicable to the derived observation sets is therefore clearly needed. Our difficulties with the problem of changing scale suggest that compa-risons of scale effects on landscape properties should be made with careful consideration of all the sampling and analysis procedures that might affect scale.

4.5 Summary

The Haralick texture measures were originally developed to quantify certain aspects of pattern in digital pictorial imagery, but they can be used as pattern indices for ecological data sets having the required format of regularly spaced gridded values. These measures are based on a statistical approach to the description of pattern, as contrasted with a structural approach based on the spacing and other attributes of primitive elements (i.e., patches). The measures are based on adjacency, i.e., the relationship of each sample value to the values of neighboring samples. The measures we examined in detail, Angular Second Movement (ASM) and Inverse Difference Moment (IDM), are both indices of homogeneity but differ in their formulation. ASM is essentially an inverse measure of interspersion and may be used either for data values representing qualitatively different types or for data that are measurements of intensity or amount. IDM is a measure of the shallowness of local gradients and is appropriate only for measurements of intensity or amount. The procedures for calculating these indices may be varied in several ways to extract information on different aspects of pattern, including directionality and spacing of repetitive patterns.

The calculation of a texture measure value is governed by several scale-related variables. These include the distance specified for adjacency and, if the data set is analyzed blockwise, the size and spacing of blocks. These variables, plus the scale-related properties of the input data set (area per sample value and area of the entire data set), give to any single texture measurement a complex dependence on scale not easily described in terms of current concepts of scale.

Analysis of the calculation procedures for these indices suggests that the derivation of an index value may be equivalent to stepwise derivation of a series of observation sets, each produced from the preceding one by a process of sampling and analysis. Each step is governed by different spatial variables and may place different limits on observable phenomena while capturing and incorporating some information on pattern at various scales. The scale of the final observation set (the texture value) is determined by the effects of all the sampling and analysis processes used, and to condense these effects into a single number for one or two aspects of scale (i.e., grain and extent) has proven difficult.

We examined the responses of ASM and IDM to varying block (window) size and of IDM to varying pixel size in a collection of remotely sensed images used as natural test patterns. The indices differed in their inherent capacity to vary with block (window) size, with IDM constrained by its mathematical formulation to be invariant. ASM decreased with block size because it is a nonlinear measure and, therefore, sensitive to variance among blocks. Thus, different indices of a given aspect of pattern might be, as these are, largely redundant when scale is held constant, yet provide very different information about change in pattern with scale.

The IDM response to varying pixel size by averaging blocks of pixels depended on the image pattern and the range of pixel sizes. IDM response to pixel size was interpreted as reflecting the contrast and size distribution of patches of similar value.

The introduction to Haralick texture measures provided in this chapter is intended to encourage more detailed studies necessary to determine whether these indices will be useful components of the landscape ecologist's "tool kit." Further explorations of these and other pattern analysis methods developed for image analysis may also serve to stimulate landscape ecologists to expand and clarify their concepts of pattern and to formulate new hypotheses relating ecological processes to complex aspects of pattern.

Acknowledgments

We thank T.F.H. Allen for suggesting this investigation and R.V. O'Neill for a helpful review of an early version of the manuscript and for freely sharing ideas and unpublished manuscripts. B.T. Milne also provided valuable advice and encouragement. We are especially grateful to the editors, M.G. Turner and R.H. Gardner, for a thorough and useful review. T. Budge performed the preprocessing of data, and we are thankful for his assistance throughout the subsequent data processing. This work would not have been possible without the help of many others on the staff of the Technology Application Center. This research was funded by NSF Ecosystems Studies Grant BSR86–06700.

References

Allen, T.F.H.; O'Neill, R.V.; and Hoekstra, T.W. 1984. Interlevel relations in ecological research and management: Some working principles from hierarchy theory. USDA Forest Service, General Technical Report RM–110.

Billingsley, F.C.; Anuta, P.E.; Carr, J.L.; McGillem, C.D.; Smith, D.M.; and Strand, T.C. 1983. Data processing and reprocessing. In *Manual of Remote Sensing*, 2nd ed., ed. R.N. Colwell, pp. 719–92. Falls Church, Va.: American Society of Photogrammetry.

Brenneman, J., ed. 1989. *Long-Term Ecological Research in the United States, A Network of Research Sites*. 5th ed. Seattle, Wash.: National Science Foundation Long-Term Ecological Research Network Office.

Brodatz, P. 1956. *Textures: A Photograph Album for Artists and Designers*. New York: Dover.

Burrough, P.A. 1981. Fractal dimensions of landscapes and other environmental data. *Nature* 294:240–42.

Curran, P.J. 1980. Multispectral remote sensing of vegetation amount. *Progress in Physical Geography* 4:315–41.

Curran, P.J. 1983. Multispectral remote sensing for the estimation of green leaf area index. *Philosophical Transactions of the Royal Society of London* A309:257–70.

Curran, P.J., and Hay, A.M. 1986. The importance of measurement error for certain procedures in remote sensing at optical wavelengths. *Photogrammetric Engineering and Remote Sensing* 52:229–41.

Earth Resources Laboratory. 1987. ELAS: Earth Resources Laboratory applications software. Earth Resources Laboratory Report No. 183. National Aeronautics and Space Administration, National Space Technology Laboratories, NSTL, Miss.

Forman, R.T.T. 1987. The ethics of isolation, the spread of disturbance, and landscape ecology. In *Landscape Heterogeneity and Disturbance*, ed. M.G. Turner, pp. 213–29. New York: Springer-Verlag.

Forman, R.T.T., and Godron, M. 1981. Patches and structural components for a landscape ecology. *Bioscience* 31:733–40.

Forman, R.T.T., and Godron, M. 1986. *Landscape Ecology*. New York: Wiley.

Halfpenny, J.C., and Ingraham, K.P., eds. 1984. *Long-Term Ecological Research in the United States, A Network of Research Sites*, 3rd ed. Corvallis, Oreg.: National Science Foundation Long-Term Ecological Research Network Office.

Haralick, R.M. 1979. Statistical and structural approaches to texture. *Proceedings of the IEEE* 67:786–804.

Haralick, R.M., and Shanmugam, K.S. 1974. Combined spectral and spatial processing of ERTS imagery data. *Remote Sensing of Environment* 3:3–13.

Haralick, R.M.; Shanmugam, K.; and Dinstein, I. 1973. Textural features for image classification. *IEEE Transactions on Systems, Man, and Cybernetics* SMC–3:610–21.

Markham, B.L., and Barker, J.L. 1986. Landsat MSS and TM post-calibration dynamic ranges, exoatmospheric reflectances and at-satellite temperatures. *Eosat Landsat Technical Notes* 1:3–8.

Meentemeyer, V., and Box, E.O. 1987. Scale effects in landscape studies. In *Landscape Heterogeneity and Disturbance*. ed. M.G. Turner, pp. 15–34. New York: Springer-Verlag.

Milne, B.T. 1988. Measuring the fractal geometry of landscapes. *Applied Mathematics and Computation* 27:67–79.

Milne, B.T.; Johnston, K.M.; and Forman, R.T.T. 1989. Scale-dependent proximity of wildlife habitat in a spatially-neutral Bayesian model. *Landscape Ecology* 2:101–10.

Nellis, M.D., and Briggs, J.M. 1989. The effect of spatial scale on Konza landscape classification using textural analysis. *Landscape Ecology* 2:93–100.

Nevatia, R. 1982. *Machine Perception*. Englewood Cliffs, N.J.: Prentice-Hall.

O'Neill, R.V.; DeAngelis, D.L.; Waide, J.B.; and Allen, T.F.H. 1986. *A Hierarchical Concept of Ecosystems*. Princeton, N.J.: Princeton University Press.

O'Neill, R.V.; Krummel, J.R.; Gardner, R.H.; Sugihara, G.; Jackson, B.; DeAngelis, D.L.; Milne, B.T.; Turner, M.G.; Zygmunt, B.; Christensen, S.W.; Dale, V.H.; and Graham, R.L. 1988. Indices of landscape pattern. *Landscape Ecology* 1:153–62.

Pielou, E.C. 1969. *An Introduction to Mathematical Ecology*. New York: Wiley-Interscience.

Remillard, M.M.; Gruendling, G.K.; and Bogucki, D.J. 1987. Disturbance by beaver (*Castor canadensis* Kuhl) and increased landscape heterogeneity. In *Landscape Heterogeneity and Disturbance*. ed. M.G. Turner, pp. 103–22. New York: Springer-Verlag.

Risser, P.G. 1987. Landscape ecology: state of the art. In *Landscape Heterogeneity and Disturbance*. ed. M.G. Turner, pp. 3–14. New York: Springer-Verlag.

Risser, P.G.; Karr, J.R.; and Forman, R.T.T. 1984. Landscape ecology, directions and approaches. Special Publication No. 2, Illinois Natural History Survey, Champaign.

Romme, W.H., and Knight, D.H. 1982. Landscape diversity: The concept applied to Yellowstone Park. *Bioscience* 32:664–70.

Rosenfeld, A., and Kak, A.C. 1976. *Digital Picture Processing*. New York: Academic Press.

Schonewald-Cox, C.M. 1988. Boundaries in the protection of nature reserves. *Bioscience* 38:480–86.

Schowengerdt, R.A. 1983. *Techniques for Image Processing and Classification in Remote Sensing*. New York: Academic Press.

Townshend, J.R.G. 1981. The spatial resolving power of earth resources satellites. *Progress in Physical Geography* 5:32–55.

Turner, M.G., ed. 1987. *Landscape Heterogeneity and Disturbance*. New York: Springer-Verlag.

Turner, M.G., and Bratton, S.P. 1987. Fire, grazing, and the landscape heterogeneity of a Georgia barrier island. In *Landscape Heterogeneity and Disturbance*, ed. M.G. Turner, pp. 85–101. New York: Springer-Verlag.

Turner, M.G.; O'Neill, R.V.; Gardner, R.H.; and Milne, B.T. 1989. Effects of changing spatial scale on the analysis of landscape pattern. *Landscape Ecology* 3:153–62.

Wiens, J.A.; Addicott, J.F.; Case, T.J.; and Diamond, J. 1986. Overview: the importance of spatial and temporal scale in ecological investigations. In *Community Ecology*, eds. J. Diamond and T.J. Case, pp. 145–53. New York: Harper and Row.

Wiens, J.A.; Crawford, C.S.; and Gosz, J.R. 1985. Boundary dynamics: A conceptual framework for studying landscape ecosystems. *Oikos* 45:421–27.

Woodcock, C.E., and Strahler, A.H. 1987. The factor of scale in remote sensing. *Remote Sensing of Environment* 21:311–32.

5. Biogeochemical Diversity and Element Transport in a Heterogeneous Landscape, the North Slope of Alaska

Gaius R. Shaver, Knute J. Nadelhoffer, and Anne E. Giblin

Terrestrial ecosystems are typically studied as discrete, independent units within a heterogeneous landscape, even though they are never completely isolated from each other. In fact, all ecosystems are more or less strongly linked to their neighbors through the transport of water, energy, organic matter, and mineral elements: in other words, one ecosystem's outputs are its neighbor's inputs. These transport processes are often extremely important in determining an ecosystem's productivity, its species composition, and the relative importance of its various element cycling processes (Odum 1971; Forman and Godron 1986). Thus in order to understand how an ecosystem works, it is necessary to understand how it interacts with its neighbors.

The first aim of this chapter is to describe one approach to determining the importance of element transport between neighboring ecosystems. Our approach is empirical, comparative, and budget oriented. Our overall question is this: How important is element transport between ecosystems relative to internal element cycling processes in support of primary production? Answers to this question are especially useful for at least three reasons. First, by knowing how ecosystems vary in the relative importance of external transport versus internal cycling, we can make better predictions about how ecosystems should also vary in their response to a disturbance of transport processes. Second, we can better understand how biogeochemical disturbances spread over a landscape and why some ecosystem types are more effective than others in limiting the spread of disturbance (i.e., through differences in their ability to immobilize element inputs in soil water).

Third, by understanding how the various units of an entire landscape are linked together, we can better predict the implications of management practices (such as agricultural fertilizer application) that affect element transport from one landscape unit to its neighbors.

Our second aim is to provide an example of the importance of spatial heterogeneity in element cycling at the landscape level. Viewing the landscape as a patchwork of internally homogeneous ecosystems linked by soil water transport, we ask this: how does a particular sequence of ecosystems determine the chemistry of soil water flowing through them, and what might be the effect of a change or interruption of that particular sequence? In this respect our approach differs from most watershed studies (e.g., Likens et al. 1977), which usually consider a watershed as a single ecosystem and view soil and stream water outputs as an overall "integrator" of element cycling. Here, we are focusing on variation *within* the watershed or landscape in an attempt to identify which components of the landscape are most important as element sources or sinks.

5.1 The Arctic as a Model Landscape

The focus of our work is on the landscape of the northern foothills of the Brooks Range, Alaska. Most people think of this part of the world as scenically spectacular but uniformly barren, biotically impoverished, wet, and cold. In fact, the arctic landscape has a number of features that make it particularly useful for research on ecosystem- and landscape-level questions. These include (1) an array of sharply different vegetation and soil types (ecosystems) with distinct boundaries; (2) a repeated series of mesotopographic gradients that include contrasting ecosystems immediately adjacent to each other in a predictable sequence; (3) low plant stature and fine-grained scale of species heterogeneity, which reduces sampling effort and makes replicated ecosystem-level experimentation feasible; and (4) continuous permafrost close to the ground surface, which prevents deep water drainage and thus greatly simplifies hydrological studies.

In the Sagavanirktok River valley near Toolik Lake, Alaska, we have identified a toposequence of ecosystems that is particularly appropriate for our research. The toposequence (Fig. 5.1) extends across a series of old floodplain terraces and includes six contrasting vegetation types with major differences in growth form composition. This sequence is typical of major river valleys of the North Slope, in particular along the Sagavanirktok, Kuparuk, and Atigun rivers near the Trans-Alaska oil pipeline (Fig. 5.2; Walker 1985). The sequence ranges from typical moist tussock tundra on the surrounding uplands to tall (1- to 1.5-m) willow stands on recently vegetated alluvial soils. In between are a well-drained heath, a deciduous shrub-*Cassiope*-lupine zone, a grass-sedge-*Equisetum* zone, and a poorly drained wet sedge tundra. The soils along the toposequence are also highly variable in their temperature, moisture, organic matter content, depth to permafrost, and nutrient availability (Table 5.1).

A great deal of background information is also available about similar toposequences; in fact, one of the classic and highly successful approaches to tundra

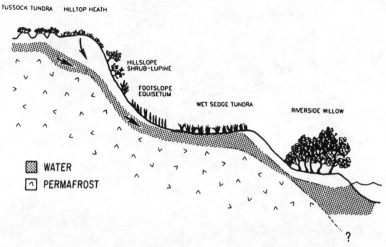

Figure 5.1. The study site in the Sagavanirktok River valley in northern Alaska, including a toposequence of six ecosystem types running from tussock tundra on the uplands, down a series of old floodplain terraces, to the river itself. A large snowdrift accumulates each year in the hillslope and footslope ecosystems, while the hilltop heath is blown free of snow. The toposequence is entirely underlain by permafrost, and during the summer the soil water flows downslope, across the top of the permafrost. In all six ecosystems, plant roots reach all the way down to the maximum depth of thaw (except perhaps in the hilltop heath, where depth of thaw exceeds 1 m). The total vertical drop is approximately 10 m, and the total horizontal distance ranges from 100 to 200 m.

plant ecology has been to analyze changes along such "mesotopographic gradients" (Billings 1973). Results are available from a number of studies in alpine as well as arctic environments (Bliss 1956, 1977; Johnson and Billings 1962; Teeri 1973; Peterson and Billings 1980; McGraw and Antonovics 1983). In past research the major environmental factors of interest have included wind, water, snow cover, and temperature. Changes in overall element cycling patterns have not been studied with the same thoroughness, although one study along a toposequence at Eagle Creek, Alaska, is especially suggestive about the importance of nutrient exchanges between ecosystems (Miller 1982). That toposequence was very similar to ours; from top to bottom, it included a *Dryas* fell-field, an evergreen heath, a deciduous shrub zone, a *Cassiope* bed, a grass-forb zone, more deciduous shrubs, and a wet sedge-moss bed. The principal conclusion was that " . . .the vegetation zones are largely associated with different levels of nitrogen and phosphorus availability rather than the length of the snow-free season, water availability, and soil pH" (Miller 1982). The most nutrient-rich sites were those at the bottom of the slope, before it flattened out, similar to our foot-slope *Equisetum* zone (Fig. 1). The most nutrient-poor sites were the heath and sedge-moss zones, similar to our heath and wet sedge zones. However, the exchanges of nitrogen or phosphorus between ecosystems were never measured, and the relative importance of internal recycling versus external exchanges remains unresolved.

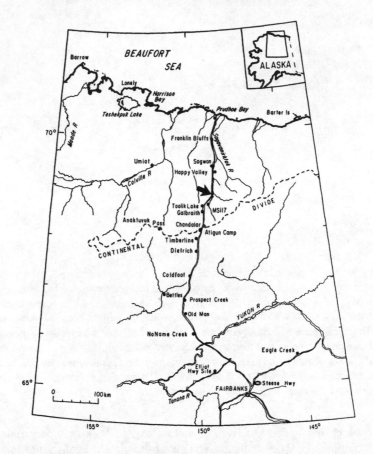

Figure 5.2. Map showing the location of the study site, approximately 30 km north of Toolik Lake, in the Sagavanirktok River Valley near the Dalton Highway and Trans-Alaska Oil Pipeline.

The six ecosystem types along our toposequence are nicely representative of the broad spectrum of ecosystem types that occur over the entire North Slope of Alaska (Britton 1966; Walker 1985; Walker et al. 1982). Thus, one of the advantages of our toposequence is that it allows us to study specific interactions between ecosystems in considerable detail and also to extrapolate our conclusions about individual ecosystem types over much broader areas.

5.2 A Conceptual Model of Landscape Element Cycling

The central theme of our research is that of element transport over the arctic landscape and especially element exchanges across ecosystem boundaries. We view this landscape as a patchwork or mosaic of contrasting ecosystem types and assume homogeneity within each ecosystem or "patch." The boundaries between

Table 5.1. Characteristics of Ecosystems Along the Toposequence[a]

	Tussock Tundra	Hilltop Heath	Hillslope Shrub-Lupine	Footslope Equisetum	Wet Sedge Tundra	Riverside Willow
Dominant vegetation	Eriophorum Dwarf shrubs	Dryas Ericads Betula-Salix	Salix Cassiope Lupinus	Equisetum Grasses Sedges	Sedges	Salix Betula Lupinus
Soil type	Pergelic Histic Cryaquept	Pergelic Ruptic Entic Cryumbrept	Pergelic Histic Cryaquept	Pergelic Histic Cryaquept	Pergelic Cryohemist	Pergelic Cryofluvent
Horizontal distance along toposequence (m)	30–100	8–15	10–30	8–15	20–50	10–100
Max depth of thaw (m)	40	200–500(?)	100–200(?)	45	55	?
Soil moisture (%)	Moist	Dry	Dry	Moist	Wet	Dry
Average of 01 horizon	415%	175%	235%	435%	530%	205%
Percentage C in 01 horizon	37	38	31	34	40	20
Percentage N 01 horizon	1.8	1.4	1.1	1.6	2.5	1.2
Average N:P ratio in bulk soils	14	8	12	11	41	12
Net primary production[b] g C/m²/y	160	90	150	100	70	320
Limiting nutrient[c] for net primary production	N (some species) P (some species)	N	?	N	P (Strong N interaction)	N?

[a] From Giblin et al. (in review).
[b] Estimated from Shaver and Chapin (1986) and Chapin and Shaver (1985). These papers were based on data collected at similar, nearby sites.
[c] Actual results of factorial nitrogen-times-phosphorus fertilization experiments in tussock and wet sedge tundra sites; for other sites based on interpretation of relative nitrogen and phosphorus availability measures (Giblin et al., in review).

these ecosystems, or landscape patches, are defined by distinct and usually abrupt changes in vegetation and soils. The two major elements we are concerned with are nitrogen and phosphorus, the most frequently limiting elements in the Arctic as well as most other terrestrial ecosystems (Chapin and Shaver 1985). However, in order to evaluate the importance of nitrogen and phosphorus transport, we must place it in the context of other major pathways of nitrogen and phosphorus cycling. To do this we have developed a conceptual model of element cycling along our riverside toposequence (Fig. 5.3).

Our approach is to view the toposequence as a 1-m-wide wide transect passing through all six of our ecosystem types. The transect consists of a sequence of individual square meters, linked through the transport of nitrogen and phosphorus either upslope or downslope. Each ecosystem type occupies a variable distance along the transect. The overall effect of each ecosystem type on the toposequence is thus dependent on both the process parameters for each square meter that it occupies and the number of square meters occupied. In the field the distance along our toposequence varies from about 30 to 100 m for tussock tundra to about 8 to 15 m for the hilltop heath and footslope *Equisetum*.

In any square meter along the transect, plants and soils may cycle nitrogen and phosphorus internally, and they may exchange nitrogen and phosphorus with each other, with the atmosphere, with the relatively inactive nitrogen and phosphorus pools deep in the soil, and with neighboring square meters (Fig. 5.3). There is also the possibility of longer-distance element transport, between nonadjacent square meters, due to animal movements, wind-blown litter, and soil water flow beneath the plant rooting zone (where depth to permafrost is greater than 1 m).

Many of the exchanges indicated by single arrows in Fig. 3 actually involve several element cycling processes, and not all involve both nitrogen and phosphorus. For example, atmospheric inputs include nitrogen fixation as well as rainfall, snow, and dry deposition. Outputs to the atmosphere are primarily gaseous nitrogen losses, such as by denitrification. The exchanges with inactive soil pools include mineralization of phosphorus from parent material, long-term chemical immobilization of both nitrogen and phosphorus, and gains or losses due to changes in the depth of permafrost.

In our conceptual model (Fig. 5.3) the primary means of lateral element transport is through soil and surface water flow. We have not considered stream flow and its associated element content, because our primary research site is a reasonably uniform hillside without any streams or springs and with no obvious channelized subsurface water flow (as might be indicated by vegetation patterning perpendicular to the slope). In practice, however, we do use stream chemistry and flow measurements from nearby small, first-order watersheds to help constrain estimates of element transport based on soil solution chemistry and soil hydrology on the primary site. For the purposes of this chapter, we also will consider other transport processes, such as animal movements or the redistribution of leaf litter by wind, to be negligible over the short term (one to two years). We know that in many other landscapes animals in particular may exert a tremendous long-term influence over element transport processes both directly and through their effects

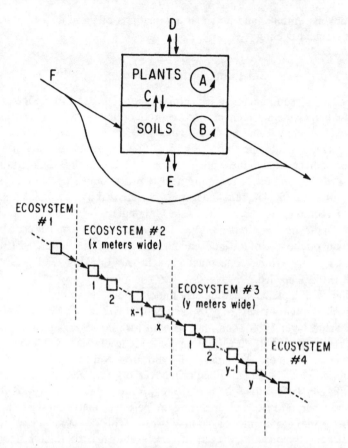

Figure 5.3. A conceptual model of element cycling along the toposequence at the study site. The sequence is viewed as essentially one-dimensional, consisting of a series of individual square meters linked by element transport in soil water. In each square meter there may be internal recycling of elements by plants and soil (A and B), there may be exchange of elements between plants and soil through uptake or litter fall (C), the plant-soil system may exchange elements with the atmosphere through processes such as nitrogen fixation or denitrification (D), and elements may be weathered from parent material or sequestered in immobile forms (E). The transport of elements down slope in soil water is indicated by (F); other means of element transport (not shown) include animal movements and movement of litter by wind or gravity.

on landscape hydrology, vegetation structure, and decomposition processes (Naiman 1988).

In this chapter we will not discuss the details of the methods used to make the actual measurements of nitrogen and phosphorus cycling processes (Fig. 5.3). All of this information is available elsewhere (Giblin et al. in review; Nadelhoffer et al. in press; Shaver and Chapin 1986 and in press). Instead, we will concentrate on the problems, pitfalls, and potential applications of these data in a study of landscape element cycling.

5.3 Plant Nutrient Budgets

The total element requirements of net primary production (NPP) are the amounts of each essential element that are used in the production of new plant biomass each year. These requirements may come from many different sources, including lateral transport between landscape units (Fig. 5.4). To determine the importance of element transport relative to these other sources, our approach is to determine first the portion of the total element requirement that is supplied by the recycling of elements within plants. The remainder must be satisfied by new uptake, and this new uptake requirement is calculated as a remainder by subtracting the amount recycled from the total requirement. The new uptake requirement can then be used as a constraint on the sum of the independent estimates of its components (Fig. 5.4) to help us identify errors, conceptual gaps in our thinking, and areas where additional work is needed.

Recycling within plants is particularly important in the Arctic, as it is in many nutrient-limited environments (e.g., Specht and Groves 1966; Miller et al. 1976; Gray and Schlesinger 1983), and may account for 50 to 90% of the nitrogen or phosphorus supply to NPP in a wide range of arctic plants and vegetation types (Shaver and Chapin in press; Chapin et al. 1980; Jonasson and Chapin 1985). Most of the recycled elements are resorbed or recovered from senescing leaves, stems, and other tissues before they die and are reallocated to new growth. If the senescence and resorption takes place during the period of active growth, the recovered elements may go directly to the new tissues. This direct recycling from old to new tissues is typical of evergreen leaves and also of rhizomes of some graminoid species, which die during the early part of the growing season, simultaneous with the period of active new growth (Shaver 1981, 1983). If the

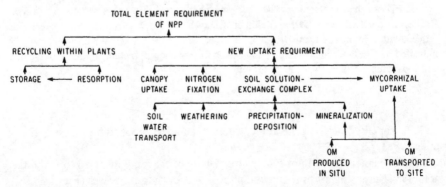

Figure 5.4. Sources of elements in support of net primary production (NPP). The individual sources are described in text. OM equals organic matter, including litter or dung.

senescence and resorption take place at the end of the growing season, on the other hand, the elements may be stored over the winter in perennial tissues. Deciduous shrubs, for example, lose their leaves at the end of each growing season and store the recovered elements in stems over the winter (Chapin et al. 1978).

Both the total element requirement of NPP and the amounts recycled to support that requirement can be estimated by sampling the vegetation at key times during the year. The NPP requirement is equal to the mass of elements in all new tissues per square meter of ground. The amounts recycled directly from senescing tissues are estimated by measuring the masses of the old leaves, stems, and rhizomes that die during the growing season and the changes in element concentration over the period of their senescence (Shaver and Melillo 1984). The amounts recycled from overwinter storage are estimated by knowing the masses of the surviving tissues and the changes in their element concentrations (Chapin et al. 1978).

This approach is much easier to apply in arctic vegetation than in most other vegetation types. The growing season of arctic plants in restricted to only 60 to 90 days, and there is only a single flush of growth each year (Wielgolaski et al. 1981). At least for all of the common or dominant species, senescence of tissues produced within the current year does not normally occur, as it does, for example, in temperate salt marshes (Houghton 1985). Finally, it is possible to separate new from old tissues and to determine the age of old tissues with accuracy and consistency (e.g., Gartner et al. 1986). Even for secondary wood production, methods have been developed based on age/mass relationships which make it possible to estimate its element requirements (Shaver 1986). Although the scheme outlined in Fig. 4 should apply almost anywhere, additional indirect methods would have to be used in other landscapes to estimate the components of plant element budgets.

To estimate the annual uptake requirements for nitrogen and phosphorus along our toposequence, we actually used detailed data on production, biomass, and element cycling in tussock and wet sedge tundras, riparian willow vegetation, and heath vegetation at several nearby sites (Shaver and Chapin 1986). We estimated the uptake requirements for the hillslope and footslope ecosystems based on data from other sites in northern Alaska and nitrogen and phosphorus analyses of plant tissues collected on our sites. The detailed sampling and analysis necessary to calculate these variables directly for all six ecosystems along our toposequence is still under way.

The final estimates of nitrogen and phosphorus uptake requirement (Table 5.2) varied by more than fivefold among the six ecosystem types, reflecting the variation in estimated NPP (Table 5.1). One feature of this variation among landscape units is particularly important with regard to the development of integrated element budgets for the entire landscape. There was no consistent trend in either NPP (and its total element requirement, not shown) or in the individual element uptake requirements along the toposequence. The lack of a regular increase or decrease in element uptake requirement makes it difficult to believe that any single microclimatic or nutritional gradient controls the productivity of the vegetation or the availability of essential elements along the toposequence. There

Table 5.2. Estimated Annual Nitrogen and Phosphorus Uptake Requirements for the Vegetation of Each Ecosystem Type, and Annual Supply (or Loss) from Various Sources

	Uptake Requirement[a]	Net Mineralization[b]	Summer Precipitation[c]	Winter Snow[c]	Soil Water Transport[d]
Nitrogen					
Tussock tundra	350	130	25	10	-16
Hilltop heath	250	470	25	0	-31
Hillslope shrub-lupine	410	120	25	50	+20
Footslope *Equisetum*	270	90	25	15	+10
Wet sedge tundra	220	550	25	10	-29
Riverside willow	1000	320	25	20	?
Phosphorus					
Tussock tundra	24	n.d.	2	0	-1.7
Hilltop heath	12	n.d.	2	0	-0.3
Hillslope shrub-lupine	25	n.d.	2	1	-0.8
Footslope *Equisetum*	17	n.d.	2	1	-3.3
Wet sedge tundra	23	n.d.	2	0	+1.5
Riverside willow	47	n.d.	2	0	?

Units are mg m^{-2} yr^{-1}. n.d. = no data.
[a] Estimates based on data collected at similar nearby sites as described in text.
[b] Three-year average (1985–1987), from Giblin et al. (in review).
[c] Two-year average (1986–1987), from Giblin et al. (in review).
[d] From Table 5.3.

was also considerable variation in the ratio of nitrogen to phosphorus required in annual uptake along the toposequence. The nitrogen-to-phosphorus uptake ratio declined from 22.5:1 at the top of the toposequence to only 9.6:1 at the wet sedge site, suggesting a relatively greater dependence on nitrogen uptake in the tussock tundra and on phosphorus uptake at the bottom. In the riverside willow site, the nitrogen-to-phosphorus uptake ratio jumped back up to 21.3:1.

5.4 Soil Element Supply and Sources

In most terrestrial ecosystems, the majority of the new uptake requirement is satisfied by uptake of elements in inorganic form dissolved in the soil solution (Kramer and Kozlowski 1979; Fig. 5.4). Plants associated with nitrogen-fixing microbes, however, may obtain much of their nitrogen uptake requirement essentially directly from the atmosphere. The importance of nitrogen fixation to these species and to the vegetation as a whole often depends on the relative rate of nitrogen supply to the soil solution and associated exchange complex (Fitter and Hay 1987; Sprent 1987). In early successional sites or other areas where there is little soil organic matter accumulation, nitrogen fixation may be quite significant on an annual basis as well as serving as an important long-term net input (Crocker and Major 1955; Marrs et al. 1983). In late successional sites, or sites with large amounts of soil organic matter, the supply of nitrogen to soil solution is usually much larger than the nitrogen-fixation rate, and nitrogen fixation normally provides only a small percentage of the annual whole-vegetation uptake requirement. Some element uptake may also occur in the vegetation canopy, by absorption of aerosols or of elements dissolved in rain (Cole and Rapp 1981; Lindberg et al. 1986). Canopy uptake is also insignificant in most ecosystems relative to uptake from soil solution on an annual basis, although it, too, may be an important long-term means of element accumulation. Finally, a portion of the uptake requirement may be met through direct mycorrhizal uptake of organic compounds, such as amino acids (Bowen and Smith 1981).

Inorganic nutrients may be supplied to the soil solution through a number of pathways (Melillo and Gosz 1983; Fig. 5.4). In most ecosystems the mineralization of elements in soil organic matter is by far the dominant pathway. Atmospheric deposition and precipitation represent another pathway that is often particularly important in the long-term accumulation and maintenance of element stocks in ecosystems such as raised bogs and perhaps arctic ecosystems, where the accumulation of organic matter on the surface has isolated the plants from elements supplied by weathering of parent material.

Because the mineralization of soil organic matter was expected to be the dominant source of nitrogen and phosphorus for plant uptake along our toposequence, we focused intensively on its measurement. The method we used was the "buried bag" method (Raison et al. 1987), in which the primary assumption is that, when soil is isolated from plant uptake, the net mineralization of elements will be reflected in an increase of their total amounts in soil solution and as exchangeable ions. The method is normally used only as an assay for net nitrogen mineralization. The method works by placing a relatively undisturbed soil core in a plastic bag, incubating it over the time period of interest in the same hole from which it was taken and then comparing the total exchangeable nitrogen concentrations before and after the incubation.

The results of the mineralization studies (Table 5.2; Giblin et al. in review) indicate that, although annual net nitrogen mineralization is at least as variable as the nitrogen uptake requirement along the transect, there is little or no correlation between these two variables. The highest nitrogen mineralization rates occurred in the two least productive sites, the hilltop heath and the wet sedge tundra (Fig. 5.1). These two sites also were perhaps the most different in terms of their basic soils characteristics, with the hilltop heath being well drained, warm, and rocky with relatively little organic matter accumulation and the wet sedge soil being water saturated, cold, and entirely organic. This lack of correlation between mineralization rate and uptake requirement suggests a great deal of variation in the relative importance of mineralization as a nitrogen source for plant uptake (Table 2). The data indicate that nitrogen mineralization can account for only about 30 to 50% of the uptake requirement in the tussock, hillslope, footslope, and riverside sites but 190 to 250% of the requirement in the heath and wet sedge sites.

We also measured the inputs of nitrogen and phosphorus in precipitation during the summertime and the distribution of nitrogen and phosphorus in snow before the start of the growing season (Table 5.2). Assuming a uniform distribution of summer precipitation (both rain and snow), the inputs of nitrogen from this source were 0.02 g m^{-2} year^{-1}, or 2 to 9% of the uptake requirements at each site. For phosphorus, the summer precipitation inputs were 0.002 g m^{-2} year^{-1}, or about 5 to 10% of the phosphorus uptake requirements. There was also a significant amount of dissolved inorganic nitrogen and phosphorus held in the 2- to 3-m-deep snowbank that formed each year over the hillslope shrub-lupine and the footslope *Equisetum* sites (Fig. 5.1). The tussock and wet sedge tundras and the riverside willow site also accumulate 30 to 60 cm of snow each year, with measurable inorganic nitrogen content, but the hilltop heath site is windblown and nearly snow free. This unequal distribution of winter snowcover (when melted) may contribute 5 to 10% of the nitrogen and phosphorus uptake requirements of the hillslope and footslope sites. Because the areas downslope of the snowbank are snowfree for one to three weeks before the snowbank begins to melt (Giblin et al. in review), the upper 5 to 10 cm of soil are thawed and the meltwater passes through the soil rather than running off on the surface. Assuming that the winter snowfall is initially distributed evenly and then blown around, the formation of this snowbank may be considered as another important mechanism of lateral element transport along our toposequence.

5.5 Soil Solution Transport

In most landscapes the importance of lateral movement of elements dissolved in soil water (Fig. 5.4) is quite difficult to estimate because it requires accurate information on the rates and volumes of lateral soil water flow, as well as the extent of contact between plant roots and soil organic matter and the moving water. Often plant roots do not penetrate to the water table. Furthermore, even if the roots are in continuous, close contact with moving soil water, it is important to know how much of the soil water is interacting with the rooting zone and how much water either passes beneath the roots without interaction or flows more deeply into the earth.

On the North Slope of Alaska the problem is greatly simplified because the entire region is underlain by continuous permafrost, which prevents deep water drainage and thus eliminates a major uncertainty in most soil water budgets. The seasonally thawed "active layer" of soil is usually less than a meter thick (often only 20 to 40 cm), with continuously saturated water flow across the top of the permafrost. The depth to the water table typically ranges from 0 to 50 cm but may be greater where the soil is thawed exceptionally deeply (up to 3 to 4 m). Roots normally penetrate all the way to the maximum depth of soil thaw (Shaver and Billings 1975; Miller et al. 1982), through the entire saturated zone, so they are capable of interaction with the entire volume of soil water.

A further simplification is possible in our case by assuming that the rate of water movement is uniform horizontally across the slope of our toposequence. In other words, we are assuming that there is no channelized water flow, either above-ground or belowground. In fact, we originally chose our site on the basis of its uniform topography and lack of any indication in the vegetation of the channelized water flow or "water tracks" that are common in some other tundra slopes (Chapin et al. 1988).

Under such conditions there are two principal ways to estimate the rates and volumes of soil water movement downslope (Freeze and Cherry 1979). One way is to measure the actual water content of the saturated portion of the soil, the hydraulic conductivity of the soil and its changes with depth, and the changes in elevation of the surface of the water table along the toposequence. A second way is to measure the total inputs of water in precipitation and the outputs in evapo-transpiration and to assume that the remainder must be either stored in the soil or transported down slope. We will use the second of these methods in this chapter because the data are already available and the calculations are simpler. Research based on the first approach is still in progress but generally supports this presentation, as do our tracer studies with conservative ions such as bromide and chloride.

By using the hydrologic budget method, we start with the average annual precipitation during the period of soil thaw (June, July, and August), which was 151 mm at our toposequence in 1986–87. Summer evapotranspiration was approximately 76 mm (estimated from Stuart et al. (1982) and checked against pan evaporation measurements on site), leaving 75 mm to be accounted for. We assume the same evapotranspiration rates at all six sites, because the vegetation cover was complete everywhere and the soil was always moist (Giblin et al. in review). Over this two-year period, we are assuming there was no net change in the total volume of water stored in the soil, and thus each year the entire 75 mm remainder must have moved through the soil, carrying elements along with it.

In the context of our conceptual model (Fig. 5.3), this net annual input of 75 mm of water translates into 75 L of water per square meter per year. The volumes accumulate as the water moves down slope, so that at the bottom of a 1-m-wide transect, down a slope 100 m long, there would be 7500 L of soil water passing through each square meter annually. These large volumes are reflected in the increasingly shallow soil water table and larger total volumes of soil water in the lower parts of the toposequence. In the wet sedge site in 1986, for example, the total soil water volume (from the surface to the maximum depth of thaw) averaged about 240 L/m^2. If we assume that the average annual duration of soil thaw is 100 days and that the soil water flows at a constant rate, then the turnover time of the soil water volume in the center of the wet sedge site would be about 3 to 4 days. This compares with a soil water turnover time of 75 to 150 days in the tussock tundra site at the top of the toposequence.

The net exchange of inorganic nitrogen and phosphorus between the soil water and the vegetation and soils above it can be estimated by calculating the changes in the amounts of these nutrients carried by soil water as it passes beneath each of the six ecosystem types along our toposequence (Table 5.3). For the sake of simplicity, we will use the three-year average concentrations of dissolved inorganic nitrogen (DIN) and phosphorus (DIP) in soil solution (Giblin et al. in review). Multiplying these average concentrations times the estimated volumes of soil water coming out the bottom of each ecosystem gives an estimate of the total outputs of DIN and DIP. Subtracting these outputs from the outputs of neighboring uphill ecosystems gives an estimate of the net uptake from or addition to the soil solution by each ecosystem type.

For DIN, there was quite a large variation in total outputs from each succeeding ecosystem, moving down the toposequence (Table 5.3). The tussock tundra and hilltop heath were both net sources of DIN to soil water, with nearly a 40% increase in total DIN occurring over just 10 m as the soil water passed beneath the hilltop heath. The hillslope and footslope ecosystems were both net sinks for DIN, removing nearly half of the DIN from the water leaving the hilltop heath over a total distance of 30 m. The amount of DIN in soil water leaving the wet sedge ecosystem then more than doubled over the next 30 m. We have not calculated the net output of DIN from the riverside ecosystem because we are uncertain of the extent to which the water in and beneath the river mixes with the soil water there.

The variation in DIP outputs from each ecosystem was much smaller, on both absolute and relative bases (Table 5.3). There was about a 15% increase in total DIP losses from the tussock tundra to the hillslope. This increase occurred despite a continuous decrease in the average DIP concentrations, because of the greater soil water volumes involved. In the footslope ecosystem, there was then a fairly rapid, 33% increase in DIP outputs occurring over 10 m, followed by a 33% decrease in the wet sedge zone.

When net DIN and DIP exchange are expressed on a meter-squared basis, they can be compared with the estimated annual nitrogen and phosphorus uptake requirements of the vegetation and with the other inputs to soil solution (Table 5.2). For DIN, exchange with moving soil water ranges from a net loss of 31 mg/m

Table 5.3. Calculation of Net Inorganic Nitrogen and Phosphorus Exchange by Each Ecosystem Type

	Mean Concentration in Soil Water (ug/l[a])				Horizontal Distance (m)	Cumulative 1 H_2O/yr[b]	Total Output[c] (mg ecosystem^{-1} y^{-1})
	1986	1987	1988	avg.			
Nitrogen							
Tussock tundra	259	140	249	216	50	3750	810
Hilltop heath	272	172	300	248	10	4500	1116
Hillslope shrub-lupine	91	81	190	121	20	6000	726
Footslope *Equisetum*	34	49	195	93	10	6750	628
Wet sedge tundra	102	162	232	165	30	9000	1485
Riverside willow	169	87	182	146	—	—	—
Phosphorus							
Tussock tundra	37	13	15	22	50	3750	83
Hilltop heath	27	11	18	19	10	4500	86
Hillslope shrub-lupine	30	10	10	17	20	6000	102
Footslopetum	30	11	18	20	10	6750	135
Wet sedge tundra	16	9	5	10	30	9000	90
Riverside willow	18	11	12	14	—	—	—

[a] From Giblin et al. (in review).

[b] "Cumulative 1 H_2O" is the estimated volume of water that passes out the bottom of each ecosystem from a 1-m-wide transect starting at the top of the tussock tundra, assuming a net annual water input of 75 L/m^{-2} (75 mm) along the transect.

[c] Total output is equal to this cumulative water volume times the three-year average concentration.

squared per year in the hilltop heath to a net gain of 20 mg/m squared per year on
the hillslope. The amounts are about equal in magnitude to nitrogen inputs in
precipitation, but variable in direction. In the hilltop heath and wet sedge zones the
DIN losses equaled 12 to 13% of the annual plant uptake requirement, while in the
hillslope and footslope the gains could account for 4 to 5% of the uptake requirement. The DIP exchanges were generally smaller, except in the footslope where
the DIP loss equaled nearly 20% of the phosphorus uptake requirement.

5.6 Putting the Pieces Together: Landscape Biogeochemistry

Our budgets for nitrogen and phosphorus supplies in support of net primary
production along the toposequence are still incomplete (Table 5.2), but we can say
that element transport in soil water between neighboring terrestrial ecosystems is
probably not a major source of elements to plants in any single year. Only two sites
showed any net uptake of DIN from soil water (the hillslope and footslope), and
only one site (the wet sedge tundra) took up DIP. This net ecosystem uptake could
have been due solely to microbial immobilization in soil organic matter without
any uptake by plants. Internal recycling of nitrogen and phosphorus within plants,
followed by net mineralization of soil organic matter to inorganic forms (at least
for nitrogen), are far more important element sources in support of productivity on
an annual basis. The magnitude of the net exchanges with soil water on an areal
basis are about the same as the inputs of nitrogen and phosphorus in summer
precipitation and/or winter snow accumulation.

On the other hand, the total amounts of DIN and DIP that pass through our
1-m-wide transect (Table 5.3) are several times larger than the annual uptake
requirements per square meter (Table 5.2). The values in the rightmost column of
Table 3 actually understate the total throughput because they are net amounts and
do not include the additions of DIN and DIP due to mineralization, precipitation,
and weathering (Fig. 5.4), which are balanced by plant uptake. What this means
is that, although all six ecosystems are fairly effective at keeping the concentrations of DIN and DIP in soil water quite low (Table 5.3), the volumes of water
passing through each successive square meter are large enough to result in significant annual element fluxes.

From the perspective of lake and stream productivity in particular, the changing
total amounts of DIN and DIP in soil water as it moves downslope are of major
importance. Ultimately, soil water must flow into a lake or stream, and we know
that the productivity of arctic lakes and streams is both strongly element limited
and much less well "buffered" by a diversity of element sources and large organic
matter standing stocks than in any of our terrestrial ecosystem types (Peterson et
al. 1983, 1985). The more than twofold variation in total DIN output from the
various ecosystems along our toposequence and the nearly 50% variation in DIP
output (Table 5.3) suggest that one critical control over aquatic productivity is the
specific sequence of ecosystem types that happen to filter the soil water before it
actually reaches a lake or stream.

We have focused on the cycling, transport, and uptake of inorganic forms of nitrogen and phosphorus because, conventionally, these are considered the only forms that can be taken up by plants in significant amounts. However, recent work near Toolik Lake, Alaska, has concluded that several of the dominant plant species along our toposequence are also capable of taking up dissolved amino acids and perhaps other forms of dissolved organic nitrogen (DON), from soil solution (Kielland 1988).

We also know that DON concentrations are frequently much higher than DIN concentrations along the toposequence (Fig. 5.5). The mechanism for direct uptake of DON by plant roots is still unclear, but mycorrhizal uptake of DON from soil solution and transport to the host plants (Fig. 5.4) could easily account for some of the gaps between our estimates of plant uptake demand and the sum of our estimates of three sources of inorganic nitrogen (Table 5.2).

Our data on DON concentrations (Fig. 5.5) have been collected on only a few scattered dates, and for this reason we have not tried to estimate the net transport of DON between ecosystems and into the river. However, the consistency higher

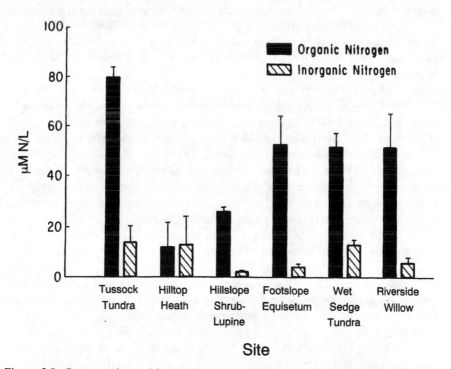

Figure 5.5. Concentrations of inorganic vs. organic nitrogen dissolved in soil solution along the toposequence (DIN vs. DON) on 30 June 1987. Units are micromoles of nitrogen per liter.

concentrations of DON (versus DIN) indicate that more nitrogen may be moving downslope in organic forms than as ammonium and nitrate. Furthermore, the changes in DON concentration between successive ecosystems suggest very different patterns of net uptake or release of DON versus DIN. For example, the hillslope and footslope, which are net sinks for DIN (Table 5.3), may be large net sources of DON (Fig. 5.5). Further investigation of the sources, sinks, and uptake pathways of DON in these ecosystems is clearly an important priority for the future.

There is little reason why our conceptual, essentially one-dimensional model (Fig. 3) cannot be scaled up to a full, three-dimensional watershed. The principal additional requirements will be an accurate map of the mosaic of ecosystem types in the watershed, and information about both the directions and rates of soil water movement between adjacent square meters. The hydrologic data will be much more complex and difficult to obtain, however, because we cannot use the same simplifying assumptions as in the one-dimensional model, covering only 100 to 200 m without any surface or channelized flow. A major difficulty will be predicting when the volumes of water become large enough that subsurface flow is inadequate to carry such volumes, and surface flow occurs. Another difficulty will be including the characteristic "water tracks" (Chapin et al. 1988), or areas of greater subsurface and intermittent surface flow, that are typical of the moist, upland tundras in the northern foothills of the Brooks Range. These "water tracks" appear as distinctive stripes of deciduous shrub-dominated vegetation, running downslope, and are much more productive and nutrient-rich than the intertrack areas.

An additional improvement, in either one dimension or in three dimensions, would be to model nitrogen and phosphorus transport by using the same basic model structure but with daily or weekly time steps rather than annual time steps, as we have done here. This would require a much more extensive data set, including not only changes in soil solution chemistry, nitrogen mineralization rates, and plant phenology but also within-season changes in soil water storage and depth of soil thaw. There may be enough data available already, either published or unpublished, to do this, but the computations would be much more complex. In the three-dimensional case, a computer-based Geographic Information System (GIS) would be required. Our primary aim in this chapter was not so much to produce such a detailed model as to demonstrate the utility of our conceptual approach.

Despite the simplicity of our assumptions, our results are both reasonable and useful. It is not surprising, for example, that the strongly phosphorus-limited wet sedge tundra should immobilize phosphorus in soil solution and be relatively inefficient at retaining nitrogen. Our work both confirms and extends the conclusions of Miller (1982) about the importance of variation in nitrogen and phosphorus cycling along such toposequences, which are strongly characteristic of alpine as well as arctic landscapes (Billings 1973). Although the importance of the overall terrestrial landscape in controlling the chemistry of lakes and streams is universally accepted, there have been relatively few studies that have attempted to separate so explicitly the roles of a sequence of terrestrial ecosystems in controlling aquatic inputs (e.g., Peterjohn and Correll 1984). Our arctic toposequence serves as a particularly useful example because of our ability to account for the total volumes of soil water, as well as variation in the relative availability and cycling rates of nitrogen and phosphorus.

5.7 Summary

Along a toposequence of six contrasting ecosystem types in northern Alaska, the relative importance of various sources of nitrogen and phosphorus in support of primary production is described and compared. Particular attention is paid to the net exchanges of inorganic nitrogen and phosphorus carried by soil water as it moves down slope, across the surface of the permafrost that underlies the toposequence. Although these exchanges are always small relative to the annual plant uptake requirements, the total amounts of inorganic nitrogen and phosphorus carried by the soil water increase and decrease greatly as the water moves down slope. Thus the inputs of inorganic forms of nitrogen and phosphorus to aquatic ecosystems are strongly affected by the specific sequence of terrestrial ecosystems that the soil water passes through before it reaches a lake or stream. This chapter offers a conceptual model for analyzing and quantifying the importance of such biogeochemical transport processes between landscape units (ecosystems) and suggests that arctic landscapes offer a number of advantages as model systems for this kind of investigation.

Acknowledgments

This research was supported by National Science Foundation Grants #BSR 85–07493 and 88–06635. Most of the hard work was done by Jim Laundre, Alexa McKerrow, Tim Yandow, and Dan Sperduto.

References

Billings, W.D. 1973. Arctic and alpine vegetation: Similarities, differences, and response to disturbance. *BioScience* 23:697–704.

Bliss, L.C. 1956. A comparison of plant development in microenvironments of arctic and alpine plants. *Biological Review* 43:481–530.

Bliss, L.C., ed. 1977. *Truelove Lowland, Devon Island, Canada: A High Arctic Ecosystem.* Edmonton: University of Alberta Press.

Bowen, G.D., and Smith, S.E. 1981. The effects of mycorrhizas on nitrogen uptake by plants. In *Terrestrial Nitrogen Cycles*, eds. F.E. Clark and T. Rosswall. Ecological Bulletins (Stockholm) 33:237–47.

Britton, M.E. 1966. Vegetation of the arctic tundra. In *Arctic Biology*, ed. H.P. Hanson, pp. 67–130. Corvallis: Oregon State University Press.

Chapin, F.S., III; Fetcher, N.; Kielland, K.; Everett, K.R.; and Linkins, A.E. 1978. Productivity and nutrient cycling of Alaskan tundra: Enhancement by flowing soil water. *Ecology* 69:693–702.

Chapin, F.S., III; Johnson, D.A.; and McKendrick, J.D. 1980. Seasonal nutrient allocation patterns in various tundra plant life forms in northern Alaska: implications for herbivory. *Journal of Ecology* 68:189–209.

Chapin, F.S., III, and Shaver, G.R. 1985. The physiological ecology of arctic plants. In *Physiological Ecology of North American Plant Communities*, eds. B. Chabot and H.A. Mooney, pp. 16–40. London: Chapman and Hall.

Cole, D.W., and Rapp, M.R. 1981. Elemental cycling in forest ecosystems. In *Dynamic Properties of Forest Ecosystems*, ed. D. Reichle. IBP Synthesis 23:341–409.

Crocker, R.L., and Major, J. 1955. Soil development in relation to vegetation and surface age, Glacier Bay, Alaska. *Journal of Ecology* 43:427–48.

Fitter, A.H., and Hay, R.K.M. 1987. *Environmental Physiology of Plants*. 2nd ed. London: Academic Press.

Forman, R.T.T., and Godron, M. 1986. *Landscape Ecology*. New York: John Wiley and Sons.

Freeze, R.A., and Cherry, J.A. 1979. *Groundwater*. Englewood Cliffs, N.J.: Prentice-Hall.

Gartner, B.L.; Chapin, F.S., III; and Shaver, G.R. 1986. Reproduction of *Eriophorum vaginatum* by seed in Alaskan tussock tundra. *Journal of Ecology* 74:1–18.

Giblin, A.E.; Nadelhoffer, K.J.; Shaver, G.R.; Laundre, J.A.; and McKerrow, A.J. Biogeochemical diversity along a riverside toposequence in arctic Alaska. Submitted to *Ecological Monographs*, in review.

Gray, J.T., and Schlesinger, W.H. 1983. Nutrient use by evergreen and deciduous shrubs in southern California. II. Experimental investigations of the relationship between growth, nitrogen uptake, and nitrogen availability. *Journal of Ecology* 71:43–56.

Houghton, R.A. 1985. The effect of mortality on estimates of net above-ground production by *Spartina alterniflora*. *Aquatic Botany* 22:121–32.

Johnson, P.L., and Billings, W.D. 1962. The alpine vegetation of the Beartooth plateau in relation to cryopedogenic processes and patterns. *Ecological Monographs* 32:105–35.

Jonasson, S., and Chapin, F.S., III. 1985. Significance of sequential leaf development for nutrient balance of the cotton sedge, *Eriophorum vaginatum* L. *Oecologia* 67:511–18.

Kielland, K. 1988. Circumvention of net nitrogen mineralization through amino acid uptake by arctic plants. *Bulletin of the Ecological Society of America* 69:191.

Kramer, P.J., and Kozlowski, T.T. 1979. *Physiology of Woody Plants*. New York: Academic Press.

Likens, G.E.; Bormann, F.H.; Pierce, R.S.; Eaton, J.S.; and Johnson, N.M. 1977. *Biogeochemistry of a Forested Ecosystem*. New York: Springer-Verlag.

Lindberg, S.E.; Lovett, G.M.; Richter, D.D.; and Johnson, D.W. 1986. Atmospheric deposition and canopy interactions of major ions in a forest. *Science* 231:141–45.

McGraw, J.B., and Antonovics, J. 1983. Experimental ecology of *Dryas octopetala*: ecotypic differentiation and life-cycle stages of selection. *Journal of Ecology* 71:879–97.

Marrs, R.H.; Roberts, R.D.; Skeffington, R.A.; and Bradshaw, A.D. 1983. Nitrogen and the development of ecosystems. In *Nitrogen as an Ecological Factor*, eds. L. Paley and D. Aspinall, pp. 315–47. London: Academic Press.

Melillo, J.M., and Gosz, J.R. 1983. Interactions of biogeochemical cycles in forest ecosystems. In *The Major Biogeochemical Cycles and Their Interactions*, eds. B. Bolin and R.B. Cook, pp. 217–28. New York: John Wiley and Sons.

Miller, H.G.; Cooper, J.M.; and Miller, J.D. 1976. Efect of nitrogen supply on nutrients in litterfall and crown leaching in a stand of Corsican pine. *Journal of Applied Ecology* 13:233–48.

Miller, P.C., ed. 1982. The availability and utilization of resources in tundra ecosystems. *Holarctic Ecology* 5:81–220.

Miller, P.C.; Mangan, R.; and Kummerow, J. 1982. Vertical distribution of organic matter in eight variation types near Eagle Summit, Alaska. *Holarctic Ecology* 5:117–24.

Nadelhoffer, K.J.; Giblin, A.E.; Shaver, G.R.; and Laundre, J.A. Effects of temperature and organic matter on C, N, and P mineralization in soils of six tundra ecosystems. *Ecology in press*.

Naiman, R.J. 1988. Animal influences on ecosystems. *BioScience* 38:750–53.

Odum, E.P. 1971. *Fundamentals of Ecology*. 3rd ed. Philadelphia: W.B. Saunders.

Peterjohn, W.T., and Correll, D.L. 1984. Nutrient dynamics in an agricultural watershed: Observations on the role of a riparian forest. *Ecology* 65:1466–75.

Peterson, B.J.; Hobbie, J.E.; Corliss, T.L.; and Kriet, K. 1983. A continuous-flow periphyton bioassay: tests of nutrient limitation in a tundra stream. *Limnology and Oceanography* 28:583–91.

Peterson, B.J.; Hobbie, J.E.; Hershey, A.; Lock, M.; Ford, T.; Vestal, R.; Hullar, M.; Ventullo, R.; and Volk, G. 1985. Transformation of a tundra river from heterotrophy to autotrophy by addition of phosphorus. *Science* 229:1383–86.

Peterson, K.M., and Billings, W.D. 1980. Geomorphic processes and vegetation change along the Meade River sand bluffs, northern Alaska. *Arctic* 31:7–23.

Raison, R.J.; Connell, M.J.; and Kanna, P.K. 1987. Methodology for studying fluxes of soil mineral-N *in situ*. *Soil Biology and Biochemistry* 19:521–30.

Shaver, G.R. 1981. Mineral nutrition and leaf longevity in an evergreen shrub, *Ledum palustre* ssp. decumbens. *Oecologia* 49:362–65.

Shaver, G.R. 1983. Mineral nutrition and leaf longevity in *Ledum palustre*: the role of individual nutrients and the timing of leaf mortality. *Oecologia* 56:160–65.

Shaver, G.R. 1986. Woody stem production in Alaskan tundra shrubs. *Ecology* 67:660–69.

Shaver, G.R., and Billings, W.D. 1975. Root production and root turnover in a wet tundra ecosystem, Barrow, Alaska. *Ecology* 56(2):401–10.

Shaver, G.R., and Chapin, F.S., III. 1986. Effect of NPK fertilization on production and biomass of Alaskan tussock tundra. *Arctic Alpine Research* 18:261–68.

Shaver, G.R., and Chapin, F.S., III. Production:biomass relationships and element cycling in contrasting arctic vegetation types. *Ecological Monographs, in press*.

Shaver, G.R., and Melillo, J.M. 1984. Nutrient budgets of marsh plants: Efficiency concepts and relation to availability. *Ecology* 65:1491–1510.

Specht, R.L., and Groves, R.H. 1966. A comparison of the phosphorus nutrition of Australian heath plants and introduced economic plants. *Australian Journal of Botany* 14:201–21.

Sprent, J.I. 1987. *The Ecology of the Nitrogen Cycle*. Cambridge: At the University Press.

Stuart, L.; Oberbauer, S.; and Miller, P.C. 1982. Evapotranspiration measurements in *Eriophorum vaginatum* tussock tundra in Alaska. *Holarctic Ecology* 5:145–50.

Teeri, J.A. 1973. Polar desert adaptations of a high arctic plant species. *Science* 179:496–97.

Walker, D.A. 1985. Vegetation and environmental gradients of the Prudhoe Bay Region, Alaska. U.S. Army CRREL Report 85–14.

Walker, D.A.; Acevedo, W.; Everell, K.R.; Gaydos, L.; Brown, J.; and Webber, P.J. 1982. Landsat-assisted environmental mapping in the Arctic National Wildlife Refuge, Alaska. CREEL Report 82–87.

Wielgolaski, F.E.; Bliss, L.C.; Svoboda, J.C.; Doyle, G. 1981. Primary production of tundra. In *Tundra Ecosystems: A Comparative Analysis*, eds. L.C. Bliss, O.W. Heal, and J.J. Moore, pp. 187–226. Cambridge: At the University Press.

6.   Thermal Remote Sensing Methods in Landscape Ecology

Jeffrey C. Luvall and H. Richard Holbo

6.1 Introduction

Plant ecologists are well acquainted with the importance of the microclimate in governing physiological processes. Temperature plays a fundamental and often limiting role in many biological processes through its control of the rate of biological-chemical reactions, i.e., Q_{10}. Because plants are poikilothermic, their temperatures are closely linked to thermal energy exchanges with the microclimate. Evaluating these energy exchanges to improve our understanding of the interdependencies between plants and their microclimates for various types of landscape vegetations is not a simple task. For example, in a forested landscape, the upper leaves of the canopy receive high levels of solar radiation and dissipate the resulting heat while maintaining leaf temperatures within a fairly narrow range, if a survivable balance between photosynthesis and respiration is to exist. In addition, the presence of a forest canopy is known to moderate the physical microclimate of all associated plant communities.

It is recognized that canopy and leaf temperatures are controlled by a combination of thermal energy fluxes and vegetation-dependent or site-specific factors. We know that the amount of solar radiation received by the forest canopy has a major influence on all other thermal energy fluxes in the forest canopy and thus on leaf temperatures. Although the microclimate of the canopy is influenced by prevailing regional weather conditions, during clear weather the amount of solar radiation received is more strongly influenced by the local topography and land management practices. Among the many site-specific factors that contribute to canopy temperature are leaf and canopy morphology, species composition, and water status. Perhaps not so surprisingly, air temperature is often a poor predictor of leaf temperature, especially when air temperature is measured by using traditional meterological shelters in different parts of the landscape.

We may expect that the structural and physiological variability that characterizes forests, in contrast to those of other landscape types like prairies and shrublands, leads to a wide range of small microclimates within a forest. In fact, the complexity of the forest canopy, the volume it occupies, stomatal controls over water loss, and the distances separating tree crowns from the soil below all interact to produce thermal energy budgets that are dramatically different from those typical of shorter vegetations or crops. For even a seemingly uniform forest type, vertical differences in leaf and air temperatures within the canopy, variations in depth of the forest canopy, leaf area index, and the crown topographies operate to preclude a generalized summary about its microclimate across the lateral extent of forested landscapes. Many forests occur in mountainous areas, adding further complications because of elevational differences, topography, highly variable soil moisture availabilities, and diversity of forest community types.

Forest edges constitute yet another source of variability, further generating complex microclimate gradients. Ultimately, the physical environments at such edges support the many interesting and dynamic associations of woody and herbaceous species characteristic of forest plant communities (Ranny et al. 1981). Nonetheless, these edges complicate the micrometeorological picture excessively, and measurements across them, which are needed to define the nature of spatial variability in forest edge microclimates and energy budgets, are generally lacking (e.g., Miller 1975; Thistle et al. 1987).

In landscape ecology both the technological means and a corresponding analytical rationale are needed to gain more information about microclimatic processes at scales which can be extended over large areas to interpret the vegetation patterns observed. For microclimatological analyses, traditional techniques of determining leaf and/or canopy temperatures have had to rely on the use of small thermocouples or hand-held infrared thermometers. The few measurements that may be taken with these techniques necessarily provide a very small sampling of the overall landscape. Practical constraints usually limit the number of samples that can be acquired across a landscape to far less than a reasonable number. But now, because of recent developments in thermal remote sensing techniques, there is a feasible way to quantitatively assess relatively small-scale microclimatic patterns and then evaluate them with respect to the larger patterns of landscape elements. Thermal remote sensing instruments are available to measure temperatures over fairly large areas at snapshot speeds.

In his article "Landscape Ecology: State of the Art," Risser (1987) appealed for new techniques designed to address specific questions concerning landscape structure and function. Accordingly, we propose that thermal remote sensing of forest canopy temperatures is such a technique and that methodologies can be derived from it to analyze the thermal functions of landscape structures and lead toward a better understanding of pattern and process in landscape ecology. In this chapter we address the topic of applying forest canopy temperatures to landscape ecology by posing these questions:

1. What are the relationships between temperature and energy budgets of forest canopies?
2. What techniques are available to measure canopy temperatures at landscape spatial scales?
3. Can our understanding of forest function and structure be improved by having landscape-scale measures of spatial variability in forest canopy temperatures and thermal response?

6.2 Flow of Thermal Energy Across Forested Landscapes

The heterogeneity of forests across the landscape results in a spatial discontinuity of microclimates. Distinct differences in air temperature and humidity are created by the forest canopy in response to the environment. As a result, horizontal transfers of energy occur between adjacent forest canopies. These lateral exchanges of energy are termed "advection."

Advective effects in forests may be grouped into three general types (Oke 1987). One type, the "clothesline effect," occurs along the edges of dissimilar vegetation types. The term *clothesline effect* describes the condition when a flow of dry air leaves a moisture-limited surface, such as a clearing, and enters the forest. Responding to the lower humidity, evapotranspiration rates temporarily increase, soil moisture is reduced, and a drier microclimate results. The counterpart to the "clothesline effect" is an advective condition called the "oasis effect." The *oasis effect* describes a situation wherein a flow of dry air passes over a moisture-unlimited surface. Because surface moisture is unlimited, evapotranspiration increases at the expense of heat energy from the air, resulting in lower temperatures and higher humidities. The third type of advective effect, accompanying either "clothesline" or "oasis" processes, is the concept of "fetch." Fetch is an index to the thickness of the layer of air in contact with a surface that is in equilibrium with the energy exchanges occurring at the surface. Typical thicknesses for equilibrium surface layers range from 20 cm for smooth desert surfaces with an inversion to 200 m or more for forests. The properties of a moving air mass at the leading edge of a new surface obviously represent the old surface it has just passed over, and it takes some distance before a new boundary layer forms that is fully in balance with the energy exchanges occurring at the new surface. The distance that the air must travel before it achieves such an equilibrium throughout a specified thickness is called fetch. Fetch is an important consideration, primarily when making micrometeorological estimates of surface thermal energy exchanges. The practical consequence of fetch is to emphasize the point that micrometeorological instrumentation must be positioned within this equilibrium layer. Otherwise, energy exchange estimates will be in error because the measurements will have inadvertently sampled properties representative of upwind surfaces. Depending on the type of surface, atmospheric stability, prevailing wind speeds, and the deployment height of the micrometeorological instruments, required fetch distances can exceed several hundred meters. These considerations make it virtually impossible to select micrometeorological measurement sites free of the potential effects of edge discontinuities.

6.3 Surface Energy and Radiation Budgets

In spite of the micrometeorological limitations of advection, which complicate the choices of sites, it has proven useful to model in thermal energy budget terms energy exchanges occurring at surfaces. The energy budget provides a framework for comparing microclimates in ways that cannot be accomplished by using other formulations involving environmental variables (e.g., temperature, humidity, wind speed, and solar radiation). By translating these variables into the components of an energy budget, all types of landscape surfaces from vegetated (forest and herbaceous) to nonvegetated (bare soil, roads, and buildings) may be compared. The units normally used to express these energies are watts per square meter (Wm^{-2}) to describe the rates of energy flow and joules per square meter (Jm^{-2}) to quantify the total amounts of energy accumulated over time.

We begin a brief review of the thermal energy budget of forests by first considering the radiation fluxes acting upon the surface, the residual of which results in the net radiation flux, Q^*. Then, the nonradiative terms of the energy budget are discussed, the sum of which must equal Q^*. Although the discussion emphasizes forests, the principles apply to surfaces of all types. For a comprehensive discussion of the thermal energy budgets, the reader may consult the excellent text by Oke (1987).

The net all-wave radiation balance, Q^*, of forest canopies may be defined by considering its shortwave (or solar) and long-wave radiation components separately. This compartmented approach helps focus on areas where remote sensing techniques are applicable.

The net flux of solar radiation, K^*, can be found from

$$K^* = (1 - \alpha) \, \phi \, K\!\downarrow \tag{6.1}$$

where α = surface albedo = $K\!\uparrow/K\!\downarrow$, ϕ = a surface slope and aspect solar gain coefficient, $K\!\downarrow$ = incoming solar radiation, and $K\!\uparrow$ = outgoing (reflected) solar radiation.

The surface angle dependency, ϕ (Garnier and Ohmura 1968), is sometimes left out of the equation when the surface in question is assumed to be essentially flat, as an entire landscape might appear to be. However, while ϕ is usually not large (<1.2), it takes on increasing importance as short-term observations are made and as spatial resolution is improved. Its influence can often explain much of the variability in thermal image data.

The net flux of long-wave radiation at the surface, L, is

$$L^* = L\downarrow - L\uparrow \qquad (6.2)$$

where $L\downarrow$ = incoming long-wave radiation from the atmosphere, and $L\uparrow$ = outgoing long-wave radiation emitted from the surface.

The long-wave radiation emitted from a surface is the variable measured by using thermal remote sensing. It is a function of surface temperature:

$$L\uparrow = \varepsilon \, \sigma \, T_s^4 \qquad (6.3)$$

where ε = emissivity, σ = Stefan-Boltzman constant ($5.7 \times 10^{-8} Wm^{-2}T^{-4}$), and T_s = surface temperature (Kelvin).

The net flux of all-wave radiation, Q^*, is the sum of Eqs. 6.1 and 6.2:

$$Q^* = K^* + L^* \qquad (6.4)$$

Net radiation (Q^*) is a very important term because it represents the amount of energy available at the surface for partitioning into the nonradiative processes, LE, H, and G, of the thermal energy budget:

$$Q^* = LE + H + G \qquad (6.5)$$

where LE = latent heat exchange, or evapotranspiration (ET), H = sensible heat exchange into the atmosphere, and G = conduction heat exchange with biomass and soil.

For a typical daytime situation during fair summer weather, Q^* will be the heat energy source, except for short periods following shading by clouds. For coniferous forests, Q^* may be about 65% of $K\downarrow$, proportionally lower for deserts, and higher for well-watered crops. LE and H are the largest dissipative energy fluxes. But it would be ill-advised to assume that LE is predominant. Coniferous canopies can exchange an amount equal to Q^* by H alone, without large increases in needle temperatures. The magnitude of G is usually small; however, during brief periods near sunrise or sunset, it may become disproportionately large.

These approximations aside, the partitioning of the terms in the energy budget depends on a great many surface characteristics. In forested landscapes, the proportioning of the surface energy budget terms depends largely on forest type and moisture conditions. Forest types that exhibit distinct energy budgets will likely have differences in plant community types, canopy biomass, or leaf area index. Moisture conditions encompass a number of factors, including phenological status, and thus stomatal activity, soil depth, rooting effectiveness, and soil moisture content.

6.3.1 Estimating Latent Heat

Remotely sensed surface temperatures can be employed in relationships used to estimate latent heat flux (Jarvis 1981):

$$LE = \rho C_p/\gamma \ (VD_s - VD_a)/R_c \tag{6.6}$$

where ρ = density of the air, C_p = specific heat of air at atmospheric pressure, γ = psychrometric coefficient, VD_s = saturated water vapor density of the air, VD_a = water vapor density of the air, and R_c = stomatal resistance per unit leaf area. In this relationship, it is possible to infer the saturated vapor density, VD_s, for moisture-unlimited surfaces from remotely sensed surface temperature observations, using

$$VD_s = [217/(273.15 + T_s)] \ 6.1078 \ e^{((17.269T_s)/(237.0 + T_s))} \tag{6.7}$$

where T_s = observed surface temperature, °C, and the term in brackets transforms Teten's equation for saturation vapor pressure over water to vapor concentration units.

6.3.2 Estimating Sensible Heat

Sensible heat flux can be written

$$H = \rho C_p \ (T_s - T_a)/R_a \tag{6.8}$$

where ρ = density of the air, C_p = specific heat of air at atmospheric pressure, T_a = air temperature above the canopy, T_s = remotely sensed surface temperature, and R_a = the heat transfer resistance between the surface and the air.

Figure 6.1 illustrates a typical late-summer diurnal energy budget for a white pine canopy during a mostly cloud-free day. The variability in fluxes during the afternoon resulted from the influence of clouds. The role of surface temperature in performing the LE and H flux estimations has been described by Eqs. 6.6 and 6.8. Remotely sensed surface temperatures can be determined on a landscape scale. Airborne sensors such as the Thermal Infrared Multispectral Scanner (TIMS) can provide calibrated surface temperature data (Appendix 6.A). The surface temperature measurements can be combined with ground-based measurements of $K\downarrow$, $L\downarrow$, air temperature, water vapor deficits, and estimates of surface resistance to determine the other terms in the equations, and for Eq. 6.4.

6.3.3 Nonradiative Terms in the Thermal Energy Budget

In developing ideas for a primarily remote-sensing means of describing the surface energy budget, we proposed (Luvall and Holbo 1989) a procedure that treated changes in surface temperature as an aggregate response of the dissipative thermal energy fluxes (LE, H, and G) and also expressed the influences of surface properties (canopy structure, amount and condition of biomass, heat capacity, and moisture). We had observed surface temperature changes in data collected by the aircraft-borne TIMS from uniform landscape elements like mature forests, plantations, and clear cuts. These temperature changes were evident when data from successive overflights separated by about 30 minutes were compared. From the

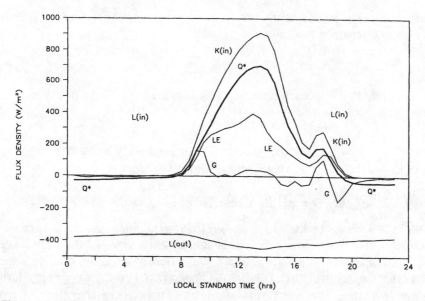

Figure 6.1. Surface energy budget for a white pine plantation at the Coweeta Hydrologic Laboratory, North Carolina. The measurements used to calculate the fluxes were taken with above canopy, tower-mounted, and ground-based instrumentation. The fluxes shown are incoming solar radiation [K↓ = K(in)], incoming long-wave radiation [L↓ = L(in)], net radiation (Q*), latent heat flux, or evapotranspiration (LE), conduction heat exchange (G), and outgoing long-wave radiation [L↑ = L(out)]. Cloudiness reduced flux densities during the afternoon.

energy budget, we knew that net radiation, Q*, constrains the total magnitude of the nonradiative fluxes, and surface-specific Q* values were used to normalize the measured temperature changes between the different types of surface. Since the rates of the thermal fluxes could be expected to change over the measurement interval, it was reasonable to use the rate of change in surface temperature as the value that reveals how the nonradiative fluxes react to changing radiant energy inputs. We then used the following formulation of the ratio between Q* and ∆T to define a surface property we called a Thermal Response Number (TRN), which would have units $Jm^{-2} \, °C^{-1}$:

$$TRN = \sum_{t_1}^{t_2} Q^* \Delta t / \Delta T \qquad (6.9)$$

where:

$$\sum_{t_1}^{t_2} Q^* \Delta t$$

represents the total amount of net radiation (Q*) for that surface over the time period between flights ($\Delta t = t_2 - t_1$), and ∆T is the change in mean temperature of that surface. The mean spatially averaged temperature for the surface elements at the times of imaging was estimated by using

$$T = 1/n \sum T_p \qquad (6.9a)$$

where each T_p is a pixel temperature (smallest area resolvable in the thermal image), and n is the number of pixels in the surface type.

6.4 Landscape Studies Using Remotely Sensed Forest Canopy Temperatures

6.4.1 Using the Thermal Response Number to Determine Forest Type

The TRN was used by Luvall and Holbo (1989) to examine thermal responses of several surface types across a forested landscape. Their approach takes advantage of TIMS capability for repeating surface temperature observations from the same site at short intervals. The H. J. Andrews Experimental Forest in Oregon provided a patchwork of various forest types covering a complex topography. Forest types ranged from old-growth Douglas-fir forests to recent clear-cuts.

This study represented the first time that short-term (28-minute) surface thermal response changes have been measured on a "landscape scale" for forests with the use of remote thermal sensors. Each cover type produced a characteristically different TRN (Table 6.1). The sites could be ranked by using the TRN value. The TRN for the forested sites was much greater than for the quarry and clear-cut sites. The TRN even differentiated between similar forested sites. For example, from Table 6.1, it is evident that one could not distinguish between the plantation and the natural regeneration sites by using only surface temperatures. The information shows that it is possible for two different types of forest canopies to have the same temperature, but the TRN indicates the existence of surface properties that result in differences in energy disposition. Thus the TRN, which quantifies the energy required to change surface temperature, also appears to distinguish between various types of surface.

Table 6.1. Surface Rankings Based on Thermal Response Number (TRN) for Douglas Fir Forest Types at the Andrews Experimental Forest, Oregon

Surface Type	Q* (Wm^{-2})	T (°C)	ΔT (°C)	TRN (KJm^{-2} °C^{-1})
Plantation	730	29.5	0.76	1631
Douglas fir forest	830	24.7	0.91	1549
Natural regeneration	771	29.4	1.66	788
Clear-cut	517	51.8	2.16	406
Rock quarry	445	50.7	4.50	168

Forested sites and forest plantations were found to have large TRNs, consistent with their tendency to exhibit moderated microclimates. Clear-cuts and barren sites had quite small TRNs, indicating little capacity to resist microclimate extremes. Sites midway in their development toward becoming forest had intermediate TRN values. The TRN seems to characterize the combined influences of surface properties controlling microclimate processes.

6.4.2 Using Remotely Sensed Surface Temperatures to Estimate Evapotranspiration

A few studies have used remotely sensed data from aircraft-borne scanners for estimating forest ET (Luvall 1988; Pierce and Congalton 1988). Luvall (1988) directly compared ET rates determined with the use of remotely sensed surface temperature from the TIMS with tower-based measurements of ET determined by using the energy budget approach (Penman-Monteith; Monteith 1973) from a white pine canopy. The initial results indicated a close agreement between the Penman-Monteith estimates and the TIMS (Table 6.2).

6.4.3 Spatial Heterogeneity of Forest Canopy Temperature

The spatial variation of canopy temperatures across the landscape produces a complex pattern. Differential solar loading due to topography can produce significant spatial variation of canopy temperatures even in monospecific canopies of uniform age and structure (Fig. 6.2). The white pine canopy temperatures ranged from 14.2°C to 18.8°C within the 14.1-ha south-facing watershed.

Observations made simultaneously across a landscape with the use of TIMS show that different surface types take on a wide range of temperatures with different surface temperature frequency distribution patterns (Fig. 6.3, from Holbo and Luvall 1989). Each type of surface appears to have uniquely influenced its pattern, apparently the result of its thermal properties and microclimates.

The distributions were strongly controlled by the presence of a plant canopy: distributions are markedly broader and warmer where a plant canopy is absent

Table 6.2. Evapotranspiration Estimated for a White Pine Canopy Using TIMS-Derived Surface Temperatures and the Penman-Monteith Equation

	Time Ending					
Latent energy flux (Wm^{-2}) (Evapotranspiration)	8:25 a.m.	8:55 a.m.	1:00 p.m.	1:25 p.m.	7:30[a] p.m.	8:00[a] p.m.
Penman-Monteith method	44	308	387	422	—	—
TIMS-based method	55	395	450	492	—	—

[a] Stomata were closed.

Figure 6.2. A thermal image composed of 5-m resolution TIMS temperature data acquired from a morning flight (8:30 a.m.) of the 28-year-old white pine plantation at Coweeta, for which the energy budget is shown in Fig. 6.1. Notice the variability in canopy temperatures within the watershed, shown by differences in shading, even though the canopy is of uniform age and structure. The variability was apparently the result of differences in incoming solar radiation because of differences in topographic position.

[e.g., the clear-cut (Fig. 6.3d)] and sensitive to the spatial uniformity of canopy elements on the site [e.g., the narrow distribution of the mature Douglas-fir stand (Fig. 6.3a) versus the broader distribution of the naturally regenerating, more sparsely occupied site (Fig. 6.3b)]. At night, although the range in temperatures is smaller, the thermal image still seems to capture subtle variations that may be as characteristic of surface type as during the daytime. In both day and night situations, the thermal images appear to have potential for helping to increase understanding of the dominant physical processes operating at the surface.

Areas of incomplete canopy coverage, as well as clear-cut areas, often exhibit extreme differences between needle temperatures and soil surface temperature. Vanderwaal and Holbo (1984) showed that Douglas-fir needle temperatures tracked air temperatures very closely (±3°C). However, sunlit soil surface temperatures directly under the seedlings were 30°C to 50°C higher than either air or needle temperatures. In such cases, one could not use an average surface temperature for the estimation of LE or H from the forest canopy without substantial errors.

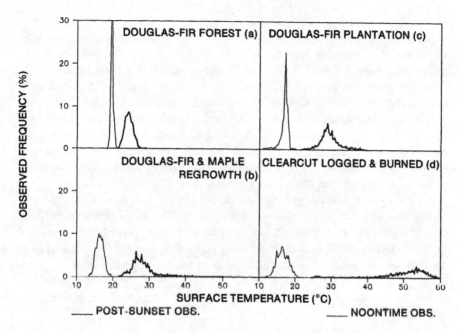

Figure 6.3. TIMS surface temperature frequency distribution patterns for four landscape types at the H. J. Andrews Experimental Forest in the Oregon Cascade Mountains: (a) a Douglas-fir forest; (b) a naturally regenerating site with an incomplete canopy coverage of about 60% Douglas fir and 40% maple; (c) a fifteen-year-old Douglas fir plantation; and (d) a clear-cut, logged and burned two years earlier. The two sets of data values for each type of surface represent night (lighter line) and day (heavier line) temperature distributions respectively.

Therefore, the average surface temperature observations may not characterize all the spatial variability needed for energy budget flux estimates. However, as demonstrated by Holbo and Luvall (1989), the frequency distributions of temperatures can be used as a powerful model in the differentiation and identification of land surface cover types and their properties. They found that a beta probability distribution can be closely fitted to surface temperature frequency distributions for a wide variety of forest landscapes. One advantage of using the beta distribution as a model is that it utilizes the pixel frequency distributions directly, and no high-order, measurement-error-magnifying statistics are used. (There is always a quantizing error in digital data that will propagate exponentially into the higher statistical moments.)

A beta probability distribution can be fitted to any closed distribution assumed to have a single mode (Selby 1968; Press et al. 1986). It is defined by two parameters (α, β) and a gamma (Γ) function of these same two parameters:

$$f(x) = \Gamma(\alpha,\beta)[x^{\alpha} (1 - x)^{\beta}] \tag{6.10}$$

for $0 < x < 1$, and $\alpha > -1$, $\beta > -1$.

Note that the distribution is a closed one, owing to the term in square brackets, which forces a return to zero at each end of the x range. The position of the maximum along the x scale is governed by the relative sizes of α to β; when $\alpha > \beta$, the distribution is shifted to the right (negatively skewed and $\bar{x} > 0.5$); when $\alpha < \beta$, the distribution is shifted to the left (positively skewed and $\bar{x} < 0.5$); if $\alpha = \beta$, the maximum and the mean of the distribution are both at 0.5 and may be made to look very much like a normal (Gaussian) distribution. Smaller α or β values spread the distribution, as used to model the clear-cut. Selecting larger α and β values narrows the distribution and increases the probable frequency of the mode, modeling observations from forest canopies well. See Fig. 6.4 and Table 6.3.

The role of the gamma function is an important one and cannot be neglected. The gamma function performs as a weighing coefficient, or third parameter in the model. It conditionally assures that the integral of the beta f(x) distribution is equal to one (unity, or 100%, probability) for any α,β values that may be required to fit the data set. Because of this condition, f(x) can truly be regarded as an estimator of the probability of x [i.e., p(x)], for data sets being fitted by a beta distribution.

For integer values of α and β, the Gamma function part of Eq. 6.10 can be written in the form:

$$\Gamma(\alpha,\beta) = [\Gamma(\alpha + \beta + 2)]/[\Gamma(\alpha + 1)\Gamma(\beta + 1)] \tag{6.11}$$

defining $\Gamma(y) = (y - 1)!$, where ! is the factorial operation, and y is the sum of values in parentheses. A description of the computational technique for determining parameters α and β is presented in Appendix 6.B.

Beta modeling has the potential to become a useful tool for describing, comparing, and classifying thermal image data sets; see Table 6.3 (from Holbo and Luvall 1989). In the table, α and β beta parameters for several forest types at the H. J. Andrews Experimental Forest are shown. With the model parameters, the surface temperature distributions can be compared. The α and β values vary widely and can be seen to relate strongly to surface type. An index, the beta index, has been devised to simplify the comparison. The index is a consolidation of the beta model's parameters into a single number that conveys the same information, although a bit of contrast stretching has been introduced (see Appendix 6.B).

6.4.4 Forest Edge Microclimates

Figure 6.5 shows two surface temperature profiles, one at noon and one after sunset, taken along a transect trending from southwest to northeast through a clear-cut on the H. J. Andrews forest (Fig. 6.6). Each profile shows the same ground, differing only in the time of imaging. The transect begins in the forest at the edge of the clear-cut, then moves across its logged and burned southwest slope,

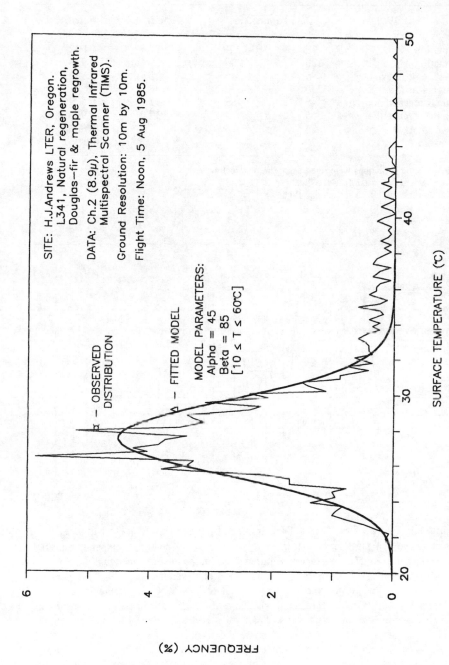

SITE: H.J.Andrews LTER, Oregon. L341, Natural regeneration, Douglas–fir & maple regrowth.

DATA: Ch.2 (8.9μ), Thermal Infrared Multispectral Scanner (TIMS).

Ground Resolution: 10m by 10m.

Flight Time: Noon, 5 Aug 1985.

— OBSERVED DISTRIBUTION

— FITTED MODEL

MODEL PARAMETERS:
Alpha = 45
Beta = 85
[10 ≤ T ≤ 60°C]

SURFACE TEMPERATURE (°C)

FREQUENCY (%)

Figure 6.4. An example of an observed surface temperature frequency distribution (irregular line) and its modeled beta distribution fit (smooth, heavier line). This example is of the H. J. Andrews natural regeneration surface type from first overflight of TIMS. The beta parameters producing the fitted curve are $\alpha = 45$ and $\beta = 85$. The major features of the surface temperature frequency distribution are well described by the beta probability distribution.

Table 6.3. Comparisons of Four Surface Temperature Distributions from the H. J. Andrews Experimental Forest, Oregon[a]

Surface Type	Beta Parameters		Beta Index	Fitted R^2
Plantation 1[b]	$\alpha = 70$	$\beta = 120$	21.5	0.942
Plantation 2[c]	120	200	34.4	0.925
Douglas fir forest 1	95	245	62.8	0.989
Douglas fir forest 2	240	570	138.9	0.984
Natural regeneration 1	45	85	17.1	0.923
Natural regeneration 2	55	95	17.2	0.925
Clear-cut 1	23	5	−7.1	0.900
Clear-cut 2	20	4	−6.3	0.896

[a] Frequency distributions were fitted using the beta technique.
[b] Number 1 signifies the first-day flight.
[c] Number 2 signifies the second-day flight.

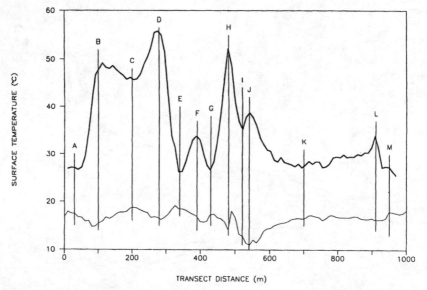

Figure 6.5. Two profiles along a 1-km transect from the H. J. Andrews Experimental Forest, Oregon, of TIMS surface temperature observations. The higher temperature profile was extracted from thermal imagery acquired at noon, the lower temperature profile from the postsunset imagery. TIMS resolution for these observations was 10 m. Temperature variability along the profiles is seen to span a range of 10°– 60°C. The figure demonstrates the unique capability of TIMS for sampling across the edges of landscape elements. Some of the surface features along the transects are identified with vertical lines on the figure: (A) edge of forest; (B) narrow road; (C) somewhere within a clear-cut; (D) a wider road; (E) one side of a small shelter wood of Douglas fir; (F) a pond within the shelter wood; (G) the other side of the shelter wood; (H) the wide road; (I) trees along road; (J) in a flatter part of the clear-cut; (K) somewhere within a 15-year-old Douglas fir plantation; (L) a trail; and (M) in an older stand of Douglas fir regrowth.

Figure 6.6. One of the TIMS images used to obtain the temperature transect data for Fig. 6.5. The white line, from lower left (SW corner) to upper right (NE corner), shows the direction of the transects.

through an isolated stand of mature Douglas fir, across a flatter part of the same clear-cut, and across a fifteen-year-old Douglas fir plantation.

From the profiles, it appears that surface temperature transitions across the known abrupt boundaries, as between forest and clear-cut, are about 30 m in length. But it must be appreciated that each observation point along the transect represents the average temperature of a 10 m by 10 m surface area (the pixel size for this data set), thus limiting edge definition to lengths longer than twice the width of a pixel in the direction of the transect. Nevertheless, it would be difficult to improve upon this with ground-based sampling.

A meteorological station on the same clear-cut (near D in Fig. 6.5) measured average air temperatures (at 2-m height) of 25°C during the noon imaging over-flight and 14°C during the evening imaging. However, all points along the transect were warmer than the noontime air temperature, with a few exceeding 55°C. Just after sunset, most points were slightly warmer than the air, with a small section along the transect (J) dropping to nearly 10°C. This region, near (J), seems to indicate the onset of cold air drainage from upslope, apparently being impeded by something in that part of the clear-cut.

6.5 Limitations of Thermal Techniques

6.5.1 Thermal Response Number

Some site-specific information is needed for calculation of the TRN for optimal discrimination of forest canopy types. Specifically, one must know incoming solar radiation at the time of the flight and determine the net radiation for a range of forest types. The relationship between $K\downarrow$ and Q^* is well established, and generally the literature values can be used (Oke 1987). Sites at which $K\downarrow$ is routinely measured are relatively rare. (Its measurement is prescribed for all Long-Term Ecological Research sites.) Also, Q^* can be calculated by using Eq. 6.5. Luvall and Holbo (1989) showed that errors in Q^* estimates would not significantly affect the TRN ranking of forest types.

At the present time the TRN technique is limited to aircraft-obtained data. Aircraft and calibrated thermal sensors are limited in number and expensive to operate. Weather conditions ideal for obtaining the measurements are rigorous. Data from sites with frequent cloud cover (e.g., tropical forests) are limited and difficult to obtain. In rapidly changing weather conditions, it may also be difficult to obtain repeated aircraft passes.

The use of TRN to describe landscape functional processes is in its infancy. Only a limited number of sites for western coniferous forest have been described (Luvall and Holbo 1989). No research has been done that describes seasonal variability in the TRN. TIMS data has been collected by the authors during 1987 and 1988 from several tropical and eastern deciduous forests. Preliminary analysis of that data supports the use of the TRN for landscape function analysis in those ecosystems.

6.5.2 Estimating Evapotranspiration

The determination of LE with remotely sensed surface temperatures through the use of Eq. 6.6 requires that several plant or surface-dependent variables be known. First, measured or good models of the canopy conductances are needed to account for the plant physiological control of water loss. This information may not be readily available for the forest or during the time frame of the flights.

Canopy conductance is one of the most difficult factors to estimate accurately. Plant species have different rates of transpiration and stomatal conductance, both daily and seasonally. In forested ecosystems, evergreen, coniferous tree species tend to have lower rates of transpiration and stomatal conductance than deciduous, broad-leaved species. A wide variation in transpiration and conductance has been reported among species. Since ET is directly related to canopy conductance, it can differ between forests or stands because of species composition, as well as the environment (Chambers et al. 1985; Federer and Gee 1976; Hinckley et al. 1981).

Second, it is difficult to determine the vapor pressure gradient to which the forest canopy is actually exposed. Usually, vapor pressure information is measured at one site within the study area and taken as representative of the area.

Since most remotely sensed data is taken at a instantaneous point in time, the measurements must be extrapolated for the whole day. Several approaches have been taken to equate the two ET estimates. Nieuwenhuis et al. (1985) proposed the following relationships in terms of surface temperature and latent heat flux (following the notation and formulations used by Gash 1987):

$$LE_d = LE_{ref.d} - B^1(\delta T_s)_i \qquad (6.12)$$

where LE = latent heat flux (evapotranspiration), d = the 24-h average values, ref = reference evapotranspiration, i = to an instantaneous midday value, B^1 = a calibration constant similar to the constant B proposed by Jackson et al. (1977), which related the difference between surface and air temperature to evaporation, and δT_s = surface temperature.

6.5.3 Surface Temperature Distributions

The information content of any distribution function is constrained by the size and number of the samples taken and is always limited by practical considerations. For the surface temperature distributions modeled here by beta probability distributions, it was fortunate that pixel sizes were 10 by 10 m. This size is near the upper limit of what could be reasonably expected to be descriptive in a forested landscape, since the dominant surface features (crowns of the trees and shrubs) frequently approximately that size. Should a subsequent examination of the same surface types on the same landscape be performed with the TIMS flown at lower altitudes to achieve a resolution of 5 by 5 m, better surface-type discrimination would be expected. And, if flown at a coarser resolution, say 20- by 20-m pixels, details would be lost to the extent that recognition of only the most dramatically different landscape elements would be obtained. As a rule, the information content of a collection of samples is strictly interpretable only to those elements in the sample space which are twice as large as the spatial resolution of the sampling process. This rule is analogous to Shannon's sampling theorem (1949), which was developed for extracting information from temporal phenomena. This is not to conclude that the smaller elements exert no influence on the individual samples, but rather to caution against attempting to extract significance from the data at scales smaller than the sampling can support.

Another factor limiting the broader applicability of beta distribution modeling to landscape elements of differing types is that only a few TIMS missions have been flown for the purpose of collecting thermal images of other landscape types. The majority of the flights have been flown to exploit the potential of TIMS to acquire imagery for interpreting surface geology across desert landscapes.

It is certainly anticipated that temporal variability, caused by changing solar angles or redistribution of components of the energy budget, will influence the distributions of surface temperatures. One example might be a water-stressed canopy, where the partitioning of Q^* shifts from LE to H as stomata close late in the day. In this instance, canopy temperatures might be expected to increase, in response to the need to dissipate a greater proportion of the energy as sensible heat. While the potential impact of such effects should not be ignored, the analyses conducted thus far suggest that beta˙ modeling will detect contrasts between different landscape elements during fair weather. Part of this success may be attributable to the fact that the energy budgets of different surface types respond to the conditions at any given time in ways that are consistent with, and characteristic of, their specific surface type.

6.6 Uses of Thermal Models in Landscape Ecology

Surface temperature is of importance to many ecological processes. Together with moisture availability, surface temperature can encourage or restrict the initiation of plant life on a site. Until the advent of spatial surface temperature sampling by instruments like TIMS, management of the microclimate of plant establishment depended largely upon relationships based on air temperature measurements. However, air temperature is far more conservatively behaved and rarely reveals the true extremes experienced by plants (and animals) at the surface. There are several areas in which thermal models can be used in landscape ecology.

6.6.1 Plantation Forestry

Plantation forestry is one area where avoidance of surface temperature extremes is critically important to the success of reforestation efforts (Childs et al. 1985). The experienced manager may have intuitive beliefs about this type of environment. However, a site-specific investigation, as exemplified in Fig. 6.5, can improve the manager's objectivity with regard to the actual conditions at the surface.

6.6.2 Evaluation of Thermal Habitats of Fish and Wildlife

Increasing utilization of the landscape has resulted in greater competition for resource values in some areas. For example, as logging activities encroach upon the boundaries of small mountain streams, it is critical to ensure adequate shading to prevent water temperatures from rising to uninhabitable levels for desired fish species. Thermal imagery could be used to gauge the effectiveness of buffer strips along riparian zones or to help locate stretches that might need special protective measures.

6.6.3 Frost Protection Planning for Nurseries and Orchards

Thermal images can reveal the presence of cold air collection areas and drainage paths when a foliar cover is present. High-elevation grasslands and clear-cuts become sources of cold air. Forests exhibit much smaller changes in temperature from day to night. Forest canopies may extend into warmer air aloft, in effect buffering the intrusion into the canopy space of colder surface air (Holbo 1983). Nighttime radiation losses of nonforested surfaces under cloud-free skies are about

50 Wm^{-2} (Holbo and Childs 1987). Since $L\uparrow = \varepsilon\sigma T^4$, one can use its derivative, $\delta L\uparrow/\delta T = 4\varepsilon\sigma T^3 \cong 5$ Wm^{-2}°C^{-1}, to estimate that each 1°C drop in surface temperature represents an additional 5 Wm^{-2} differential in the radiation balance, progressively accelerating the heat loss rate and increasing the chance of freezing.

6.6.4 Predicting Nighttime Air-Shed Smoke Impact

The cold air drainage patterns also identify areas that tend to entrain smoke originating with controlled burning activities, such as slash burning done following logging and field burning of grass stubble following seed harvesting. These activities are subject to increasing regulation, especially near population centers. Studies of thermal imagery before scheduling a burn might help managers avoid violations of regulations and could be used for mapping areas likely to provide routes for nighttime air drainage (Gossmann 1986).

6.6.5 Estimating Evapotranspiration

Water is a critical resource. Although concern for its conservation has waned, there will be a renewal of interest in improved evapotranspiration estimation techniques if current trends continue. A substantial amount of effort has already been expended on the development of evapotranspiration estimation methods that use remotely sensed thermal data as a primary input variable (e.g., Abdellaoui et al. 1986; Heilman et al. 1976; Ho 1985; Reginato et al. 1985; Seguin and Itier 1983; Soer 1980; Stone and Horton 1974). With the improved spatial resolution and surface temperature accuracy offered by the TIMS, those methods can now be made site specific and have greater relevance to landscape management practices (Luvall 1988).

6.6.6 Detection and Delineation of Microtopographic Terrain Features

The direct linkage between thermal emissions from a surface and its physical properties (e.g., thermal diffusivity) permits the extraction of information about features too subtle to appear in unenhanced thermal images (Pelletier 1985; Pelletier et al. 1985). Features that can be located in this way include soil conservation measures, minor drainages, and archaeological structures.

6.6.7 Assessment of Vegetatively Moderated Urban Microclimates

Microclimatic issues are important to urban planners. Structural materials within the urban environment often result in hostile microclimates. By using remotely sensed thermal image data, the influence of various amounts and arrangements of vegetation in those environments can be evaluated and mediating alternatives recommended (Quattrochi and Rowntree 1988).

6.6.8 Recognition of Landscape Type with Thermal Image Data

Thermal remote sensing offers many possibilities for studying landscapes because there is virtually no other means by which the spatial character of the surface temperature field in those systems can be observed. This spatial character tends to be diagnostic of the type of surface from which the thermal image is taken. For example, Fig. 6.3 contains observed (nighttime and daytime) frequency distributions of surface temperature of four landscape types. It can be seen that each surface type exhibits a different pattern of dispersion, which is especially obvious in the daytime.

By using a beta probability distribution as a model, these patterns can be described in terms of just two parameters (Holbo and Luvall 1988). Different landscape types yield unique combinations of these parameters. It should be feasible to implement this technique for landscape recognition. Furthermore, because of the microclimatic information also available from this data, its association with specific landscape types should improve regional evapotranspiration estimation methods.

6.6.9 Predicting Microclimatic Change Due to Landscape Type Conversion

The interrelationships among the four components of a surface's thermal energy budget (radiation, convection, conduction, and latent heat (i.e., evapotranspiration) make it difficult to predict the consequences of converting from one landscape type to another. These components respond to both external and internal factors, but it is the internal biophysical properties of the site that are changed by management practices. When complicated even more by multiple land management practices across a landscape, assessing the net impact is beyond the scope of traditional approaches. A new approach, arising out of the study of short-term changes in surface temperature and made with the use of TIMS, has been proposed. It is based on a derived surface property called the TRN (Luvall and Holbo 1989).

6.7 Summary

Thermal remote sensing offers special and unique qualities to the landscape ecologist that cannot be obtained from other kinds of remote sensing data products. Thermal remote sensing instruments measure a very fundamental surface phenomenon, namely, temperature, a primary expression of the physical environment. Because temperature data can be used to evaluate functional relationships, such as the energy budget, thermal remote sensing can become a valuable tool for improving our understanding of patterns and processes in landscape ecology. This is especially true for instruments like the TIMS, which offer measurement accuracy and spatial resolution far better than previously attainable.

Immediate applications for remotely sensed thermal imagery and the corresponding temperature data include avoiding or managing stressful microclimates (high temperature/low moisture) in plantation silviculture or cropland agriculture; frost protection planning for nurseries and orchards; watershed, crop, or forest evapotranspiration estimation; evaluation of thermal habitats of fish and wildlife (buffer strip effectiveness along mountain streams or in riparian zones); predicting nighttime air-shed smoke impact regions during controlled burns; thermal recognition and classification of landscape types; predicting microclimatic change due to landscape type conversion (fire, logging); detection and delineation of microtopographic terrain features (soil erosion control structures, locating archaeological sites); and assessment of vegetatively moderated urban heat islands.

The analytical models and image processing techniques that have been presented here are in their initial stages of development. The applications examples posed are just a few of those which come to mind when we consider the large potential for landscape applications that this remote sensing technology appears to have. Improvements are already being planned for the instrumentation, so that the next generation of TIMS will have better spatial resolution and measurement precision. Work is also under way to make thermal data processing easier, with more of the image-handling tasks being transferred to microcomputers and thus becoming accessible to more researchers.

Appendix 6.A: Obtaining Calibrated Remotely Sensed Canopy Temperatures

The Thermal Infrared Multispectral Scanner (TIMS) is an aircraft-mounted instrument (Palluconi and Meeks 1985). It was used to obtain the surface temperature measurements needed for investigating forest canopy thermal response and for the development of an analytical model of surface temperature distributions. The TIMS has six channels in the thermal wavelength region from 8.2 to 12.2 μm (Fig. 6.7). It scans a swath 60° wide along the flightline to capture images of the surface in raster style. The TIMS is continuously calibrated through the use of on-board reference blackbody cavities, which results in a noise equivalent temperature (NEΔT) as low as 0.09°C for channel 2 and a precision of 0.2°C over the temperature range of 10 to 65°C. Its spectral responsivity and blackbody radiance calibration coefficients are determined from preflight optical bench calibrations. The on-board low and high temperature blackbodies are read at the beginning and end of each raster's scan line. The blackbody temperatures are selected to bracket the expected range of surface temperatures to optimize TIMS' temperature resolution. Spatial resolution can be varied from 5 to 30 m, depending on aircraft altitude.

It is necessary to account for atmospheric radiance to obtain accurate surface temperature data. Although TIMS' channels fall within the so-called atmospheric window for long-wave transmittance (8.0 to 13.0μm), the maximum transmitance there is actually only about 80%, meaning that some 20% of the radiance received by TIMS' thermal sensing element may originate in the path between the surface to the sensor. Ground-based measurements show that the incoming (down-welling) long-wave energy flux density ($L\downarrow$) from the atmosphere varies from 300 to 425 Wm^{-2}, of which some 60 to 85 Wm^{-2} might be interpreted as caused by long-wave emission from the surface. Atmospheric radiance is mostly a function of the amount and temperature of water vapor.

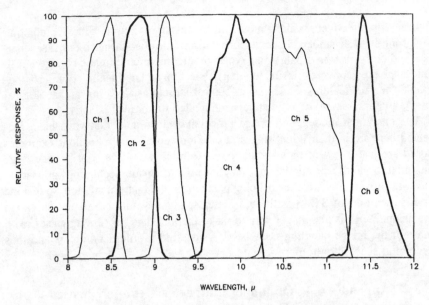

Figure 6.7. The spectral response curves for the six channels of TIMS. Channel 2 appeared to offer the best accuracy and was used to obtain surface temperature data presented here.

Atmospheric radiance can be determined in one of two ways: (1) measuring L↓ by using an pyrgeometer (typical passband of 5 to 50 μ), which then is adjusted according to the specific wavelengths used by TIMS (Sweat and Carroll 1983); or (2) measuring the atmospheric profiles of air temperature and water vapor content and then estimating the amount of radiance correlation from these measurements. At many meteorological stations throughout the United States, such profiles are routinely taken using radiosondes twice daily (commonly at 6:00 a.m. and 12:00 midnight). The authors used the second approach, but employed on-site, TIMS flight-time concurrent radiosonde launches to obtain the profile data. The atmospheric profiles were then incorporated in the LOWTRAN6 model for calculation of atmospheric radiance (Kneizys et al. 1983). Atmospheric long-wave radiance values calculated by an earlier version, LOWTRAN5, have been shown to be in excellent agreement with measured atmospheric radiance values (Sweat and Carroll 1983). Wilson and Anderson (1986) endorsed the validity of LOWTRAN5 for atmospheric radiance corrections of aircraft thermal data collected over atmospheric path lengths similar to those used here.

The output from LOWTRAN6 is then combined with TIMS' spectral response curves, using the module TRADE on the ELAS system at NASA's Stennis Space Center, Science and Technology Laboratory (Graham et al. 1986). TRADE produces a look-up table for pixel temperatures as a function of TIMS values (Anderson 1985).

Appendix 6.B: Modeling Observations with a Beta Probability Distribution

A TIMS data set can be modeled with a beta probability distribution. The following describes the steps used:

1. Scale the TIMS temperature values from the observed frequency distribution data set to range from 0 to 1:

$$x = (T_j - T_{low})/(T_{high}) - T_{low})$$

where low and high are the extreme TIMS temperature values, and

$$T_{low} \le T_j \le T_{high}$$

2. Choose values for α and β and compute for each x:

$$x^\alpha (1 - x)^\beta$$

Near 0 and 1 of x, this beta product function will produce very small numeric values, so small that computational difficulties may be experienced. Such situations arise when a frequency distribution is narrow and has large frequency classes, thus requiring large values of α and β.

3. Compute the integral function of the values obtained by step 2 for successive x_i over the interval $0 \le x \le 1$:

$$f(x_i) = \int_0^{x_i} x^\alpha (1 - x)^\beta \delta x$$

If the data intervals are equivalent for all data sets, this may be done by successively summing the results of step 2 for x_i over its 0 to 1 range.

4. Normalize the function described by step 3 according to its value at $x = 1$. In step 3, this maximum value will likely be a small number. Dividing it into all values of the function produced in step 3 will result in a function produced by this step (step 4), which is a cumulative beta distribution with an integral value of unity at $x = 1$. This step performs an equivalent operation to the coefficient $\Gamma(\alpha, \beta)$, which is otherwise troublesome to compute.

5. Differentiate the resultant function of step 4 at the intervals of x. This produces the beta probability density function (PDF) of α, β, or $p(x)$.

6. Compare $p(x)$ with the observed frequency distribution. Loop iteratively from step 2 (selecting new trial values of α and β) through step 6 until an acceptable level of comparability is found. For the cases shown here, a combination of least squares fit and visual (graphical) matching at the modal frequency was used. This combination approach helped avoid choosing an α, β parameter set that may yield a high regression coefficient but that does not simultaneously represent the modal class of the observed distribution.

The regression test was made a part of the fitting procedure. The fitting procedure utilized only those frequencies within the TIMS-observed pixel temperature range, excluding the tails of the beta distribution, which were beyond the low-high range of the data set. By using the regression, a succession of α and β values were tried until a maximum R^2 was found. Tabulating the resulting R-squared values in the form of an $[\alpha,\beta]$ matrix helped in the process of choosing the best parameter combination. With the use of such a matrix table, the optimal α and β values were chosen from along what appears as a diagonal "ridge" of higher correlation coefficients. This iterative technique was also assisted by viewing graphs of the "fit" on a graphics monitor, since the convergence of α and β values toward an optimum could be visually evaluated, especially during the early stages of fitting. Additionally, viewing the fit helped ensure against selecting α and β values that may happen to yield a high R^2 but fail to satisfactorily produce the frequency near the principal mode of the distribution. This was often the situation for the more irregular data sets. Also, in view of the wide range in α and β combinations representing all observed surface temperature distributions in the Andrews data sets, the regression-fitting technique seemed to provide fairly adequate discrimination among the sites examined.

To make ranked comparisons between sites, the two beta parameters can be combined into a single value, or beta index. This index contains information about both the parameter ratio and parameter magnitudes. The fitted α, β values for the sites had absolute ratios of less than 1 to above 3, at the same time ranging over three orders of magnitude: $1 < \alpha/\beta$ (or β/α) < 1000, or more. The geometric mean (GM) of α, β, $\sqrt{\alpha\beta}$ was selected to represent magnitude information. The logarithm of the β/α ratio (LR), $\log_{10}(\beta/\alpha)$, was chosen to convey the relationship between α and β. The advantage of this logarithmic form is twofold: first, it indicates model asymmetry in a manner similar to the third moment of the data set, the second, it offers greater discrimination between distributions having ratios close to unity, essentially stretching the contrast level for scenes with narrow and/or nearly symmetrical temperature frequency distributions, i.e., the nighttime data sets. The beta index was formed as the product of these two terms: GM times LR.

References

Abdellaoui, A.; Becker, F.; and Olory-Hechinger, E. 1986. Use of Meteosat for mapping thermal inertia and evapotranspiration over a limited region of Mali. *Journal of Climate Applied Meteorology* 25:1489–1506.

Anderson, J.E. 1985 Thermal infrared data: its characteristics and use. American Society for Photogrammetry and Remote Sensing Technical Paper 1:143–55.

Chambers, J.L.; Hinckley, T.M.; Cox, G.S.; Metcalf, C.L.; and Aslin, R.C. 1985. Boundary-line analysis and models of leaf conductance for four oak-hickory forest species. *Forest Science* 31:437–50.

Childs, S.W.; Holbo, H.R.; and Miller, E.L. 1985. Shadecard and shelterwood modification of the soil temperature environment. *Soil Science Society of America Journal* 49:1018–22.

Federer, C.A., and Gee, G.W. 1976. Diffusion resistance and xylem potential in stressed and unstressed northern hardwood trees. *Ecology* 57:975–84.

Garnier, B.J., and Ohmura, A. 1968. A method of calculating the direct shortwave radiation income of slopes. *Journal of Applied Meteorology* 7:796–800.

Gash, J.H.C. 1987. An analytical framework for extrapolating evaporation measurements by remote sensing surface temperature. *Internation Journal of Remote Sensing* 8:1245–49.

Gossman, H. 1986. The influence of geography on local environment as inferred from night thermal infrared imagery. *Remote Sensing Review* 1:249–75.

Graham, M.H.; Junkin, B.G.; Kalcic, M.T.; Pearson, R.W.; and Seyfarth, B.R. 1986. ELAS-Earth Resources Laboratory Applications Software, rev. January 1986. NASA/NSTL/ERL Report Number 183.

Heilman, J.L.; Kanemasu, E.T.; Rosenberg, N.J.; and Blad, B.L. 1976. Thermal scanner measurement of canopy temperatures to estimate evapotransportation. *Remote Sensing of Environment* 5:137–45.

Hinckley, T.M.; Teskey, R.O.; Duhme, F.; and Richter, H. 1981. Temperate hardwood forests. In *Water Deficits and Plant Growth*, vol. 6,pp. 153–208. New York: Academic Press.

Ho, D. 1985. Soil thermal inertia and sensible and latent heat fluxes by remote sensing. In *Proc. 4th Thematic Conference: Remote Sensing for Exploration Geology*, pp. 635–43. San Francisco, April 1–4, 1985. Ann Arbor: Environmental Research Institute of Michigan.

Holbo, H.R. 1983. Wind Patterns in a clearcut and an adjacent forest. FIR Report 5(2). Oregon State University Cooperative Extension Service.

Holbo, H.R., and Childs, S.W. 1978. Summertime radiation balances of clearcut and shelterwood slopes in southwest Oregon. *Forest Science* 33:504–16.

Holbo, H.R., and Luvall, J.C. 1988. Thermal responses and surface temperature distributions of forested landscapes. In *Remote Sensing for Resource Inventory, Planning and Monitoring*, ed. J.D. Greer, pp. 308–17. Falls Church, Va.: American Society for Photogrammetry and Remote Sensing.

Holbo, H.R., and Luvall, J.C. 1989. Modeling surface temperature distributions in forest landscapes. *Remote Sensing of Environment* 27:11–24.

Jackson, R.D.; Reginato, R.J.; and Idso, S.B. 1977. Wheat canopy temperatures: a practical tool for evaluating water requirements. *Water Resources Research* 13:651–56.

Jarvis, P.G. 1981. Stomatal conductance, gaseous exchange and transpiration. In *Plants and Their Atmospheric Environment*, eds. J. Grace, E.D. Ford, and P.G. Jarvis, pp. 175–204. Oxford: Blackwell.

Kneizys, F.X.; Settle, E.P.; Gallery, W.O.; Chetwynd, J.H., Jr.; Abreu, L.W.; Selby, J.E.A.; Fenn, R.W.; and McClatchey, R.A. 1983. Atmospheric transmittance/radiance: computer code Lowtran-6. Air Force Geophysics Laboratory Report AFGL-TR83–0187. Optical Physics Division, Hanscom Air Force Base.

Luvall, J.C. 1988. Using the TIMS to estimate evapotranspiration from a white pine forest In *Remote Sensing for Resource Inventory, Planning and Monitoring*, ed. J.D. Greer, pp. 90–98. Falls Church, Va.: American Society for Photogrammetry and Remote Sensing.

Luvall, J.C., and Holbo, H.R. 1989. Measurement of short-term thermal responses of coniferous forest canopies using thermal scanner data. *Remote Sensing of Environment* 27:1–10.

Miller, D.R. 1975. Microclimate at an oak forest–asphalt parking lot interface. In *Proceedings of the 12th Conference on Agricultural and Forest Meteorology*, pp. 35–36. Tucson, Arizona, April 14–16, 1975, American Meteorological Society.

Monteith, J.L. 1973. *Principles of Environmental Physics*. London: Edward Arnold.

Nieuwenhuis, G.J.A.; Schmidt, E.H.; and Thunnissen, H.A.M. 1985. Estimation of regional evapotranspiration of arable crops from thermal infrared images. *International Journal of Remote Sensing* 6:1319–34.

Oke, T.R. 1987. *Boundary Layer Climates*. New York: John Wiley and Sons.

Palluconi, F.D., and Meeks, G.R. 1985. Thermal Infrared Multispectral Scanner (TIMS): an investigator's guide to TIMS data. Jet Propulsion Laboratory Publication 85–32.

Pelletier, R.E. 1985. Identification of linear features in agricultural landscapes through spatial analysis of thermal infrared multispectral data. In *Proceedings Technical Papers American Society Photogrammetry*, pp. 381–90. Washington, D.C., March 10–15, 1985.

Pelletier, R.E.; Ochoa, M.C.; and Hajek, B.F. 1985. Agricultural applications for thermal infrared multispectral scanner data. In *Proceedings 11th International Symposium Mach. Process. of Remotely Sensed Data*, pp. 321–28. West Lafayette, Ind., June 25–27, 1985.

Pierce, L.L., and Congalton, R.G. 1988. A methodology for mapping forest latent heat flux densities using remote sensing. *Remote Sensing of Environment* 24:405–18.

Press, W.H.; Flannery, B.P.; Teulosky, S.A.; and Vetterling, W.T. 1986. *Numerical Recipes*. Cambridge: Cambridge University Press.

Quattrochi, D.A., and Rowntree, R.A. 1988. Potential of thermal infrared multispectral scanner (TIMS) data for measurement of urban vegetation characteristics. In *Remote Sensing for Resource Inventory, Planning and Monitoring*, ed. J.D. Greer, pp. 344–51. Falls Church, Va.: American Society for Photogrammetry and Remote Sensing.

Ranney, J.W.; Bruner, M.C.; and Levenson, J.B. 1981. The importance of edge in the structure and dynamics of forest islands. In *Forest Island Dynamics in Man-Dominated Landscapes*, eds. R.L. Burgess and D.M. Sharpe, pp. 67–95. New York: Springer-Verlag.

Reginato, R.J.; Jackson, R.D.; and Pinter, P.J., Jr. 1985. Evapotranspiration calculated from remote multispectral and ground station meteorological data. *Remote Sensing of Environment* 18:75–89.

Risser, P.G. 1987. Landscape ecology: state of the art. In *Landscape Heterogeneity and Disturbance*, ed. M.G. Turner, pp. 3–14. New York: Springer-Verlag.

Seguin, B., and Itier, B. 1983. Using midday surface temperature to estimate daily evaporation from satellite thermal IR data. *International Journal of Remote Sensing* 4:371–83.

Selby, S.M., ed. 1968. *Standard Mathematical Tables*. 16th ed. Cleveland, Ohio: Chemical Rubber Co.

Shannon, C.E. 1949. Communication in the presence of noise. *Proceedings of the IRE* 37:10–21.

Soer, G.J.R. 1980. Estimation of regional evapotranspiration and soil moisture conditions using remotely sensed crop surface temperatures. *Remote Sensing of Environment* 9:27–45.

Stone, L.R., and Horton, M.L. 1974. Estimating evapotranspiration using canopy temperatures: field evaluation. *Agronomy Journal* 66:450–54.

Sweat, M., and Carroll, J.E. 1983. On the use of spectral radiance models to obtain irradiances on surfaces of arbitrary orientation. *Solar Energy* 30(4):373–77.

Thistle, H.W.; Miller, D.R.; and Lin, J.D. 1987. Turbulent flow characteristics at the edge of a foliated and unfoliated deciduous canopy. In *Proceedings of the 18th Conference on Agricultural and Forest Meteorology*, pp. 179–80. West Lafayette, Ind., Sept. 15–18, 1987.

Vanderwaal, J.A., and Holbo, H.R. 1984. Needle-air temperature differences of Douglas-fir seedlings and relation to microlimate. *Forest Science* 30:635–44.

Wilson, S.B., and Anderson, J.M. 1986. The applicability of LOWTRAN 5 computer code to aerial thermographic data correction. *International Journal of Remote Sensing* 7:379–88.

7. Intelligent Geographic Information Systems for Natural Resource Management

Robert N. Coulson, Clark N. Lovelady, Richard O. Flamm,
Sharon L. Spradling, and Michael C. Saunders

7.1 Introduction

Landscapes are the functional management units for service agencies charged with husbandry of natural resources. Natural resource management concerns the orchestrated modification of landscape structure (components of the landscape and their linkages and configurations), function (quantities of flows of energy, materials, and species within and among landscape elements), and rate of change (alteration in the structure and function of the ecological mosaic over time). As with studies of basic issues in landscape ecology, management of natural resources involves integration and interpretation of various forms of knowledge, e.g., simulation results, historical data bases, technical reports, and heuristic information. One of the principal technologies available for investigations of landscapes is the geographic information system (GIS). In addition to representing landscape features, a GIS can be used in management to predict the consequences of a contemplated action, evaluate the results of actions that have been taken, and compare alternative actions. The utility of a conventional GIS can be greatly expanded by adding artificial intelligence (AI) methodologies. The resulting product is an intelligent GIS or IGIS (Coulson et al. 1987). An IGIS application is a blend of methodologies associated with several different academic specialties. Our goal in this chapter is to illustrate how an IGIS can facilitate representation, analysis, and

[1]Texas Agricultural Experiment Station Pa. TA No. 24768.

interpretation of spatially referenced data and thereby improve landscape management planning and practice. The specific objectives are (1) to examine GISs and IGISs in the context of a suite of computer-based tools for problem solving and decision making, (2) to describe the IGIS concept and method, and (3) to illustrate the application of IGISs in natural resource management decision making and problem solving.

7.2 GISs and AI in the Context of Computer-Aided Decision Making and Problem Solving

The expanded knowledge base available for natural resource management has necessitated computer applications for problem solving and decision making. Several different types of computer-based systems facilitate analysis, integration, interpretation, and delivery of knowledge (Fig. 7.1). The tasks performed by the systems increase in complexity in the following order: file management systems, data base management systems, management information systems, knowledge-based systems, and the knowledge system environment (Coulson and Saunders 1987; Coulson et al. 1989). At various steps in the progression, additional technologies have been introduced to extend or enhance the utility of a particular type of system. Examples of the supplemental technologies include simulation models, expert systems, intelligent data base management systems, GISs, and IGISs (Fig. 7.1). It is noteworthy that these computer applications can be used as "stand-alone" systems. In a general sense, the evolution of complexity in the systems parallels the development of the six levels of comprehension identified by Bleecker (1987). These systems also reflect a progression of innovation associated with technical elements of computer science and engineering, e.g., hardware, languages, operating systems, communication, and networking. We are particularly interested in GISs, AI, and the relation of these two technologies. The significant message embedded in Fig. 7.1 is that the concept and methodology of the IGIS, described in this chapter, developed as a function of the evolution of computer science technology and a human need for more sophisticated control over problem solving and decision making for natural resource management.

7.2.1 GISs

A GIS is a computerized mapping system for capture, storage, management, analysis, and display of spatial and descriptive data. Burrough (1988) provides a comprehensive review of GIS principles and uses for land resource assessment. The basic elements of a GIS are illustrated in Fig. 7.2. All GISs have four fundamental components: (1) a data input subsystem to collect and process spatial and descriptive data from maps, aerial photographs, remote sensors, and other sources; (2) a database management subsystem to store and retrieve data; (3) an analysis subsystem to interpret within and among data themes; and (4) a reporting subsystem to display maps and reports. In addition, most GISs have a facility for communication with independent data bases, mathematical models, evaluation

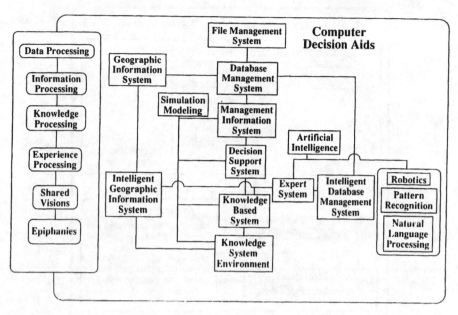

Figure 7.1. Relation of IGIS to other types of computer decision aids. Note that the decision aids increase in complexity from file management systems to the knowledge system environment. The progression in complexity parallels the levels of human comprehension (Knowledge Engineering Laboratory, Texas A&M University, and Laboratory for Artificial Intelligence Applications, Pennsylvania State University).

functions, and statistical analysis systems (Parker 1988). This capacity for communication permits blending of AI practices with GIS functions.

Use of the GIS as a tool for natural resource management has been greatly expedited through several recent technological advancements (Croswell and Clark 1988; Haefner 1987; Iverson 1988; Johnston 1987; Johnston et al. 1988; Walsh 1985; Welch et al. 1988). First, GIS ports to microcomputers have substantially reduced hardware costs associated with the development and delivery of applications. The GIS installation on microcomputers is possible because of increased memory, storage space, and speed of operation now available on this class of machine. Second, peripheral devices for data input (digitizing tablets, scanning devices, global positioning instrumentation) and output (printers, painters) have been refined. Third, map and image-processing systems for data base assembly and editing have been greatly improved. Fourth, use of GIS software has been simplified while functional utility has been expanded. Fifth, networking and communication software has been developed that permits centralized data storage while still allowing for a distributed data processing environment (Fig. 7.3).

A variety of public domain (e.g., GRASS and MOSS) and proprietary (e.g., ARC/INFO and ERDAS) GISs are now available for use on microcomputers. In general, each system was developed for a specific type of application or need, e.g.,

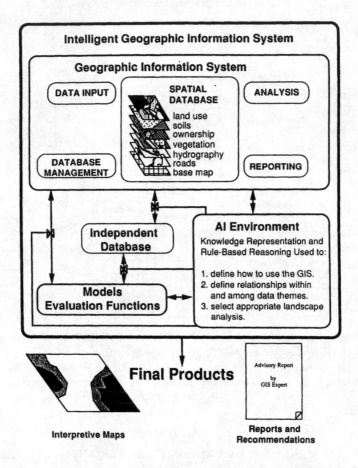

Figure 7.2 Schematic of the basic elements of a GIS. All GISs have facilities for data input, data base management, analysis, and reporting. An IGIS includes an artificial intelligence environment for knowledge representation and rule-based reasoning (Knowledge Engineering Laboratory, Texas A&M University).

automated cartography or thematic mapping versus landscape analysis. There are two fundamental GIS designs: vector-based and raster (cell)-based. These two systems differ in the way in which spatially referenced data are represented (Fig. 7.4). In a vector-based GIS, map data are represented as coordinates for points, lines, or polygons. The vector data structure is an efficient way to store spatially referenced data in a computer. This approach is useful for the management and display of discrete classes of map data (Fig. 7.4). However, it is not suitable for analysis and modeling where spatially contiguous values for data are necessary. In a raster-based system, numeric values for map data are represented in a grid containing rows and columns of cells of a prescribed size. Each cell corresponds to a fixed area in real space. Both categorical and continuous map information can

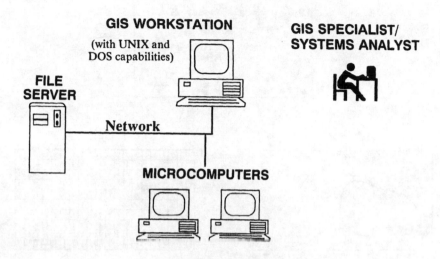

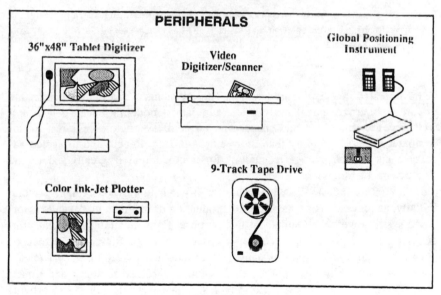

Figure 7.3. Elements of a GIS laboratory. In addition to the GIS workstation, peripheral devices include a tablet digitizer, video digitizer/scanner, painter, tape drive, and perhaps global positioning instrumentation (Knowledge Engineering Laboratory, Texas A&M University).

Base Map

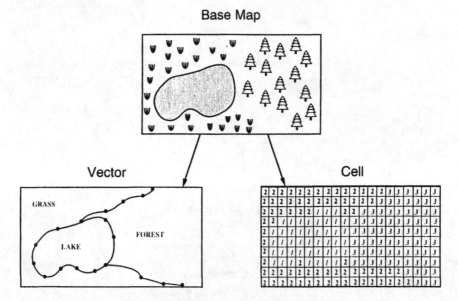

Figure 7.4. Illustration of the two principal methods for representing map information in a GIS: vector-based representation and raster (cell)-based representation (modified from Kvamme and Kohler 1989).

be incorporated in a raster-based system. These systems are particularly amenable to various types of spatial analysis and simulation modeling because each unit (cell) has the same size and shape. The raster data structure requires greater computer storage capability than the vector approach. Recent and projected advancements in disk storage technology for microcomputers greatly reduce the significance of this issue.

Both vector-and raster-based GISs have utility in natural resource management. Initially, major emphasis was placed on production of thematic maps, and vector-based systems were well suited for this purpose. Formalization of general principles of landscape ecology (Forman and Godron 1986; Risser 1987) focused attention on the need for spatial analysis and modeling. Raster-based systems are particularly suited for these purposes. Burrough (1988) and Kvamme and Kohler (1988) examine and contrast vector- and raster-based systems; they also provide recommendations for appropriate uses for each type of data structure. Logical partnerships between landscape ecologists, natural resource managers, geographers, and GIS developers are just beginning to be recognized and defined. No single approach to a GIS will satisfy the needs of all users, but it is reasonable to anticipate enhanced communication among the various sytems as networking procedures are developed and refined.

7.2.2 AI

AI is an interdisciplinary subject area that draws from several different academic specialties, including computer science, engineering, linguistics, mathematics, physics, and psychology. The issues addressed by AI include (1) definition and classification of principles of intelligent behavior; (2) design and development of computer systems (hardware and software) capable of mimicking intelligent behavior; and (3) use of such systems to solve problems of perception, analysis, and adaptation (Coulson et al. 1987). There is considerable interest in AI applications for business, medicine, geology, and natural resource management.

The areas of basic research in AI include inference, knowledge representation, search, and pattern matching. The major topics for AI application are generally defined to include expert systems, vision and image processing, robotics, and natural language processing. Intelligent data base management and IGISs are recent additions to the list. In Fig. 7.1 we have indicated that the simplest AI/GIS integration involves expert system concepts (Robinson and Frank 1987). In particular, the AI component of the IGIS approach described below centers on principles associated with rule-based reasoning. Note also from Fig. 7.1 that an expert system belongs to a broader category of computer decision aids known as a knowledge-based system. Furthermore, the essential distinction between a knowledge-based system and a knowledge system environment is the addition of a facility that permits use of spatially referenced data, i.e., the IGIS (Coulson et al. 1989).

Before examining the blend of AI with the GIS, it is necessary to have a general understanding of expert systems and rule-based reasoning. An expert system is a computer program designed to simulate or emulate the problem-solving behavior of a human who is an expert on a subject. General texts on AI and expert systems include Charniak and McDermott (1987), Firebaugh (1988), and Luger and Stubblefield (1989). There are several different types of expert systems. We are interested in rule-based systems in which knowledge is represented in the form of rules expressed as "if-then" statements. Rules are "antecedent-consequent" clauses; e.g., if there is evidence that A and B are true, then conclude there is evidence that C is true. The rules form the knowledge base, and they can be modified or changed as needed. The knowledge base is a separate element of the expert system. The "control program" for a rule-based expert system is termed an "inference engine." This program is the supervisor of the system's "reasoning process." Two principal control strategies are used in an expert system to chain a set of rules together for solution of a problem: forward chaining (data-driven) and backward chaining (goal-driven). The essential distinctions between these two procedures are illustrated in Fig. 7.5. In the forward chaining procedure, data that are known (in the example, "A") drive the inference from left to right in the rules. The rules are chained together to deduce a conclusion ("C" in the example). In the backward chaining procedure, the system starts with the statement of a desired

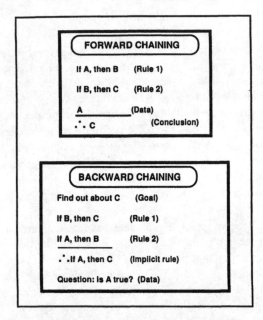

Figure 7.5. Illustration of the two principal control strategies used to chain a set of rules together for problem solving in a rule-based expert system: forward chaining (data-driven) and backward chaining (goal-driven) (modified from Coulson and Saunders 1987).

goal and works "backward" through inference rules, i.e., from right to left, to find the data that establish the goal (Buchanan and Shortliffe 1985). Both procedures have utility in natural resource management.

Firebaugh (1988) reviews in detail a spectrum of both public domain and proprietary AI software systems. Included in the discussion are the various types of appropriate languages (C, LISP, Pascal, etc.), knowledge engineering environments (KEE, ART, Knowledge Craft, etc.), and tools (M.1, EXSYS, CLIPS, etc.). In the following section we examine how rule-based reasoning can be used with a GIS to produce an IGIS.

7.3 Elements of an IGIS

Research on GISs and AI can be viewed from both basic and applied perspectives. Goals of research range from providing new understanding of a fundamental scientific issue, such as knowledge representation, to discovering innovative ways to use existing tools, such as developing a decision aid for natural resource management. Emphasis in this section is placed on issues related to natural resource management rather than the fundamental research areas of GISs or AI. The following topics are examined: (1) specific ways that rule-based reasoning can be used to improve GIS functionality, (2) methodology for developing·an IGIS application, and (3) benefits derived from blending the two technologies.

7.3.1 Rule-Based Reasoning and GIS Functionality

An IGIS is a computer application that consists of a conventional GIS with the addition of an AI environment for knowledge representation and rule-based reasoning (Fig. 7.2).[2] The AI environment can enhance the functionality of a GIS in three principal ways: (1) automation of interpreting relations within and among landscape data themes, (2) selection of appropriate analytical solutions for natural resource management and landscape ecological problems, and (3) guidance in use of the system.

7.3.1.1 Interpreting Relations Within and Among Landscape Data Themes

The use of rule-based reasoning to automate the interpretation of information represented in multiple data themes has been described by Coulson et al. (1987, 1988) and Graham et al. (1989). The goal of the approach is to formulate a map or develop an advisory report that illustrates known relations that exist for different spatially referenced data themes. For example, a forester might be charged with the task of evaluating the impact of a proposed timber harvest on a resident population of elk (*Cervus canadensis*Erxleben). Pertinent facts about elk (defined through research by wildlife biologists) include the following: they prefer certain plant species for food; they spend considerable time foraging in areas where the favored food plants occur; they seek refuge in a forest type different from the foraging area; they travel between forage and refuge sites; and they are game sought after by hunters. In judging the probable impact of the timber harvest, the forester might consider the following facts ("rules"): if the food resources are depleted below a critical biomass level, then elk will starve; if the refuge area is reduced below a critical size, then mortality due to exposure will occur; if the elk have to travel across the harvest site in moving between the foraging area and the refuge area, then they will be exposed to hunters; if hard-surface roads are near the harvest site, then a larger than normal number of hunters will come to the area. GIS data themes for vegetation types, roads, and elevation provide values for the variables needed for the impact assessment. The vegetation theme contains information on the location, biomass, and size of foraging and refuge areas relative to the harvest site. The elevation theme indicates probable routes the elk would take in traveling between foraging and refuge areas. The road theme identifies access routes and indicates how difficult it would be for hunters to approach the site.

In this example, the expertise (judgment) of the wildlife biologist is used to

---

[2]In this discussion, we represent the IGIS as a computer application that consists of a conventional GIS with the addition of an AI environment for knowledge representation and rule-based reasoning. However, there is another type of IGIS application that blends AI, GIS, and object-oriented programming techniques to model animal behavior. The general approach is described in Coulson et al. (1987) and discussed in detail by Saarenmaa et al. (1988). Eventually the approach will be useful in wildlife and fisheries management.

develop a rule base dealing with knowledge of elk populations and timber-harvesting practices; GIS data themes supply information about the specific forest landscape being examined; the logical relations among the rules in the data base are identified by the inference engine associated with the AI environment, and the final product of the exercise is an advisory report. A skilled GIS technician, wildlife biologist, and forester working together could develop the same report without using the expert system approach described above. The AI component permits encoding of the knowledge of these specialists into a computer program that can be used by practitioners to solve a complex problem in natural resource management.

7.3.1.2 Selection of Appropriate Analytical Solutions and Data Base Preparation

Formalization of general principles of landscape ecology has provided a framework needed for classifying analytical procedures based on a standard nomenclature (Risser 1987; Toth 1988). However, the nomenclature of landscape ecology, natural resource management, and geography is often quite different, and logical relationships among the disciplines are not always evident. For example, most natural resource managers, at the present time, do not view landscape structure in terms of concepts associated with matrices, patches, corridors, and networks. Certain of the GISs have well-developed functions suitable for detailed landscape analyses, in addition to procedures for description. However, these functions have not been examined and classified as general principles associated with the different academic specialties, i.e., there is little cross-referencing between the disciplines. For example, Forman and Godron (1986) and Turner (1987) describe analytical procedures used by ecologists to examine landscape heterogeneity. The relation of these procedures to analytical functions explained in a user's guide for a GIS would not necessarily be obvious. Rule-based reasoning is an approach that can be used to organize analytical and descriptive procedures in the context of landscape ecology, natural resource management, and geography. This utility would serve to guide selection of methods appropriate for a specific task.

Rule-based reasoning also can be used to guide data base preparation. In this application, the spatially referenced data are ordered for subsequent use in analyses associated with the GIS or with models and evaluation functions located peripherally to the system. Examples of these approaches include Band (1989), Coughlan and Running (1989), and DeMers (1989).

7.3.1.3 Simplification of GIS Use

GISs are complex computer programs. Proficient application in natural resource management and landscape ecology involves a commitment to training and practice by the user. None of the GISs would be considered "user friendly" by a human factors engineer. One obvious improvement in existing GISs would be to apply rule-based reasoning to guide operation of the user interface. Furthermore, since

GISs are inherently object oriented, they could be structured by using a combination of nonprocedural programming and rule-based reasoning techniques (Coulson et al. 1987). The goals of such an approach are to guide the user and to simplify application of the GIS. New practitioners from the ecological and natural resource management communities will provide incentive for developers to enhance ease of use and improve utilities associated with GISs.

7.3.2 Methodology for Development of an IGIS Application

Several steps are involved in the development of an IGIS application for natural resource management. We consider some of the important issues associated with (1) system specification, (2) problem definition, and (3) knowledge base development. In the following discussions we illustrate each of these steps using the infestation forecasting system for the southern pine beetle, *Dendroctonus frontalis* Zimn. (Coleoptera:Scolytidae), (Coulson et al. 1988) as an example. This system uses information on tree species, forest stand structure, lightning distribution and abundance, landforms, and past insect activity to forecast, at a regional landscape scale, where new infestations will occur.

7.3.2.1 System Specification

In developing an IGIS it is necessary to identify a class and type of computer for delivery of the application, a GIS, and a specific expert system tool for rule processing. There are a variety of choices for each of the elements. In some cases, agency or corporate policy and standards may limit the options available. In other cases, personal preference or cost may guide selection. Because applications that blend GISs with AI are just beginning to be developed, the design philosophy for the IGIS should emphasize flexibility, compatibility, and peripheral device independence. Some of the important considerations are (1) a microcomputer delivery system, (2) a GIS with good analytical utilities and communications capabilities that can be ported to a variety of hardware platforms, and (3) a well-supported expert system tool with communication capabilities. It may also be important to have access to source codes for the GIS and expert system tool. The southern pine beetle infestation forecasting system was developed for delivery on a SUN 386i computer using the GRASS (Geographic Resource Analysis Support System, developed by the U.S. Army Construction Engineering Research Laboratory) as a GIS, and CLIPS ("C" language production system, developed by Artificial Intelligence Section, NASA/Johnson Space Center) as an expert system tool. Both GRASS and CLIPS are public domain systems with the desirable characteristics enumerated above.

7.3.2.2 Problem Definition

Within an agency or organization, there is normally a hierarchy of users each with different information needs for a particular management problem. It is therefore necessary to identify at the onset the intended users of the application. Once this task is completed, a specialist, known as a knowledge engineer, works with

domain experts to (1) define the bounds for a specific problem, (2) identify
relevant types of information, and (3) organize the information to solve a problem
or make a decision. The process is known as knowledge engineering, and it
consists of a formalized set of techniques and tools. The initial product of the
process is a function model that can be used to develop diagrams, known as
dependency networks, that illustrate problem-solving logic.

For the southern pine beetle forecasting system, the goal was to identify forest
stands by their hazard, or relative vulnerability to herbivory by the insect. Previous
research by a variety of entomologists and foresters indicated that hazard was a
function of the interaction of several variables: (1) vegetation cover (representing
the presence of suitable host type for the insect), (2) forest stand conditions
(composition, age, and density), (3) lightning centers (classed by season of the
year), (4) background populations of insects (derived from aerial survey data on
infestations), and (5) landform.

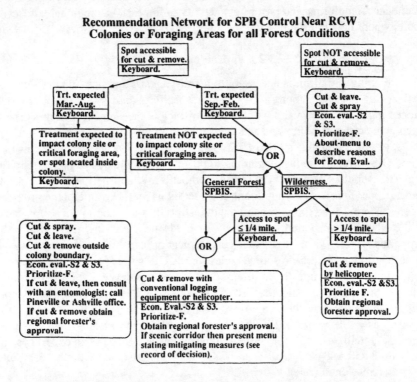

Figure 7.6. A dependency network illustrating the problem-solving logic used for selecting
a suppression tactic for the southern pine beetle (SPB). The selection process is complicated
by the presence of an endangered species, the red-cockaded woodpecker (RCW), *Picoides
borealis*. This dependency network is part of the southern pine beetle integrated expert
system (ISPBEX) developed for the USDA Forest Service, Forest Pest Management,
Region 8.

Figure 7.6 is an example of a dependency network developed from knowledge engineering on a subject related to the forecasting system. It illustrates the problem-solving logic used in the southern pine beetle integrated expert system (ISBPEX)[3] to select a suppression tactic for a pest insect. This particular example involves the consideration of an endangered species, the red-cockaded woodpecker (RCW), *Picoides borealis* (Vieillot).

7.3.2.3 Knowledge Base Development

Following development of the dependency network (or networks), the next step is to specify the knowledge base that will be used for problem solving or decision making. The knowledge representation scheme emphasized here centers on rules. There are three principal rule types: (1) data base, (2) map, and (3) heuristic (Fig. 7.7). Data base rules are used to evaluate numerical information derived from simulation models or stored in on-line data bases; i.e., they deal with nonmap

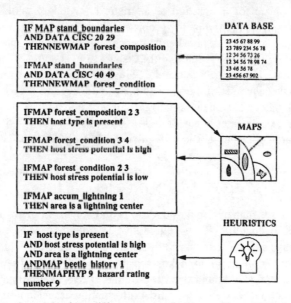

Figure 7.7. Illustration of the three different types of rules used in an IGIS: data-base rules, map rules, and heuristic rules (Knowledge Engineering Laboratory, Texas A&M University).

[3]ISPBEX is a computer application developed by the Knowledge Engineering Laboratory and Knowledge Based Systems Laboratory at Texas A&M University for the USDA Forest Service, Forest Pest Management, Region 8. The application concerns integrated pest management decision making and problem solving applied at the landscape level of organization.

information. For example, in the southern pine beetle forecasting system, one of the important data types is forest condition. The USDA Forest Service in the southern United States maintains a stand-level data base for each national forest, CISC (continuous inventory of stand conditions). This data base contains numerical information about stands, e.g., age, basal area, and site index. Data-base rules are used to modify an existing map (Fig. 7.7). Map rules, the second type, are used to evaluate categorical information associated with an on-line map. For example, since the southern pine beetle forecasting system was developed on GRASS, a raster-based GIS, map data are stored as a matrix. Each grid cell has associated with it a numeric value that represents a specified category, e.g., loblolly pine vegetation = category 1, and shortleaf pine = category 2. Map rules are used to construct new maps or to assign values to variables in a map (Fig. 7.7). Heuristic rules, the third type, are used to evaluate knowledge of domain experts. Information for the rules can be based on map categories or nonmap numerical data (Fig. 7.7).

An IGIS application would contain all three types of rules. The significant message regarding rule types and rule-based reasoning is that the approach provides a mechanism for integration of heuristic knowledge, spatially referenced data, and tabular (nonmap) data for problem solving and decision making. In the specific case of the southern pine beetle forecasting system, the rule base required to integrate knowledge for predicting the distribution, abundance, and location of infestations of the insects is quite simple. Figure 7.8 illustrates the projected hazard zones versus the actual distribution of southern pine beetle sites for 1983. Results are presented in three hazard classes: high, medium, and low. Within these classes, 8 (1.9%) of the beetle sites occurred in low hazard sites, 189 (44.5%) occurred in medium hazard cells, and 228 (53.6%) occurred in high hazard cells.

7.3.3 Benefits Derived from an IGIS

A natural resource manager makes decisions and solves problems by integrating various types of quantitative and qualitative information. Most problems dealt with do not have exact solutions, and final actions taken are often judgmental. The GIS component is the tool that makes possible the analysis and representation of quantitative spatial and tabular data. It provides access to simulation models and evaluation functions as well as technical information associated with data bases. However, qualitative information, represented as expert opinion, has traditionally guided problem solving and decision making at operational levels. The methods used for analysis and description of quantitative data are not suited for heuristic knowledge. The AI component provides the means for using qualitative information. In particular, it facilitates (1) incorporation of qualitative domain-specific information into the natural resource management process, (2) treatment of problems that do not respond to algorithmic solutions, (3) use of metalevel knowledge to effect more sophisticated control of problem-solving strategies, and (4) merging different representations of knowledge. The IGIS application is a particularly

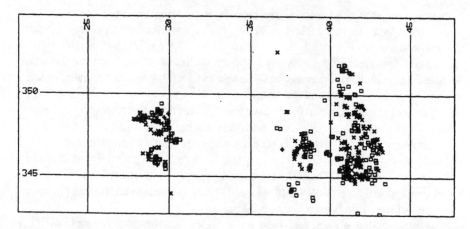

+ - Low Hazard

x - Medium Hazard

☐ - High Hazard

Figure 7.8. Projected hazard zones vs. the actual distribution of southern pine beetle sites for 1983. Results are presented in three hazard classes: high, medium, and low. Within these classes, 8 (1.9%) of the beetle sites occurred in low hazard sites, 189 (44.5%) occurred in medium hazard cells, and 228 (53.6%) occurred in high hazard cells.

significant tool because it blends methodologies for representation, analysis, and interpretation of quantitative, spatially referenced data with heuristic knowledge of experts.

7.4 Application of an IGIS in Natural Resource Management Problem Solving and Decision Making

In the example that follows, we illustrate how an IGIS can be used to investigate a landscape management problem concerning pesticide use and protection of endangered species. The intent of the example is to demonstrate the utility of an IGIS as both a management decision aid and a research tool. The example is a typical resource management and landscape ecology problem in the sense that resolution involves the use of qualitative and quantitative data.

In 1987 the U.S. Environmental Protection Agency (EPA) initiated the Endangered Species Protection Program. The goal was to protect endangered species from further adverse effects from pesticides. The program was a response to the Endangered Species Act (ESA), which provides legal protection for endangered species, and to the Federal Insecticide, Fungicide, and Rodenticide Act (FIFRA), which provides for registration and controlled use of pesticides. The strategy was to regulate pesticide use through label restrictions in areas (states and counties) containing endangered species. The program would change regulation and use of pesticides in agriculture, forestry, human and animal disease vector control, and aquatic and right-of-way vegetation control. The ESA was a significant legislative initiative that contained provisions to ensure the intent of the law is not compromised for financial or other motives. Numerous federal, state, and private agencies and organizations are involved in implementation of the act.

The ecological impact of the Endangered Species Protection Program cannot be easily assessed using traditional analytical tools, because a great deal of the information available on the subject exists as the qualitative knowledge of experts. Decision making relative to the program is largely judgmental and therefore subject to contention. The IGIS methodology provides a means for organization, integration, and interpretation of qualitative and quantitative information on the impact of pesticide use on endangered species.

Assuming that a GIS data base for a region has been developed, an IGIS approach can be used to evaluate the Endangered Species Protection Program. The analysis can be structured to include four parts: (1) definition of the probable range of the endangered species, (2) delineation of an enforcement zone to protect the endangered species, (3) optimization of pesticide application relative to natural history of the endangered species, and (4) enumeration of alternative practices to pesticide use (Fig. 7.9). Investigation of each of these activities involves establishing a specific goal, developing a function model, constructing dependency networks, and formulating a rule base, as described above. Each of the three types of rules is used.

The first step in the evaluation is to define the probable range of the specific endangered plant or animal of interest. Operational data available on the location of many endangered species are based on cultural (roads, power transmission lines, etc.) and physical (rivers, lakes, etc.) features of the landscape, not on actual habitat conditions needed by the organism. However, the natural history of most endangered species has been studied in sufficient detail, at least at the generic taxonomic level, to permit construction of a rule base that describes actual habitat requirements. Definition of the range of the organism then becomes a matter of searching the GIS data themes to locate the habitat with the defined set of characteristics for a particular region. Standard GIS commands are used for this purpose. Coincidence of endangered species habitat and location of pesticide use can be identified simply by overlaying the two data themes (Fig. 7.9).

Following identification of locations where pesticide use is a hazard to the endangered species, the second step in the evaluation is to define a zone of sufficient size around the habitat to protect the organism. Therefore, the goal is to delineate an enforcement zone where pesticide use will not be permitted. The rule base developed for this purpose would integrate two types of information: essential habitat requirements for the organism and pesticide use patterns within the landscape (Fig. 7.9). This evaluation is important because it will ensure protection of the endangered species and minimize the landscape area removed from agricultural production, ranching, and forestry.

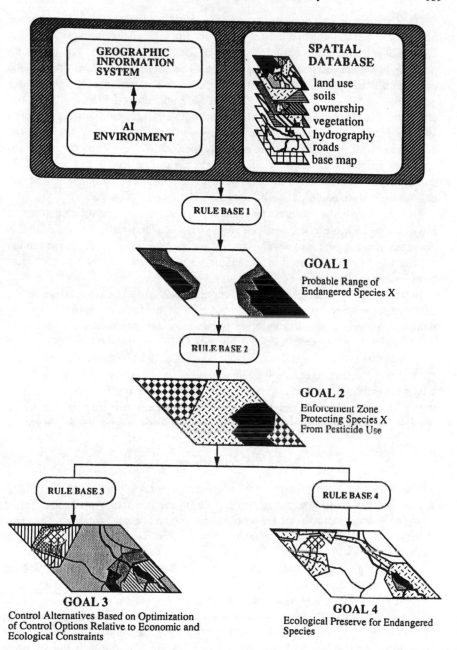

Figure 7.9. Illustration of an intelligent geographic information procedure for evaluation of the Endangered Species Protection Program. Note that there are four goals and four rule bases (Knowledge Engineering Laboratory, Texas A&M University).

Once an enforcement zone has been established, it may be possible to adjust the boundary by examining habitat requirements for the endangered species in the context of pesticide use pattern, toxicology of the pesticide, and the actual cultural practice that requires pesticide application. The goal of the third step in the evaluation is to optimize land use for the protection of endangered species. An additional rule base is required.. Knowledge engineering for this goal would consider specific details of the interaction between pesticide and endangered species. Is the pesticide toxic to the endangered species? Does it persist in the environment? Is it present during sensitive periods of development of the endangered species? For example, if the pesticide use pattern is not temporarily coincidental with activities of the endangered species, the boundaries of the enforcement zone could be modified to reflect this fact (Fig. 7.9).

The final step in the evaluation is included to address the circumstance in which it will not be possible to use pesticides and protect endangered species. Given this circumstance, the goal is to develop an alternative plan. Knowledge defined in the previous steps could be used in formulating a new rule base that also considers principles associated with ecological preserves.

In this example we have considered four approaches to a complicated landscape and natural resource management problem. The fundamental aim of the Endangered Species Protection Program is clear. Achieving the goal of providing protection for endangered species from pesticide use involves an analysis of land-use patterns relative to habitat conditions. The final judgments made by the natural resource manager involve the consideration of scientific principles, legislative mandates, and landscape use patterns. The IGIS approach is a methodology that can be used to solve a complex landscape-level problem for which spatially referenced data, tabular data, and heuristic knowledge must be integrated.

7.5 Summary

In the preceding discussion we examined the concept and practice of the IGIS as a tool for natural resource management. An IGIS can be used to integrate different representations of knowledge for problem solving and decision making. Development of IGIS applications is an interdisciplinary activity that requires technical expertise in GISs, AI, landscape ecology, geography, and management practices. The technology is particularly useful for interpreting relations within and among data themes, guiding use of a GIS, and selecting appropriate landscape analyses. We emphasized how rule-based reasoning and expert system concepts can be used with GIS software to solve problems in landscape management. Three different types of rules are used to develop a knowledge base for an IGIS application: data base rules, map rules, and heuristic rules. This simple approach is extremely significant because it permits integration of both quantitative information and qualitative knowledge of experts for problem solving and decision making. Uses of AI methodologies, standard computer science techniques, and GISs are just beginning to be identified.

Acknowledgments

We acknowledge Ms. A. M. Bunting for technical assistance in the preparation of the manuscript. We are especially grateful to Dr. Pamela C. Case (USDA Forest Service), Dr. W. F. Limp (Arkansas Archaeological Survey), and the two anonymous reviewers who provided thorough and candid critique of the manuscript. We thank Dr. V. B. Robinson, Associate Editor of AI Applications for Natural Resource Management, for providing access to prepublication manuscripts from a forthcoming edition of the journal that deals specifically with AI applications and GISs. Developmental projects in the Knowledge Engineering Laboratory dealing with IGISs were supported by the USDA Forest Service, Forest Pest Management, Region 8, and the USDA Competitive Grants Program G6961. The opinions expressed herein are those of the authors.

References

Band, L.E. 1989. Automating topographic and ecounit extraction from mountainous forested areas. *AI Applications for Natural Resource Management* 3:1–11.

Bleecker, S.E. 1987. Rethinking how we work: the office of the future. *The Futurist*, July-Aug., pp. 34–40.

Buchanan, B.G., and Shortliffe, E.H. 1985. *Rule-Based Expert Systems*. Reading, Mass.: Addison-Wesley.

Burrough, P.A. 1988. *Principles of Geographical Information Systems for Land Resources Assessment*. Oxford: Clarendon Press.

Charniak, E., and McDermott, D. 1987. *Artificial Intelligence*. Reading, Mass.: Addison-Wesley.

Coughlan, J.C., and Running, S.W. 1989. An expert system to aggregate biophysical attributes of a forested landscape within a geographic information system. *AI Applications for Natural Resource Management* 3:35–43.

Coulson, R.N.; Folse, L.J.; and Loh, D.K. 1987. Artificial intelligence and natural resource management. *Science* 237:262–67.

Coulson, R.N.; Graham, L.A.; and Lovelady, C.N. 1988. Intelligent geographic information system for predicting the distribution, abundance, and location of southern pine beetle infestations in forest landscapes. In *Integrated Control of Scolytid Bark Beetles*, eds. T.L. Payne and H. Saarenmaa. Proceedings, IUFRO Working Party and XVII International Congress of Entomology Symposium Vancouver, British Columbia, Canada, July 4, 1988.

Coulson, R.N., and Saunders, M.C. 1987. Computer-assisted decision-making as applied to entomology. *Annual Review of Entomology* 32:415–37.

Coulson, R.N.; Saunders, M.C.; Loh, D.K.; Oliveria, F.L.; Drummond, D.; Barry, P.J.; and Swain, K.M. 1989. Knowledge system environment for integrated pest management in forest landscapes: the southern pine beetle. *Bulletin of Entomological Society of America* 35:26–32.

Croswell, P.L., and Clark, S.R. 1988. Trends in automated mapping and geographic information systems hardware. *Photogrammetric Engineering and Remote Sensing* 54:1571–76.

DeMers, M.N. 1989. Knowledge acquisition for GIS automation of the SCS LESA model: An empirical study. *AI Applications for Natural Resource Management* 3:12–22.

Firebaugh, M.W. 1988. *Artificial Intelligence: A Knowledge-Based Approach*. Boston: Boyd and Fraser.

Forman, R.T.T., and Godron, M. 1986. *Landscape Ecology*. New York: Wiley.

Graham, L.A.; Coulson, R.N.; and Lovelady, C.N. 1989. Intelligent geographic information systems. In *Resource Technology 88, Proceedings*, ed. G.J. Buhyoff. Falls Church, Va.: American Society of Photogrammetric and Remote Sensing.

Haefner, H. 1987. Assessment and monitoring of renewable natural resources: Concepts and applications. *Applied Geography* 7:7–15.

Iverson, L.R. 1988. Land-use changes in Illinois, USA: the influence of landscape attributes on current and historic land use. *Landscape Ecology* 2:45–61.

Johnston, C.A.; Datenbeck, N.E.; Bonde, J.P.; and Niemi, G.J. 1988. Geographic information systems for cumulative impact assessment. *Photogrammetric Engineering and Remote Sensing* 54:1609–15.

Johnston, K.M. 1987. Natural resource modeling in the geographic information system environment. *Photogrammetric Engineering and Remote Sensing* 53:1411–15.

Kvamme, K.L., and Kohler, T.A. 1988. Geographic information systems: technical aids for data collection, analysis, and display. In *Quantifying the Present and Predicting the Past: Theory, Method, and Application of Archaeological Predictive Modeling*, eds. W.J. Judge and L. Sebastian. Denver, Colo.: U.S. Department of the Interior, Bureau of Land Management Service Center.

Luger, G.F., and Stubblefield, W.A. 1989. *Artificial Intelligence and the Design of Expert Systems*. Redwood City, Calif.: Bengamin/Commings.

Parker, H.D. 1988. The unique qualities of a geographic information system: A commentary. *Photogrammetric Engineering and Remote Sensing* 54: 1547–49.

Risser, P.G. 1987. Landscape ecology: state of the art. In *Landscape Heterogeneity and Disturbance*, ed. M.G. Turner. New York: Springer-Verlag.

Robinson, V.B., and Frank, A.U. 1987. Expert systems for geographic information systems. *Photogrammetric Engineering and Remote Sensing* 53:1435–42.

Saarenmaa, H.; Stone, N.D.; Folse, L.J.; Packard, J.M.; Grant, W.E.; Makela, M.E.; and Coulson, R.N. 1988. An artificial intelligence modelling approach to simulating animal/habitat interactions. *Ecological Modelling* 44:125–41.

Toth, R.E. 1988. Theory and language in landscape analysis, planning, and evaluation. *Landscape Ecology* 1:193–201.

Turner, M.G., ed. 1987. *Landscape Heterogeneity and Disturbance*. New York: Springer-Verlag.

Walsh, S.J. 1985. Geographic information systems for natural resource management. *Journal of Soil and Water Conservation* 40:202–5.

Welch, R.; Remillard, M.; and Slack, R. 1988. Remote sensing and geographic information system techniques for aquatic resource evaluation. *Photogrammetric Engineering and Remote Sensing* 54:177–85.

8. Methods for Analyzing Temporal Changes in Landscape Pattern[1]

Christopher P. Dunn, David M. Sharpe, Glenn R. Guntenspergen,
Forrest Stearns and Zhao Yang

8.1 Introduction

Landscapes, like other ecological units of study, are dynamic in structure, function, and spatial pattern. Just as communities are composed of species populations, landscapes are assemblages of habitats, communities, and land use types. The spatial configuration of these landscape elements can be attributed to a combination of environmental correlates and human forces (Forman and Godron 1986). In some cases the disturbance regime is dominated by natural disturbance, e.g., fire (Heinselman 1981). In others, land use practices predominate (Sharpe et al. 1987), so that changes in a landscape are due to changes in management practices and the social, political, and economic forces controlling land use (di Castri and Hadley 1988). The interaction among natural and modified elements creates complex patterns of change (e.g., Peterjohn and Correll 1984; di Castri and Hadley 1988). Temporal changes in most landscape patterns are dictated by a combination of natural and human influences, each of which operates at different spatial and temporal scales, so that temporal changes or rates of change may themselves by "patchy." Within a single landscape, several types of patches may be experiencing

[1]This research was supported by the National Science Foundation, Ecology Program, through Grant No. DEB–8214792 to University of Wisconsin-Milwaukee, and Grant No. DEB–8214702 to Southern Illinois University at Carbondale.

changes at different temporal scales. For example, woodlots in an agricultural matrix may be undergoing secondary succession, even though the gross patterns of land uses are not changing at a noticeable rate. Finally, the relevance of patch structure and landscape change to particular biota depends upon factors such as habitat requirements, dispersal capability, and longevity; perception of landscape pattern and its changes are in the eye of the beholder.

This chapter focuses on issues that relate to temporal change in landscape patterns and the ecologic impact of such changes. We review ways to identify changes in landscape patterns; methodologies that have been used, including data sources and analytic procedures; and insights about ecological dynamics of landscapes that have resulted. We illustrate these points using analyses of Cadiz Township, Wisconsin. For a discussion of specific models of landscape change, see Baker (1989).

8.2 Analyzing Change in Landscape Pattern

Landscape pattern refers to the number, size, and juxtaposition of landscape elements or patches (e.g., land use and land cover types), which are important contributors to overall landscape pattern and to interpretation of ecological processes (Gardner et al. 1987; O'Neill et al. 1988). Patch size can influence floral and faunal composition and richness (chapters in Burgess and Sharpe 1981; Ambuel and Temple 1983). For example, small patches of forest tend to have a greater proportion of edge to interior than larger patches have and thus are more likely to harbor exotic or weedy species (Levenson 1981; Dunn and Loehle 1988). Patch size affects bird species composition (Galli et al. 1976; Lynch and Whigham 1984). Furthermore, smaller patch size might contribute to local and regional extinction of sensitive species adapted to interior conditions.

Patch shape (as measured by perimeter-to-area ratio or fractal dimension) can have important implications at the landscape scale. Patches of simple shape with straight boundaries might have more abrupt transitions (of biotic and abiotic features of the landscape) to the adjacent patch (e.g., forest to field). Complex patches (e.g., convoluted edge), by contrast, might have different sets of environmental conditions (and thus species) in close proximity to each other (Forman and Milne 1986).

Connections between like and unlike patches (e.g., hedgerows or proximity) can facilitate the movement of individuals and propagules across a landscape (Middleton and Merriam 1983; Johnson and Adkisson 1985). If connections are absent, distance between patches can determine the effectiveness of dispersal (Merriam 1988).

In sum, landscape pattern has important consequences regarding the effective dispersal of organisms (or propagules) among landscape elements and the spread of disturbance across the landscape (Turner and Bratton 1987; Turner et al. 1989). The landscape pattern at any given moment is a stage on which dynamic processes occur. The fact that landscapes are dynamic requires that time, or temporal changes, be considered in quantitative landscape studies.

8.2.1 Temporal Change in Landscapes

Temporal changes in a landscape could include changes in (1) patch number, (2) patch size, (3) number and type of corridors, (4) number and type of dispersal barriers (e.g., agricultural fields and roads), and (5) probability and spread of disturbance (Turner 1987b; Turner et al. 1989). The proximate consequences of changes in landscape pattern relate to changes in dispersal capabilities of organisms, among other factors. Some likely ultimate consequences would include changes (either positive or negative) in population viability and in genetic diversity.

The proximate consequences of temporal changes have been the major focus of temporal landscape studies. If species cannot move from patch to patch, the probability of local extinction can increase with time (Hamabata 1980; Merriam 1988). Populations of woodland fauna that are relegated to habitat fragments may be more susceptible to local extinction than populations in larger contiguous patches (Merriam 1988; Lynch and Whigham 1984). Many bird species respond negatively to smaller and/or more isolated habitat patches, even though total forest area may remain the same (Galli et al. 1976). At the scale of the patch itself, greater forest fragmentation can, with time, result in the degeneration of patches into a nonforested condition (Hill 1985).

Discerning the ultimate consequences of landscape changes, however, is considerably more challenging. In studying temporal changes, we need to be aware of their evolutionary/genetic, regional biodiversity, and management implications. The overriding issue in studying temporal changes is developing methodologies to understand the importance of landscape patterns, the changes in these patterns, and the mechanisms underlying these changes. At least five aspects or parameters of change are important in this regard.

First, in any study of change, some starting point must be established. The presettlement (i.e., pre-European) landscape, such as portions of North America (e.g., Auclair 1976; Moss and Hosking 1983; Whitney and Somerlot 1985; Whitney 1987; Sharpe et al. 1987), has been used as an initial condition. It is tempting to view the presettlement landscape normatively as a benchmark or as the standard for comparison with the current landscape. However, this viewpoint has two flaws: (1) landscape pattern (e.g., vegetation mosaics) and species composition would change even in the absence of human influence (Shugart and West 1980; Seagle and Shugart 1985); and (2) virtually all landscapes are, in fact, cultural landscapes that retain vestiges of past management practices, as well as current practices (Birks et al. 1988); so the issue of human impact on landscapes is one of degree, not kind. However, historical data on landscapes are often limited; therefore, descriptions and analyses of change inevitably will be incomplete. In many cases, data on the landscape before a recent wave of settlement are not available, and more recent data such as early aerial photography must substitute as the initial conditions, even though the history of the landscape is being truncated (Turner 1987a; Sharpe et al. 1986; Lowell and Astroth 1989).

Second, the trajectory (direction) of change could reveal a great deal about temporal changes. This approach has been used commonly in ordination studies of vegetation succession (Austin 1977; Grieg-Smith 1983; Dunn and Sharitz 1987). Although trajectories are not explicitly predictive tools or models, they do summarize historical changes. This temporal summary can be at several spatial scales. In vegetation succession, each sample (e.g., a stand) usually includes a value (cover and biomass) for each species. In a landscape, each sample (year) would include values (e.g., land area) for each type of element. A single landscape or several landscapes could be followed through time.

Third, rates of change in landscape patterns can have several implications. For example, very rapid changes might drive sensitive species to local or regional extinction, altering regional biodiversity by eliminating native species and encouraging the spread of exotic species (Hamabata 1980). Rates can be estimated from trajectories (i.e., greater distances between successive samples suggest more rapid change than shorter distances) or can be calculated as gains or losses during a specified time period. Sharpe et al. (1982; 1986) calculated rates of change (in terms of hectares per year per 100 km^2) for several land use/cover types in southeastern Wisconsin at ten-year intervals.

The fourth parameter of change is predictability; namely, is there a characteristic sequence of changes, either of the entire landscape(s) or of specific elements within landscapes? Agricultural development inevitably involves replacing native vegetation with crops and the infrastructure of an agricultural landscape. The interplay between cultural and natural factors, such as topography and pattern of soil fertility, may lead to landscapes with characteristic features and patch configurations (Johnson 1976; Bowen and Burgess 1981).

Finally, given that changes are occurring, are these changes persistent? In other words, does a change of a unit of land surface from one land cover type to another preclude or make less probable any future changes in that unit? In two southeastern Wisconsin landscapes, changes from natural vegetation to urban uses were never reversed, whereas an ongoing exchange existed between cropland, pasture, and natural vegetation in the rural landscape (Sharpe et al. 1982, 1987).

8.2.2 Characteristic Patterns of Temporal Change

In North America, the emphasis in studies of landscape change has been on supplanting natural vegetation, usually forest, first by agricultural land uses and associated cover types, then by urban land uses (Curtis 1956; Burgess and Sharpe 1981; Auclair 1976; Moss and Hosking 1983; Whitney and Somerlot 1985; Whitney 1987; Sharpe et al. 1987). A few studies have recognized that landscape change is ongoing and that return of native vegetation is characteristic of many landscapes (Johnson and Sharpe 1976; Turner 1987a; Turner and Ruscher 1988). That is, removal of natural vegetation is not always inexorable and irreversible.

The trends of deforestation in the vicinity of Harvard Forest in New England, the Piedmont of Georgia, and Cadiz Township in south-central Wisconsin (Fig. 8.1) underscore the ongoing nature of landscape dynamics. Each of these regions was rapidly deforested in conjunction with European occupation of North America. New England was settled and cleared of forest in the eighteenth and early

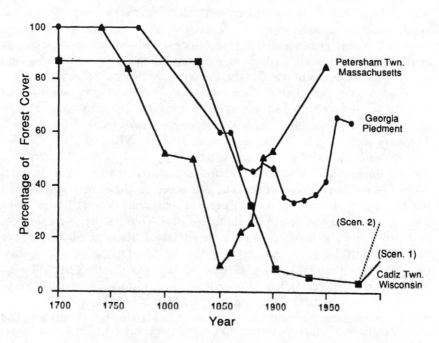

Figure 8.1. Deforestation and reforestation of landscapes in New England (Petersham Township, MA), and southeastern U.S. (Upper Piedmont, GA), and the actual and potential trends for the upper Midwest (Cadiz Township, Green County, WI).

nineteenth centuries. Reforestation in the nineteenth and early twentieth centuries was equally dramatic, and the region is now 60% forest (Irland 1982).

Likewise, the forests of the southeastern United States were cleared in the early 1800s. It is estimated that over 80% of the forests of the lower Piedmont of Georgia was cleared for agriculture at one time or another (Bond and Spillers 1935), although abandonment of agricultural land kept the percentage of woodland above 30% (Johnson and Sharpe 1976; Sharpe and Johnson 1981), even at the height of agricultural development. Forest has more than doubled in area in the past seventy years and now occupies 65% of the Piedmont, in spite of periods of renewed agricultural and urban development.

Thus, these landscapes in the eastern United States experienced rapid deforestation that lasted 50–100 years, followed by equally dramatic reforestation. However, the record of agricultural development in a representative landscape in the upper Midwest, Cadiz Township, Green County, Wisconsin, appears to contradict the conclusion that all landscapes are dynamic (Fig. 8.1). Deforestation was as rapid and agricultural development as extensive in southern Wisconsin as in New England and the Southeast (Curtis 1956; Burgess and Sharpe 1981). However, land management in the past several decades has been characterized by gradual clearing of residual forest for agriculture with little counteracting reforestation.

Does this indicate that some landscapes are stabilized over long periods of time? We think not. The specific impetus for land use change varies from one region and moment in history to another. However, episodes of wholesale landscape change in the United States frequently have followed changes in governmental land policy, the most recent of which is the Conservation Reserve Program (CRP) of the Food Security Act of 1985. Under the CRP, the federal government is empowered to contract with farm managers to retire 40–45 million acres (16–18 million hectares) of highly erodible land throughout the United States by 1990. These are lands with land capability classes of IV to VIII and with excessive average annual rates of erosion, as designated by the U.S. Soil Conservation Service. (In general, Class I soils have few limitations on agricultural land use, while Class VIII soils have limitations that preclude commercial agriculture.) Farm managers who enroll in the CRP agree to convert the land to a less intensive use, e.g., pasture, permanent grass, shrubs, or trees. Tree planting is encouraged; at least one-eighth of enrolled land is to be reforested. Thus, the conditions have been established for significant reallocation of rural land to less intensive purposes. Some authorities (Moulton and Dicks 1987) predict that the CRP will stimulate the largest reforestation program in the nation's history. The potential exists for a significant reversal in the trend toward deforestation of this region. Given this notion, we anticipate that the landscape in the upper Midwest is far from static, and we use Cadiz Township as an example to explore the implications for landscape patterns of enrolling land in the CRP.

8.2.3 Scenarios for Landscape Change

While it is impossible to predict which eligible lands in particular landscapes such as Cadiz Township will be enrolled in the CRP by farm managers, it is clear that the impacts of the CRP will accumulate over decades as vegetation responds to changing land use. We can extend the history of landscape change for Cadiz Township (shown in Fig. 1) by developing scenarios that express the potential impact of the CRP. We have analyzed two scenarios whose implications for landscape patterns are discussed below: (1) only the more highly erodible land (land capability classes VI–VIII) is enrolled, and (2) all eligible land (land capability classes IV–VIII) is enrolled. We have further assumed that only eligible land in blocks of 4 hectares (10 acres) or more will be enrolled in the CRP when disjunct from current forest, but eligible land in blocks of any size that is adjacent to current forest land will be enrolled. That is, we assume that farm managers will enroll fields, not portions of fields, unless they are adjacent to current forest. This region was platted into townships (36 mi^2) and sections (1 mi^2 or 640 acres) by the U.S. Rectangular Land Survey so that fields tend to be multiples of 10 acres, e.g., 10, 40, or 80. Conversion to forest vegetation through tree planting or old-field succession is assumed for both scenarios. The methods for the analysis are discussed below.

8.3 Methods for Assessing Landscape Change

8.3.1 Sources of Data

Methods to detect, describe, and analyze temporal landscape changes seem crude in relation to the complexity of landscape-scale processes. Even the more sophisticated methods of obtaining and summarizing data can require extensive "ground truthing" or field verification (Driscoll et al. 1987; Jensen 1986). Types of data are relatively few (although the sources may be many) and fall into three main categories.

8.3.1.1 Aerial Photography

Until recently, the most widely used type of data has been aerial photography (usually black and white panchromatic) from which land use/land cover types can be obtained directly (either by planimetry or by gridding the area with an overlay and counting "cells"). Many conclusions about temporal changes and rates of change can be drawn from a series of aerial photographs without using sophisticated cartographic software or computer technology. Photointerpretation currently provides the most accurate classification (90% or higher) of temporal landscape changes (Lindgren 1985; Jensen 1986).

Nevertheless, there are several potential drawbacks to the use of aerial photography. First, it is rare to find aerial photographs taken before the mid- to late-1930s. Thus, detection of changes over long time periods is often impossible, although other data sources might be used to draw generalized conclusions about earlier land uses. A second problem is that assessment of changes is limited to those dates on which aerial photographs were taken, which might not include some pivotal events in the process of landscape change. Third, the quality of the photographs might be of limited value for distinguishing certain land use/land cover types. Fourth, problems with registration and distortion may be such that successive years cannot be readily superimposed. Finally, detection of change using aerial photographs is generally acknowledged to be time-consuming and cumbersome (Lindgren 1985; Jensen 1986).

Nonetheless, aerial photography has provided valuable assessments of temporal change, especially when combined with other types of data. For example, Auclair (1976) monitored land cover changes in Iowa and Dane counties, Wisconsin, using a combination of aerial photography, United States government and state of Wisconsin censuses, and soil surveys. Principal components analysis, canonical correlations, and Simpson's dominance index were then employed to illustrate the contribution of several ecological parameters to land use/land cover changes associated with the development of intensive agriculture.

Similarly, Sternitzke (1976) used USDA Forest Service timber resource inventories (based on panchromatic aerial photographs and ground measurements) to document the effect of land use changes on forests of the lower Mississippi

River Valley. More recently, Ilberry and Evans (1989) used aerial photography
and Ordnance Survey maps to estimate temporal changes in agricultural and urban
land used on the outskirts of Birmingham, England.

8.3.1.2 Digital Remote Sensing

Remote sensing technology (exclusive of traditional aerial photography) has advanced rapidly and is accessible to many researchers around the world. The several
Landsat satellites have propelled the analysis of landscape change into a new
technological era. The entire surface of the earth is now photographed by the
satellites' multispherical scanners (MSS), providing very useful digital data for
monitoring and quantifying landscape changes (Naveh and Lieberman 1984) and
a continuity not previously available with traditional aerial photography (di Castri
and Hadley 1988). Sequences of satellite data (e.g., months and years) can be used
to derive estimates of transition frequencies between land use categories (Hall et
al. 1988). For instance, the output from a multiyear analysis can be in the form of
a "transition map" that depicts rates of change from one category to another. Hall
et al. (1988) refer to rates of change as "transition frequency" and the proportion
of landscape elements remaining in a particular state as "retention frequency."
Newer satellites (e.g., SPOT) provide much finer resolution than Landsat. Unfortunately, the scale of resolution (pixel size) of satellite imagery is too large for
many ecological applications (79×79 m for Landsat MSS; 30×30 m for Landsat
thematic mapper; 10×10 m for SPOT panchromatic; Jensen 1986).

Many researchers are now turning to airborne scanners because of the greater
resolution available. The most promising of these are data from NASA's experimental Airborne Imaging Spectrometer (AIS). These data can not only provide
transition frequencies but also begin to uncover "invisible" landscape processes
not previously detected by other remote methods. For example, Westman (1987)
discusses the possible use of AIS to estimate forest productivity, decomposition
rates, and rates of nutrient release and assimilation. This technology has been used
to demonstrate a strong relationship between tree leaf lignin concentration and
nitrogen availability and can thus be used to estimate rates of nitrogen-cycling
across forested landscapes (Wessman et al. 1988).

8.3.1.3 Published Data and Censuses

Census data can be extremely useful, especially as ancillary data. These include
data from such agencies as the U.S. Bureau of the Census (the Census of Agriculture, Census of Population), U.S. General Land Office Survey data (for presettlement vegetation descriptions), and the British annual Agricultural Census.
Again, these have been used, commonly complemented with photography, to
describe types and rates of changes in landscape patterns (e.g., Auclair 1976;
Ilberry and Evans 1989).

The annual Agricultural Censuses have been used extensively in Great Britain
to estimate rates of agricultural land loss (Ilberry and Evans 1989). Although the
census is primarily used by the government to track trends in crop production, it

also includes "total holding area," that is, the total agricultural area in a civil parish. By comparing these total holding areas from one year to another, temporal changes can be estimated. The U.S. Census of Agriculture provides similar information, but at five-year intervals. However, these data do not seem to be reliable at the scale of small political units, e.g., the individual parish. In comparing the estimates of agricultural land loss at the urban fringe of Birmingham, Ilberry and Evans (1989) concluded that the Agricultural Census is reliable over large areas and is best used in conjunction with aerial photography and map data.

8.3.2 A Classification System for Patches

Investigation of historic changes in landscape patterns requires a classification system that can be applied uniformly over the time series. However, the time spans over which landscape change has been assessed make it necessary to use a variety of data sources and technologies, each of which may be suited to a different classification system. A flexible classification system that is not bound to a specific technology is needed. The land use and land cover (LULC) classification (Anderson et al. 1976, see Table 8.1) is an example of such a system. It is widely used and has been adopted as a standard for U.S. governmental agencies. It was developed for use with remote sensing data (e.g., aerial photography) and provides a flexible framework for developing special-purpose classifications. For example, the LULC data can be provided in digital form for analysis in geographic information systems (Fegeas et al. 1983).

The LULC classification is hierarchical. It has first-level categories based on general land uses, and second-, third-, and fourth-level categories that provide increasing resolution. The first- and second-level categories are well established and may be applied with little modification, whereas the higher level categories are intended to be defined to meet special purposes. For example, forest cover types (Eyre 1980) can be added as level III or level IV classes. A hierarchical classification system such as the LULC is also useful when it is necessary to collapse the taxonomy (i.e., classify patches at a lower level) to deal with differences in the resolution of data during a time series. For example, maps of the distribution of woodlots are available for selected areas of the United States in the late 1800s (Burgess and Sharpe 1981; Whitney and Somerlot 1985), but forest type or specifics on agricultural land (level II classes) are missing. The more detailed information available from aerial photography and digital remote sensing, which may be classified at higher resolution, can be collapsed to be compatible with earlier data. In addition, studies of landscape change have used other classification systems that can be mapped onto the LULC classification so that information about several landscapes can be compared (see Turner and Ruscher 1988).

The LULC classes used to develop a spatial data base for Cadiz Township, Wisconsin, are underlined in Table 8.1. The resolution of level II was adopted, except for the forest patches, which were of primary interest. Distinctions were made between stand densities in order to record the impact of prolonged grazing and other disturbances in woodlots.

Table 8.1. USGS Land Use/Land Cover Classification System[a]

Level I	Level II	Sample Level III
1. Urban or built-up land	*11. Residential*	
	12. Commerical or services	
	13. Industrial	
	14. Transporation, communication or utilities	
	15. Industrial and commerical	
	16. Mixed urban or built-up	
2. Agricultural land	*21. Cropland and pasture*	
	22. Orchards, groves, vineyards, horticulture	
	23. Confined feeding operations	
	24. Other agricultural land	
3. Rangelands	*31. Herbaceous rangelands*	
	32. Shrub and brush rangelands	
	33. Mixed rangelands	
4. Forest land	*41. Deciduous forest land*	*411,421. < 10% Tree cover*
	42. Evergreen forest land	*412,422. 10-50% Tree*
	43. Mixed forest land	*413,423 > 50% Tree cover*
5. Water	*51. Streams and canals*	
	52. Lakes	
	53. Reservoirs	
	54. Bays and estuaries	
6. Wetland	*61. Forested wetlands*	
	62. Nonforested wetlands	
7. Barren land	71. Dry salt flats	
	72. Beaches	
	73. Sandy areas except beaches	
	74. Bare exposed rock	
	75. Strip mines, quarries, gravel pits	
	76. Transitional areas	
	77. Mixed barren land	
8. Tundra		
9. Perennial snow or ice		

[a] Anderson et al. (1976). Classes used in Cadiz Township Data Base are italicized.

8.3.3 Spatial Data Base

Studies of landscape change frequently require that the data be spatially explicit. As discussed above, data are often derived from a variety of sources. Overlaying data from different sources raises a number of issues. One major issue is the accuracy and compatible resolution or detail in the data (Bailey 1988). A second is consistent characterization of the phenomenon being mapped, e.g., classification of vegetation (Domon et al. 1989). Additionally, maps are produced at a variety of scales and on various projections, which call for transforming the data to a common spatial referencing system.

Aerial photography and published maps have been used to develop a spatial data base of Cadiz Township. The spatial data base for Cadiz Township is a raster data base of land use and environmental data. The resolution is 1.0 hectare, i.e., each cell is 1.0 hectare in size, and the township has approximately 10,000 hectare. Data were derived from a number of sources, including maps of presettlement vegetation (Finley 1976); published maps of the distribution of forest tracts > 4 hectare for 1882 and 1902 (Shriner and Copeland 1904); aerial photographs for 1937, 1963, and 1978; a soil survey (Glocker 1974); and topographic maps. These original data sources were converted to a uniform scale (1:24,000) and digitized by using a mylar overlay with the grid cells.

8.3.4 Field Survey of Forest Patches

Field surveys have not typically been a component of studies of landscape change. It is laborious to assess the innumerable patches in a landscape. Selected components of the landscape have been surveyed, e.g., fencerows (Willmot 1980; Forman and Baudry 1984) and woodlands (Peterken and Game 1984). However, historic reconstruction of even one stand is time consuming (Oliver and Stephens 1977); so retrospectives are frequently limited to the earliest regional scale vegetation studies, provided by sources such as the U.S. General Land Office surveys (Sharpe et al. 1987).

To assess current vegetation, we surveyed most forest tracts in Cadiz Township during the summers (May to August) of 1983 and 1984. One or more cells in each forest tract were sampled by the stratified random line strip method (Lindsey 1955). Plots were located along transects across each 1.0-hectare cell, beginning at the forest edge to assure that both edge and interior species were recorded. Transects parallel to an edge were maintained 25 m from the edge. Sampling intensity in each forest patch (most of which are farm woodlots) depended on size, topographic and vegetational heterogeneity, and ownership patterns. All layers (trees, saplings, shrub, and ground layer) were surveyed. Trees (≥ 10 cm diameter breast height [dbh] and saplings (2.5 to 9.9 cm dbh) were recorded by species and diameter breast height in 10 m × 25 m plots along the transect, with four plots per 100 m transect in each cell. Species names follow Gleason and Cronquist (1963). We have surveyed approximately 1700 plots in 380 cells in 98 forest patches in Cadiz Township and adjacent Clarno Township.

The importance values (an average of relative basal area and relative density) of the tree layer were summarized by forest tract. These data were entered into the spatial data base for each forest tract: that is, each cell of the forest tract is given the same importance values for each species since the basic unit for our spatial analysis is the forest tract, not the cell. The vegetation data were generalized to presence/absence and analyzed using a geographic information system software program, IDRISI, as discussed below.

8.3.5 Data Analysis

Spatial data can be analyzed with a geographic information system (GIS) (Berry 1986; Jensen et al. 1986), permitting the generation of computer maps, spatial statistical analysis, analysis of temporal changes in one or more land use/land cover types, and cartographic modeling (Berry 1986). The success of early GIS programs, e.g., IMGRID (Sinton 1977), has led to development of a number of other systems, including MAP (Tomlin 1983; Berry 1986, ARC/INFO for both mainframe and personal computers (Environmental Systems Research Institute 1987), IDRISI (Eastman 1988), and a number of others.

The Cadiz spatial data base was analyzed with a raster-based GIS, IDRISI (Eastman 1988). Four types of analyses were conducted. The first involves grouping cells in the data base on the basis of their proximity. A raster data base considers the cell (not a group of like cells, e.g., a forest patch) as the basic unit of analysis. Contiguous cells are grouped to define an area of cells with the same attribute. The forest patches can be identified in this way.

The second analysis involves combining layers in the spatial data base (a layer is a file of one variable, e.g., land use in 1978) in order to define juxtapositions of features. For example, location of forest patches in 1978 was combined with data on land capability classes from the soil data base to identify areas of potential reforestation in response to the CRP.

The third analysis involves spatial analysis. Distance from a target forest patch (e.g., a seed source) to all other cells in the Cadiz Township data base can be identified, and then overlaid with the map of forest patches to compute their respective distances from the target. Another spatial analysis involved producing buffer zones of specified widths around each forest patch to show other patches in its vicinity. These were used to measure distances between woodlots.

The fourth analysis involves logical and mathematical operations. For example, cells with combinations of attributes (e.g., forest in 1978 and the presence of a selected tree species) can be identified. Overlaying the spatial analysis of forest patches with locations of other features e.g., patches of highly erodible land, allows the distance from the forest to each patch of erodible land to be computed.

8.4 Cadiz Township—A Dynamic Landscape

8.4.1 Trends in Deforestation and Reforestation

Given the potential for an ongoing dynamic in Cadiz Township, we can reconstruct the history of deforestation and the potential for its reversal (Table 8.2). This region was dominated by forest prior to European settlement but was deforested rapidly. Two-thirds of the forest had been cleared by 1882, and only 5% remained by 1978, not including small clumps of trees scattered throughout the landscape. The impact of the CRP under these two scenarios is to increase the area in natural vegetation from a current 5% of Cadiz Township to 13% (scenario 1) or to 27% (scenario 2) (Table 2).

These dynamics have fundamental impact upon landscape structure. Deforestation led to the creation of distinct patches, predominantly farm woodlots, with the number of woodlots increasing as deforestation progressed. The current landscape has 84 woodlots, about 2.3 per square mile (Table 8.2). The CRP would increase the number of discrete patches to 97 (scenario 1) or 103 (scenario 2), with a concurrent increase in average patch size from 5.6 hectare to 12.8 hectare for scenario 1, or to 25.5 hectare for scenario 2. In short, the number of forest patches would increase moderately, but their average size would increase twofold to fourfold.

Other parameters have been devised to characterize the impact of fragmentation of natural vegetation on selected biota. One is the amount of edge that develops at the interface between forest and agricultural land uses (Ranney et al. 1981). The total amount of edge has decreased in the past century from about 242 km in 1882 to about 94 km in the current landscape, but the amount of edge per unit area of forest has increased (Table 8.2). Addition of CRP land to current forest would more than double the edge but would decrease the edge/area ratio.

Another general parameter that has been identified is the area of interior forest (Levenson 1981). By defining interior forest as the area within a forest that is at least 100 m from an edge, we used the Cadiz Township spatial data base to delineate interior forest. This analysis suggests that the area of interior forest in

Table 8.2. Historic and Prospective Changes in Landscape Patterns in Cadiz Township, Green County, Wisconsin[a]

Variables	1831	1882	1978	Scenario 1	Scenario 2
Total forest area (ha)[b]	8724	3339	473	1223	2623
Number of forest islands	1	47	84	97	103
Average forest island size (ha)	8724	58	6	13	26
Total perimeter or edge (km)	—	242	94	222	366
Average interisland distance (m)	—	153	439	351	288
Edge/forest area (m/ha of forest)	—	72	199	178	139
Interior forest area (ha)	—	1295	11	109	359
Forest islands with interior forest	—	44	3	17	13

[a] See Burgess and Sharpe (1981) for 1831 and selected 1882 data.
[b] The abbreviation ha designates hectares.

1882 was about 1295 hectare, distributed among forty-four of the forty-seven forest patches. That is, the forest patches existed as large blocks, so that about 38% of the forest was interior. The area of interior forest declined to about 11 hectare in only three of the eighty-four woodlots in 1978. The impact of reforestation under the CRP would be to increase interior forest tenfold if the more severely erodible land is enrolled in the CRP (scenario 1), or over thirtyfold if all eligible land were enrolled (scenario 2): that is, some areas of CRP land would allow the conversion of current forest edge to interior forest while, through time, larger blocks of CRP land would themselves develop into interior forest.

A third parameter relates to increases in interpatch distances and loss in connectivity between patches (Johnson et al. 1981). Interpatch distances have increased from an average of about 150 m in 1882 to 440 m in the current landscape. Addition of the CRP land to current forest would decrease the average distance between patches by 20% (scenario 1), or by 35% if all eligible land were enrolled (scenario 2).

In summary, the forest patches in Cadiz Township have become small, isolated, and dominated by edge in the past century. Thus, the landscape is "fragmented." Addition of the CRP land has the potential to allow the development of a less fragmented landscape with larger forest patches that are less isolated with a greatly increased area of forest interior.

8.4.2 The Significance of Landscape Change for Species Dispersal

The above analysis of landscape change in Cadiz Township has been based on characteristics that have been found to be generally useful in describing landscapes. However, current landscape patterns and ongoing landscape dynamics interplay with each species differently. Interaction between dispersal capability of selected tree species and landscape patterns in Cadiz Township makes this point. We compare dispersal capabilities of three species with the distance between woodlots in Cadiz Township.

The impact of current and potential patterns of land use differs between species. For example, *Acer saccharum* (sugar maple) has limited dispersal capability. Johnson (1988) found an exponential decrease in *A. saccharum* seedlings away from an isolated seed source and from a line source along a fencerow in Wisconsin. Johnson's data suggest that the distance over which seed fall decreases by one-half (d_h, see McClanahan 1986) is 10 m. Simulation of *A. saccharum* seed dispersal using seed fall velocities and wind-rose data (Johnson et al. 1981; Sharpe and Fields 1982) yields a similar conclusion concerning the limited dispersal capability of *Acer* spp. Most seeds are windblown only a few tens of meters from the seed source; few are dispersed over 100–200 m. Ignoring the specific locations of *A. saccharum* within woodlots and the exponential decay in seed fall with distance, we will approximate an effective dispersal distance for *Acer saccharum* as 200 m, which is a multiple of the cell size (100 m) in the Cadiz Township data base. We assume that woodlots farther than 200 m apart would not exchange significant number of *A. saccharum* seeds.

Field data suggest that *Fraxinus americana* (white ash) is dispersed farther than *A. saccharum* slightly (Johnson 1988). The former has a d_h of 30 m, compared with 10 m for *A. saccharum*. Conservatively, we assume the limit of dispersal for *F. americana* to be 400 m, i.e., four cell widths in the Cadiz Township data base. The greater dispersal distances for *Fraxinus* give it the capability to invade woodlots farther away from seed sources. Consequently, we would expect that fewer woodlots are isolated in relation to *F. americana* than to *A. saccharum*. However, even interisland distances of several hundred meters, which are characteristic of this landscape (see Table 2), exceed the dispersal capabilities of *F. americana*.

By contrast, *Quercus spp.* are present in few woodlots in this landscape. However, they have the potential to be dispersed over kilometer-long distances by birds (Johnson and Adkisson 1985). Nominally, we considered *Quercus spp.* to be capable of being dispersed 1 km, and use *Quercus rubra* as an example of a zoochore with a dispersal capability that exceeds the dispersal capabilities of windblown seeds.

8.4.2.1 Isolated Versus Proximal Woodlots and Regenerating Forest

The current landscape can be classified into four categories of woodlots: proximal and isolated seed sources (woodlots with a species present) and proximal and isolated invasion sites (woodlots with a species absent). Attributes for each of these categories are outlined below:

1. Proximal seed source: a woodlot that contains the species of interest and whose distance to another woodlot is within the dispersal distance of the species.
2. Isolated seed source: a woodlot in which the species is present, but the distance to another woodlot is greater than the dispersal distance of the species.
3. Proximal invasion site: a woodlot in which the species is absent, but the distance to a seed source is less than the dispersal distance of the species.
4. Isolated invasion site: a woodlot in which the species is absent, and the distance to a seed source is greater than the dispersal distance of the species.

The woodlots in Cadiz Township were classified into these four categories, based upon the field survey, which determined the presence/absence of each species, and spatial analysis using the spatial data base mentioned above.

The classification for the current landscape is shown in Table 8.3 and Fig. 8.2.a, c, and e for the three species. There are three woodlots in which *A. saccharum* is present that are close enough to adjacent woodlots to be effective seed sources and twenty-eight in which interpatch distances are too great. Likewise, woodlots in which *A. saccharum* is absent generally are too isolated from potential seed sources; four are within the dispersal distance of *A. saccharum*, whereas forty-nine are too distant. Likewise, many *F. americana* seed sources are generally isolated; only ten woodlots are proximal seed sources, whereas eleven are isolated. Reciprocally, most potential invasion sites are isolated from seed sources. By contrast,

Table 8.3. Summary of Proximal and Isolated Seed Sources and Invasion Sites for Selected Species for 1978 and for Landscape with Potential Effect of the Conservation Reserve Program (Scenario 1) Taken into Account[a]

	Landscape in 1978			
	Seed Sources		Invasion Sites	
Species	Proximal	Isolated	Proximal	Isolated
Acer saccharum	24 (3)	173 (28)	17 (4)	240(49)
Fraxinus americana	60(10)	58(11)	66(12)	270(51)
Quercus rubra	172(20)	12 (2)	178(10)	92(52)

	Landscape with Potential Effects of the Conservation Reserve Program					
	Seed Sources		Invasion Sites		Scenario 1	
Species	Proximal	Isolated	Proximal	Isolated	Proximal	Isolated
Acer saccharum	113(12)	84(19)	29 (8)	228(45)	210(30)	559(56)
Fraxinus americana	86(14)	32 (7)	76(16)	260(47)	306(25)	463(61)
Quercus rubra	175(21)	9 (1)	195(41)	75(21)	588(67)	181(19)

[a] Values are hectares; numbers in parentheses are number of woodlots or patches of CRP land. Some patches in scenario 1 are composites of woodlot and CRP land.

the seed sources for *Q. rubra* are generally within the dispersal capability of this species. Twenty are proximal, whereas two are isolated. The invasion sites generally are isolated, small patches, as indicated by the small total area involved.

Consideration of the changing land use patterns described in scenarios 1 and 2 adds another dimension to the distribution of seed sources and invasion sites. Enrollment of eligible agricultural land in the CRP in Cadiz Township would set in motion old-field succession over large areas. The different dispersal capabilities of the three sample species would lead to differential access to the areas undergoing old-field succession, which will become either proximal or isolated invasion sites.

The impact of the areas eligible for the CRP to the current woodlots is shown for scenario 1 for the three species in Table 8.3 and Fig. 8.2 b, d, and f. One major impact is to decrease interpatch distances so that seed sources that are isolated in the current landscape would again become functional. The number of isolated seed sources declines for all species. However, significant numbers and areas of CRP

Figure 8.2. Interaction between forest patches for *Acer saccharum, Fraxinus americana,* or *Quercus rubra* considering the presence/absence of the respective species in each patch and its dispersal capability. The left panel shows the landscape with 1978 woodlots, and the right panel shows Scenario 1.

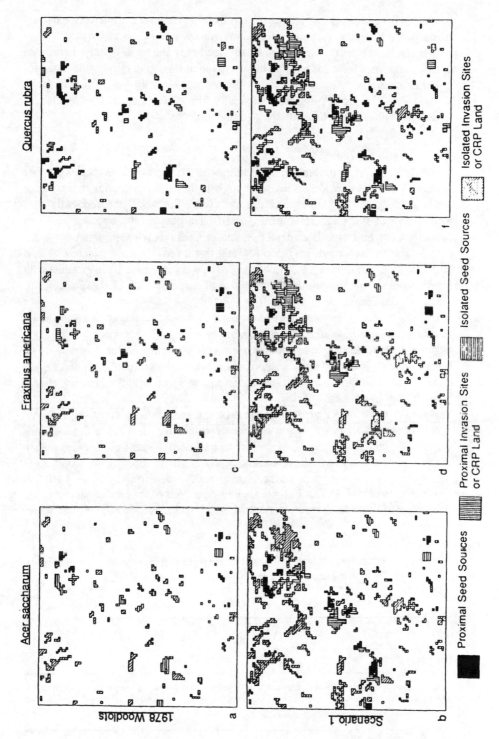

Acer saccharum   Fraxinus americana   Quercus rubra

1978 Woodlots   a   c   e

Scenario 1   b   d   f

■ Proximal Seed Sources   ▥ Proximal Invasion Sites or CRP Land   ▦ Isolated Seed Sources   ▨ Isolated Invasion Sites or CRP Land

patches may be created that are too distant from seed sources for effective natural regeneration to occur, especially for *A. saccharum* and *F. americana.*

It is ironic that the two wind-dispersed species are the most widely distributed in this landscape, yet woodlots lacking these species are generally too isolated for effective interisland dispersal to occur. By contrast, *Q. rubra* is present in fewer woodlots, but its greater dispersal capability brings more current woodlots and CRP land within reach than for the two anemochores.

8.4.2.2 Clustering of Woodlots

The distribution of woodlots and of these species among woodlots creates another pattern—clusters of woodlots close enough to have the potential to exchange seeds. Although average interpatch distances (e.g., Table 8.2) are frequently used to characterize the fragmentation of forested landscapes, the spacing of forest patches is not uniform. The impact of this in Cadiz Township, using the three sample species, is shown in Table 8.4 and displayed for *Acer saccharum* and *Fraxinus americana* in Fig. 8.3. In this analysis, woodlots have been grouped with other woodlots that are within the dispersal distance of each of these species to create the clusters shown.

In 1882 the forty-seven distinct forest patches created twenty-six clusters using the dispersal capability of *A. saccharum* but only five clusters when the dispersal distance is increased to 400 m, as for *F. americana.* The greater dispersal capability of *Q. rubra* created one cluster of woodlots, none of which exceeded 1.0 km from another woodlot. However, we do not know which woodlots in each cluster actually contained these species.

By 1978 the number of distinct clusters had increased to sixty-seven for *A. saccharum* but only twenty-eight for *F. americana.* By identifying the locations of woodlots containing these two species, as indicated by our field survey, we can identify those clusters which have each species in one or more of the woodlots in the cluster. Only twenty-nine of the clusters had *A. saccharum,* and only ten had *F. americana* (Table 4); that is, there are entire portions of the landscape composed of clusters of proximal woodlots, none of which includes seed sources and which

Table 8.4. Summary of Clusters of Woodlots for 1882, 1978, and for Potential Forest (1978 Forest Plus CRP Land), shown in Fig. 8.3

Species	1882	1978	Scenario 1	Scenario 2
Acer saccharum	26	67(29)	48(23)	27(13)
Fraxinus americana	5	28(10)	21 (6)	10 (5)
Quercus rubra	1	6 (3)	1	1

[a] Values are the number of clusters for each year or scenario. Numbers in parentheses are clusters with one or more current woodlots with the species present. Species presence/absence in 1882 is unknown.

Figure 8.3. Clusters of woodlots in 1882 and 1978 and for Scenario 1, based on the dispersal capability of *Acer saccharum* and *Fraxinus americana.*

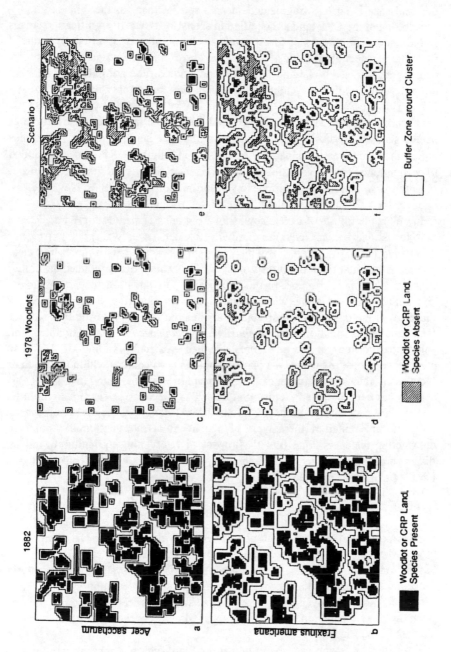

are isolated from seed sources in other clusters. There were six clusters for *Q. rubra*, three of which were located along the western border of Cadiz Township, and whose proximity to woodlots in the adjacent township is unknown. However, the landscape is far less fragmented for this species than for the others.

Addition of the CRP land as specified in scenario 1 leads to significant changes in the number of clusters of forest patches. The number is significantly reduced from the perspective of both *A. saccharum* and *F. americana* as interpatch distances are decreased by addition of CRP land, and corridors of CRP land connect current woodlots, to forty-eight for *A. saccharum* and twenty-one for *F. americana*. However, a minority of these clusters have woodlots that can serve as seed sources. By contrast, addition of CRP land returns the landscape to a fully connected state for *Q. rubra*; i.e., the maximum distance between any forest patches is less than the dispersal capability of oaks. If all eligible CRP land were added to current forest (scenario 2), there would be fewer clusters for *A. saccharum* and *F. americana*, and a higher proportion would harbor seed sources. Nevertheless, about half of all clusters would still lack a seed source.

In effect, the forests of Cadiz Township are more fragmented for *A. saccharum* than for *F. americana*. Given the number of clusters (twenty-six) that existed for *A. saccharum* in 1882, some of which probably lacked a seed source, it is clear that the fragmentation of the forests of Cadiz Township quickly reduced the ability of *A. saccharum* to disperse in this landscape. The situation for *F. americana* appears to be less extreme, whereas the landscape appears to have provided for dispersal of the oaks until deforestation attained current levels. Addition of CRP land ameliorates isolation of *A. saccharum* and *F. americana* nominally; however, substantial portions of the new forest patches created by adding CRP land to current woodlots are themselves distant from the seed sources within the patch or the cluster. The slow accretion of *A. saccharum* to woodlots being invaded by this species has been modeled (Johnson et al. 1981). Decades or centuries would be required for each species to become established throughout the clusters in which they are represented if there were no management intervention. By contrast, considering distance alone (not the impact of intervening agricultural land on dispersal patterns for such bird-dispersed species as the oaks), Cadiz Township is hardly fragmented for oaks.

8.5 Summary

Landscape ecologists have developed a number of descriptors for landscape pattern, such as patch size, patch shape, and interpatch distance. Concurrently, data bases from remote sensing technologies, classification systems, and analytic procedures, especially those provided by geographic information systems, have greatly facilitated analysis. We have reviewed these concepts and tools and applied them to a typical landscape in the upper Midwest of the United States, Cadiz Township, Wisconsin.

While the landscape patterns in this agricultural region commonly are viewed as stable, we argue that they are in fact quite dynamic in ecologically relevant time frames. Agricultural development of Cadiz Township over the past 150 years has led to a decrease in the total area of forest, with concomitant decreases in patch size and area of interior forest and increases in edge/area ratio and interpatch distances. Change was rapid in the late nineteenth century, but the landscape seemingly has stabilized. However, this upper Midwest landscape has the prospect to undergo a rapid increase in natural vegetation, just as New England and the southeastern United States have. Currently, the major impetus for change in agricultural landscapes in the United States is the Conservation Reserve Program (CRP), administered by the U.S. Department of Agriculture. Scenarios of plausible landscape change resulting from the CRP suggest that as much as 20% of the agricultural land could be reforested. This would lead to a fourfold increase in average patch size and the return of forest interior, which has virtually disappeared from the small patches in Cadiz Township.

While deforestation in Cadiz Township has led to a landscape of isolated forest patches, the impact varies with species. We demonstrated this in two ways. First, we classified patches into proximal and isolated seed sources and invasion sites, and applied the classification to *Acer saccharum, Fraxinus americana,* and *Quercus rubra.* Using field survey data on the location of these species in the woodlots in Cadiz Township, we mapped the distribution of patches on the basis of the classification. Even though *A. saccharum* is widely distributed among the woodlots in Cadiz Township, the vast majority of seed sources are isolated from potential invasion sites. For *A. saccharum,* Cadiz Township is a fragmented landscape with little opportunity for dispersal from one patch to another. By contrast, *Q. rubra* is not present in many patches, but its greater dispersal capability means that the majority of woodlots that lack this species have the potential to be invaded by virtue of interpatch distances. In effect, this landscape is less fragmented for the oaks than for many wind-dispersed species.

Likewise, the ongoing change in landscape patterns alters the isolation of species differentially. Addition of CRP land to the forest land base would reduce interpatch distances, so that a greater proportion of the new or enlarged patches would be within the dispersal capabilities of the sample species. Using scenario 1, which assumes a conservative response to the CRP, to simulate the impact of landscape change, we found that the area of potential invasion sites increases for all species. However, *A. saccharum* still remains relatively isolated, with only about 20% of the patches that are potential invasion sites within its dispersal capabilities. By contrast, over 75% of the potential invasion sites for *Q. rubra* are within range of a seed source, in spite of its more sparse distribution in the landscape.

A second view of landscapes and landscape change in Cadiz Township is in terms of how clusters of forest patches develop and change through time. We considered members of a cluster of patches as patches within the dispersal capability of each species. Thus, a different pattern of clusters develops for each species in the landscape. The dispersal capability of *Q. rubra* meant that all the patches in Cadiz Township were grouped into one cluster in 1882, but this increased to six by 1978. Addition of CRP land would return this landscape to one large cluster for the oaks. By contrast, deforestation quickly created a large number of clusters of patches that were isolated from each other for *A. saccharum;* this isolation would be ameliorated by the CRP program, but the landscape still would be fragmented for this species. The current landscape would be more or less fragmented for other members of the biota, and the spatial expression of land use change will have a different effect on each species.

8.6 Conclusion

Change is characteristic of virtually all landscapes, but it is episodic and driven by social, economic, and political factors. Predicting landscape change is an ongoing challenge in many disciplines. In forestry, the challenge is to combine stand dynamics models with land use dynamics models to project timber supplies (Parks and Alig 1988). Reciprocally, agricultural economists are attempting to model how complex social, economic, and political forces affect trends in agricultural acreages and their allocation to crops, pasture, natural vegetation (e.g., forest), and urban land uses (see Alig et. al. 1988). Other investigators are assessing the dynamics of land use change at the rural-urban interface (Vesterby 1988; Ilbery and Evans 1989). The challenge in landscape ecology is to translate the predictions about land use change made by experts in these other fields into assessments of ecologic impact. Simulations can use "plausible futures" as points of departure to assess how general trends might be allocated in the landscape and to assess the sensitivity of populations, communities and ecosystems to these allocations. Management strategies can be developed from analyses that identify the role of individual patches in the landscape.

Given the ongoing nature of landscape change, we believe that the focus of research in landscape ecology needs to be on how landscape dynamics interacts with species tolerances in time and space. If, for example, deforestation in an era of agricultural expansion is followed by periods of reforestation in many landscapes, how do lags in species response to fragmentation compare with observed rates and levels of fragmentation with given durations? The assumption that land use patterns are static in a given landscape fosters application of equilibrium models. The implicit assumption that populations are in equilibrium with current landscape patterns is probably wrong (Pickett 1980). This statement appears trite, but we make it explicit because we find that it is assumed that some landscapes, especially predominantly agricultural areas, have stable land use patterns that persist over long periods. This assumption is especially apparent when it is assumed that deforestation in one period of history and economic development cannot be followed by reestablishment of natural vegetation in the near future. The implications of this assumption extend to the political stance that ecologists take with respect to regional development and the way in which fundamental research questions are posed.

Acknowledgments

The authors wish to thank the Green County Forestry Education Association, Monroe, Wisconsin, for donating research facilities during the field study of woodlots in Cadiz Township, and Craig Strus, Department of Geography, SIUC, for assistance in developing the graphics.

References

Alig, R.J; White, F.C.; and Murray, B.C. 1988. Economic factors influencing land use changes in the south central United States. U.S. Dept. Agric. Forest Service, Southeast Forest Expt. Sta. Research Paper SE–272, Asheville, N.C.

Ambuel, B., and Temple, S.A. 1983. Area-dependent changes in the bird communities and vegetation of southern Wisconsin forests. *Ecology* 64:1057–68.

Anderson, J.R.; Hardy, E.E.; Roach, J.T.; and Witmer, R.E. 1976. A land use and land cover classification system for use with remote sensor data. United States Geological Survey, Professional Paper 964.

Auclair, A.N. 1976. Ecological factors in the development of intensive-management ecosystems in midwestern United States. *Ecology* 57:431–44.

Austin, M.P. 1977. Use of ordination and other multivariate descriptive methods to study succession. *Vegetatio* 35:165–75.

Bailey, R.G. 1988. Problems with using overlay mapping for planning and their implications for geographic information systems. *Environmental Management* 12:11–17.

Baker, W.L. 1989. A review of models of landscape change. *Landscape Ecology* 2:111–33.

Berry, J.K. 1986. GIS: learning computer-assisted map analysis. *Journal of Forestry* Oct. 1:39–43.

Birks, H.H.; Birks, H.J.B.; Kaland, P.E.; and Moe, D. eds. *The Cultural Landscape—Past, Present and Future.* New York: Cambridge University Press.

Bond, W.E., and Spillers, A.R. 1935. Use of land for forests in the lower Piedmont region of Georgia. South. Forest Expt. Stn. Occas. Paper 53.

Bowen, G.W., and Burgess, R.L. 1981. A quantitative analysis of forest island patterns in selected Ohio landscapes. ORNL/TM#7759. Oak Ridge National Laboratory, Oak Ridge, Tenn.

Burgess, R.L., and Sharpe, D.M. eds. 1981. *Forest Island Dynamics in Man-Dominated Landscapes.* New York: Springer-Verlag.

Curtis, J.T. 1956. The modification of mid-latitude grasslands and forests by man. In *Man's Role in Changing the Face of the Earth,* ed. W.L. Thomas, Chicago: University of Chicago Press. pp. 721–36.

di Castri, F., and Hadley, M. 1988. Enhancing the credibility of ecology: Interacting along and across hierarchical scales. *GeoJournal* 17:5–35.

Domon, G., Gariepy, M.; and Bouchard, A. 1989. Ecological cartography and land-use planning: Trends and perspectives. *Geoforum* 20:69–82.

Driscoll, R.S.; Betters, D.R.; and Parker, H.D. 1978. Land classification through remote sensing—techniques and tools. *Journal of Forestry* 76:656–61.

Dunn, C.P.,Loehle, C. 1988. Species-area parameter estimation: testing the null model of lack of relationship. *Journal of Biogeography* 15:721–28.

Dunn, C.P., and Sharitz, R.R. 1987. Revegetation of a *Taxodium-Nyssa* forested wetland following complete vegetation destruction. *Vegetatio* 72:151–57.

Eastman, J.R. 1988. *IDRISI: A Grid-Based Geographic Information System.* Worcester, Mass.: Clark University, Graduate School of Geography.

Environmental Systems Research Institute (ESRI). 1987. *ARC/INFO User's Guide.* Redlands, Calif.: ESRI.

Eyre, F.H., ed. 1980. *Forest Cover Types of the United States and Canada.* Society of American Foresters: Washington D.C.

Fegeas, R.G.; Claire, R.W.; Guptill, S.C.; Anderson, K.E.; and Hallam, C.A. 1983. Land use and land cover digital data. U.S. Geological Survey, Geological Survey Circular 895–E.

Finley, R.W. 1976. Original vegetation cover of Wisconsin from U.S. General Land Office Notes (map). U.S. Forest Service, North Central Forest Experiment Station, St. Paul, Minn.

Forman, R. T. T., and Baudry, J. 1984. Hedgerows and hedgerow networks in landscape ecology. *Environmental Management* 8:495–510.

Forman, R.T.T., and Godron, M. 1986. *Landscape Ecology*. New York: Wiley.

Forman, R.T.T., and Milne, B.T. 1986. Peninsulas in Maine: woody plant diversity, distance, and environmental patterns. *Ecology* 67:967–74.

Galli, A.E.; Leck, C.F.; and Forman, R.T.T. 1976. Avian distribution patterns within different sized forest islands in central New Jersey. Auk 93:356–64.

Gardner, R.H.; Milne, B.T.; Turner, M.G.; and O'Neill, R.V. 1987. Neutral models for the analysis of broad-scale landscape pattern. *Landscape Ecology* 1:19–28.

Gleason, H.A., and Cronquist, A. 1963. *Manual of Vascular Plants of Northeastern United States and Adjacent Canada.* New York: Van Nostrand.

Glocker, C.L. 1974. *Soil Survey of Green County, Wisconsin.* U.S. Dept. Agric. Soil Conserv. Serv., Washington, D.C.

Grieg-Smith, P. 1983. *Quantitative Plant Ecology.* 3rd ed. Berkeley: University of California Press.

Hall, F.G.; Strebel, D.E.; and Sellers, P.J. 1988. Linking knowledge among spatial and temporal scales: vegetation, atmosphere, climate and remote sensing. *Landscape Ecology* 2:3–22.

Hamabata, E. 1980. Changes of herb-layer species composition with urbanization in secondary oak forests of Musashino Plain near Tokyo—studies on the conservation of suburban forest stands. I. *Japanese Journal of Ecology* 30:347–58.

Heinselman, M.L. 1981. Fire and succession in the conifer forests of Northern North America. In *Forest Succession: Concepts and Application,* D.C. West, H.H. Shugart, D. B. Botkin, eds, pp. 374–405. New York: Springer-Verlag.

Hill, D.B. 1985. Forest fragmentation and its implications in central New York. *Forest Ecology and Management* 12:113–28.

Ilbery, B.W., and Evans, N.J. 1989. Estimating land loss on the urban fringe: A comparison of the agricultural census and aerial photograph/map evidence. *Geography* 74:214–221.

Irland, L.C. 1982. *Wildlands and Woodlots.* Hanover, N.H.: University Press of New England.

Jensen, J.R. 1986. *Introductory Digital Image Processing.* Englewood Cliffs, N.J.: Prentice-Hall.

Jensen, J.R.; Mackey, H.E., Jr.; Tinney, L.R.; and Sharitz, R. 1986. Remote sensing inland wetlands: a multispectral approach. *Photogrammetric Engineering and Remote Sensing* 52:87–100.

Johnson, H.B. 1976. *Order Upon the Land: The U.S. Rectangular Land Survey and the Upper Mississippi Country.* New York: Oxford University Press.

Johnson, W.C. 1988. Estimating dispersibility of *Acer, Fraxinus,* and *Tilia* in fragmented landscapes from patterns of seedling establishment. *Landscape Ecology* 1:175–87.

Johnson, W.C., and Adkisson, C.S. 1985. Dispersal of beech nuts by blue jays in fragmented landscapes. *American Midland Naturalist* 113:319–24.

Johnson, W.C.; and Sharpe, D.M. 1976. An analysis of forest dynamics in the northern Georgia Piedmont. *Forest Science* 22:307–22.

Johnson, W.C.; Sharpe, D.M.; DeAngelis, D.L.; Fields, D.E.; and Olson, R.J. 1981.Modeling seed dispersal and forest island dynamics. In *Forest Island Dynamics in Man-Dominated Landscapes,* R.L. Burgess and D.M. Sharpe, pp. 215–39. New York: Springer-Verlag.

Levenson, J.B. 1981. Woodlots as biogeographic islands in southeastern Wisconsin. In *Forest Island Dynamics in Man-Dominated Landscapes,* eds. R.L. Burgess and D.M. Sharpe, pp. 13–39. New York: Springer-Verlag.

Lindgren, D.T. 1985. *Land Use Planning and Remote Sensing.* Dordrecht, Netherlands: M. Nijhoff Publishers.

Lindsey, A.A. 1955. Testing the line-strip method against full tallies in diverse forest types. *Ecology* 36:485–95.

Lowell, K.E., and Astroth, J.H., Jr. 1989. Vegetative succession and controlled fire in a glades ecosystem. A geographical information system approach. *International Journal of Geographical Information Systems* 3:69–81.

Lynch, J.F.; and Whigham, D.F. 1984. Effects of forest fragmentation on breeding bird communities in Maryland, USA. *Biological Conservation* 28:287–324.

McClanahan, T.R. 1986. Seed dispersal from vegetation islands. *Ecol. Modelling* 32:301–9.

Merriam, G. 1988. Landscape dynamics in farmland. *Trends in Ecology and Evolution* 3:16–20.

Middleton, J., and Merriam, G. 1983. Distribution of woodland species in farmland woods. *Journal of Applied Ecology* 20:625–44.

Moss, M.R., and Hosking, P.L. 1983. Forest associations in extreme southern Ontario ca 1817: a biogeographical analysis of Gourlay's Statistical Account. *Canadian Geographer* 27:184–93.

Moulton, R., and Dicks, M. 1987. Implication of the 1985 Farm Act for forestry. In *The Blue and the Gray: Proc. of the 1987 Joint Meeting of Southern Forest Economics Workers and the Mid-West Economists*, eds. R.L. Busby, J. deSteigner, J. Edwards, and W.B. Kurtz. April 8–10, 1987.

Naveh, Z., Lieberman, A.S. 1984. *Landscape Ecology: Theory and Application*. New York: Springer-Verlag.

Oliver, C.D., and Stephens, E.P. 1977. Reconstruction of a mixed-species forest in central New England. *Ecology* 58:562–72.

O'Neill, R.V.; Krummel, J.R.; Gardner, R.H.; Sugihari, G.; Jackson, B.; DeAngelis, D.L.; Milne, B.T.; Turner, M.G.; Zygmunt, G.; Christensen, S.W.; Dale, V.H.; and Graham, R.L. 1988. Indices of landscape pattern. *Landscape Ecology* 3:153–62.

Parks, P.J., and Allg, R.J. 1988. Land base models for forest resource supply analysis: A critical review. *Can. J. For. Res.* 18:965–73.

Peterjohn, W.T., and Correll, D.L. 1984. Nutrient dynamics in an agricultural watershed: Observations on the role of a riparian forest. *Ecology* 65:1466 75.

Peterken, G.F., and Game, M. 1984. Historic factors affecting the number and distribution of vascular plant species in the woodlands of central Lincolnshire. *Journal of Ecology* 72:155–82.

Pickett, S.T.A. 1980. Non-equilibrium coexistence of plants. *Bull. Torrey Bot. Club* 107:238–48.

Ranney, J.W.; Bruner, M.C.; and Levenson, J.B. 1981. The importance of edge in the structure and dynamics of forest islands. In *Forest Island Dynamics in Man-Dominated Landscapes*, eds. R.L. Burgess, and D.M. Sharpe, pp. 67–95. New York: Springer-Verlag.

Seagle, W.W., and Shugart, H.H., Jr. 1985. Faunal richness and turnover on dynamic landscapes: A simulation study. *Journal of Biogeography* 12:499–508.

Sharpe, D.M., and Fields, D.E. 1982. Integrating the effects of climate and seed fall velocities on seed dispersal by wind: A model and application. *Ecol. Modelling* 17:297–310.

Sharpe, D.M.; Guntenspergen, G.R.; Dunn, C.P.; Leitner, L.A.; and Stearns, F. 1987. Vegetation dynamics in a southern Wisconsin agricultural landscape. in *Landscape Heterogeneity and Disturbance*, ed. M.G. Turner, pp. 137–55. New York: Springer-Verlag.

Sharpe, D.M., and Johnson, W.C. 1981. Land use and carbon storage in Georgia forests. *Journal of Envir. Mngmt.* 12:221–33.

Sharpe, D.M.; Stearns, F.; Burgess, R.L.; and Johnson, W.C. 1982. Spatio-temporal patterns of forest ecosystems in man-dominated landscapes of the eastern United States. In *Perspectives in Landscape Ecology*, eds. S.P. Tjallingii and A.A. de Veer. Proceedings of the International Congress. Centre for Agricultural Publ. and Documentation, Wageningen, Netherlands.

Sharpe, D.M.; Stearns, F.; Leitner, L.A.; and Dorney, J.R. 1986. Fate of natural vegetation during urban development of rural landscapes in southeastern Wisconsin. *Urban Ecology* 9:267–87.

Shriner, F.A., and Copeland, E.B. 1904. Deforestation and creek flow about Monroe, Wisconsin. *Bot. Gaz.* 37:139–43.

Shugart, H.H., Jr., and West, D.C. 1980. Forest succession models. *Bioscience* 30:308–13.

Sinton, D.F. 1977. *The User's Guide to I.M.G.R.I.D.: An Information Manipulation System for Grid Cell Data Structures.* Harvard University, Dept. Landscape Architecture, Graduate School of Design, Cambridge, Mass.

Sternizke, H.S. 1976. Impact of changing land use on Delta hardwood forests. *Journal of Forestry* 74:25–27.

Tomlin, C.D. 1983. Digital cartographic modeling techniques in environmental planning. Doctoral dissertation, Yale University, School of Forestry and Environmental Studies, New Haven, Conn.

Turner, M.G. 1987a. Spatial simulation of landscape changes in Georgia: a comparison of 3 transition models. *Landscape Ecology* 1:29–36.

Turner, M.G., ed. 1987b. *Landscape Heterogeneity and Disturbance.* New York: Springer-Verlag.

Turner, M.G., and Bratton, S.P. 1987. Fire, grazing and the landscape heterogeneity of a Georgia barrier island. In *Landscape Heterogeneity and Disturbance,* ed. M.G. Turner, pp. 85–101. New York: Springer-Verlag.

Turner, M.G.; Gardner, R.H.; Dale, V.H.; and O'Neill, R.V. 1989. Predicting the spread of disturbance across hetereogeneous landscapes. *Oikos* 55:121–29.

Turner, M.G.; and Ruscher, C.L. 1988. Changes in landscape patterns in Georgia, USA. *Landscape Ecology* 1:241–51.

Vesterby, Marlow. 1988. Land use change in fast-growth counties: analysis of study methods. U.S. Dept. Agric., Economic Research Service, Resources and Technology Division. Staff Report No. AGES880510.

Wessman, C.A.; Aber, J.D.; Peterson, D.L.; and Melillo, J.M. 1988. Remote sensing of canopy chemistry and nitrogen cycling in temperate forest ecosystems. *Nature* 335:154–56.

Westman, W.E. 1987. Monitoring the environment by remote sensing. *Trends in Ecology and Evolution* 2:333–37.

Whitney, G.G. 1987. An ecological history of the Great Lakes forest of Michigan. *Journal of Ecology* 75:667–84.

Whitney, G.G., and Somerlot, W.J. 1985. A case study of woodland continuity and change in the American Midwest. *Biological Conservation* 31:265–87.

Willmot, A. 1980. The woody species of hedges with special reference to age in Church Broughton Parish, Derbyshire. *Journal of Ecology* 68:269–85.

Wilson, J.P., and Ryan, C.M. 1988. Landscape change in the Lake Simcoe-Couchiching Basin, 1800–1983. *Canadian Geographer* 32:206–22.

9. Lessons from Applying Fractal Models to Landscape Patterns

Bruce T. Milne

9.1 Introduction

Landscape structure affects the spread of disturbance and regulates the movements of resources, organisms, and energy (Risser et al. 1984; Forman and Godron 1986; Turner 1987; Kolasa and Pickett 1991; Milne 1990). Many familiar ecological interactions and dynamics have spatial components, such as dispersal (McDonnell and Stiles 1983; Gaines and Roughgarden 1985), genetic neighborhoods (Levin and Kerster 1971), spatial distributions of growth potentials (Wu et al. 1985), historical events that create canopy gaps of many sizes (Pickett and White 1985; Hubbell and Foster 1986), herbivory (Senft et al. 1987), and moisture relations that vary with scale (Neilson and Wullstein 1983; Wiens et al. 1986). The quantification of spatial relationships is a common focus in ecology.

Yet in making predictions, landscape ecologists face the daunting task of quantifying patterns of great temporal and spatial complexity. The complexity stems from biotic interactions that generate patterns (Rykiel et al. 1988) or from physical processes that alter landscape structure. For example, the distribution of moisture (and, consequently, forests) may be affected by topographically controlled thunderstorms initiated at the broad scale of mountain ranges and at the finer scale of individual peaks (Barker Schaaf et al. 1988). Likewise, the juxtaposition of different terrestrial surfaces alters local wind and temperature distributions (Pielke 1984), thus producing a connection between landscape structure and the atmosphere. Landscape patterns modify processes that regulate landscape structure.

GOVERNORS STATE UNIVERSITY
UNIVERSITY PARK
IL 60466

Topography, clouds, and many landscape patterns are quantified by the methods of fractal geometry (Mandelbrot 1983). Fractals represent many kinds of patterns, including density, diversity, dendritic stream networks, mountainous terrain, and size distributions of islands (Mandelbrot 1983; Peitgen and Saupe 1988). Recent studies have included measures of the fractal geometry of landscapes (Burrough 1981; 1983a, b; Krummel et al. 1987; Gardner et al. 1987; Milne 1988; O'Neill et al. 1988a; De Cola 1989; Wiens and Milne, 1989), thereby increasing ecologists' awareness of scale and its effects (see also Meentemeyer and Box 1987).

The goals of this chapter are threefold: (1) to clarify the definitions, utility, and generality of fractals; (2) to illustrate the wide variety of situations to which they may be applied, and (3) to encourage an intuitive understanding of the origin of fractal patterns and the equations used to describe them. Fractal geometry is a continuously developing calculus of heterogeneity; the application of several basic principles may reveal novel features of landscape patterns and the dynamics of organisms living in fractal landscapes.

9.1.1 The Importance of Scale

Patterns in space (Barnsley 1988) and time (e.g., music; Voss and Clarke 1975) possess remarkable and intriguing combinations of regularity and randomness seen at many scales simultaneously. Structure at multiple scales is the hallmark of a fractal or, more generally, of patterns with characteristics that vary strongly with scale. To illustrate, the coast of North America clearly has several major features visible at broad scales, such as the Florida, Seward, and Baja peninsulas. Upon closer inspection, many smaller spits and peninsulas, such as Cape May, Cape Cod, and Monterey, are apparent.

In fact, 174 peninsulas ≥ 0.5 km long are apparent at a 1:24,000 scale along the coast of Maine (Fig. 9.1). The probability of finding a peninsula of a given length decreases as peninsular length is raised to the -1.93 power ($R^2 = 0.96$). Thus, there is a probability of ≈ 0.2 of observing a 2-km-long peninsula but only a chance of ≈ 0.005 of finding a 15-km-long peninsula. Most peninsulas are short, and a diligent beachcomber would find many "peninsulas" that were shorter than a meter and even more peninsulas that were only a decimeter long. Cartographers conveniently ignore the truly small peninsulas when producing maps, but the existence of peninsulas of many lengths challenges scientists from many fields to accommodate such details in models of nature.

Thus, to answer the question "How long is the coastline?" Mandelbrot (1983) replies, "Coastline length depends on the scale of measurement." Ambiguous geometric characteristics like coastline length are described as "scale-dependent." Scale-dependent coastlines have an exponentially greater amount of littoral habitat available to organisms that differentiate space at the 1-cm scale, compared to humans, who use maps with coarser scales. A major contribution of fractal geometry is the quantification of scale dependence in ways that allow predictions to be made in the face of ambiguity.

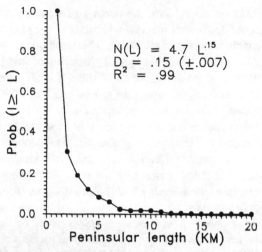

Figure 9.1. The preponderance of small peninsulas along the coast of Maine is described by a decreasing probability of observing peninsulas > L km, the length scale. The function with a fractal exponent $D = 0.15$ describes the average number of peninsulas with lengths ≥0.5 km occurring between two points on the coastline separated by L km.

9.1.2 Ecological Implications of Scale Dependence

It is helpful to consider some ecological consequences of scale before we discuss how to measure scale-dependent structure in landscapes. A major lesson from studies conducted at multiple scales is that organisms operating at different scales may perceive vastly different densities and arrangements of resources (Milne et al. 1989). Mammals explore their home ranges at rates that increase proportionally to body mass (kg) raised to the 0.36 power (Swihart et al. 1988). Thus, massive species (e.g., small antelopes of 40 kg) require ~15 h to cover their home range, while a species that is 1/100th the mass requires just 3 h. Here, scale-dependence stems from the fact that home range area increases as body mass raised to the ~1.4 power (Swihart et al. 1988). It would be naive to assume that the juxtaposition and density of resources is effectively the same for all species.

The ecological consequences of scale dependence may be divided conveniently into two classes: biotic and metric. Biotic scale dependence originates from the differential responses of organisms to the abundance of a resource. Metric scale dependence results when physical processes produce statistically similar aggregations of abiotic quantities such as water, minerals, or energy.

Biotic scale dependence occurs in insect populations. Morse et al. (1985) show that body size and metabolic rate are related to insect density, which in turn is related to the scale-dependent structure of host plants. At coarse resolution, plants are somewhat linear objects, but at finer scales branches and their surface area become more apparent (Loehle 1983). Consequently, a tenfold decrease in the body size of insects reflects an exponential increase in available surface area of plants, a presumed resource. Thus, the density of small insects is exponentially higher than that of large insects in a given habitat (Morse et al. 1985). Other biotic effects of scale include the direct relationship between body mass and home range area (Brown 1981; Peters 1983), the spatial clustering of suitable habitat for deer (Milne et al. 1989), and the modification of the diffusion patterns of tenebrionid beetles by fractal networks of bare soil (Wiens and Milne 1989). Wiens (1989) reviews other examples of biotic scale dependence.

Metric scale dependence concerns the translation of measurements obtained at one scale to other scales. For example, global circulation models (GCMs) are used to model the earth's climate for various scenarios of climatic change. Computational limitations restrict the spatial resolution of GCMs to ≈250,000 km² cells stratified over three to nine levels of altitude. Attempts to include interactions between the atmosphere and biota in the cells at ground level would benefit from representations of photosynthesis and evapotranspiration. However, little is known about the translation of traditional ecological measurements made at meter scales to the 500-km-wide scales of GCMs. The goal of scale translation is to preserve information about details in surface features of the earth (e.g., Pielke and Avassar 1990), while maintaining a relatively coarse, economical gridwork of cells within the analysis. Metric scale dependence is of concern when one is characterizing the nature of physical factors.

The distinction between biotic and metric scale dependence highlights a principle that is useful in the development of new models of nature. At one level, the physical environment presents a template upon which organisms and ecological systems operate. Thus, attention to metric scale dependence is fundamental. At a second level, the organismal responses to the physical environment may exhibit patterns that vary between species and are constrained by the geometry of the environment (Wiens and Milne 1989). Thus, landscape structure and its effects may be partitioned into two components, and the responses of organisms may be described by parameters that are sensitive to scale-dependent constraints in the environment and to biological responses of other organisms.

It is convenient to adopt several conventions regarding the concept of scale. The scale at which data are represented can be partitioned into two components, namely, extent and grain (Fig. 9.2). *Extent* describes the areal or temporal breadth of a study, a data set, or a subset of data from which a statistic is obtained. *Grain* is the minimum spatial or temporal resolution of the data.

The other aspects of scale are important during analysis, i.e., *window size* and *lag distance* (Fig. 9.2). Windows are used to aggregate grains and effectively modify the grain of the data during analysis. By definition, windows must be greater than or equal to the size of grains and are composed of a contiguous series of grains. Lag distances, or lags, are the distances between pairs of windows under comparison and are typically greater than or equal to the length of one grain (e.g., Palmer 1988).

Definitions of grain and extent are context sensitive. For instance, in some analyses, a spatial subset may be obtained, and statistics may then be acquired for a series of windows within each subset, e.g., the mean quantity of the grains

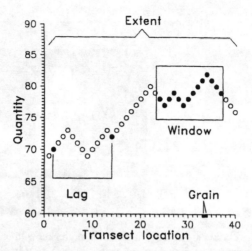

Figure 9.2. Components of scale include lag, window size, the spatial or temporal extent of the observed quantities, and the grain at which observations are resolved. During analysis, pairs of variates separated by a lag distance may be compared, or data may be aggregated within windows that are positioned within the extent.

enclosed by the windows. In such cases the statistics are representative of the extent encompassed by each subset rather than the extent of the entire data set. Thus, extent is the spatial or temporal domain from which a statistic is obtained. The undesirable alternative to this convention is to adopt a long series of words to accommodate each level of subdivision of the data. The concepts of extent, grain, window, and lag help to clarify operations performed during the fractal analysis of landscapes.

9.2 Definitions and Characteristics of Fractals

As with any mathematical representation of nature, fractal models are necessarily abstractions. Stanley (1986) distinguishes two broad classes of mathematical objects or sets, both of which qualify as fractals. Stanley divides fractals into "exact" or highly regular fractals, and "statistical" fractals, or those which exhibit regularity when average properties are examined. Exact fractals are highly artificial and are not expected to appear in nature, but they are very useful for describing several basic principles.

9.2.1 Lessons from Exact Fractals

Exact fractals are geometrically regular, just as circles and triangles are regularly shaped. The regularity stems from a strict relationship between the total length of the set and the number of units composing it. The apparent regularity of the "Sierpinski carpet" (Fig. 9.3a) can be viewed in a 3 by 3 array of cells with eight

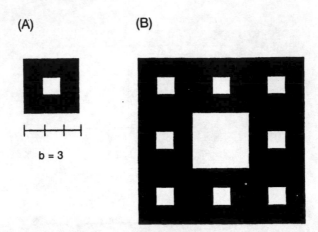

Figure 9.3. The Sierpinski carpet. (A) The smallest aggregate with $b = 3$ segments along one edge, a hole in the center, and eight unit cells (shaded). (B) An aggregate of eight of the smallest aggregates, giving a total of sixty four shaded cells.

of the nine available cells shaded (Mandelbrot 1983). It is helpful to envision the shaded cells as having "mass" and the blank areas as being "empty," like holes in a carpet. Thus, the set is $b = 3$ units long and is composed of $N = 8$ shaded cells. For calculations of the total mass of the carpet, each shaded cell is assigned a mass of 1 and is of unit area. The density of the carpet is calculated as mass/area = 8/9 = 0.888.

A larger, more complex carpet is generated recursively, much as a quilt is made by sewing small squares together to form a larger rectangle. Copying the small carpet of Fig. 3a seven times and then arranging the eight versions of the carpet produces a new carpet with a large central hole and eight small peripheral holes. The new carpet has $b = 9$ cells along one edge (Fig. 9.3b), giving a total of 8 × 8 = 64 shaded cells from the total of 81 possible cells. Of course the number of shaded cells and the mass have increased, but the density of cells has decreased (mass = 64, density = 64/81 = 0.790). Reiteration of this procedure would yield a very ornate carpet with similar structure at all scales, thereby representing the essence of exact fractals. A fractal set is called "self-similar" if a small part of it can be used to generate the whole of a larger version.

At each step of the carpet construction, the mass of the carpet may be calculated by tallying the number of shaded cells. If the mass of a given carpet with b cells along one edge is denoted as $M(b)$, then a change from b to $3b$ cells along one edge results in an eightfold increase in mass, or

$$M(3b) = 8\,M(b) = 3^{D_s}M(b) \tag{9.1}$$

and the number of shaded cells, or the total mass $M(b)$, varies with b as:

$$M(b) = b^{D_s} \tag{9.2}$$

D_s is a "fractal dimension" describing the relationship between mass and length scale. By taking logarithms of both sides,

$$D_s = \ln[M(b)]/\ln(b) \tag{9.3}$$

For example, substituting real numbers for $M(b)$ and b of the Sierpinski carpet gives $D_s = \ln(8)/\ln(3) = 1.8928$. Similarly for Fig. 9.3b, $\ln(64)/\ln(9) = 1.8928 = D_s$.

The most significant result is that the exponent remains constant even though the total mass of the carpet changes as b increases. Notice also that density decreases from 0.888 to 0.790 as b changes from 3 to 9. Thus the slope of the relationship between logarithmically transformed density and $\ln(b)$ is $[\ln(0.888)-\ln(0.790)]/[\ln(3) - \ln(9)] = -0.106$, which (within rounding error) is equal to $D_s - 2$, where 2 is the dimension of the plane within which the Sierpinski carpet resides. Thus, the fractal dimension describing the scale-dependent mass of the carpet can be used to estimate density at various scales.

The significance of the noninteger scaling exponent, D_s is seen by contrasting the above calculations with an alternative method by which the mass of the carpet could grow. If the carpet were continuous "wall-to-wall" carpeting without holes, then the mass of the carpet would grow strictly as the square of b, or the width of the floor. The fact that b is squared for wall-to-wall carpeting is often taken for granted, but squaring is appropriate when the entire plane of the floor is covered; the plane is a two-dimensional space, and thus an exponent of 2 reflects the two-dimensional nature of the carpet. In contrast, sets such as the Sierpinski carpet occupy less than a two-dimensional space, and noninteger exponents such as $D_s = 1.8928$ characterize the relationship between mass and the length scale at which measurements are made (Mandelbrot 1983).[1]

The density of carpeting changes regularly with scale and is a function of both D_s and d, where d is the dimension of the space within which the carpet resides, i.e., $d = 2$. The density (i.e., mass/area) of the Sierpinski carpet is

$$p(b) = M(b)/b^2 = b^{D_s}/b^D = b^{D_s-d} \tag{9.4}$$

which provides a general relationship that may be applied to density estimates for other fractals.

Ecologists often lack a complete map of the landscape but instead have data from isolated locations. Strictly speaking, the variates are not sampled from a

---

[1]The reader may wish to confirm that $D_s = 1.585$ for an exact fractal constructed from equilateral triangles that are $b = 2$ units long with $M(2) = 3$ shaded triangular cells out of four possible cells; i.e., the triangle has a central triangular hole, just as the Sierpinski carpet has a square hole (Fig. 9.4).

(A)   (B)

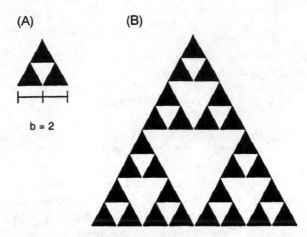

b = 2

Figure 9.4. The Sierpinski triangle. (A) The smallest aggregate has just two segments along the base and three shaded triangles within. (B) Expansion of the small aggregate yields a triangle that is eight units long.

two-dimensional universe. Rather, the sampling sites occupy a space of lower dimension. Fractal density may be adjusted for the sparse or incomplete coverage of sampling locations by first determining the fractal dimension of the locations and then estimating density at a given scale by the ratio b^{D_s}/b^{D_l}, where D_l is the fractal dimension of the sampling locations.

Density calculations that rely on fractal geometry are appropriate when pro-rating estimates obtained at one scale to broader scales. For example, assume that the population size of a rare plant species is estimated to be twenty individuals per hectare of habitat. Suitable habitat (e.g., bogs) may occupy just 2% of the landscape (when measured at a very high resolution!), and may be distributed in a fractal manner, as are the plants within the bogs. A suitable estimate of plant density over a 100-ha landscape would be made from the ratio of the number of plants and the fractal distribution of habitat:

$$p(b) = A/K \ b^{D_f - D_g} \qquad\qquad (9.5)$$

where A and K are coefficients obtained from the regressions of the logarithms of the number of plants (or bogs) versus $\ln(b)$, D_f is the dimension of the number of plants, and D_g is the "grid" fractal dimension of bogs in the landscape. The restriction of the plants to a particular subset of the two-dimensional landscape makes their density a function of the distribution of bogs.

The grid dimension is obtained by superimposing a series of grids over a map of a patch mosaic and then counting the number of grid cells of length L that are occupied by any portion of the mosaic (Fig. 9.5). The number of occupied cells increases as a function of the extent of the map divided by L, e.g., $(1/L)^D$ (Voss

(A) Patch Mosaic

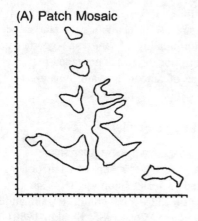

(B) Low resolution

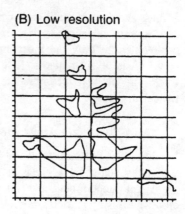

(C) High Resolution

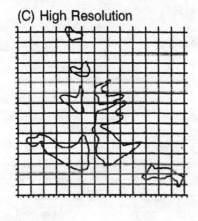

(D)

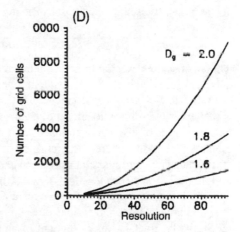

Figure 9.5. The dispersion of patches in the geographical plane is measured by counting the number of grid cells occupied by the mosaic. (A) The mosaic. (B) Occupancy of large grid cells provides a low-resolution estimate of dispersion. (C) Resolution is higher for smaller grid cells (D) The fractal dimension is the slope of the curve from the double-logarithmic regression of the counts of grid cells versus resolution. Mosaic of *Clematis fremontii* after Erickson (1945).

1988). Thus, if the mosaic fills the plane completely, the number of cells grows as the square of $1/L$ and will grow at slower rates if the mosaic covers the plane incompletely. The grid dimension is sensitive to the variance, or dispersion, of the mosaic.

In summary, exact fractals exhibit very regular structure because they are constructed on the basis of a simple and constant rule. Fractal characteristics such as mass and density are described precisely by the exponent D_s, which is a constant over a range of length scales. Exact fractals are useful heuristics for landscape ecologists, but only the most intentional manipulation of landscape geometry could create exact patterns. Rather, the description of real landscapes requires fractal models that accommodate some degree of randomness and thus are statistical in nature.

9.2.2 The Statistical Nature of Fractals

Stanley (1986) discusses several "statistical" fractals whose regular shape is revealed by statistical analysis rather than by direct inspection of each object. Statistical fractal models have been constructed to depict natural patterns such as clouds, diffusing particles, complexly folded molecules, mountainous terrain, and plants (Barnsley 1988; Lovejoy and Schertzer 1985; Voss 1988; Meakin 1988). The growing number of examples from nature suggests that complexity may be described precisely and that new fractal models may be generated virtually at will for as yet unforeseen applications in many fields.

9.2.3 Fractal Dimensions

Dimensions are essential in both Euclidean and fractal geometry for representing characteristics such as area, density, and perimeter length. For example, when the area of a disk $= \pi r^2 = \pi r^d$ is calculated, the radius is squared because the disk is a planar, or two-dimensional, set of points (i.e., $d = 2$). Although the geometrical interpretation of the exponent is often taken for granted, the implications of noninteger exponents are at the crux of fractal geometry.

Fractals have various dimensions, just as points, lines, planes, and solids have topological dimensions $d = 0$, 1, 2, and 3, respectively. Any of these familiar sets may be envisioned as a set of points residing in a higher dimensional space. Thus, a single point with $d = 0$ resides on a line of $d = 1$, or perhaps in a plane of $d = 2$. Similarly, fractals can be thought of an subsets of higher-dimensional spaces. For example, the fractal Sierpinski carpet with its noninteger $D_s = 1.8928$ resides in a higher-dimensional space of $d = 2$. However, the converse is not true, that any subset of a high-dimensional space is fractal.

A generic notion of fractal dimension may be described as the relationship between a quantity Q, and the length scale, L, over which Q is measured (Stanley 1986). Thus,

$$Q(L) = L^{D_q} \qquad (9.6)$$

where D_q represents a fractal dimension for the quantity Q. For example, the fractal dimension of a beetle's trail or walk of length l_w is measured by plotting the mean distance traveled between two points that are separated by a reference distance, L (Fig. 9.6). In practice, the reference distance L is the distance from the origin of the walk to the beetle's location after t time intervals. The mean distance traveled is calculated from many walks lasting t time intervals. Notice that a biased

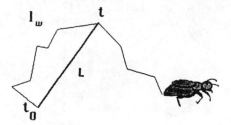

Figure 9.6. Estimation of the fractal dimension of a beetle's path. The beetle walks distance l_w between times t_0 and t. The reference length scale is the bee-line distance L.

estimate of D_w is obtained by using walks of different duration because some beetles would have a longer time to increase their path length than others would.

If a beetle walks in a very straight line, then the distance traveled equals the reference distance (i.e., $l_w = L^{D_w} = L^1$; $D_w = 1$ if the walk is linear). However, if the walk is curvy, then it occupies a greater portion of the plane and $1 < D_w < 2$. A calculation of D_w shows that beetles such as *Eleodes longicollis* follow a somewhat tortuous path (Fig. 9.7a and b) with $D_w = 1.1$, whereas a randomly walking individual has $D_w = 2$ if the landscape lacks obstructions (Stanley 1986). Clearly, real beetles do not behave randomly.

Two principles follow from this observation and may guide future inventions of new fractal dimensions. First, any D_q represents the way in which the quantity Q varies with scale. Since there are many possible quantities (e.g., area, perimeter, mass, and speed) that yield a rich variety of dimensions, it is vague to simply discuss "the fractal dimension." Rather, there are many potential dimensions pertaining to a given set.

Second, since $D_q \leq d$, a fractal may be thought of as a subset obtained by extracting some portion of the space within which it resides (see Barnsley [1988] or Morgan [1988] for discussions related to geometric measure theory). Some process may concentrate the mass of the set within a subregion of the Euclidean space within which it resides. In landscapes, erosion may remove parts of the surface in a statistically consistent fashion, leaving ridges and valleys in a fractal pattern (Gupta and Waymire 1989). As discussed by Stanley (1986) and Meakin (1988), operations may be performed on an existing fractal and the geometry of the set obtained from the operation may also be fractal (e.g., fractal density obtained from the mass of a fractal, or fractal spread of fire through fractally distributed forests). Thus, other dimensions or exponents that describe scale dependence obtain directly from the geometry of the initial set. These related dimensions are direct measures of the geometrical "consequences" of processes operating in a fractal environment.

How might these principles be applied by landscape ecologists? For example, fires spread across landscapes along a connected network of fuel (for now, fire that

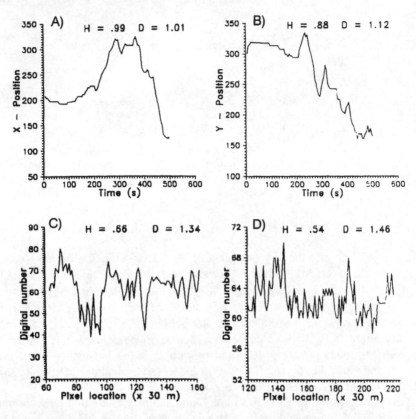

Figure 9.7. Self-affine functions are exemplified by: (A) The x-axis location of a flightless *Eleodes longicollis* beetle during a 500-*s* time interval. (B) The y-axis location of the same beetle. (C) The radiance of Thematic Mapper band 4 (near-infrared) along a transect through a mountainous region of the Sevilleta National Wildlife Refuge, New Mexico. (D) Near-infrared radiance along a second transect across a grassland in the Sevilleta. The parameters H and D describe the curvilinearity of the functions.

"jumps" or is blown about as firebrands must be viewed as a means by which isolated fuels effectively are connected by the process of burning). Rarely could fuels in a forest or grassland be envisioned as uniformly occupying the plane. Rather, a subset of the plane is rich in fuels and may exhibit consistent changes in mass or density with scale. If fire spreads by direct contact between contiguous fuels, then Stanley (1986) predicts the velocity, v, of the moving front to be

$$v \approx (p - p_c)^{(D_{min}-1)} \tag{9.7}$$

where p is the percentage of the landscape occupied by fuel, p_c is the critical probability above which the fuel forms a contiguous cluster across the landscape (see Gardner et al. 1987; O'Neill et al. 1988b), and D_{min} is the fractal dimension of the shortest path between any two points residing on the contiguous cluster. As for many quantities, D_{min} is obtained from an analysis of

$$l_{min} \approx L^{D_{min}} \qquad (9.8)$$

where l_{min} is the length of the shortest path between two points on a contiguous cluster; the two points are separated by the straight-line distance L. The critical probability p_c is "critical" because it is a threshold above which flows increase dramatically (substituting several values of p below and above $p_c = 0.5928$ in Eq. 9.7 will demonstrate this). The dimension D_{min} is theoretically sufficient to predict the velocity of the spread of fire, although the aggregation of similar patches in natural landscapes causes empirical values of p_c and D_{min} to deviate from the expectations for random maps ($D_{min} = 1.1$ when $d = 2$; Stanley 1986). Similar models of spatial propagation may apply to gene flow, seed dispersal to safe sites (Harper 1977), or the movement rates of pollutants, energy, and nutrients. The control of velocity by the critical probability p_c is an example of a "critical phenomenon," or threshold effect.

The study of critical landscape phenomena could provide a common framework for discussions of apparently unrelated processes. If, for example, p_c were known for each of several processes (e.g., the spread of fire, genes, and nutrients), then the geometry of a given landscape could be interpreted regarding its ability to propagate flows of each type. The exact values of p_c would vary among processes because factors like wind, hydrological flow, and the activity of pollinators affect the flow rates differently. Interactions between the flows might occur at broad scales only if $p > p_c$ for each of the processes. The interactions could be predicted from knowledge of p_c and D_{min} for each kind of flow. Thus, quantification of landscape geometry in an ecologically meaningful way requires information about proximal factors that regulate flows.

9.2.4 The Scaling of Fractal Moments

Voss (1988) presents a general notion of fractal dimension with a single formula to interrelate several moments of a fractal's statistical properties. Each moment, analogous to the mean, variance, skewness, and higher moments of a frequency distribution, is sensitive to particular aspects of the pattern's geometry.

Voss (1988) envisions a fractal as a set of S points residing within a larger space of dimension d (Fig. 9.8). The spatial dependence among the points is studied by measuring the probabilities $P(m,L)$ of observing m points within d-dimensional regions of size L (squares or cubes) centered on individual points from among the S points (Fig. 9.8). In the analysis of landscapes, the set of S points may be pixels denoting a particular landscape cover type, such as forest.

On a two-dimensional map, $P(m,L)$ is measured by centering a window of size L^2 on each pixel and counting the number of pixels of that type within the window. After normalizing the probabilities to a cumulative probability of 1, moments of

B.T. Milne

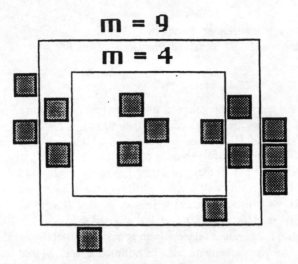

Figure 9.8. Methodology for measuring the probability of finding m cells within a window of size L. Windows of length L are centered on each cell in turn, and the number of shaded cells is counted within each window. In general, small windows enclose fewer cells ($m = 4$) than large windows ($m = 9$), and the rate at which the mean number of cells increases with L is a function of the fractal dimension of the set.

the spatial distribution can be obtained from a measure of the mass $M^q(L)$ of the S points at a given L:

$$M^q(L) = \sum_{m=1}^{N(L)} m^q P(m, L) \qquad (9.9)$$

Here, q is an index of the moment of the fractal distribution on the map, and $N(L)$ is the number of different values of m observed for a given L. The measurement of $P(m, L)$ is a departure from the sampling methods used by many ecologists in that the "quadrats" are purposefully placed directly on the objects to be measured, rather than thrown randomly. Consequently, $P(m, L)$ is a measure of the spatial dependence of the objects of interest, as viewed using a particular window size.

The probabilities $P(m, L)$ were calculated for areas with 10—19% bare soil in the Sevilleta National Wildlife Refuge, New Mexico (Milne 1990). As expected, the shape of the distribution of $P(m, L)$ versus m varied strongly as a function of L (Fig. 9.9). Taking the first moment for $q = 1$, $M^q(L)$ indicates the mean number of points observed in windows of length L. Since $M^q(L)$ can be considered a "quantity," as in Eq. 9.6, it is expected to vary with L according to a power law, where the exponent is the fractal dimension of the mosaic. Thus $M^1(L) = KL^{Dm}$ is the central tendency or the mass of the set at a particular scale, just as the mean of a frequency distribution describes its average tendency (K is a constant).

Equation 9.9 was used to solve for D in Fig. 9.1, which represents a set of coastal peninsulas relative to their length. The peninsulas were compared according to the one-dimensional attribute of length, and consequently the fractal dimen-

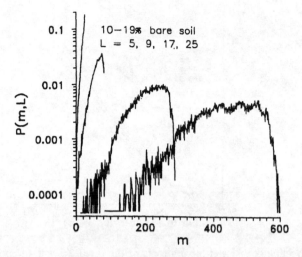

Figure 9.9. Probability density functions $P(m,L)$ from an analysis of pixels representing 10-19% bare soil in the Sevilleta National Wildlife Refuge. Curves obtained using windows that were 5, 9, 17, and 25 pixels long progress from left to right.

sion D_c must range from 0 to 1. The value of D_c = 0.15 indicated that the number of peninsulas observed simultaneously along the coast of Maine increased by this power as larger expanses of coastline were examined.

The second moment of a fractal pertains to its variance in space. The second moment is obtained by substituting $m^q = m^2$ in Eq. 9.9 (Voss 1988), thus providing a scaling exponent for the spatial variance according to $<M^2(L)>^{1/2} = K L^{D_v}$. When a two-dimensional landscape map is considered, the exponent of the second moment is sensitive to the variance, or dispersion, of the set in the plane, just as the second moment of a frequency distribution measures dispersion from the mean. When D_v = 2, the second moment has a value of $(L^2)^2 = L^4$ because $P(m,L)$ takes the value of 1.0 at $m = L^2$, and 0 for all other m. The different interpretations of Eq. 9.9 as q varies from 1 to 2 indicate the potential for tuning measurements of fractals to emphasize various properties. Interested readers are referred to Voss (1988, p. 66) for a full account of this topic. Examples appear in Lovejoy and Schertzer (1985).

Theoretically, exact fractals have identical fractal dimensions for each moment. However, natural features, like Mount Everest, are not composed of miniature copies of themselves and therefore are not exact fractals. Rather, the scaling exponents of successive moments of statistical fractals vary predictably from one to another (Meakin 1988) because of the random processes that generate natural patterns, or because of constraints that operate strongly at broad scales. Lovejoy (1982) showed that the empirical ratio between the area and perimeter of clouds and the rain shadows they cast has a constant slope when examined over six orders of magnitude. However, the thinness of the atmosphere limits the height of clouds and introduces a constraint on cloud geometry at very broad scales (Lovejoy and

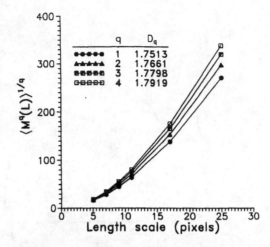

Figure 9.10. Scaling behavior of the qth roots of the q = 1,2,3 and 4th moments of the probability density function $P(m,L)$ for areas containing 0-9% bare soil in the Sevilleta National Wildlife Refuge. The fractal dimension, D_q, describing the moment as a function of length scale (i.e., window size) changes regularly with q, as is expected for naturalistic statistical fractals.

Schertzer 1985, Schertzer and Lovejoy 1985). Natural processes generate statistical, rather than exact, scale-dependent variation.

Remotely sensed measurements of the Sevilleta grasslands in New Mexico reveal areas with 0 to 9% bare soil that exhibit striking differences in the scaling exponents of the first through fourth moments (Fig. 9.10). These vegetated patterns represent aeolian deposits, arroyos, and bajadas that have remained ungrazed by cattle since 1974. Apparently the mixture of soil conditions, differential responses of plant species to the cessation of grazing, and variation in the rates of erosion and deposition contribute to the statistical fractal character of plant distribution. Thus, the lack of a constant scaling exponent for each moment indicates that adequate models of landscape formation require several processes, perhaps operating at several scales, as suggested by Delcourt et al. (1983) and Urban et al. (1987).

In summary, a fractal dimension indicates the ability of a set to fill the Euclidean space in which it resides. In landscape research, fractal measurements provide information about the space-filling properties of a mosaic of patches at all scales. If the constraints on a process (e.g., cloud formation) vary with scale but the same mechanism of pattern generation holds at all scales, the processes may amplify characteristics of the pattern differentially with scale (e.g., length versus height). In such cases, functions describing the pattern may exhibit statistical, rather than exact, scale-dependence.

9.2.5 Self-Affinity

Ecologists often collect data describing a variable, like soil moisture, as a function of spatial location or time. Random variables as functions of independent variables may exhibit a special form of fractal scale-dependence, namely self-affinity (Hurst et al. 1965). By definition, an affine transformation is one for which a point y residing in a coordinate system $(x_1, \ldots x_d)$ is transformed to become point y' with coordinates $(r_1 x_1, \ldots, r_d x_d)$, where the scaling coefficients are not all equal (Feder 1988). Variables such as soil moisture are affine in that soil moisture at a particular location x can be represented by transforming x by a factor r. A familiar example of an affine transformation is $y = mx + b$, where m is a scaling coefficient that transforms x to equal $y - b$.

Self-affinity occurs when a function exhibits small changes in the y variable for small changes in x and predictably larger variation in y over large changes in x. Examples include the spatial locations of a diffusing beetle, soil mineral content along a transect, and modeled terrain (e.g., Barnsley 1988). In these cases, effects of scale occur in the random variable for which there is one value for each combination of independent variables.

To illustrate, I obtained two data sets describing (1) the location of an individual beetle of the genus *Eleodes* as it moved within a 5-by-5-m plot in New Mexico (Wiens and Milne, 1989), and (2) the radiance of light in the near-infrared region of the spectrum along two transects extending ~2.9 km across the Sevilleta National Wildlife Refuge, New Mexico. Near-infrared radiance is used in remote sensing to monitor the amount of green vegetation present (Goetz et al. 1985). Both data sets (Fig. 9.7) involve measures of random variables (i.e., beetle location and radiance) as functions of fixed variables (i.e., time and space, respectively).

To study temporal self-affine fractals, one varies the window size and observes changes in the range of the random variable Y. The time scale is purposely manipulated by aggregating the raw observations originally separated by t time units into windows that are rt units long. Thus, N/r is the number of windows covering the N observations, and r is a scale factor for constructing windows on the abscissa within which the behavior of the random variable is measured.

In self-affine fractals, the mean range of the dependent variable within a window of length r increases with window length by the factor r_H (the symbol H is in honor of Hurst [1951]; Gefen et al. 1983; Mandelbrot 1983; Orbach 1986; Voss 1988). The scaling exponent H is related to the fractal dimension of the function by $D = 2 - H$. Thus, analysis over many intervals of length rt in the independent variable yields a measure of the mean range of Y:

$$<\max(Y)_{rt} - \min(Y)_{rt}> = K(rt)^H \qquad (9.10)$$

in which the mean range, denoted by inequality signs, increases with the time scale as the power H (Fig. 9.11).

In practice, the dependent and independent variables are rescaled to fit on the

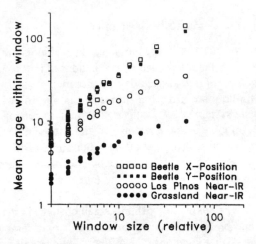

Figure 9.11. The mean range of self-affine functions observed within a window increases as a power of *H*, the slope of the curve. Explained variance is > 95% in all cases. Results based on data from Fig. 9.7, in which *H* and *D* = 2 − *H* are reported.

unit inverval (i.e., all values range from 0 to 1, inclusive). Then the independent axis is divided into *N/r* adjacent windows for each of several values of *r* used in the analysis, and each window is examined to determine the range of the random variable in the window. The mean range for each value of *rt* is then regressed against *rt*, and the slope of the double-logarithmic relationship provides an esti-mate of *H*. The range is desired for measures of *H* because it indicates the magnitude of fluctuation that occurs over windows of size *rt*. The Hurst exponent has an expected value of 0.5 for Brownian functions in which the Gaussian random variable has equal probabilities of increasing or decreasing at all scales.

In general, self-affinity pertains when a pattern has one form of scaling on one axis and a second form of scaling on a second axis, owing to the different qualitative "meanings" of the axes. Whereas a coastline meanders similarly with regard to latitude and longitude (i.e., the reference axes), diffusion patterns, for example, vary one way with time and another way with location. For this reason, estimating the dimension of self-affine functions by a grid counting method (Fig. 9.5) results in erroneous estimates of *D* because the orthogonal sides of the grid squares are assumed to have identical relationships to variation in the fractal pattern on both axes (see Feder 1988, p. 184).

Variation in the fractal dimensions obtained from self-affine functions can be used to detect edges and ecotones. As an illustration, an ≈5.1-km transect was made across part of the Sevilleta National Wildlife Refuge, New Mexico, spanning from an extensive grassland southeastward to the Los Pinos Mountains (Fig. 9.12a). Radiance measurements were obtained from the Thematic Mapper Landsat sensor in the near-infrared region and mapped according to the positions of the 30-m pixels (i.e., grains). The transect was then divided into many overlapping

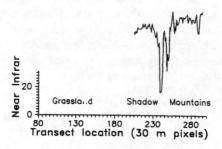

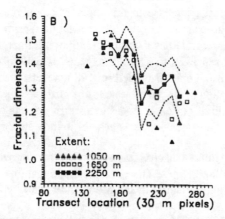

Figure 9.12. Ecotone detection by fractal analysis. (A) Radiance (digital number) in the near-infrared channel of the Thematic Mapper (TM) sensor along a 5.1-km transect through the Sevilleta National Wildlife Refuge, recorded 10 September 1987. The major dip in TM band 4 represents a shadow cast by Los Pinos mountain range at the high end of the transect. (B) Fractal dimensions of spectral data for subsets of the transect. Subsets varied in length from 35 pixels (i.e., ≈1050 m) to 75 pixels (≈2250 m); the centers of subsets were located on the transect as indicated by points on the graph. Dashed lines enclose two standard errors of the fractal dimensions estimated in subsets of ≈2250 m.

subsets of differing lengths (i.e., 1050 to 2250 m). The subsets represented smaller extents within which H and the fractal dimension were estimated.

Mapping the fractal dimension according to the position of the subset along the transect revealed a striking change from $D \approx 1.5$ to $D \approx 1.3$ (Fig. 9.12b; R^2 of regressions used to estimate D ranged from 0.86 to 0.97). The change in dimension at about pixel number 210 corresponds to the interface between the grassland and the shadow cast by the mountain. Analyses made within small extents detected fine-scale ecotones between grass swards living on different soils or alluvial fans. In contrast, analyses within relatively broad extents revealed ecotones stemming from broad-scale features (e.g., mountain ranges). These findings suggest a general strategy for partitioning variation in landscape structure according to the scale at which constraints on landscape patterning operate. By changing extent and measuring variation within each extent, maps of landscape texture or fractal structure may be constructed for each extent. The detection of ecotones at multiple scales is consistent with predictions of hierarchy theory (Allen and Starr 1982; O'Neill et al. 1986) because the appearance of hierarchical entities, like patches, is presumed to vary with the scale at which observations are made.

9.3 The Diversity of Fractal Models

Mandelbrot (1983), Barnsley (1988), and Peitgen and Saupe (1988) read like catalogs of naturalistic and whimsical fractals. Each fractal has a dimension, D, describing the exponential increase in the apparent complexity of the whole as finer scales are used to examine the set. The fractal dimension from a particular model should not be interpreted as a universal number. Rather, the model and the corresponding D are sensitive to specific aspects of pattern or structure, e.g., density, area occupied, and mass (Mandelbrot 1983; Schertzer and Lovejoy 1985). Consequently, application of fractal models to the analysis of patterns requires an awareness of the sensitivity of each model to specific aspects of structure. When one is reporting a fractal dimension, it is helpful to indicate the geometric aspect represented.

Burrough (1986) gathered useful fractal models that are the cornerstone of techniques applied to landscapes. The ecological applications, assumptions, and computational limitations of some of the models are discussed below.

9.3.1 Area Versus Perimeter Relationships

Landscape patches are readily characterized by area and perimeter lengths, which affect species utilization of patches (e.g., Forman et al. 1976; Van Dorp and Opdam 1987). Fractal models of area and perimeter are appropriate given that patch shape varies with patch size (Krummel et al. 1987).

In Euclidean geometry the area A of a patch is related to the diameter L by

$$A = BL^2 \tag{9.11}$$

For disks, B is a constant equal to $\pi/4$. The relationship is generalized to fractal patches by the equality between area and patch length:

$$A = \beta L^{D_a} \tag{9.12}$$

This relationship is similar to the generic equation for fractal scaling (Eq. 9.6), in that patch length provides the reference length scale (L), and area is a dependent quantity. Thus D_a is the fractal dimension of area given the distance between the two most distant points on the perimeter. The area dimension is sensitive to the lobes and convexity of the patch, with high dimensions ($D_a \sim 2$) indicating an obtuse shape. Patches with peninsular lobes have lower area dimensions because such patches resemble wide, albeit convoluted, lines (D_a ~1.2). The area is low, for example, given the patch length.

Landscape ecologists could use the area dimension to predict species interactions. For example, brown-headed cowbirds are effective nest parasites of woodland birds (Brittingham and Temple 1983). Female cowbirds explore the interiors of forest patches to locate nests of other species in which to deposit an egg, leaving it for the host female to incubate and rear. The search for host nests is balanced by the need to remain near to patch edges where food for cowbirds is relatively dense. Thus, maximal penetration distances for cowbirds are limited to ~300 m from the forest edge (Brittingham and Temple 1983). Area dimensions estimated from Eq. 9.12 could be useful for modeling the vulnerability of nests to cowbird parasitism. Low-dimensional patches have an exponentially higher proportion of area represented by "edge" and should be subject to intense parasitism throughout.

The general relationship between a quantity and the length scale (Eq. 9.6) suggests that a different dimension can be constructed to describe patch perimeter relative to the diameter or length of the patch. Taking diameter as the length scale L, indicates that

$$P = \tau L^{Dp} \tag{9.13}$$

where D_p is a dimension describing the perimeter relative to the diameter, and τ is a constant. If perimeter grows rapidly with L (i.e., D_p ~2), then the perimeter must be very convoluted for a given patch diameter. Expressing patch area (Eq. 9.12) in terms of perimeter and the perimeter dimension (Eq. 9.13) gives

$$A = \beta \left[\frac{P^{1/D_p}}{\tau} \right]^{D_a} \tag{9.14}$$

$$= \beta (P/\tau)^{(D_a/D_p)}$$

This relationship holds for familiar objects like disks, which have $D_a = 2$ and $D_p = 1$. Substituting the values $\beta = \pi/4$, $\tau = \pi$, and $P = \pi L$ into Eq. 9.14 confirms that the area of a disk $= \pi(L/2)^2$. Ecologically, this relationship allows the synthesis of information about the scale-dependent behavior of patch area and patch perimeter. In general, the area of fractal patches can be expressed as a function of perimeter raised to an exponent, which is the ratio of the area and perimeter dimensions.

9.3.2 Density

Percolation theory (Stauffer 1985; Orbach 1986; Gardner et al. 1987; O'Neill et al. 1988b) is the study of spatially random processes. Originally applied in studies of porous materials such as ceramics and metallic films, percolation theory has found application in landscape ecology as a statement of the expected structure of landscapes under the assumption of randomness. Percolation theory provides random expectations for the aggregation of patches and the movement of animals or disturbances.

Percolation theory considers the implications of casting points onto the plane with some probability p. In studies of so-called site percolation, a lattice is formed and then the cells of the lattice are turned on with probability p. Surprisingly, if $p > 0.5928$ there is an asymptotically high probability that the cells on a square lattice will form a continuous "infinite" cluster spanning from one edge of the map to the other (Orbach 1986). Be definition, a "cluster" of connected cells on a square lattice consists of cells having either vertical or horizontal neighbors. The possibility that an organism could potentially move across the entire map without leaving the infinite cluster led to the term "percolating network" to describe the cells. The infinite cluster exhibits scale dependence in its mass and density with an area that has a fractal dimension of ≈ 1.92 (Orbach 1986). The process of mapping randomly chosen points affects the arrangement of boundaries between the infinite cluster and neighboring clusters of finite size (Milne 1987). Maps (or landscapes) that are analyzed on triangular or hexagonal lattices (Stauffer 1985) may also exhibit infinite clusters, but p_c will differ from $p_c = 0.5928$.

Although the density of occupied sites on the infinite cluster varies with window size, some properties, such as the likelihood of spanning the window, are essentially constant for large maps. The density of occupied cells on the cluster satisfies

$$p(L) = BL^D/CL^d \qquad (9.15)$$

where $p(L)$ is the scale-dependent density at length scale L, d is the dimension of the space containing the fractal, and B and C are constants (Orbach 1986). The density of points within the cluster decreases as L^{D-d}, because there is relatively more open space at broad length scales.

9.3.3 Diffusion

The dispersal ability of organisms effectively alters the consequences of landscape structure such that vagile species (e.g., Kareiva 1983) perceive a highly connected landscape mosaic, while poor dispersers may perceive highly dissected conditions within the same landscape (M.G. Turner and R.H. Gardner, personal communication). Thus, diffusion rates depend on the fractal geometry of landscape pattern and on the diffusion ability of organisms (Wiens and Milne, 1989).

Classical diffusion models describe motion in homogeneous environments in which the distribution of barriers to movement is assumed to be constant from one subregion to another and across all spatial scales. These assumptions do not apply over short time periods in percolating networks (Fig. 9.13) where the mean-squared displacement (i.e., a measure of the distance moved) varies with time raised to an exponent (Gefen et al. 1983). On the infinite cluster, the exponent equals 0.71, and thus the rate of change in mean-squared displacement decreases with time. In networks having $D = 2$, i.e., a homogeneous plane, the fractal diffusion equation reduces to the classical formula (e.g., Berg 1983), and mean-

Figure 9.13. Example of a percolation cluster (shaded) and a path through the cluster. Side branches of the path indicate reversals in the walk, which progressed generally from left to right.

squared displacement increases linearly with time. For example, the expected mean-squared displacement for a particle diffusing on a surface with $D = 2$ and with a diffusion coefficient of 10 will be 50 m² over five time units and 500 m² over fifty time units. However, on the percolating cluster, the same particle would move only 31.3 m² in five time units and 160.7 m² in fifty time units, a decrease of 67.8% compared to diffusion on the continuous plane. Of course, if an organism is sufficiently vagile, the small bumps of the landscape vanish from the viewpoint of the organism. Diffusion in a "heterogeneous" landscape may behave according to classical theory (e.g., Okubo 1980; Berg 1983) for some organisms and according to fractal theory for others (Wiens and Milne, 1989).

The slowing of diffusion within fractal networks can be explained by imaging an organism that moves only within one particular kind of patch. Assume that the patch, interspersed with holes of all sizes, is highly irregular and that it is possible to move from any point in the patch to another, albeit via a highly convoluted pathway. Assuming that the organism is small relative to the narrowest constriction in the patch, the organism will be relatively free to move in a straight line over small distances. Over very small distances the patch appears to be uniform, with $D = 2$. However, as we consider even larger displacements (as the crow flies), the organism will find it increasingly difficult to make the traverse without encountering convolutions in the interior boundaries of the patch. On average, the spatial complexity within fractal networks damps diffusion most strongly at large spatial and temporal scales.

9.3.4 Semivariance

Fractal analysis shares characteristics with other techniques for evaluating spatial patterning. Burrough (1981) and Palmer (1988) estimated fractal dimensions by using the semivariogram, a mainstay of geostatistics (Davis 1986; Robertson 1987). Semivariograms provide measurements of variance at many scales by comparing the values of a random variable at two points separated by a given lag distance (Fig. 9.2). Given points in space, each with a value for some random variable X, the semivariance is defined as:

$$g(h) = \tfrac{1}{2} N(h) \sum_{j=1}^{N(h)} (X_j - X_{j+h})^2 \qquad (9.16)$$

where $g(h)$ is the semivariance at lag h, $N(h)$ is the number of pair-wise comparisons at lag h, and X_j is the random variate at position j.

Autocorrelated patterns within a statistically homogeneous region have an expected semivariance that increases linearly over relatively short lags but then reaches as asymptote at long lags (Burroughs 1983a). An estimate of the fractal dimension of the variable is obtained via transformation of the slope m of the double logarithmically transformed semivariogram over its linear extent: $D_b = (4 - m)/2$ (Burrough 1981, 1983a). Analyses of semivariance indicate that anomalies or drift obscure the expected shape of the semivariogram, thus limiting the range of lags over which the fractal dimension is constant (Burrough 1983b; Palmer 1988). Known limits to the length scales over which constant fractal dimensions are found (Orbach 1986) enable the spatial or temporal responses of variables to be partitioned into domains of scale-dependence versus scale-independence (Wiens 1989).

Simulation models may be subject to underestimation of the role of extreme events. Mandelbrot and Wallis (1969) provide examples of river discharges that exhibit regular increases in variance through time (Mandelbrot 1983). As discharge records are observed over ever-increasing windows in time, the range of values increases steadily, suggesting that ecologists involved with long-term studies (Likens 1989) will also observe more extreme events as time goes on. Ecosystem models driven by external factors (e.g., river flows, storm events, or pollutant inputs) may be made more realistic by first establishing the time-scale dependent dynamics of the forcing factors (e.g., as in Eq. 9.10) and then assuming that the range in values of the forcing factor will increase steadily through time. Thus, long-term simulations could include fluctuations of large magnitude that exceed any previously observed fluctuations. Feder (1988) provides further discussion.

9.3.5 Mixed Fractal Models

Some ecological interactions, such as nest parasitism, may be regulated simultaneously by two or more geometric effects. For example, if female cowbirds readily penetrate 300 m into forest patches, then we might expect patch perimeter dimensions to be of little use for predicting penetration success in patches < 600 m in diameter ~28–36 ha. However, the perimeter length of larger patches will be a predictor of parasitism because lobes are more easily invaded than patch interiors. Thus, two dimensions may be needed to model parasitism, i.e., a dimension sensitive to the distribution of patch area and a dimension of patch perimeter.

Dual fractal dimensions may enhance predictability in landscapes where the dimensions vary with patch area (e.g., Krummel et al. 1987) or location (O'Neill et al. 1988a; Fig. 9.12b).

Mandelbrot (1983) describes relationships between the number of islands in an archipelago and the diameters of the islands. Island geometry was partitioned into two components that allowed fragmentation per se to be treated independently of the fractal dimension of the islands' coastlines (Milne 1988). Partitioning the geometric characteristics of an archipelago would be of use if coastline shape and island number were hypothesized to control ecological interactions differently.

Mixed models could apply equally well to the dynamics of forest fires and seed dispersal. The spread of forest fires depends on fuel loads and the spatial patterning of the fuel (Kessell 1979). The first moment from Eq. 9.9 could be used to estimate the expected mass of fuel at each scale. Gaps in fuel would be found at fine scales, but the total mass of fuel would increase as broader areas were studied. Consequently, the fire will burn somewhere within the study region (broad scale) but will be missing from small areas within the region due to a local lack of fuel. Regarding the spread of fire, the clustering of fuel may be analyzed by the grid dimension (Fig. 9.5), which characterizes the spatial variation throughout the extent of the study independently of the fuel mass. If the study region is partitioned into subregions and the grid dimension of fuel is estimated within each subregion, then the fire should be expected to spread most easily within subregions having high grid dimensions for fuel. Thus, the dynamics of fire throughout a region may be functions of both the mass and spatial variance of fuel, which can be parameterized across a range of scales by fractal models.

Harper (1977) suggests that plant abundance is regulated by the number of seeds available for germination, the suitability of microsites for successful germination, and predation. Seed dispersal and safe-site distribution could be addressed by a mixed model. The density of dispersed seeds (mass dimension, D_m) and the distribution of safe sites (grid dimension, D_g) could be combined to study the interaction of these factors across a range of scales. Hypothetically, ruderal species that disperse well would exhibit seed density with $D_g \approx 2$ but would be limited by the distribution of safe sites which have a lower fractal grid dimension and are encountered infrequently by seeds (e.g., Murray 1988).

Alternatively, seed dormancy could be viewed as a strategy for integrating the instantaneously fragmented distribution of safe sites into a uniform and complete coverage through time. The "shadow theorem" (Barnsley 1988) explains how such a system could work (Fig. 9.14). If we envision a dry sidewalk suddenly pelted with raindrops, there will be a sparse coverage of drops just seconds after the cloudburst begins (i.e., fractal dimension of the raindrops is $D_g \ll 2$). Soon thereafter, drops will have covered the surface completely (i.e., $D_g = 2$). Thus, the effective fractal dimension approaches two when the raindrops are integrated through time.

Similarly, germination sites wink on and off in the landscape (Marks 1974) such that any seed has a probability of < 1 of occupying a germination site at any one time. However, because there is a shifting mosaic of germination sites through

equal to A. The maximum area-perimeter dimension for any cluster of A pixels can be found by the following algorithm. First, a quantity A_s is defined as the largest perfect square that is $\leq A$:

$$A_s = [\text{INT}(A^{\frac{1}{2}})]^2 \qquad (9.17)$$

Then, the maximum possible fractal dimension is found by considering how D_a relates A to the perimeter. As before,

$$A = (P/4)^{D_a} \qquad (9.18)$$

(Gardner et al. 1987) and after substitution of the perimeter in terms of A_s,

$$D_{max} = \frac{\log(A)}{\log\left[\dfrac{A_s^{\frac{1}{2}} + C}{4}\right]} \qquad (9.19)$$

where
$$C = \begin{cases} 0 \text{ if } A = A_s \\ 2 \text{ if } A - A_s \leq (A_s)^{\frac{1}{2}} \\ 4 \text{ if } A - A_s > (A_s)^{\frac{1}{2}} \end{cases}$$

These rules can be envisioned by drawing a perfect square on graph paper and adding pixels along one edge of the square (Fig. 9.16). If the number of new pixels is $\leq (A_s)^{\frac{1}{2}}$, then $C = 2$ because the only perimeter in addition to that of A_s corresponds to the two sides of the pixels at the end of the row of additional pixels. A total of $C = 4$ new pixel sides is used in Eq. 9.19 when the number of additional pixels wrapped around two sides of the perfect square increases to $> (A_s)^{\frac{1}{2}}$.

9.4.2 Ensembles of Objects

In contrast to the analysis of single patches, there are several philosophical and pragmatic reasons for working with groups of patches. First, Mandelbrot (1983) considers the fractal geometry of sets with $\gg 1$ objects and shows that one of the special contributions of fractal geometry is the ability to treat numerous objects simultaneously. By examining suites of islands , ensembles of mountain slopes, or clusters of galaxies, complex sets can be treated holistically, rather than as individuals. Second, estimates of fractal dimensions based on a single object may be

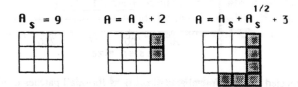

Figure 9.16. Characteristics regulating the constant C in Eq. 9.19. Left, a perfect square, $A_s = 9$, $C = 0$; center, the addition of $\leq (A_s)^{1/2}$ cells, $C = 2$; right, the addition of $\geq (A_s)^{1/2}$ cells, $C = 4$.

subject to bias, as shown above, and may provide no estimate of the error in the estimate of *D*.

Krummel et al. (1987) measured the perimeter dimension for groups of forest patches rather than for individuals. Two thousand patches were ranked from smallest to largest, and then subsets of 200 patches of similar rank were selected for analysis. The fractal perimeter dimension of the smallest 200 patches was estimated by regressing the logarithm of perimeter on the logarithm of area. Successive subsets of increasing mean rank were chosen, and the dimension of each ensemble of 200 patches was graphed as a function of mean patch area. A discrete change in the fractal dimension at 60 to 73 ha was hypothesized to reflect a change in the factors regulating patch shape (i.e., agriculture versus geomorphic features).

Changes in the fractal dimension are affected by the window size or the extent used for analysis. Variation in Fig. 9.12b indicated that the fractal dimension was affected somewhat by the extent over which the dimension was estimated. If we tune the extent used to estimate *D*, dimensions that are sensitive to local processes could be partitioned within a landscape, perhaps in accordance with the predictions of hierarchy theory (Urban et al. 1987). The patterns of vegetation within the Sevilleta National Wildlife Refuge (S. Turner et al., Chapter 2) suggest that a cascade of patterns exists at scales below the resolution of the spectral data in Fig. 9.12a. The complexity that appears with decreasing grain size challenges ecologists to consider how the high-resolution patterns of plant abundance relate to the relatively coarse resolution patterns apparent in spectral data. In summary, dimensions may be measured for single objects and for suites of many objects. The latter approach yields error estimates and best reflects the spirit of fractal geometry.

9.5 Intuitive Development of Scale-Dependent Models

Models of complex patterns, like snowflakes and dendritic features of flowing fluids, are statistical mechanical explanations of the origin of fractal patterns (Nittmann et al. 1985; Nittmann and Stanley 1986). Hypothetically, generative forces and constraints work in opposition to create pattern. In models applicable to the growth of plant roots, Meakin (1983) assumes that the generative process of cell addition is favored by diffusion gradients. Small irregularities on the surface of existing roots bias the addition of future cells to the growing mass, although the probability of adding cells varies with resource concentration outside the root. Interestingly, knobs or protuberances in the young structure are magnified through time to create macroscopic structures (Nittmann et al. 1985, Sander 1986. Fig. 9.17). The fractal dimension remains constant (within limitations of Eq. 9.19) through time, even in the face of noise (Nittmann and Stanley 1986) because the process regulating cell addition is constant. In landscapes, interactions between process and constraint seem apparent in stream erosion, where gravity and the second law of thermodynamics dictate the eventual downward travel of material. However, local impediments to material flow bias the cutting of stream channels,

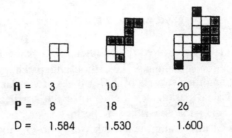

Ꭺ =	3	10	20
P =	8	18	26
D =	1.584	1.530	1.600

Figure 9.17. Simulated growth of a cluster of cells. Area (A) and perimeter (P) increase from left to right, while the fractal dimension relating area and perimeter remains steady. Shaded cells represent new growth.

resulting in small irregularities that are propagated to broad scales, as is observed in automata (Wolfram 1986). Streams exhibit scale dependence (Gupta and Waymire 1989) because the volume of water (i.e., the power that moves material) is low in the upper reaches of the stream so that only small rivulets are created, but power increases downstream as water is accumulated and more concerted cutting of stream banks occurs. Downstream reaches appear more linear (over a given spatial extent) because sufficient power has accumulated to overcome relatively small obstructions.

The change in geometry as a function of constraint suggests a general axiom for interpreting fractal dimensions. Virginia Dale (ORNL, personal communication) quantified the fractal boundaries of patches created by aphids in the Fraser fir forest of the Smoky Mountains. During the initial outbreak, patch perimeter dimensions were nearly 1.0, indicating smoothness. As the aphid population began to saturate the fir forest, the dimension of the aphid patch adopted that of the elevational isopleth below which Fraser fir did not grow. Thus, the topographic boundary of the food resource increased the perimeter dimension of the aphid patch, indicating greater constraint on the spread of the population. The fractal dimension may be indicative of constraints that restrict flows.

In the interpretations of fractal dimensions observed throughout eastern North America (O'Neill et al. 1988a), patch perimeter shape was affected by topography in Edmonston, Maine, and agricultural practices in Iowa. As with aphids, patch perimeter shape was attributed to the dominant constraints in the landscape, and in each case patch perimeter assumed the dimension of the constraints (i.e., edges of plowed fields, topography, and elevational isopleths).

Two expectations may be stated here regarding constraint and the perimeter dimension. First, a dynamic process will spread across the landscape until physical constraints are reached (Fig. 9.18). The dimension of the growing patch may not equal the dimension of the constraining landscape feature until the system becomes constrained. While spread is limited by rate processes early on, thus allowing for smooth perimeters and obtuse polygons, spread eventually is limited by the availability of suitable environments. Second, at equilibrium the dimension

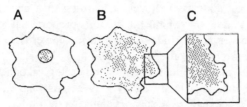

Figure 9.18. Expansion of a growing patch. (A) An expanding patch (stippled) grows without external constraint of the surrounding patch. (B) As the patch grows, it fills the constraining perimeter. (C) The filling of complex perimeters may require exponentially longer periods of time, compared to smooth perimeters.

of the patches created will equal that of the constraint. If the constraining features have higher perimeter dimensions, there will be exponentially more perimeter for the growing patch to fill, and the rate of filling may taper off asymptotically with time. These expectations suggest relationships between patterns and processes at many scales.

9.6 Conclusions

Each application of fractals in landscape research has been rewarded with new insights gained by conducting measurements across a wide range of scales, but limitations are also apparent. Exact fractal patterns are unlikely to occur in land-scapes because contemporary patterns are the results of several processes that dominated in the past. So far, little is known about how the legacies of different processes (e.g., erosion versus deposition) accumulate to produce existing fractal landscape patterns. As in the transition from grassland to mountainous terrain along a transect, fractal dimensions may vary spatially if the constraints that affect pattern vary throughout the landscape (Fig. 9.12b). Mandelbrot (1983) recognizes the nonstationary aspect of natural patterns, and Burrough (1983b) offers an alternative way of incorporating discrete changes in the fractal patterns of soil properties.

Additional limitations result from interactions between the grain and extent of the data and the scale of analysis (i.e., lag and window size, Fig. 9.2). For example, in the area-perimeter relationship Eq. 9.14, very fine-grain data are necessary to reduce bias created by a few pixels in small patches. In general, calculations of fractal dimensions may be bounded by characteristics of scale other than the length-scale variable in the fractal model.

The distinction between self-affine scaling and statistically self-similar patterns is a more subtle aspect of natural patterns. In general, analyses of self-affinity are appropriate if a random variable is a function of an independent variable, e.g., biomass versus transect location. An example of a potentially confusing case is the mapping of a random variable as a function of both latitude and longitude. Although both independent axes are geographical, the random variable will not be (e.g., biomass). Thus, biomass will vary according to a generalized model of Brownian motion, and it will exhibit fluctuations proportional to the square root of the area of the observation window raised to the power H in Eq. 9.10. Affine functions are common and require special attention.

Long-term studies may observe fluctuations in ecological behavior as functions of both space and time, in which case H may assume one value for temporal changes and another for changes in space. Falsifying the null hypothesis that H is the same in time and in space would enable variation in the long-term record to be partitioned according to the source of long- versus short-term trends at multiple scales in time and space. Differences between the spatial and temporal scaling behavior would provide measures of the degree to which space-for-time substitution is possible (see Pickett 1989). For instance, if temperature is measured both in time and space, an affine parameter for fluctuations in time may be found ($H_t = 0.5$; $D = 2 - H_t = 1.5$), while $H_s = 0.5$ may be found for spatial variation. These equal values would indicate that Brownian motion was sufficient to model temperature fluctuations in time and space. However, a change in the temporal value to $H_t = 0.9$ would indicate a correlation in temperature through time despite the random fluctuation in space. Thus, spatial patterning and temporal dynamics are similar only to the extent that the respective values of H are equal.

If yet another definition of disturbance could be tendered here, it would be that "disturbance" occurs when the affine scaling parameters in time and space converge and the affine variation in time resembles that in space. Undisturbed systems may be those with $H_t > H_s$, which means they appear stable because spatial variation changes slowly.

Fractal geometry allows ecologists to view landscapes at multiple scales and thereby achieve predictability in the face of complexity. However, much of the theoretical rigor of fractal geometry has not been exploited as in other fields. For instance, Barnsley (1988, p. 147) states that "two sets which are metrically equivalent have the same fractal dimension," although the converse need not hold. Metric equivalence occurs when one set is stretched without ripping, folding, or extreme stretching to yield a second set, much as a drawing on a toy balloon is transformed by inflating the balloon. Although the absolute (i.e., Euclidean) distance between two points changes as the balloon inflates, a new distance metric can be defined for the inflated balloon, such that each of the new locations of the points correspond one-to-one with the positions before inflation. Such a transformation ensures metric equivalence. Remarkably, the fractal dimension of the pattern before and after inflation will be the same if the calculations are made using the appropriate metrics.

This example suggests two falsifiable hypotheses for landscape ecologists interested in developing a comprehensive theory of landscape structure. First, identical fractal dimensions for forests in two landscapes represented by the same kind of data (e.g., photographically interpreted at 1:24,000 scale, digitized with grain sizes of 1 ha and extent of 30,000 ha), and therefore residing in the same metric space, would indicate an extraordinary geometric similarity. Second, and much less trivially, we could envision two metrically equivalent sets (e.g., a stream and a cow pasture residing in appropriate metric spaces) that exhibit identical fractal dimensions of nutrient distribution. Under Barnsley's (1988) definition, we expect that if the two systems are indeed metrically equivalent, then they share an identical fractal dimension. Thus, disparate systems can be equated by appropriate transformations of the metric space.

An alternative use of fractal dimensions involves the relationship between theory and empirical study. In theoretical models of dendritic growth (e.g., Meakin 1983), the models yield an expected fractal dimension (e.g., of snowflakes or root patterns), given the explicit assumptions about the process involved. Empirical tests of the model in metrically equivalent systems must yield similar dimensions. Falsification of the model by empirical data challenges the theorist to develop new models.

Landscapes throughout the eastern United States have patch perimeter dimensions in the range of 1.23 for Dubuque, Iowa, to 1.44 for Edmundston, Maine (O'Neill et al. 1988a). The low dimension reflects the straight patch boundaries of agricultural fields, while the higher dimension results from irregular topographic or coastal boundaries. Here, empiricism leads theory, and models are needed that incorporate processes (e.g., agricultural land use and erosion) and constraints (topography, barriers to development) working in opposition to create landscapes with particular fractal dimensions. Modeled landscapes will be sufficient representations of nature when both exhibit similar dimensions.

Fractal geometry is rich in yet undiscovered applications and will benefit by the linking of its inherently descriptive capabilities with explanations of the origin of fractal patterns by means of dynamic processes. The potential contribution of fractal theory to ecology is perhaps as fundamental as that of Whittaker (1967), who formalized the theory of gradients put forth by Gleason (1926), and the contribution of hierarchists (Simon 1973; Rosen 1977; Allen and Starr 1982; O'Neill 1986) who espoused the nearly discrete organization of systems. Similarly, fractal theory offers a fundamentally different view of the world and ultimately may enable ecologists to integrate systems across levels of organization at many temporal and spatial scales.

9.7 Summary

The complexity and ubiquity of natural patterns beg for quantification in ways that enhance ecologists' ability to make predictions. Much of the complexity stems from interactions between the differential responses of organisms operating at various scales and the heterogeneous patterns of resource abundance or fragmented spatial distributions of habitat. Fractal geometry includes a rich collection of models that apply to diverse quantities such as density, radiance, movement locations, patch mosaics, and peninsular lengths, all of which exhibit consistent changes with scale. Equally diverse are the scaling parameters, or fractal dimensions, that are used to relate various quantities to the length scale, or resolution, at which measurements are made. A general probability density function provides a link between concepts of abundance, spatial variance, and fractal dimensions, thereby giving a context within which many disparate aspects of pattern may be treated. Several aspects of scale affect estimates of fractal dimensions and offer opportunities for tuning a spatial analysis to highlight landscape structures such as ecotones, which exist at many scales simultaneously. Fractal models of landscape structure provide the elements of a calculus for quantifying and predicting the multiscale dynamics of landscape processes.

Acknowledgments

Ongoing discussions with V. Dale, R.H. Gardner, A. Johnson, N. Kotliar, R. Lathrop, Y. Marinakis, R.V. O'Neill, M.G. Turner, and J.A. Wiens have helped to identify the potential contributions of fractals to landscape ecology. R.H. Gardner first showed me the fluctuation in the expected fractal dimension of patches. S. Andrews, S. Loftin, and T. Morse assisted with the beetle observations, and C. Crawford provided beetle identifications. B. Musick is a constant help in matters of remote sensing. Funding provided by NSF Grants BSR–8806435, BSR–8614981, and DOE Grant DE–FG04–88ER60714. Sevilleta LTER contribution no. 5.

References

Allen, T.F.H. and Starr, T.B. 1982. *Hierarchy: Perspectives for Ecological Complexity.* Chicago: Univ. of Chicago Press.

Barker Schaaf, C., Wurman, J., and Banta, R.M. 1988. Thunderstorm-producing terrain features. *Bull. Am. Meteorol. Soc.* 69:272–277.

Barnsley, M.F. 1988. *Fractals Everywhere.* New York: Academic Press.

Berg, H.C. 1983. Random Walks in Biology. Princeton, N.J.: Princeton University Press.

Brittingham, M.C. and Temple, S.A. 1983. Have cowbirds caused forest songbirds to decline? *BioScience* 33:31–35.

Brown, J.H. 1981. Two decades of homage to Santa Rosalia: toward a general theory of diversity. *Am. Zoo.* 21:877–88.

Burrough, P.A. 1981. Fractal dimensions of landscapes and other environmental data. *Nature* 294:241–43.

Burrough, P.A. 1983a. Multiscale sources of spatial variation in soil. I. Application of fractal concepts to nested levels of soil variations. *J. Soil Sci.* 34:577–97.

Burrough, P.A. 1983b. Multiscale sources of spatial variation in soil. II. A non-Brownian fractal model and its application in soil survey. *J. Soil Sci.* 34:599–620.

Burrough, P.A. 1986. Principles of Geographical Information Systems for Land Resources Assessment. Oxford: Clarendon Press.

Davis, J.C. 1986. *Statistics and Data Analysis in Geology.* 2nd ed. New York: J. Wiley and Sons.

DeCola, L. 1989. Fractal analysis of a classified Landsat scene. *Photogram. Eng. and Remote Sensing* 55:601–10.

Delcourt, H.R., Delcourt, P.A., and Webb, T. III. 1983. Dynamic plant ecology: the spectrum of vegetational change in space and time. *Quat. Sci. Rev.* 1:53–175.

Erickson, R.O. 1945. The Clematis Fremontii var. Riehlii population in the Ozarks. *Ann. Missouri Bot. Garden* 32:413–60.

Feder, J. 1988. Fractals. New York: Plenum Press.

Forman, R.T.T., Galli, A.E., and Leck, C.F. 1976. Forest size and avian diversity in New Jersey woodlots with some land use implications. *Oecologia* 26:1–8.

Forman, R.T.T. and Godron, M. 1986. *Landscape Ecology*. New York: Wiley.

Gaines, S. and Roughgarden, J. 1985. Larval settlement rate: a leading determinant of structure in an ecological community of the marine intertidal zone. *Proc. Nat. Acad. Sci. (USA)* 82:3707–11.

Gardner, R.H., Milne, B.T., Turner, M.G., and O'Neill, R.V. 1987. Neutral models for the analysis of broad-scale landscape pattern. *Landscape Ecol.* 1:19–28.

Gefen, Y., Aharony, A., and Alexander, S. 1983. Anomalous diffusion on percolating clusters. *Phys. Rev. Lett.* 50:77–80.

Gleason, H.A. 1926. The individualistic concept of the plant association. *Bull. Torrey Bot. Club* 53:7–26.

Goetz, A.F.H., Vane, G., Solomon, J.E. and Rock, B.N. 1985. Imaging spectrometry for earth remote sensing. *Science* 228:1147–53.

Gupta, V.K. and Waymire, E.D. 1989. Statistical self-similarity in river networks parameterized by elevation. *Water Resour. Res.* 25:463–76.

Harper, J.L. 1977. *Population Biology of Plants* New York: Academic Press.

Hubbell, S.P. and Foster, R.B. 1986. Canopy gaps and the dynamics of a neotropical forest. In *Plant Ecology*, ed. M.J. Crawley, pp. 77–96. Boston: Blackwell Scientific.

Hurst, H.E. 1951. Long-term storage capacity of reservoirs. *Transactions of the American Society of Civil Engineers* 116:770–808.

Hurst, H.E., Black, R.P., and Simaika, Y.M. 1965. *Long-term Storage, An Experimental Study*. London: Constable.

Kareiva, P. 1983. Local movement in herbivorous insects: applying a passive diffusion model to mark-recapture field experiments. *Oecologia* 57:322–27.

Kessell, S.R. 1979. Gradient Modeling. New York: Springer-Verlag.

Kolasa, J. and Pickett, S.T.A., eds. 1991. *Ecological Heterogeneity*. New York: Springer-Verlag.

Krummel, J.R., Gardner, R.H., Sugihara, G., O'Neill, R.V., and Coleman, P.R. 1987. Landscape pattern in a disturbed environment. *Oikos* 48:321–24.

Levin, D.A. and Kerster, H.W. 1971. Neighborhood structure in plants under diverse reproductive methods. *Am. Nat.* 105:345–54.

Likens, G.E., ed. 1989. *Long-Term Studies in Ecology: Approaches and Alternatives*. New York: Springer-Verlag.

Loehle, C. 1983. The fractal dimension and ecology. *Speculations Sci. Technol.* 6:131–42.

Lovejoy, S. 1982. Area-perimeter relations for rain and cloud areas. *Science* 216:185–87.

Lovejoy, S. and Schertzer, D. 1985. Generalized scale invariance in the atmosphere and fractal models of rain. *Water Resour. Res.* 21:1233–50.

McDonnell, M.J. and Stiles, E. W. 1983. The structural complexity of old field vegetation and the recruitment of bird-dispersed plant species. *Oecologia* 56:109–16.

Mandelbrot, B. 1983. *The Fractal Geometry of Nature*. New York: W.H. Freeman and Co.

Mandelbrot, B. and Wallis, J.R. 1969. Some long-run properties of geophysical records. *Water Resour. Res.* 5:321–40.

Marks, P.L. 1974. The role of pin cherry (*Prunus pensylvanica*) in the maintenance of stability in northern hardwood ecosystems. *Ecol. Monogr.* 44:73–88.

Meakin, P. 1983. Diffusion-controlled deposition on fibers and surfaces. *Phys. Rev.* A27:2616–33.

Meakin, P. 1983. Diffusion-controlled deposition on fibers and surfaces. *Phys. Rev. Ab* 27:2616–33.

New York: Academic Press.

Meentemeyer, V. and Box, E.O. 1987. Scale effects in landscape studies. In *Landscape Heterogeneity and Disturbance*, ed. M.G. Turner, pp. 15–34. New York: Springer-Verlag.

Milne, B.T. 1987. Hierarchical landscape structure and the forest planning model: discussant's comments in FORPLAN: An Evaluation of a Forest Planning Tool. USDA Forest Serv. Gen. Tech. Rept. RM–140.

Milne, B.T. 1988. Measuring the fractal geometry of landscapes. *Appl. Math. and Comput.* 27:67–79.

Milne, B.T. 1991. Heterogeneity as a multiscale characteristic of landscapes. In *Ecological Heterogeneity*, eds. J. Kolasa and S.T.A. Pickett. New York: Springer-Verlag, in press.

Milne, B.T., Johnston, K., Forman, R.T.T. 1989. Scale-dependent proximity of wildlife habitat in a spatially-neutral Bayesian model. *Landscape Ecol.* 2:101–10.

Morgan, F. 1988. *Geometric Measure Theory: A Beginner's Guide.* New York: Academic Press.

Morse, D.R., Lawton, J.H., Dodson, M.M., and Williamson, M.H. 1985. Fractal dimension of vegetation and the distribution of arthropod body lengths. *Nature* 314:731–34.

Murray, K.G. 1988. Avian seed dispersal of three neotropical gap-dependent plants. *Ecol. Monogr.* 58:271–98.

Neilson, R.P. and Wullstein, L.H. 1983. Biogeography of two southwest American oaks in relation to atmospheric dynamics. *J. of Biogeogr.* 10:275–97.

Nittmann, J., Daccord, G., and Stanley, H.E. 1985. Fractal growth of viscous fingers: quantitative characterization of a fluid instability phenomenon. *Nature* 314:141–44.

Nittmann, J. and Stanley, H.E. 1986. Tip splitting without interfacial tension and dendritic growth patterns arising from molecular anisotropy. *Nature* 321:663–68.

Okubo, A. 1980. *Diffusion and Ecological Problems: Mathematical Models.* New York: Springer-Verlag.

O'Neill, R.V., DeAngelis, D.L., Waide, J.B., and Allen, T.F.H. 1986. *A Hierarchical Concept of Ecosystems.* Princeton, N.J.: Princeton University Press.

O'Neill, R.V., Krummel, J.R., Gardner, R.H., Sugihara, G., Jackson, B., DeAngelis, D.L., Milne, B.T., Turner, M.G., Zygmunt, B., Christensen, S.W., Dale, V.H., and Grahm, R.L. 1988a. Indices of landscape pattern. *Landscape Ecol.* 1:153–62.

O'Neill, R.V., Milne, B.T., Turner, M.G., and Gardner, R.H., 1988b. Resource utilization scales and landscape pattern. *Landscape Ecol.* 2:63–69.

Orbach, R. 1986. Dynamics of fractal networks. *Science* 231:814–19.

Palmer, M.W. 1988. Fractal geometry: a tool for describing spatial patterns of plant communities. *Vegatatio* 75:91–102.

Peitgen, H.O. and Saupe, D., eds. 1988. *The Science of Fractal Images.* New York: Springer-Verlag.

Pentland, A.P. 1984. Fractal-based description of natural scenes. *IEEE Transactions Pattern Analysis Machine Intelligence*, Vol. PAMI–6 No. 6:661–74.

Peters, R.H. 1983. *The Ecological Implications of Body Size.* New York: Cambridge University Press.

Pickett, S.T.A. 1989. Space-for-time substitution as an alternative to long-term studies. In *Long-Term Studies in Ecology: Approaches and Alternatives,* ed. G.E. Likens, pp. 110–35. New York: Springer-Verlag.

Pickett, S.T.A. and White, P.S., eds. 1985. *The Ecology of Natural Disturbance and Patch Dynamics.* New York: Academic Press.

Pielke, R.A. 1984. *Mesoscale Meteorological Modeling.* New York: Academic Press.

Pielke, R.A. and Avissar, R. 1990. Influence of landscape structure on local and regional climate. *Landscape Ecol.,* 4:133–55.

Risser, P.G., Karr, J.R., and Forman, R.T.T. 1984. Landscape ecology: directions and approaches. Illinois Natural History Survey Spec., Publ. No. 2, Champaign.

Robertson, G.P. 1987. Geostatistics in ecology: interpolating with known variance. *Ecol.* 68:744–48.

Rosen, R. 1977. Complexity as a system property. *Gen. Sys.* 3:227–32.

Rykiel, E.J., Jr., Coulson, R.N., Sharpe, P.J.H., Allen, T.F.H., and Flamm, R.O. 1988. Disturbance propagation by bark beetles as an episodic landscape phenomenon. *Landscape Ecol.* 1:129–39.

Sander, L.M. 1986. Fractal growth processes. *Nature* 322:789–93.

Schertzer, D. and Lovejoy, S. 1985. The dimension and intermittency of atmospheric dynamics. In *Turbulent Shear Flow 4,* 3ed. B, Launder, pp. 7–33. New York: Springer-Verlag.

Sneft, R.L., Coughenour, M.B., Bailey, D.W., Rittenhouse, L.R., Sala, O.E., and Swift, D.M. 1987. Large herbivore foraging and ecological hierarchies. *BioScience* 37:789–99.

Simon, H.A. 1973. The organization of complex systems. In *Hierarchy Theory*, ed. H.H. Pattee, pp. 3–27. New York: Braziller.

Stanley, H.E. 1986. Form: an introduction to self-similarity and fractal behavior. In *On Growth and Form: Fractal and Non-Fractal Patterns in Physics,* eds. H.E. Stanley and N. Ostrowski, pp. 21–53. Boston: Martinus Nijhoff.

Stauffer, D. 1985. *Introduction to Percolation Theory.* London: Taylor and Francis.

Swihart, R.K., Slade, N.A., and Bergstrom, B.J. 1988. Relating body size to the rate of home range use in mammals. *Ecology* 69:393–99.

Turner, M.G., ed. 1987. *Landscape Heterogeneity and Disturbance. Ecological Studies 64.* New York: Springer-Verlag.

Urban, D.L., O'Neill, R.V., and Shugart, H.H. 1987. Landscape ecology. *BioScience* 37:119–27.

van Dorp, D. and Opdam, P.F.M. 1987. Effects of patch size, isolation and regional abundance on forest bird communities. *Landscape Ecol.* 1:59–73.

Voss, R.F. 1988. Fractals in nature: from characterization to simulation. In *The Science of Fractal Images,* eds. H.O. Peitgen and D. Saupe, pp. 21–70. New York: Springer-Verlag.

Voss, R.F. and Clarke, J. 1975. '1/f noise' in music and speech. *Nature* 258:317–18.

Whittaker, R.H. 1967. Gradient analysis of vegetation. *Biol. Rev.* 42:207–64.

Wiens, J.A. 1989. Spatial scaling in ecology. *Functional Ecol.,* 3:383–97.

Wiens, J.A., Addicot, J.F., Case, T.J., and Diamond, J. 1986. Overview: the importance of spatial and temporal scale in ecological investigations. In *Community Ecology,* eds. J. Diamond and T.J. Case, pp. 145–53. New York: Harper and Row.

Wiens, J.A. and Milne, B.T. 1989. Scaling of 'landscapes' in landscape ecology, or, landscape ecology from a beetle's perspective. *Landscape Ecol.,* 3:87–96.

Wolfram, S. 1986. *Theory and Applications of Cellular Automata.* World Scientific.

Wu, H., Sharpe, P.J.H., Walker, J., and Penridge, L.K. 1985. Ecological field theory: a spatial analysis of resource interference among plants. *Ecol. Modell.* 29:215–43.

3. Model Development and Simulation

10. The Development of Dynamic Spatial Models for Landscape Ecology: A Review and Prognosis

Fred H. Sklar and Robert Costanza

10.1 Introduction

We are at the dawn of a new era in the mathematical modeling of ecological systems. The advent of supercomputers and parallel processing, together with the ready accessibility of time series of remote sensing images, have combined with the maturing of ecology to allow us to finally begin to realize some of the early promise of the mathematical modeling of ecosystems. The key is the incorporation of space as well as time into the models at levels of resolution that are meaningful to the myriad ecosystem management problems we now face. This explicitly spatial aspect is what motivates landscape ecology.

All this is occurring at a time when the need for better tools for ecosystem understanding and management has never been more acute. Global climate change, acid precipitation, toxic wastes, sea level rise, and a host of other issues have focused attention on our dependence on the environment and highlighted the need to live in harmony with our ecological life-support system.

In this chapter we trace the roots of dynamic spatial modeling as it is beginning to be used in landscape ecology. There is a relatively long history of spatial model development in fields related to landscape ecology. This paper reviews models developed in a variety of disciplines, including geography, physics, biology, and ecology, which have formed the historical precedents for extant landscape models. We intend to show the types of problems researchers have solved with spatial models and hope to illustrate the different methodologies used. We also review a few existing landscape models that integrate across these fields and begin to provide the integrative synthesis that is the goal of landscape ecology. Finally, we speculate on the future of landscape modeling.

Models in landscape ecology can be used to (1) quantitatively describe spatial landscape level phenomena, (2) predict the temporal evolution of landscapes, and (3) integrate between and among spatial and temporal scales. Most ecological modeling work to date has focused on temporal changes. They tend to simulate a point in space and extrapolate the findings for an entire landscape by assuming that the landscape is "homogeneous." In other words, most models in ecology have little, if any, spatial articulation (Costanza and Sklar 1985). It is clear, however, that space needs to be more explicitly included if ecological models are to be truly useful tools for understanding and predicting the behavior of real ecosystems (Risser et al. 1984). Larger spatial scales (from regional to global) need explicit and concentrated attention from ecological modelers. One long-term goal of the scientific community is to develop ecosystem components to link with existing regional and global atmospheric and hydrologic models, creating comprehensive global geosphere-biosphere models. As the U.S. Committee for an International Geosphere-Biosphere Program (1986) points out, "It is only through simulations with comprehensive models that we can hope to discern the impact of man's activities on the environment."

For this review, we define a dynamic spatial model as any formulation that describes the changes in a spatial pattern from time t to a new spatial pattern at time $t + m$, such that

$$\mathbf{X}_{t+m} = f(\mathbf{X}_t, \mathbf{Y}_t) \tag{10.1}$$

where $\mathbf{X}_t$ is the spatial pattern at time t and $\mathbf{Y}_t$ is a vector or scalar set of variables that may affect the transition. This definition is the same as that given by Griffith and MacKinnon (1981) except we include scalar controls (i.e., variables with spatial coordinates).

This definition of a spatial dynamic model forces us to accept the notion that space and time are intertwined and cannot be reduced to two independent components. It demands that we look for the processes at time t that control processes at time $t + m$. It implies time lags between inputs and outputs, and it requires some cybernetic expression (i.e., feedback mechanisms) within, at least, the ecosystem level (Margalef 1968) of a landscape model.

This simple definition will also guide our review of spatial models and help us to focus on those systems with dynamic behavior. There are four major disciplines that use spatial models within the broad topic of ecology: geography, hydrology, biology, and ecosystem science. Within geography we can further subdivide spatial models into geometric, demographic, and network models. In hydrology there are basically two types of spatial models of fluid dynamics: finite element models, which we call hydrodynamic models, and finite difference models, which we call general circulation models. Spatial models in biology include growth models, population models, and point-averaged ecosystem models. Finally, we look at two types of spatial ecosystem models—the stochastic landscape models and the process-based landscape models. Before starting with the geography models, we look briefly at statistical models.

10.2 Spatial Statistics

Statistics represent a broad class of well-accepted mathematical procedures designed to evaluate the uncertainty of inductive inferences and to estimate the values of parameters in models that produce the "best fit" between the model and empirical data (Steel and Torrie 1960; Cliff and Ord 1973; Bennett 1979). As tools in the analysis of ecological processes, statistics are used to evaluate models. In general, spatial statistics are used to compare spatial data with expected spatial probability distributions. They are used to test for spatial differences (ANOVA, MANOVA); spatial concurrence and contiguity (correlations, regressions, spectral analysis, spatial autocorrelation); or spatial similarities (principle component analysis, ordination). For example, principle component analysis and ordination can distinguish spatial attributes along transects (Boesch 1977) by grouping hundreds of organisms, which vary with distance, into 'clusters' with similar characteristics. A host of new statistical methods are evolving to analyze spatial patterns from a landscape perspective (see S.J. Turner et al., Chapter 2), including fractals, spatial predictability, contagion analysis, and multiple resolution techniques (Costanza 1989; Turner et al. 1989).

The scope of the field of spatial statistics used to evaluate and analyze spatial patterns is too large to review within this chapter. Readers are referred to Patil et al. (1971), Cliff and Ord (1973), Pielou (1977), Sokal (1979), S.J. Turner et al. (Chapter 2), and Turner et al. (1989) for more information. We concentrate instead on landscape models themselves, with passing reference to statistical tests and methods as appropriate.

10.3 Geographic Models

Geographers have developed spatial models to evaluate cultural, social, population, and economic changes, which can be subdivided into three categories: (1) demographic models, (2) network models, and (3) geometric models. Some of these models are deterministic, while others are purely statistical. Some combine probability theory with empirically derived probabilities to create stochastic simulations, while others are purely mathematical rules for spatial patterns and development. The geographic literature is filled with so many models that simulate, predict, and explain spatial phenomena that we barely scratch the surface of this body of work in our review. The research of Hägerstrand (1953), Haggett and Chorley (1967), Cliff and Ord (1973), Isard (1975), Strahler (1980), and Huggett (1980) helped geographers adopt a 'model-based paradigm,' (Willmott 1984) and as a result, models are now commonplace within every subdiscipline of geography.

10.3.1 Geometric Models

A geometric model is one that examines distance and space to create some geometric paradigm. Geometric models use shape and design to explain spatial distributions, to hypothesize or create an emergent geometry, and to produce a generalized theory of spatial structure. These types of models usually view the landscape as some stable recursive geometric design that changes only in size. Thünen models (Chisholm 1962; Paynter 1982) are probably the best example. They are used by geographers to understand the 'space around a point,' by economists to predict peripheral land use, and by archaeologists to analyze prehistoric settlement patterns. Thünen models are based on the concept of economic rent. This rent is really a measure of potential surplus production as a function of transportation costs. Different activities respond differently to distance such that some must be close to an institutional center for them to be profitable (high surplus), while others can succeed in a more peripheral location (lower economic rent). Estimates of the costs to overcome distance for different activities produce a series of concentric circles (Fig. 10.1) that have become known as idealized

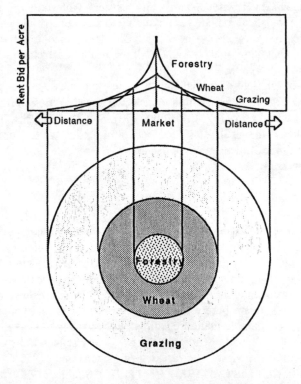

Figure 10.1. Idealized Thünen landscapes. When used to estimate the cost to overcome distance for different activities, a series of concentric circles is produced (Hoover 1975, fig. 6.1, p. 104).

Thünen landscapes. These concentric circles have been modified into stars, asymmetrical circles, and hyperbolas as a function of transportation routes and barriers (Selkirk 1982).

Geometric models have also been used to organize and categorize natural landscape features. For example, streams and distributaries in large river basins have been found to exhibit a mathematical geometric progression (Horton's law; $y = aR^x$) as a function of stream size (Horton 1945, Schumm 1956). Similarly, geometric attributes such as river basin bifurcation and area ratios (Woldenberg 1972) have been used, along with erosion equations by Sprunt (1972), to simulate drainage basin development (Fig. 10.2), while systems of towns, river subbasins, or branches on a tree have been organized (Christaller's laws) into a hierarchy of hexagons for the optimum partitioning of space (Woldenberg 1972). Although these tend to be descriptive models with no dynamic feedbacks, they can be very useful tools when incorporated into more complex dynamic spatial models. For example, the spatial hierarchical order of stream systems was part of the function used to parameterize water flow characteristics between different habitats in the Coastal Ecosystem Landscape Spatial Simulation (CELSS) model developed by Costanza et al. (1986) and Sklar et al. (in press) (see section on process-based landscape models).

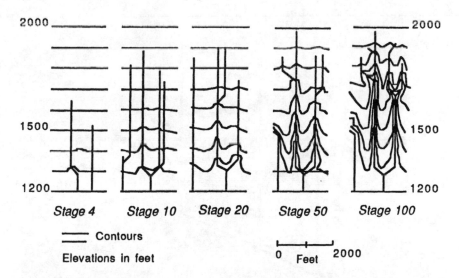

Figure 10.2. Simulated development of a drainage basin on an initially uniform plane. Contours are shown in feet. At the end of 100 iterations, 20 first-order basins, 6 second-order basins, and 1 third-order basin were formed (modified from Sprunt 1972, fig. 14.4, p. 387).

10.3.2 Demographic Models

Demographic models have been used to predict urban development (Clark 1969; Batty and March 1978; Getis and Boots 1978), change in human age structure (Haggett 1966), spread of epidemics (Cliff et al. 1981), and location of settlements (Curry 1964; Hudson 1969). Fig. 10.3, from Getis and Boots (1978), illustrates a conceptualization of the most basic demographic process for a point in space. As time progresses from t to $t + n$, a segment of the population within a grid cell will die (negative flow), will give birth (positive flow), and will diffuse or agglomerate (positive or negative flow). Obviously, such demographic models have the potential to be very dynamic, especially if cybernetic feedbacks are included.

Spatial processes in geography are thought of as a sequence of events resulting from certain physical, social, or economic forces that transform the environment. Most geographers tend not to be directly interested in the forces per se; rather, they focus on spatial aspects of these processes and the patterns to which they give rise (Getis and Boots 1978). As a result, geographers have spent much of their time analyzing and mapping two forms of diffusion: (1) expansive (those which spread to new areas but don't abandon old areas); and (2) relocative (those which leave the area where they originated) (Cliff et al. 1981). Most of the studies of diffusion in geography originated with Torsten Hägerstand's *Innovation Diffusion as a Spatial Process* (1953). Hägerstrand focused on human behavior (i.e., the acceptance of innovative farming methods) as a spatial process that moves across a landscape in "waves." These diffusion waves were seen to occur in four stages (Fig. 10.4a) as a function of the distance from the original source of the innovation and the proportion of adopters of the innovation. The study of these waves by

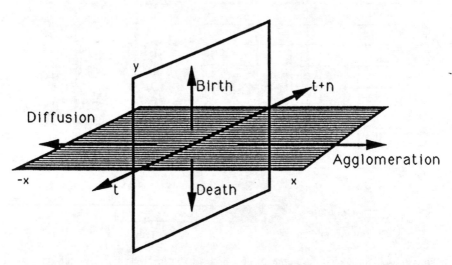

Figure 10.3. A generalized geography diagram for viewing spatial demographic processes (modified from Getis and Boots 1978, fig. 1.1, p.3).

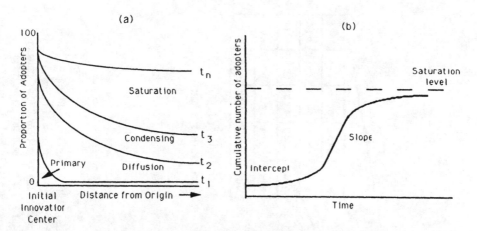

Figure 10.4. (a) Hypothetical four-stage model for the passage of what Hägerstrand (1953) termed "innovation waves" but which are more generally called "diffusion waves." The primary stage is the initial condition. The diffusion stage is the rate at which an idea is adopted. Condensing is where the relative increase in the numbers accepting an item is equal in all locations, and saturation is cessation of the diffusion process (modified from Cliff et al. 1981, fig. 2.9, p. 17). (b) The cummulative effect of diffusion waves is the logististic curve.

many subsequent investigators (Griliches 1957; Mansfield 1961; Tornqvist 1967; Casetti 1969; Cliff and Ord 1975) has resulted in the acceptance of the logistic curve (Fig. 10.4b) to denote the cumulative diffusion of demographic properties across a geographic landscape (Cliff et al. 1981).

Despite the nonlinear characteristics of logistic diffusion, many geographers have attempted to simulate diffusion using linear, stochastic techniques (Hägerstrand 1953; Dacey 1964; Hudson 1969; Alves 1974; Cliff and Ord 1975; Mollison 1977; Thomas 1977; Huff 1984) with modifications for population size, scale, neighborhood effects, and central place hierarchical effects (see Thomas and Huggett [1980] and Cliff et al. [1981] for review). At first, diffusion was modeled as a Markovian process. That is, spatial transition probabilities were established based upon movement characteristics such that:

$$M_{ij} = f[a_i, b_j, c_{ij}] \tag{10.2}$$

where M_{ij} is the probability of migration as a function of origin characteristics a_i (e.g., population density or size); destination characteristics b_j (e.g., resources); and some relationship between origin and destination c_{ij} (e.g., distance). Hägerstrand (1953) was the first to adopt this technique (Cliff et al. 1981) by developing a mean information field (MIF) composed of 25 1-km^2 cells that "floated" over a landscape (Fig. 10.5). The MIF contained the probability of information transfer

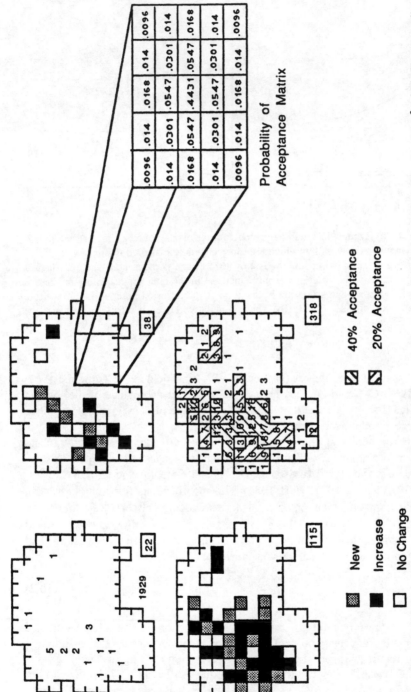

Figure 10.5. Spatial pattern of acceptance of a new agricultural technique using the Hägerstrand model. The 5 × 5 km² grid is Hägerstrand's mean information field (MIF), which "floats" across the landscape and calculates the probability of the acceptance of a new agricultural innovation for each cell in the grid. The top left map is the actual configuration of adopters used to start the model in 1929. Model iterations are annual. For 1932, the bottom-right map, the numbers of adopters are given, and cells are shaded if 20% or 40% of the total population in them accepted the innovation. Numbers in rectangles at map corners give the total number of adopters on each map for 1929, 1930, 1932, reading from right to left (modified from Cliff et al. 1981, p. 23).

(i.e., the acceptance of a new agricultural technique in Sweden) as a function of distance such that, when placed over an origin of an idea (or an adopter of an idea), an adopter located in the center of the MIF had approximately a 44% chance of passing on the information within a radius of about 2.5 km, which decreased to less than 1% at a distance of about 14 km (along the diagonals) (Hägerstrand 1967). Similarly, Sonis (1981) created an "aging wave" by using a Markov process to define the changing proportions of old and young populations across Europe. The process included a transition matrix of (1) the passage of young into old, (2) birth, (3) nonproportional mortality by age group, and (4) nonproportional migration for each group. The result was a velocity-of-change table for each country.

Another way geographers describe spatial interactions and diffusion is by using "gravity models." The original gravity models are said to date back to the nineteenth century (Ravenstein 1885 cited in Wilson 1981). Geographic gravity models are simple modifications of Newton's equation for gravity such that the spatial interaction between two bodies declines in proportion with the square of the distance:

$$T_{ij} = \frac{kW_iW_j}{d_{ij}^2} \qquad (10.3)$$

where T_{ij} = the size of the predicted flow from i to j; k = the scaling constant (converts equation into units of flow, is analogous to Newton's universal gravity constant); W_i = a measure of the capacity to leave i; W_j = a measure of the attraction capacity for each j; and d = the distance between the origin, i, and the destination, j.

Most geographic models allow distance to be raised to an empirically derived value a, since there is no a priori justification for expecting flows to decline as a square of the distance (Thomas and Huggett 1980). Solving the gravity equation is also dependent upon finding k, and solving for k can range from simple matrix algebra to complex spatial autocorrelations. The simple solution, if one knows the total migration or flux for the region (T), is T divided by the sum of all unscaled interactions, $T_{ij} = W_iW_j/d_{ij}$ (Thomas and Huggett 1980):

$$k = T/\sum_i^n\sum_j^n T_{ij} \qquad (10.4)$$

Demographic probability models in combination with gravity models have produced a wide variety of "statistical" space-time models (Bennett 1974; Batty and March 1978; Sokal and Wartenberg 1981). For example, Batty and March (1978) designed an information-minimizing urban model to estimate the probability of people locating in the Reading subregion of England. In this model the idea of gravity (i.e., distance) was modified as travel costs, and the transition probabilities were based on temporal changes in population density. What is interesting about this model is that it takes the Shannon formula for information (1948):

$$H = \sum_{i=1}^{n} p_i \ln p_i \qquad (10.5)$$

and transforms it into a function of the expected values of population density, p_i^t, p_i^{t+1} in zone i at times t and $t + 1$ respectively:

$$I_1 = \sum_i p_i^{t+1} \ln[p_i^{t+1}/p_i^t] \qquad (10.6)$$

subject to

$$\sum_i p_i^{t+1} c_i^t = <\overline{c^t}> \qquad (10.7)$$

where the objective function is to minimize I (the relative gain or difference in information between zones and time periods) as a function of c, the travel cost. It was later realized that travel costs were probably underestimated, as the model predicted too little activity in the center and too much at the periphery (Batty and March 1978).

Many demographic models, like those of Batty and March, use regressions to parameterize their models and define concepts such as travel costs. The incorporation of spatial stochastic parameters into demographic models was a technique developed by geographers at Cambridge (Cliff and Ord 1969, 1971, 1972, 1973). The classic example is the epidemiological study in southwest England that predicted the spread of measles across a 9,000-km² area as a function of spatial autocorrelation, "distance" between population centers, and the frequency of outbreaks for each region (Cliff et al. 1975). The regularity of measles outbreaks was calculated for each region by decomposing each regional time-series into its periodic sine and cosine functions (i.e., Fourier analysis), then calculating the contribution of each function to the overall variance of the series (i.e., the power spectrum). The result was a "fundamental" frequency of measles in each region. These methods have been used by economists in business forecasting and by climatologists in examining precipitation cycles (Sklar 1983). The standard reference is Box and Jenkins (1970). The incorporation of movement into this model is of particular interest because it turns out not to be based on any demographic principles but rather on weighted "spatial lag" coefficients calculated from the spatial autocorrelation of measles frequency. Spatial autocorrelation, as defined by Cliff and Ord (1973, p. 2) is this: "If, for every pair of counties, i and j, in the study area, the drawings [areas to which the frequencies refer] which yield x_i and x_j are uncorrelated, then we say that there is no spatial autocorrelation in the county system on X. Conversely, spatial autocorrelation is said to exist if the drawings are not all pairwise uncorrelated." In general, when high values of a variable in one region are associated with high values of that variable in another region, it is said that the regions exhibit positive spatial autocorrelation. When autocorrelation coefficients are weighted by population densities and distance measures (i.e., "spatial lag" coefficients) to increase predictive accuracy, techniques developed at Cambridge, they begin to resemble spatial network models (see section 10.3.3).

This type of statistical model fitting marked the beginning of a fundamental change in demography as more and more geographic modelers replaced demographic parameters with statistical parameters.

10.3.3 Network Models

Spatial network models can be thought of as complex gravity models where many simultaneous flows (T_{ij}) are dependent upon numerous locations (nodes), each with separate measures of capacity to leave (production) and to attract (demand). Network models are typified by traveling salesman problems, commodity flow problems, and transportation problems. They are referred to as either transportation problems (Isard 1975) or spatial allocation models (Henderson 1958, Thomas and Huggett 1980). The purpose of these models is generally to optimize a distance or cost objective function, such as finding the shortest route through all cities or minimizing production costs.

The simplest of the network models is the traveling salesman problem, where the objective is to find the shortest path through a network. In Fig. 10.6 a small five-node network and its distance matrix illustrate this type of model. To find the shortest path, one recursively solves for combinations of nodes. The equation takes the form:

$$L_k(A_i) = \text{Min } \{L(A_i, A_x) + L_{k-1}(A_x) \,|_x = 1, \ldots, m|\} \tag{10.8}$$

where L is path length, k is the step constraint (i.e., the number of nodes through which you can pass, maximum is $m - 1$), A_i is the node of origin, and A_x is the destination node (Werner 1985). For the example listed in Fig. 10.6, going from W_1 to W_5 with a step of 2 will constrain the answer to ten lengths, while a step of $m - 1$ results in a path length of nine (i.e., $W_1 \rightarrow W_3 \rightarrow W_4 \rightarrow W_2 \rightarrow W_5$).

Network models attempt to answer the question, "What could be the best solution?" rather than "What are the possible interactions?" Solutions to these models are matrices that are optimal according to some predefined criterion. In general, these types of models fall within the area of mathematics known as linear programming. The originator of linear programming is the American mathematician Dantzig, who designed the simplex algorithm in 1947 (Thomas and Huggett 1980).

The essence of a linear programming network model is explained with reference to the example problem listed in Fig. 10.7. The problem is composed of a set of $n = 3$ locations, each producing O_i units of some commodity and $m = 3$ locations, each demanding D_i units of the commodity. For each $O_i D_i$ interaction there is a measure of the transportation costs per unit of commodity. For a network of flows, the total transportation costs (Z) are then the sum of each individual flow times the unit cost $(T_{ij} \times c_{ij})$. The primary objective is to find the particular set of values (T_{ij}) that minimizes the total cost of all flows. This is written as:

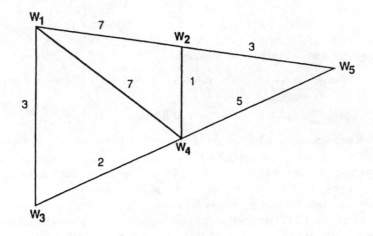

Origin	Destination W1	W2	W3	W4	W5
W1	0	7	3	7	0
W2	7	0	0	1	3
W3	3	0	0	2	0
W4	7	1	2	0	5
W5	0	3	0	5	0

Figure 10.6. Simple five-node network model (top) and its matrix of distance value: values (bottom).

$$Z = \text{Min } \{\sum_{i}^{n} \sum_{j}^{m} T_{ij}c_{ij}\} \qquad (10.9)$$

The values of T_{ij} that minimize the objective function must satisfy a set of constraints for a solution to be found. First, flows between locations cannot be negative. The second constraint asserts that the total flow from location i must be less than or equal to the production capacity, O_i, of the source,

$$\sum_{j=1}^{m} T_{ij} \leq O_i \qquad (10.10)$$

The third constraint asserts that the total flow to j must be exactly equal to the total demand at location D_j,

$$\sum_{i=1}^{n} T_{ij} = D_j \qquad (10.11)$$

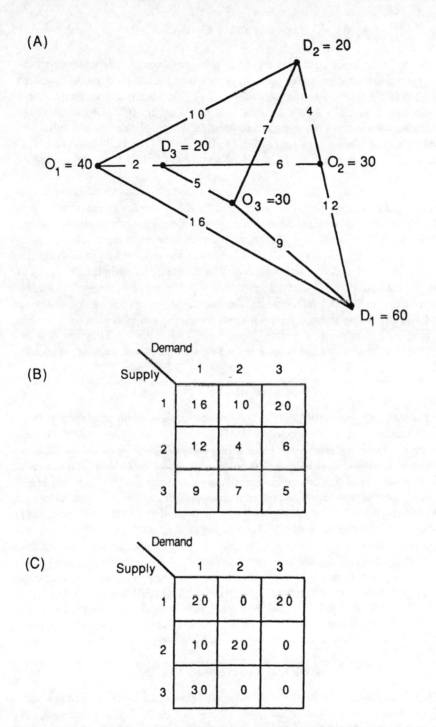

Figure 10.7. An example of a transportation-type network model. (A) Map of supply nodes (O_i), demand nodes (D_j), and unit flow cost (c_{ij}). (B) Matrix representation of unit flow costs constraints. (C) Solution matrix that minimizes network transportation costs ($T_{ij} \times c_{ij}$) (modified from Thomas and Huggett 1980, fig. 6.1, p. 172).

Taken together, these equations represent a linear program. The linear programming solution (i.e., the simplex method) proceeds by successively modifying some initial feasible solution so that the values of $\{T_{ij}\}$ gradually converge on a final optimum solution. The solution matrix is included in Fig. 10.7. The details of the methods of linear programming are beyond the scope of this review. Interested readers should consult Dantzig (1963), Dorfman et al. (1968), Thomas and Huggett (1980), and Werner (1985) or any of a number of other linear programming texts.

Network models are the most widely used and the best mathematically understood of all spatial models. They have been modified to predict the efficiency of added streets (i.e., links) and intersections (i.e., nodes) for transportation problems (Quandt 1960). In sociology they have been used to restructure school busing zones (Yeates 1963; Maxfield 1972). Manufacturing has used them to maximize production rates, and they were used in World War II to optimize the flows of weapons and food to the Allies. The most serious limitation of network models is the assumption that all relationships are linear. Also, accounting for changing flows as a function of scale does not exist, feedbacks are ignored, and flows are assumed to be infinitely divisible in an attempt to optimize. They are also restricted spatially in that the areas surrounding the nodes and links are ignored.

10.4 Models of Fluid Dynamics

The most computationally sophisticated spatial models (with relevance to landscape ecology) to be developed in the last two decades are models of fluid dynamics. These include a large number of water and gas flow models, such as air pollutant diffusion models, global atmospheric-climate models, weather models, storm-water runoff models, and global ocean-circulation models. Models of fluid dynamics are all applications of the fundamental physical continuity equations for the conservation of mass, momentum, and energy (Orlob 1983). They differ in (1) the spatial and temporal extent and resolution of their application, (2) the degree of simplification of the fundamental equations, and (3) the computational methods used to approximate the solution of the equations. We briefly discuss two classes of fluid dynamics models that were chosen for their relevance to landscape ecology: (1) general atmospheric circulation models (GACMs) because of their ability to simulate climate patterns, which are, in turn, landscape forcing functions; and (2) hydrodynamic models of watersheds and midsized water bodies because of their ability to move material through lotic and estuarine systems.

10.4.1 Hydrodynamic Models

Hydrodynamic models that incorporate spatial processes and fluid dynamics use a "link-node" design to simulate space. These designs look like geographic network models (Fig. 10.6) but are slightly different because they can have many more pathways of flow (i.e., links) converging or diverging on any single point (i.e., node). The link-node approach predominates in small and midscale applications where short-term hydrodynamics (hours to days) are the main objective of the modeling study. The Storm Water Management Model, SWMM (EPA 1975a, 1975b), which is used to simulate storm-water runoff from urban drainage systems, is a good example of such a link-node model. SWMM is a network of interconnected conduits of flows or pipelines where the "pipe" dimensions and head differences control flow rates between nodes. Dynamics are a function of the network design and the simultaneous solution of the continuity and momentum equations at every node in the network. Constraints upon this type of model include (1) the time-step stability criterion (Mahmood and Yevjevich 1976; Wang et al. 1985), which requires extensive computer processing and tends to limit simulations to very short time scales; and (2) the inability to depict spatial processes in "nonlink" areas (i.e., areas adjacent to pipelines).

We found three ways to add spatial realism to these types of models (Fig. 10.8). The first technique calculates the link size as a function of the area surrounding adjacent nodes (Fig. 8a). This works well for homogeneous landscapes but is complicated by the large number of calculations required at each node. Another technique is to use a second set of parallel pipelines to capture the dynamics of the areas adjacent to the links. This design was used by Hopkinson and Day (1980) to separate the turbulent flow characteristics within a wetland channel from the laminar sheet flow within the wetland itself (Fig. 8b). Similarly, a third technique separates channels from wetlands (Wang et al. 1985) by designating the areas around links as storage reservoirs (Fig. 8c). The link-node hydrodynamic models have been very popular and very effective for modeling open water circulation in a wide range of aquatic systems. There are however, other ways to simulate hydrodynamic space, the most popular of which uses a fixed rectangular grid approach. Grid-based models are less spatially constrained than link-node models since (1) they do not assume a particular hydrologic structure as link-node models, and (2) they are used to model substantially larger areas and longer time spans (but generally with less accuracy). Most global ocean circulation models and GACMs are of the fixed-grid format.

10.4.2 General Atmospheric Circulation Models

GACMs have existed since the late 1950s (Phillips 1956; Kasahara and Washington 1967, 1971; Williams et al. 1974; Gates and Schlesinger 1977; Washington and Williamson 1977) and have gradually evolved in complexity and spatial and temporal resolution (Washington et al. 1979; Potter et al. 1979; Washington and Meehl 1984; Boer and Lazare 1988; Schlesinger and Zhao 1989). They simulate spatial processes by dividing the atmosphere into regular, three-dimensional grid cells along latitude and longitude lines. Most versions use an 8° or 10° latitude and longitude grid and two to six vertical layers and treat the ocean as exogenous inputs (Simmons and Bengtsson 1984). More sophisticated versions use 2 to 5° latitude and longitude resolution and ten or more vertical layers of variable depth based on pressure boundaries and are linked to global ocean models forming global climate systems (Washington and Meehl 1984; Schlesinger and Zhao

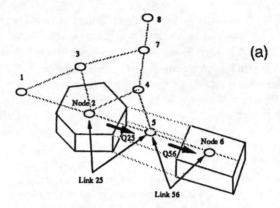

Physical Abstraction of Lac Des Allemands
Hydrologic Regime

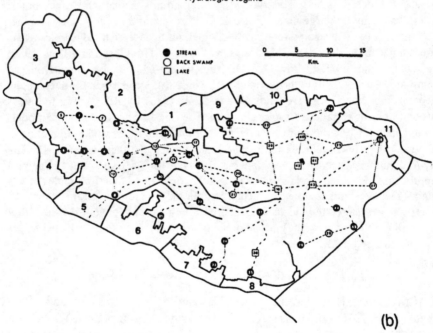

Figure 10.8. (a) Example link-node network for hydrodynamic modeling in the San Francisco Bay (modified from Huggert 1980). (b) Example of parallel links in a network (modified from Hopkinson 1978). (c) Example of reservoir nodes in a network (modified from Wang et al. 1985).

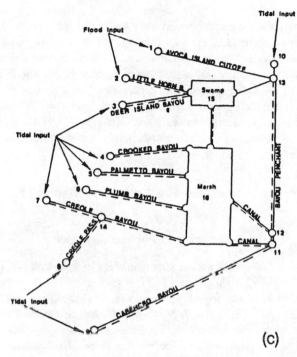

Figure 10.8. *Continued.*

1989). Unfortunately, 2°-resolution models use so much central processing time that even with supercomputers, calculations cannot be made in a reasonable length of time for the entire globe. Instead, as is done with the National Center for Atmospheric Research NCAR models, small regions of interest are simulated at 1° or 2° resolution, while the rest of the atmosphere is simulated at a 5° or 10° resolution (Allen 1989). Physical processes typically included in GACMs are absorption and reflection of solar energy; cooling and heating due to infrared radiation; cloudiness and sensible and latent heat exchanges due to water vapor, horizontal and vertical diffusion of momentum; and the hydrologic cycle.

Typical GACM results are shown as contour maps of zonally averaged model predictions. In general, predicted surface-air temperatures agree well with the data, while precipitation predications agree less well (Washington et al. 1979). This is because the hydrologic cycle is more complex than heat transfer processes and because most GACMs do not have the resolution to discern small-scale topographic features such as the relatively small Sierra Madre Mountains of California, which can influence the landward penetration of moist ocean winds and produce desert environments in the shadow of rain events (Nigam et al. 1986; Nigam et al. 1988). There have been some attempts to link terrestrial models with GACMs (Dickinson 1984; Sellers et al. 1986; Wilson et al. 1987) by using land surface conditions as forcing functions for the atmospheric processes. Current biosphere models are, however, noninteractive, because they lack landscape-level feedbacks

(Hall et al. 1988; Baker 1989). With improved algorithms and increases in computer processing time, we can expect not only better precipitation predictions, but also better integration of GACMs with other global models and data bases into what is now being called "earth system science" (Earth Systems Science Committee 1988). The stated goal of earth system science is "to obtain a scientific understanding of the entire earth system on a global scale by describing how its component parts and their interactions have evolved, how they function, and how they may be expected to continue to evolve on all timescales." Landscape ecology, by describing earth surface interactions, certainly has a major role to play in developing this scientific understanding.

10.5 Biological and Ecological Models

10.5.1 Growth and Population Models

Biological models with some spatial characteristics include competitive interaction models (Gause 1934; Hutchinson 1947; Ayala et al. 1973), patch dynamics (Paine and Levin 1981), interference and exploitation (MacArthur 1972; Schoener 1974), and niche space (Elton 1927; May and MacArthur 1972; Whittaker et al. 1973). These models tend to view the landscape as an independent, exogenous variable. Although populations in these models respond to spatially dependent variables, feedbacks from the populations to the spatial structure of the landscape are often excluded.

Animal population models in ecology are very similar to the demographic models in geography. Both deal with birth, death, and diffusion. Ecologists, however, have spent more time on temporal density-dependent interactions such as the effects of population growth on birth rates and death rates, and relatively less time on spatial density-independent interactions such as population dispersion as a function of spatial patterns (see Chapters 16 and 17 by Merriam et al. and Fahrig). Often spatial homogeneity is assumed initially so that density-dependent spatial patterns can emerge (Levin 1979). In an excellent review of population modeling, Hutchinson (1978) points out that population modeling started in 1845 with the development of the Verhulst logistic growth equation:

$$dN/dt = rN [(K - N) / K] \tag{10.12}$$

where the growth of a population (dN) increases to an upper asymptote (K) as a function of its own density (N) and some rate of increase (r). Since r can be thought of as a birthrate minus a death rate, it too has been studied as a density-dependent variable. The rediscovery of the Verhulst logistic, first by Pearl and Reed (1920) and later by Gause (1934), has since produced thousands of books and articles on animal demographics and has become the foundation for many theories on competition (Gause 1932; Frank 1957; Hardin 1960; Park 1962; Levin 1970; Ayala et al. 1973; Schoener 1974) and predator-prey interactions (Lotka 1925; Volterra

1926; Huffaker 1958; Slobodkin 1961; Paine 1966; Slobodkin 1968; Finerty 1971; Paine and Levin 1981).

It is common sense to us to suggest that competition and predator-prey interactions must include spatial information. For example, for a predator to attack its prey, the prey must remain within the "predator landscape" long enough to be detected. It is surprising that the widely accepted and simulated interactions between prey and predator (i.e., the Lotka-Volterra oscillations) does not include a landscape component. The Lotka-Volterra equations state that prey densities (N_{prey}) are controlled by an intrinsic rate of increase r, predator densities (N_{pred}), capture efficiency ($\alpha_{prey}/\alpha_{pred}$), and predator mortality m. The equations for this interaction,

$$dN_{prey}/dt = rN_{prey} - \alpha_{prey} N_{pred} N_{prey}$$

$$dN_{pred}/dt = \alpha_{pred} N_{prey} - mN_{pred}$$

(10.13)

produce cyclical trajectories when numbers of prey are plotted against numbers of predators. Hall (1988) believes these oscillations are never realized in nature. The classic experiments by Gause (1932), where the flour mite *Aleurglyphus agilis* is preyed upon by the carnivorous mite *Cheyletus eruditus*, produced trajectories that only slightly resembled the ideal Lotka-Volterra model. In general, when spatial parameters have been ignored, experimental verification of the Lotka-Volterra model has been unsuccessful. When spatial heterogeneity is introduced, as in experiments by Huffaker (1958) on mites living on oranges, oscillatory behavior can exist but only for a short time (eventually predators die off due to a lack of prey, leaving a few undiscovered prey to produce a new unpredated population). Nachman (1987a) designed a more stable model by incorporating spatial realism and synergistic feedbacks such as a large number of spatially separated patches (e.g., cucumber plants), dispersal characteristics for both species as a function of local resources, behavioral responses (feeding and reproduction) to food quality (i.e., leaf damage index), and an allowance for demographic stochasticity. The results of this model indicated how population peaks travel like waves across a twenty-four plant universe as they did in the Hägerstrand models. Nachman (1987b) used this same model to simulate a series of predator-prey oscillations as a function of the number of plants and clearly demonstrated the importance of spatial attributes to the persistence of the system (Table 10.1). Even without synergistic feedbacks, other more theoretical treatments of this subject have shown that a minimum proportion of suitable habitat distributed throughout a landscape is required for population persistence (Okubo 1980; Lande 1987).

Since the experiments by Gause (1932) and Huffaker (1958), there have been many experimental and theoretical systems devised to study dispersal (i.e., diffusion) and the importance of spatial patterning and patch dynamics (Weins 1976; Ripley 1981; Paine and Levine 1981; Nisbet and Gurney 1982; MacArthur 1972, Schoener 1974; Pickett and White 1985). For example, Levin (1979) incorporated the concepts of diffusion into equations of competition and demonstrated that nonuniform coexistence was possible if dispersion of species i from patch υ to μ

was less than the diffusion of species *i* within patches υ or μ (see Levin 1974, 1979 for details). Theories of island biography have applied much of these same demographic principles to account for how migrating species coexist or go extinct when "diffusing" to new habitats (Gilpin and Diamond 1980; Wilcox 1980; Diamond and May 1981).

Plant ecologists have long recognized the correlation between abundance and spatial heterogeneity (Zedler and Zedler 1969; Whittaker 1975; Tilman 1982). Although plant ecologists also use demographic principles to model spatial dynamics (Harper and McNaughton 1962; Harper 1977; Pacala 1987), they have a tendency to emphasize community-level rather than population-level interactions, and they have attempted to deal more explicitly with spatial-time processes than animal ecologists have. The models of Pacala and Silander (1985), for example, examine how spatial heterogeneity effects the outcome of competition among plant species by giving each species in the community a location parameter along a transect and a set of radii around the species "neighborhood." The radii was made to vary as a function of each plant's ability to make seeds, disperse the seeds, and germinate. The results were similar to those found by animal ecologists. That is, in some cases coexistence is possible only if environmental patches are larger than the neighborhood radius (Pacala 1987). Like animal population models, these models still do not allow for the possibility that the abundances and spatial distributions of resources (e.g., soil nutrients) are affected by the abundances and spatial distributions of the species using those resources.

The possibility that plants may cause the very spatial heterogeneity that allows them to coexist is of particular interest to the landscape modeler. Plant ecologists have only recently developed spatial models of growth to examine this issue. The first to model plant growth as a process affected by spatial features (e.g., crowding from adjacent trees) was Botkin et al. (1972). His forest growth model JABOWA

Table 10.1. A Summary of the Results of the Nachman (1987a, b) Model of Acarine Predator-Prey Stability

Number of Plants	Persistence of Predator-Prey System (DAYS)	Maximum Number of Prey	Minimum Number of Prey	Maximum Number of Predators	Minimum Number of Predators
2	200	13,000	0	150	0
6	350	16,000	0	110	0
12	600	15,000	0	140	0
24	>600	12,000	2	105	0.5
72	∞	11,000	1	100	1

The model demonstrated that persistence was dependent upon the number of cucumber plants used by the spider mite *Tetranychus urticae* as food and refuge from the predator mite *Phytoseiulus persimilis*. A 24-plant universe was found to be persistent but not as stable as a 72-plant universe. Numbers of prey and predators are reported as individuals per 1000 cm^2 of leaf area.

simulated plant-space interactions in a square 50-m cell as a function of growth dynamics. In JABOWA-type models, an individual tree has an optimal growth proportional to the amount of photosynthates that can be produced by leaf tissues and indirectly proportional to respiration. The rest of the photosynthate goes into the production of new plant tissue. These models are spatial in a vertical sense rather than a horizontal sense. Light penetration through the canopy is the primary feedback loop and the cartesian coordinates of individual trees are not tracked. Many modifications to JABOWA now exist (see reviews by Shugart 1984 and Dale et al. 1985). Defined as "community dynamics" models (Dale et al. 1985), these types of tree models are used to examine the impacts of feedbacks between tree species and the environment on forest development. In these community models there are numerous species in each cell and each species has certain intrinsic population parameters, including rates of birth, growth, and death. Succession occurs within a cell as trees compete with each other for resources. All the community dynamic forest models currently in use, such as JABOWA-FORET (Shugart and West 1980), FORNUT (Harwell and Weinstein 1982), FORENA (Solomon 1986), LINKAGES (Pastor and Post 1986), CLIMAX (Pastor et al., 1987), and SWAMP (Phipps 1979) are based on similar ideas. Their differences are attributed to growth, reproduction, and death descriptors as a function of environmental parameters such as soil type, nitrogen availability, shade tolerance, elevation, slope, climate, flooding regime, and disturbance frequency.

An excellent example of a JABOWA-type model, used to predict succession at a landscape level, is given in Pearlstine et al. (1985). Pearlstine et al. (1985) modified the JABOWA-FORET model to study the impact of an altered hydrologic regime on the growth and succession of a coastal forested floodplain by making germination, organic export, and tree production partially dependent upon water depth, calculated by a FLOOD subroutine (Fig. 10.9). Each cell of the model was linked to land elevation data for the Santee Basin, South Carolina, on a geographic information system (GIS). Simulations were run for each 15-cm contour of elevation to produce a set of habitat maps for predicting landscape changes from a proposed river diversion project (Fig. 10.10). Unlike models before it, this particular model allowed the plants to influence the organic/mineral content of the soil (see the conceptual diagram, Fig. 10.9), thereby affecting germination of new tree species. The only things missing from this design from a landscape perspective are the possible effects of organic exports to the next cell (downstream) and the feedback effects of organic production on the height of the water table (accretion).

10.5.2 Ecosystem Models

It has been well over 50 years since Sir Arthur Tansley (1935) defined "ecosystem" as:

> the whole system (in the sense of physics) including not only the organism-complex, but also the whole complex of physical factors forming what we call the environment of the biome—the habitat factors in the widest sense. Though the

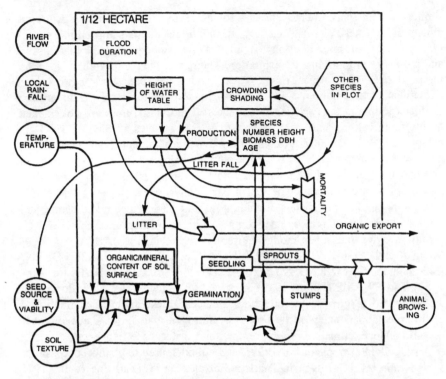

Figure 10.9. An overview of a modified FORFLO model, with Odum (1983) notation, used to simulate the succession of a forest floodplain in South Carolina. Note the internal feedback between species production and the germination of viable seeds within a half hectare cell (Pearlstine et al. 1985, fig. 3, p. 286).

organisms may claim our primary interest, when we are trying to think funda-mentally we cannot separate them from their special environment, with which they form one physical system. It is the systems so formed which, from the point of view of the ecologist, are the basic units of nature on the face of the Earth.

The major difference between the growth and population models described in the previous section and the ecosystem models in this section is that the former tend to emphasize the biological components within an ecosystem while the latter tend to emphasize the interactions among all components within an ecosystem by implementing the "whole system" concept. The primary trend in ecosystem mod-eling, since Lindemans's classic paper on trophic flows (1942), has been the condensation and aggregation of complex ecosystems into a relatively small number of differential equations (Patten 1971, 1972, 1975, 1976; Odum 1983). Most ecological models aggregate spatial components into temporally dissimilar state variables by conceptualizing the system in a variety of ways, including black boxes (Patten and Matis 1982), input-output matrices (Costanza et al.; 1983 Ulanowicz 1986), digraphs (Puccia 1983), computer diagrams (Littlejohn 1977),

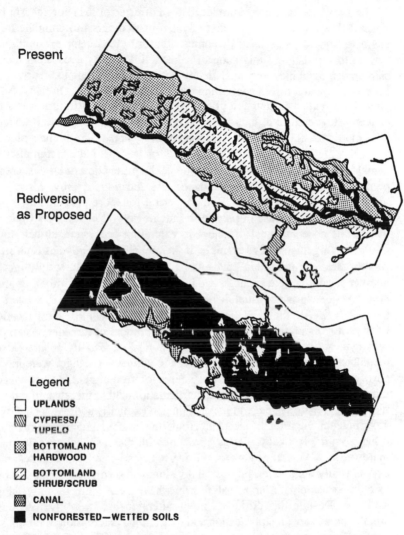

Figure 10.10. The model shown in Fig. 10.9 was used in conjunction with a GIS of elevation changes to predict a 97% loss of bottomland hardwoods associated with a plan to divert water from one river basin to another; (top) present; (bottom) proposed diversion (Pearlstine et al. 1985, fig. 7, p. 294).

industrial diagrams (Forrester 1961), and energy circuit models (Odum 1971). All these simulation environments tend to aggregate by components with similar flow characteristics. Groupings are such that animals with the same feeding behavior or soils with the same redox potentials are simulated as single-state variables. Once conceptualized in this way, the rates of change and the flows in and out of the state variables can be modeled by using any number of mathematical or statistical tools (Hall and Day 1977; Shoemaker 1977; Odum 1983; Swartzman and Kaluzny 1987; D'Elia 1988).

The symbolic language introduced by Forrester (1961) for the study of industrial flows was one of the first adopted by ecologists to simulate the spatial characteristics of ecosystem dynamics. Radford (1972), for example, used this system to simulate the migration of salmon. Like many ecosystem models, the salmon demographics were excluded in favor of a more detailed representation of the environmental forces controlling the spatial distributions. In this case, salmon moved upstream as a function of estuarine salinity, which in turn was a function of water-flow control at man-made weirs. A more implicit spatial model of fish migration was developed by Swartzman and Zaret (1983). They modeled the territorial advance of the fish *Cichla ocellaris* through Lake Gatun after its accidental introduction in 1966 as a function of the production of large schools of the predatory juveniles and the abundance and location of prey. Movement was implicitly represented by the increasing need for more prey and was predicted from the rate of increase in prey in sediments not yet invaded by *Cichla*.

There is no conceptual difficulty in extending ecosystem models to include spatial heterogeneities. For example, if the modeling objective of Radford (1972) was to track salmon within ponds along a river, one would simply increase the number of salmon state variables and include parameters to control migration from pond to pond. In fact, spatial distinctions are implicit in most ecosystem models because many of the modeling compartments are identified by location (i.e., habitat). For example, in Sklar et al.'s (1982) model of the aquatic carbon flows of a cypress swamp, the duckweed community is spatially separated from the benthic community by a layer of water, and vertical exchanges were modeled, in part, as a function of distance (i.e., depth of flooding). Like most ecosystem models, however, the vertical-spatial articulation within this swamp model serves more to aggregate system components than to explicitly address spatial dynamics.

Although it is perfectly acceptable to define modeling components by location, it has been impractical to design a system with much spatial articulation for two major reasons. First, the numbers of flows to be estimated in the model tend to rise exponentially with increasing spatial division, and second, the computer resources to track and compute large spatial arrays have been, until recently, largely unavailable. Despite this difficulty, much attention has been given to the spatial simulation of vertical stratification and longitudinal succession. Examples include nutrient exchanges across sediment-water interfaces (Gayle 1975; Kamp-Nielsen 1983; Odum 1983), plankton community changes downstream (Odum 1957), movement of biologically active radioactive materials through soil layers (White et al. 1978), nutrient spiraling (Newbold et al. 1982), and vertical distributions of plants and animals (Taylor 1988; Focardi et al. 1989).

Vertical and longitudinal models divide the ecosystem into homogeneous compartments such that each compartment can be simulated separately. If the processes within a compartment are independent of the other compartments, the simulation of a spatial model is relatively easy and is similar to population and growth models. If, however, there are dynamic interactions and feedbacks between compartments, the simulations quickly become very complex (Noy-Meir 1981). Ecosystem models tend to increase spatial complexity by simplifying or neglecting the internal processes within or between compartments. For example, White et al. (1978) simulated tritium kinetics as a process affected by thirty layers of 1-cm-thick marsh sediment. Transfers between layers were modeled as diffusion gradients proportional to concentrations within each layer, and any complicating biological interactions were omitted. This is not to say that the tritium model was incorrect or even incomplete. The objectives were not concerned with the biology of tritium uptake. They were more concerned with the permeability of marsh soils to tritium. As a physical model of how marsh sediments impede or enhance tritium movement, the White et al. (1978) model performs well. As a biological model of marsh tritium movement and uptake, the model doesn't perform at all. We have drifted slightly from our discussion of spatial models but have arrived at a junction that serves to make an important point: the goals and objectives of spatial models are intricately associated with their structure and design.

Until recently ecosystem models were limited by the computational complexity of added spatial realism. Spatial processes make ecosystem models difficult to validate and very computer intensive. For example, Taylor (1988) used the same layer approach as White et al. (1978) to model the vertical distribution of phytoplankton. Here, too, the vertical diffusion of matter was proportional to the difference in concentrations between layers. It was not possible, however, to account for phytoplankton distributions based solely on diffusion of nutrients, as did White et al. (1978) for tritium. The equations had to be modified to include irradiance, self-shading, recycling efficiencies, grazing, mortality, and Michaelis-Menten uptake kinetics. This made the internal complexity so great that the number of layers had to be limited to three. In general, when biological processes are included in vertical and longitudinal models, there is a strong tendency to reduce the spatial articulation (i.e., resolution). The reason for this trend is well illustrated in Table 10.2, where the number and complexity of the equations of various sediment phosphorus models increases dramatically with increasing spatial compartmentalization. As we shall see, however, the newest generation of supercomputers, new modeling techniques, and the development of GIS software is making this complexity more manageable and allowing the evolution of more realistic landscape models.

10.6 Landscape Models

The advent of bigger and faster computers, remote sensing, and the detection of long-term changes in ecosystem patterns has allowed the development of a new class of dynamic spatial ecosystem simulation models. These simulations more closely fit the definition of spatial modeling by Griffith and MacKinnon (1981); take up the ideas of landscape ecology as introduced by Naveh and Lieberman (1984), Risser et al. (1984), and Forman and Godron (1986); and explore ecological scales never before attempted. Landscape models have the potential to (1) map the flows of energy, matter, and information; (2) designate source, sink, and receptor areas; (3) predict succession in two- and three-dimensional space; (4)

Table 10.2. Models of Sediment/Water Exchange of Phosphorus (P) in an Aquatic System Increase in Complexity as the Number of Pools (i.e., Sediment Sources) Increases and as the Number of Layers (i.e., Spatial Resolution) Increases[a]

Conceptual Model	Model Complexity
Steady State	$\dfrac{\delta P_s}{\delta t} = K_1 P_s + R$
One Sediment Pool	$\dfrac{\delta P_S}{\delta t} = K_1 P_S + K_2 P_E V_S$
Several Sediment Pools	$\dfrac{\delta P_S}{\delta t} = K_1 P_S + K_2(P_S + P_1)$
	$\dfrac{\delta P_E}{\delta t} = K_1 P_S - K_3 P_E V_S \sigma$
	Equations for first layer
Multi-layered Pools	$\dfrac{\delta P_S}{\delta t} = K_1 P_S + K_2(P_S - P^1)$
	$\dfrac{\delta P_E^{\,1}}{\delta t} = K_1^{\,1} P_S - K_3^{\,1} P_E^{\,1} V_S^{\,1} \sigma^t - K_1^{\,1} P_E$
	$\dfrac{\delta P_1^{\,1}}{\delta t} = K_2^{\,1}(P^1 P^2) + K_3^{\,1} P_E^{\,1} V_S^{\,1} \sigma^t - K_2(P_S - P^1)$

[a] Modified from Kamp-Nielsen (1983).

determine cumulative thresholds for anthropogenic substances; and (5) address questions of scale (Reiners 1988).

Most of the models discussed thus far can be modified, extended, or combined to produce landscape simulations. The incorporation of feedback loops, neighborhood influences, and spatial exports and imports converts a one-dimensional model into a spatially articulate landscape model. The chapters to follow in this book illustrate some of the best ideas in the evolution of this new field. We have grouped these ideas into two (not necessarily mutually exclusive) schools of thought, the stochastic school and the process-based school. The stochastic school is looking for ways to combine spatial structure, patterns, and information with probabilistic distributions. The mechanistic, process-based school is trying to build models with as much realism as current computers will allow.

10.6.1 Stochastic Landscape Models

Although transition probability models have frequently been used to predict changes in vegetation patterns (Horn 1976; Debussche et al. 1977; Van Hulst 1979; Usher 1981; Lippe et al. 1985) or land use (Hett 1971; Burnham 1973;

Johnson 1977) through time, they are rarely spatially explicit. There are particular challenges in simulating changes in landscape patterns using transition probability models (Turner 1987). First, landscape changes are not strictly Markovian; that is, the change of state of a cell is not independent but is influenced by surrounding cells and other factors. Second, the rates of transition are not constant through time. Third, causation of land-use transition may be largely economic rather than natural (Burnham 1973, Alig 1986). There is also the practical difficulty of defining transition states and obtaining the transition probabilities by independent measurement, rather than inference (Lippe et al. 1985).

Despite these problems, Turner (1987, 1988) developed spatially influence land use transition models for a 12,698-ha landscape in the Georgia piedmont. Simulations were done on five land use types overlaid by square 1-ha cells, at random (i.e., a function of the transition probabilities) and with spatial influences. Spatial influences were simulated in two ways; (1) when four adjacent neighbors influenced the transition; and when eight neighbors (adjacent plus diagonal cells) influenced the transition of a cell. A transition index calculated cell change as a function of the number of neighbors of state $j(n_j)$ and the probability of i going to $j(p_{ij})$, such that the maximum value was n_j*p_{ij}, where $j = 1, \ldots$, number of states. The random simulation model produced highly fragmented landscape patterns that were quite different from the actual landscape, while the models, including spatial adjacency influences, simulated the clustering of certain land uses such as cropland and forests, reasonably well. This difference supports the non-Markovian nature of landscape changes, suggesting there are contagion effects. The nature of these types of models is explored in more detail in Chapter 11 by Gardner and O'Neill.

It has been suggested that a more realistic transition matrix is possible if, by using multiyear sequences of remote sensing data to provide estimates of transition frequencies, a climate-dependent transition matrix could be generated (Hall et al. 1988). Most transition probability models assume that climate and other changes have no effect on the transition rates. Hall et al. (1988) believe that landscape transition models can be coupled with GACMs by observing transitions in vegetation over a climatic gradient in space to calculate matrices as a function of climate change in time, by linking these matrices to spatial simulations of precipitation and temperature, and by having the fluxes of latent and sensible heat returned to the atmospheric model as a function of vegetation. While such a model has yet to be developed, it does illustrate the potential of stochastic landscape models to incorporate feedbacks and acquire a dynamic property. As they do, so they begin to overlap more and more with process-based landscape simulation models.

10.6.2 Process-Based Landscape Models

A process-based landscape model simulates spatial structure by first compartmentalizing the landscape into some geometric design and then describing flows within compartments and spatial processes between compartments according to location-specific algorithms. Four categories of flows can be identified in ecosystems (Ulanowicz 1986): (1) inputs from outside the system; (2) transfers within the system; (3) outputs of useful material from the system; and (4) energy dissipation. If the modeling of these flows, both abiotic and biotic, can be made spatially explicit, then an ecosystem model can qualify as a landscape model.

Examples of process-based, spatially articulate landscape models include wetland models (Sklar et al. 1985; Costanza et al. 1986; Kadlec and Hammer 1988; Boumans and Sklar, 1990; Costanza et al., 1990), oceanic plankton models (Show 1979), pest infestation models (Onstad 1989), coral reef growth models (Maguire and Porter 1977), and fire ecosystem models (Kessell 1977). These types of landscape models are analogous to the grid-type GACMs discussed earlier in that they simulate two- or three-dimensional flows over large areas, transport materials as a function of mass balance in combination with numerous climatic forcing functions, and require extensive computer time.

The model developed by Show (1979) to describe the distribution of *Acartia tonsa* in a tidal lagoon is a good example of a simple process-based landscape model and how one equates water movement, behavioral responses, and population processes to explicit spatial coordinates. Show (1979) compartmentalized the lagoon into rectangular solids with 1-m horizontal and 250-m vertical dimensions (Fig. 10.11). The advective transport of *Acartia* (Z) was then calculated as the sum of all U (vertical) and W (horizontal) transport coefficients, as shown here for compartment two:

$$dZ_2/dt = U_{21} Z_1 + W_{24} Z_4 + W_{26} Z_6 + P_2 Z_2 - U_{12} Z_2 - W_{42} Z_2 - W_{62} Z_2 \quad (10.14)$$

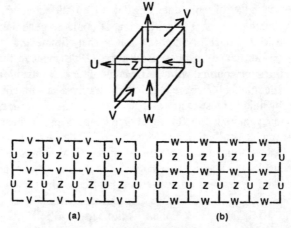

Figure 10.11. Graphic representation of some of the compartments, horizontal transport coefficients (U and V), vertical transport coefficients (W) and zooplankton densities (Z) in a spatial model that predicts *Acartia tonsa* distributions as a function of behavior, population dynamics, and environmental gradients (modified from Show 1979, figs. 3 and 5, p. 433).

where the first subscript on the transport coefficient indicates the compartment to which the flow is directed. The total Z in each cell was then multiplied by a series of analytical equations that calculated the proportion of *Acartia* swimming through a cell as a function of tide, temperature, salinity, light, and population density. The result was a time series of plankton changes for each cell.

The CELSS model, a process-based spatial simulation model consisting of almost 2,500 interconnected cells, each representing 1 km², constructed for a marsh/estuarine complex in south Louisiana (Sklar et al. 1985; Sklar and Costanza 1986; Costanza et al., in press), uses the same approach used by Show (1979) except the vertical articulation was reduced in favor of greater horizontal range. Each cell in the CELSS model contained a dynamic, nonlinear simulation model with eight state variables; was assigned one of seven habitat types, each of which had unique parameter settings; and was potentially connected to each adjacent cell by the exchange of water and materials (Fig. 10.12). The volume of water crossing from one cell to another was a function of water storage (W) and connectivity (K). Connectivity was a function of landscape characteristics, including habitat type, drainage density, waterway orientation, and levee height. The balance between sediment deposition and erosion, as influenced by the variables shown in Fig. 10.12, and thought to be critical to habitat succession and the productivity of the area, was effected by $W \cdot K$ interactions at each cell boundary. State variables were

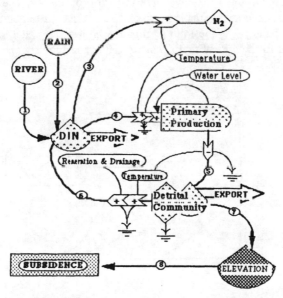

Figure 10.12. Graphical representation of the physical and biological processes controlling the internal storage (tanks) and exchange rates (lines) of carbon within a single 1-km² cell in the CELSS model. DIN = dissolved inorganic nitrogen.

monitored in each cell to see if the "environment" (i.e., salinity, elevation, productivity, etc.) was appropriate for the vegetation. Habitat succession occurred when these values became inappropriate for its designated habitat type. Succession meant that the cell habitat type and all the associated parameter settings were switched to a set of more representative parameters. For example, if salinity in a cell that was initially fresh marsh went beyond a threshold value of 3 ppt and remained at this high level for more than forty-five weeks, then the cell was converted to brackish marsh with all its associated parameters, including a habitat-specific function for primary production (Fig. 10.13). Currently the model explains about 88% of the 1978 ecosystem-type calibration data and predicted about 83% of the 1983 verification data. Figure 10.14 shows some sample output indicating the real and simulated habitat changes from 1956 to 1983, while Table 10.3 indicates the range of management options and past and future climate scenarios that the model has been used to analyze.

Fire ecosystem models are probably the most widely used of all landscape models. Predicting the direction and spread rate of a forest fire is difficult because of confounding winds, errors in the process equations, and poor spatial information. Nevertheless, a fire model has most of the characteristics of a good landscape model. Designed to help manage small to medium forest fires, the Glacier National Park Basic Resource and Fire Ecology Systems Model (Kessel 1975) uses (1) a GIS database on vegetation species and density per hectare, (2) a gradient model that predicts fuel characteristics for each hectare, (3) weather information, and (4) a fire-behavior model. The rate of fire spread, the height of the flames, and other variables are based upon the law of conservation of energy. The heat flux from burning timber becomes available to pass into and raise the temperature of the fuel in the next hectare. It is the ratio of this available heat, called the propagation intensity (IP), to the critical fuel density (RHO) and ignition energy (Q_{ig})

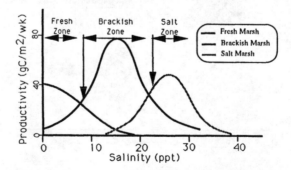

Figure 10.13. In the CELSS model, primary production was simulated as a normal distribution function along some environmental gradient (in this case salinity) for each habitat type.

Figure 10.14. Results of the CELSS model from 1956 to 1983. Each pixel represents 1 km^2 of salt marsh, brackish marsh, freshwater marsh, hardwood swamp, upland, or open water. CELSS was calibrated to the 1978 data (goodness-of-fit = 88%) and validated to the 1983 data (goodness-of-fit = 83%).

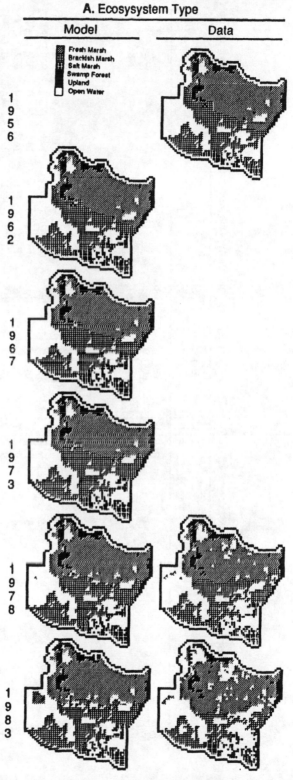

Table 10.3. Number of Square Kilometers of Each Ecosystem Type for the Three Years for Which Data Are Available, and for the Year 2033 for Various Scenarios[a]

	Swamp	Fresh Marsh	Brackish Marsh	Saline Marsh	Upland	Total Land	Open Water
1956	130	864	632	98	13	1737	742
1978	113	766	554	150	18	1601	878
1983	116	845	347	155	18	1481	998
2033 Scenarios:[b]							
Climate Scenarios:[c]							
Climate run 3 (base case)	84	871	338	120	10	1423	1056
Climate run 1	79 (−5)	874 (+3)	337 (−1)	127 (+7)	10 (0)	1427 (+4)	1052 (−4)
Climate run 4	85 (+1)	900 (+29)	355 (+17)	130 (+10)	10 (0)	1480 (+57)	999 (−57)
Climate run 5	83 (−1)	891 (+20)	332 (+6)	126 (+6)	10 (0)	1442 (+19)	1037 (−19)
Mean climate	94 (+10)	974 (+103)	402 (+64)	136 (+16)	11 (+1)	1617 (+194)	862 (−194)
Weekly average climate	128 (+44)	961 (+90)	813 (+475)	300 (+180)	11 (+1)	2213 (+790)	266 (−790)
Management Scenarios:							
No levee extension (base case)	100	796	410	123	15	1444	1035
Two reach levee extension	98 (−2)	804 (+8)	399 (−11)	123 (0)	15 (0)	1439 (−5)	1040 (+5)
CLF marsh management	102 (+2)	798 (+2)	409 (−1)	123 (0)	15 (0)	1447 (+3)	1032 (−3)

Falgout weir	104 (+4)	799 (+3)	403 (−7)	122 (−1)	16 (+1)	1444 (0)	1035 (0)
Full six reach levee extension	103 (+3)	790 (−6)	362 (−48)	122 (−1)	15 (0)	1392 (−52)	1087 (+52)
Fresh water diversion (FWD)	103 (+3)	803 (−7)	404 (−6)	123 (0)	15 (0)	1448 (+4)	1031 (−4)
FWD and Palmetto Wier	102 (+2)	802 (+6)	407 (−3)	123 (0)	15 (0)	1449 (+5)	1030 (−5)
FWD and Superior Wier	104 (+4)	799 (+3)	404 (−6)	123 (0)	15 (0)	1445 (+1)	1034 (−1)
FWD, Superior and Palmetto Weirs	104 (+4)	792 (−4)	407 (−3)	123 (0)	15 (0)	1441 (−3)	1038 (+3)
FWD, Superior and Falgout Weirs	104 (+4)	803 (+7)	407 (−3)	122 (−1)	15 (0)	1451 (+7)	1028 (−7)
Boundary Scenarios:[d]							
EPA low sea level rise	104 (+4)	800 (+4)	411 (+1)	124 (+1)	15 (0)	1454 (+10)	1025 (−10)
EPA high sea level rise	89 (−11)	794 (−2)	396 (−14)	131 (+8)	15 (0)	1425 (−19)	1054 (+19)
Historical Scenarios:[e]							
No original Avoca Levee	84	951	350	126	13	1524	955
No impacts	130	863	401	144	12	1550	929

[a] From Costanza et al. (1990). Changes from the base case (in km²) are indicated in parentheses.

[b] The summary maps and this table indicate the "dominant" habitat type for each cell (i.e., the ecosystem type that was present in the cell for the largest amount of time during the year). Alternately, we could have added the total number of cells of each ecosystem type for each week of the simulated year and divided the totals by 52. While this gives a somewhat more accurate picture of the habitat distribution, it is inconsistent with the totals from the maps.

[c] The climate analysis scenarios used a slightly different set of parameters for the model than the other scenarios.

[d] EPA low scenario is 50-cm rise by the year 2100. We used 0.46 cm/yr, which is double the historical rate of eustatic sea level rise in the study area of 0.23 cm/yr. Subsidence in the study area varies horizontally from 0.57–1.17 cm/yr, giving historical rates of apparent sea level rise (eustatic rise plus subsidence) of 0.8 to 1.4 cm/yr. EPA high scenario is 200 cm rise by the year 2100 (1.67 cm/yr eustatic or 2.24–2.84 cm/yr apparent). Base case for comparison was the no-levee extension case.

[e] No comparisons with a base case are given for the historical scenarios since these runs started in 1956 rather than 1983.

$$\text{Spread Rate} = \frac{IP}{RHO \cdot Q_{ig}} = \frac{\text{Heat Flux (kcal·m}^2\text{·min}^1\text{)}}{\text{Heat Sink (g·cm}^3\text{)(kcal·g}^1\text{)}} \qquad (10.15)$$

that allows for the prediction of the fire spread rate. The only limitation of this fire-behavior model is that it assumes uniform horizontal and vertical spatial distribution of fuels in each 1-ha cell (Kessell 1977), which doesn't allow the model to predict the impacts of some fire-fighting tactics (eg., fire breaks) or natural fire breaks (eg., rivers) because most of them operate on a scale smaller than 1 ha. We see here an example of how scale can influence model utility.

Both the stochastic and mechanistic approaches to dynamic spatial modeling usually use some uniform cell size. Uniform cell size is easy to deal with mathematically and is necessary if the spatial structure of the system is expected to change and one wishes to model that process. In CELSS, for example, updating the constantly changing man-made network of oil-access canals was simply a matter of reading a new location matrix file every year. In cases where the spatial structure is not expected to change, one might save time and effort by using irregular-shaped polygons.

Many current GISs are polygon-based. Might a more accurate or efficient landscape model be designed to directly access such a data base? This approach was explored recently by Boumans and Sklar (1990) to evaluate wetland succession patterns in Louisiana. Five nonuniform polygons (i.e., within each polygon there are many habitats) were used to compartmentalize a wetland system impacted by roads and levees (Fig. 10.15). Each interface had a different length and degree of connectivity with adjacent polygons; therefore, each polygon had a set

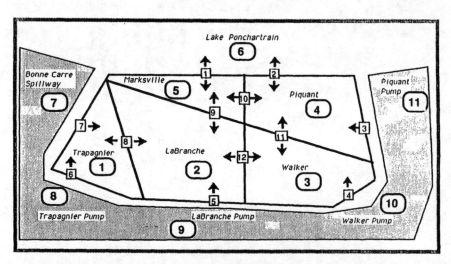

Figure 10.15. Internal polygons (1–5) and external drainage basins (6–11) were chosen to represent the different hydrological drainage units found in a Louisiana wetland. The arrows represent the direction of water flow between the polygons.

of unique flow coefficients. Habitat succession occurred within each polygon as a function of competition for resources (i.e., dry land) over time. If a polygon became too inundated with water, the ratio of marsh to open water decreased according to the population competition equations of May and MacArthur (1972). A regression analysis, used as a measure of the goodness-of-fit between predicted and actual habitat change, resulted in an r^2 of 0.56 with a $p < 0.01$. Decomposed into its separate habitat components, the r^2 indicated that the model could account for 62.5%, 50%, and 41.9% on the average of the spatial variation over time for swamp, marsh, and open water, respectively. Although this polygon approach needs more development, especially if one wants to simulate uniform polygon succession, it represents the first step in combining dynamic landscape models with popular polygon-based GIS software. We see enhancements, such as adding cells within each nonuniform polygon or incorporating amoebalike growth characteristics to uniform polygons, as fruitful avenues for further research.

Another approach to dynamic landscape modeling involves the use of rule-based simulations. These are often mislabeled as artificial-intelligence (AI) based approaches. Current applications have very little in the way of true artificial intelligence, however, and are better viewed as a variation of the process-based approach. The main distinction is that the processes are handled in a more discrete, rule-based context rather than by the more usual continuous equation approach (Coulson et al. 1987; Loehle 1987). Most existing process-based models contain at least some discrete rules, and there is a continuum over which one can move from totally continuous equation-based models to totally rule-based models.

There are only a few rule-based landscape models. Using rule-based techniques, Graham (1986) modeled the behavior of feral horses on the island of Shackleford Banks, North Carolina, Wilkie and Finn (1988) predicted land use by horticultural tribes in the tropical forests of Zaire, and Saarenmaa et al. (1988) predicted moose movements in Scandinavia. The moose-habitat model (Saarenmaa et al. 1988) is a good illustrative example of rule-based process models because moose foraging depends upon landscape characteristics such as tree species, neighborhood stands, predators, resting sites, etc., while moose landscape knowledge is developed from investigative behavior, is stored in associative memory, and is used to manifest a feasible (although not necessarily optimal) foraging strategy. The landscape sends out messages (Fig. 10.16a) that each object, in this case moose, can interpret differently according to a hierarchical tree of attributes (i.e., the age of the moose, its sex, its hunger, etc.). The rules link the landscape with the behavioral modes of the moose (Fig. 10.16b), feedbacks occur across spatial and temporal scales, and moose migrate through a landscape "tree" (Fig. 10.16c). Rules and links don't have to be biotic. One could simulate the "behavior" of some biochemical flow in a similar manner.

10.7 Prognosis

Spatial modeling, or any modeling, is as much art as science. The conceptualization phase of model development cannot be reduced to a set of systematic rules

Table 10.4 Summary of the Characteristics of Models According to Categories Within Major Disciplines[a]

Discipline	Category	Model Objective	Typical Time & Space Scales[b]	Typical Modeling Variables	Basic Characteristic	Advantages	Disadvantages	References
Geography	Geometric	Define a landscape by a system of shapes	TF=N/A dt=N/A SF=kilometers ds=meters	Shape Size Distance	Shape and design can explain spatial distributions	Used to organize and categorize natural landscape features; creates an emergent geometry related to theories of spatial structure	Descriptive model with little dynamic behavior; generalized shapes don't match reality	Horton 1945 Schumm 1956 Chisholm 1962 Selkirk 1982
	Demographic	Predict the flow from point i to point j	TF=years dt=days SF=continents ds=kilometers	Diffusion Transition rates Density Information Birth Death	A spatial pattern emerges from the influence of origin and destination characteristics on birth, death, and migration	Applicable to many situations; focus is on diffusion; easy to develop; the first to use statistical parameters such as power spectra and spatial autocorrelation.	Tends to be descriptive, linear, and non-dynamic; focus is on pattern of diffusion without concern for process; no feedbacks; spatial attributes are few and point-averaged	Hägerstrand 1953 Tornqvist 1967 Cliff and Ord 1975 Batty and March 1978 Getis and Boots 1978 Thomas and Hugget 1980
	Network	Find the shortest distance (or cost) when going from point i to point j	TF=N/A dt=N/A SF=continents ds=kilometers	Flow rates Birth (production) Death (demand) Distance	Flows between "nodes" can be optimized based upon each node's capacity to produce and consume.	Widely used and mathematically well understood; flows are a function of distance; finds the best solution; used in operations research and transportation problems.	All interactions are linear, feedbacks are ignored; flows are assumed to be infinitely divisible in an attempt to optimize; nondynamic behavior.	Henderson 1958 Dantzig 1963 Isard 1975 Thomas and Hugget 1980

	Model	Objective	Scales	Variables	Process description	Strengths	Weaknesses	References
Fluid Dynamics	Hydrodynamics	Predict the velocity and mass of a spatial fluid field	TF=days ct=seconds SF=100's of km cs=kilometers	Momentum Acceleration Depth Friction	Flows between "nodes" are based upon the equations of motion and momentum bounded by 'link' characteristics	Accurate estimates of flow for short time scales; has some feedbacks; mathematically well understood.	Requires extensive computer processing; data intensive; network structure is fixed; only 'links and nodes' are simulated	EPA 1975 Hopkinson and Day 1980 Wang et al. 1985
	GACM's	Predict the velocity, mass and direction of flows in the atmosphere/ocean system	TF=10's of yrs dt=days ds=global ds=5°lat° and long.	Momentum Turbulence Temperature Moisture (air) Salinity(oceans) Density Pressure	Space is divided into three-dimensional grid cells and flows are based upon equations of motion, momentum and M mass balance	Spatially explicit; simulates flows for large regions; incorporates many temporal and spatial feedbacks.	Very complex; requires a supercomputer; little interaction with the biosphere; coarse spatial scales	Phillips 1956 Williams et al. 1974 Washington et al. 1979 Simmons and Bengtsson 1984 Nigam et al. 1988
Ecology	Growth & Population	Predict the number or size of populations i,j,...m	TF=years dt=days SF=meters Ls=cm	Migration (diffusion) Density Information Birth Death Resources (nutrients, light, etc.)	A spatial pattern is the result of populations having different rates of growth, birth and death as a function of competitive interactions for resources.	Widely used; has simple linear mathematics; focus is on temporal density-dependent interactions; examines diffusion as a biological process.	Spatial feedbacks are ignored; landscapes are independent, exogenous variables	Huffaker 1958 Botkin et al. 1972 Levin 1979 Diamond and May 1981 Dale et al. 1985 Pearlstine Nachman 1987a, 1987b Pacula 1987
	Ecosystem	Predict resource distribution and the number or size of populations i,j,...m	TF=years dt=hours SF=N/A ds=N/a	Migration (diffusion) Density Information Birth	A spatial pattern is the result of communities having different rates of energy and mate-	Versatile and widely used; focus is on "whole system" components and flow rates; easily in-	Mathematics can get complex; functional aggregations don't explicitly address spatial dynamics	Forrester 1961 Patten 1971, 1972, 1975 Odum 1983

(Continued)

Table 10.4 *Continued.*

Discipline	Category	Model Objective	Typical Time & Space Scales[a]	Typical Modeling Variables	Basic Characteristic	Advantages	Disadvantages	References
				Death Growth Resources (nutrients, light, etc.)	rial flows as a function of biological and environmental interactions.	corporates spatial variables.		
Landscape Ecology	Stochastic	Predict changes in a spatial pattern	TF=10's of yrs dt=years SF=counties ds=10's of meters	Density Transition rates Distance Habitat	The probability of transition of a cell in a landscape is a function of intrinsic change and neighbor characteristics	Mathematics is simple and well understood; combines geometric, demographic, and population models; modifies Markovian models to include spatial resolution; finds emergent landscape structure.	Transition probabilities are constant; environmental feedbacks don't exist; cannot relate cause and effect; still in the development stage.	Hett 1971 Hall et al. 1988 Turner 1987, 1988
	Process Based	Predict flows from point i to point j and mechanisms of change in a spatial pattern.	TF=10's of yrs dt=weeks SF=counties ds=100's of meters	Diffusion Mass Density Information Growth Birth Death Momentum Turbulence Resources (nutrients, light, etc.)	Space is compartmentized and flows are based upon equations of motion and mass balance, modified by the information, energy, and material flows within each compartment.	Spatially explicit; combines all types of spatial models; incorporates many temporal and spatial feedbacks; has realistic physical and biological processes; can link cause and effect; useful for evaluating specific managment alternatives.	Uses complex nonlinear mathematics; requires extensive CPU time; coarse spatial and temporal scale; still in the development stage.	Kessell 1977 Show 1979 Sklar et al. 1985 Wilkie and Finn 1988 Costanza et al, 1990

[a] Categories represent historical developments and are not meant to be mutually exclusive

[b] TF=time frame of the model, dt=time step, SF=space frame, ds=space step

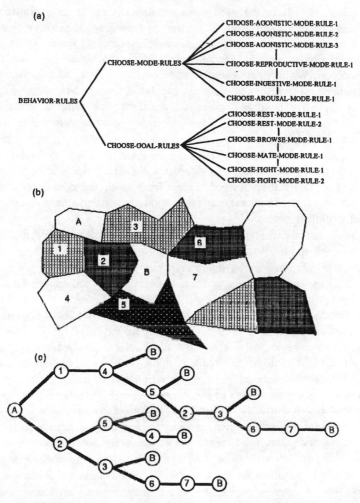

Figure 10.16. Decision-making rules and goals are used by moose (a) and linked to a hypothetical forest (b), shown in shading to indicate different object attributes, such as stand density. The tree representation (c) indicates all possible nonrepeating paths for a moose to travel from A to B (modified from Saarenmaa et al. 1988, figs. 3 and 5).

(Sklar et al., in press). The art is in finding the most appropriate variables and hierarchical level of organization for the modeling objective(s) at hand (Allen and Starr 1982). *Appropriate* in this context means some balancing of model construction costs and data requirements against the benefits gained from using the model. We see future landscape models as hybrids of previous designs, developed as a suite of useful models applicable to a range of purposes.

In the past, each discipline developed mathematical strategies more or less independently to model spatial dynamics. We have summarized these strategies in Table 10.4 to illustrate the advantages and disadvantages of these historical approaches and to show the similarities and differences across disciplines. Birth, death, and diffusion were variables common to demographic, network, population, ecosystem, and process-based landscape models. They are found in many process-based models but especially in those concerned with biological flows and the rates of production. When variables such as resources, growth, and information are added to these models, they become more realistic, more complex, and, in general, spatially explicit with more landscape-level feedbacks. What is apparent from Table 10.4 is the inverse relation between widely used models and models with many variables. Mathematically well-understood models (i.e., network, hydrodynamic, and stochastic) tend to have simple linear equations and relatively few variables. Models with more than five variables tend to include nonlinear feedbacks and are difficult to develop, except for growth and demographic models, which are widely used because they focus upon density-dependent variables rather than on spatial processes.

What is also apparent from Table 10.4 is the relation between the model objectives, which clearly separate one model from another, and their associated disadvantages. When viewed from a spatial landscape modeling perspective, many of the the objectives appear limited in scope. They are concerned only with flows from point i to point j (demographic, network, hydrodynamic and GACM) or distribution of materials among points $i, j, \ldots m$ (geometric, population, ecosystem, and stochastic). As a result, the former are inflexible (i.e., network design is fixed) and lack biological or ecosystem feedbacks, while the latter ignore spatial feedbacks and are more descriptive than dynamic. Only process-based landscape models have the advantages of being spatially explicit, realistic, and dynamic.

There are three basic ways to incorporate spatial characteristics or processes into a model: geometrically, statistically, and mechanistically. In the past these techniques were rarely combined. Geometric progressions were used to describe systems, statistics were used to analyze systems, and mechanistic models were used to simulate processes. The development of landscape ecology has set the stage for combining these techniques. Process-based landscape models work best because their objectives and techniques are unconstrained and because the tools to actually perform landscape modeling are advancing at a rapid rate. For example, supercomputers and parallel processors decrease the time needed to manipulate large, realistic, spatial arrays, spatial data from remote sensing sources are available for calibrating and verifying landscape models, and GIS data bases make organizing the mass of spatial data necessary for landscape modeling feasible. The use of GISs is becoming commonplace, and enhancing their ability to interact with dynamic spatial simulations is one obvious way to encourage the development and use of landscape models.

A modeling environment that links a flexible modeling language (that can include both continuous equation and rule-based algorithms) to a GIS data management system and a parallel computer hardware and software system for rapid execution of spatial arrays will probably be the landscape simulation environment of the future. There is still a long way to go, however, before we understand enough landscape "rules" to implement such a system on a truly large scale. We might be running out of time, because as we enter the 21st century the competition for earth resources on a spatial scale will truely intensify, and landscape-level destruction will likely threaten ecosystem structure and function on a global scale (e.g., CO_2 increase and climate change). Ecologists need to find ways for civilization and nature to evolve in harmony. This book is evidence that many scientists believe that methods in landscape ecology are powerful new tools in the quest for finding large-scale solutions.

10.8 Summary

We surveyed the use of spatial models in geography, hydrology, biology, and ecology as a way to introduce the concept of landscape modeling. We found large differences in the way space was used to structure model designs. Some models were based upon networks of links and nodes, some used two- or three-dimensional grid cells, and others described landscapes as unique geometric shapes. Geographers and anthropologists were the first to create spatial models. Many of the concepts developed in geography, such as diffusion, logistic growth, migration, and spatial autocorrelation, have been used in growth and population models as well as in stochastic and process-based landscape models. Differences in complexity and degree of spatial feedbacks vary significantly among models. Hydrodynamic models were found to be very complex but insufficient for landscape modeling since network structure cannot change, and much of the landscape "space" is not modeled. GACMs were also found to be complex but much better at including space. As GACMs become modified to include biosphere feedbacks, they will be very effective in addressing global landscape change. Probabilistic landscape models have become important as descriptive tools but need further development if they are to become dynamic predictive tools. Process-based landscape models (which incorporate some combination of historical and extant spatial models) are our most realistic landscape simulators at present. They are spatially explicit and relate cause and effect in an attempt to create effective landscape management tools. This new field, called landscape modeling, will evolve by combining the principles and designs summarized in this chapter. The excitement is in finding the right combination(s) to best exploit our rapidly improving data and computing environment.

Acknowledgments

This research was supported by National Science Foundation Grant No. BSR–8906269, titled "Landscape modeling: the synthesis of ecological processes over large geographic regions and long time scales," F.H. Sklar and R. Costanza, Principal Investigators, and the NSF LTER Program at North Inlet, South Carolina, under Grant No. BSR–8514326. The authors wish to thank M. Turner, R. Gardner, J. Finn, and two anonymous reviewers for their insightful remarks. S. Ring and S. Hutchinson assisted with the library research. This article is contribution number 783 of the Belle W. Baruch Institute for Marine Biology and Coastal Research at the University of South Carolina.

References

Alig, R.J. 1986. Econometric analysis of the factors influencing forest acreage trends in the southeast. *Forest Science* 32: 119–34.

Allen, T.F.H., and Starr, T.B. 1982. *Hierarchy*. Chicago: University of Chicago Press.

Allen, W. 1989. Supercomputers warming up to improve greenhouse models. *Supercomputing Review* 2:26–30.

Alves, W.R. 1974. Comment on Hudson's 'Diffusion in a central place system.' *Geographical Analysis* 6:303–8.

Ayala F.J., Gilpin, M.E., and Ehrenfeld, J.G. 1973. Competition between species: theoretical models and experimental tests. *Theoretical Population Biology* 4:331–56.

Baker, W.L. 1989. A review of models of landscape change. *Landscape Ecology* 2:111–33.

Batty, M., and March, L. 1978. Dynamic urban models based on information-minimising. In *Towards the Dynamic Analysis of Spatial Systems*, eds. R.L. Martin, N.J. Thrift, and R.J. Bennett, pp. 127–204. London: Pion Limited.

Bennett, R.J. 1974. Process identification for time series modelling in urban and regional planning. *Regional Studies* 8:157–74.

Bennett, R.J. 1979. *Spatial Time Series: Analysis, Forecasting, Control*. New York: Pion Press.

Boer, G.J., and Lazare, M. 1988. Some results concerning the effects of horizontal resolution and gravity-wave drag on simulated climate. *Journal of Climate* 1:789–806.

Boesch, D.F. 1977. Application of numerical classification in ecological investigations of water pollution. EPA–600/77–033. U.S. Environmental Protection Agency, Corvallis, Oreg.

Botkin, D.B., Janak, J.F., and Wallis, J.R. 1972. Some ecological consequences of a computer model of forest growth. *Journal of Ecology* 60:849–72.

Boumans, R.M.J., and Sklar, F.H. 1990. A polygon-based spatial model for simulating landscape change. *Landscape Ecology* 4(2):83–97.

Box, G.E.P., and Jenkins, G.M. 1970. *Time Series Analysis, Forcasting and Control*. San Francisco: Holden-Day.

Burnham, B.O. 1973. Markov intertemporal land use simulation model. *Southern Journal of Agricultural Economics* 5:253–58.

Casetti, E. 1969. Why do diffusion processes conform to logistic trends? *Geographic Analysis* 1:101–5.

Chisholm, M. 1962. *Rural Settlement and Land Use*. Chicago: Aldine.

Clark, W.A.V. 1969. Applications of spacing models in intra-city studies. *Geographical Analysis* 1:391–99.

Cliff, A.D.; Haggett, P.; Ord, J.K.; Bassett, K.A.; and Davies, R.B. 1975 *Elements of Spatial Structure: A Quantitative Approach*. Cambridge: At the University Press.

Cliff, A.D.; Haggett, P.; Ord, J.K.; and Versey, G.R. 1981 *Spatial Diffusion: An Historical Geography of Epidemics in an Island Community*. Cambridge: Cambridge University Press.

Cliff, A.D., and Ord, J.K. 1969. The problem of spatial autocorrelation. In *London Papers in Regional Science*, vol. 1, *Studies in Regional Science*, ed. A.J. Scott, pp. 25–55. London: Pion.

Cliff, A.D., and Ord, J.K. 1971. Evaluating the percentage points of a spatial autocorrelation coefficient. *Geographical Analysis* 3:51–62.

Cliff, A.D., and Ord, J.K. 1972. Testing for spatial autocorrelation among regression residuals. *Geographical Analysis* 4:267–84.

Cliff, A.D.; and Ord, J.K. 1973. *Spatial Autocorrelation*. New York: Pion Press.

Cliff, A.D., and Ord, J.K. 1975. Model building and the analysis of spatial pattern in human geography. *Journal of the Royal Statistical Society*, B 37:297–348.

Committee for an International Geosphere-Biosphere Program. 26.1. S. 1986. *Global Change in the Geosphere-Biosphere: Initial Priorities for an IGBP*. Washington, D.C.: National Academy Press.

Costanza, R. 1989. Model goodness of fit: a multiple resolution procedure. *Ecological Modelling* 47:199–215.

Costanza, R.; Neill, C.; Leibowitz, S.G.; Fruci, J.R..; Bahr, L.M.; and Day, J.W., Jr. 1983. Ecological models of the Mississippi deltaic plain region: data collection and presentation. U.S. Fish and Wildlife Service, Division of Biological Services. Washington, D.C. FWS/OBS–82/68.

Costanza, R., and Sklar, F.H. 1985. Articulation, accuracy, and effectiveness of mathematical models: a review of freshwater wetland applications. *Ecological Modelling* 27:45–68.

Costanza, R.; Sklar, F.H.; and Day, J.W., Jr. 1986. Modeling spatial and temporal succession in the Atchafalaya/Terrebonne marsh estuarine complex in south Louisiana. In *Estuarine Variability*, ed. D. A. Wolfe, pp. 387–404. New York: Academic Press.

Costanza, R.; Sklar, F.H.; and White, M.L. 1990. Modeling coastal landscape dynamics. *Bioscience* 40(2):91–107.

Coulson, R.N.; Folse, L.J.; and Loh, D.K. 1987. Artificial intelligence and natural resource management. *Science* 237:262–67.

Curry, L. 1964. The random spatial economy: an exploration in settlement theory. *Annals of the Association of American Geographers* 54:138–46.

Dacey, M.F. 1964. Modified Poisson probability law for point patterns more regular than random. *Annals of the Association of American Geographers* 54:559–65.

Dale, V.H.; Doyle, T.W.; and Shugart, H.H. 1985. A comparison of tree growth models. *Ecological Modelling* 29:145–69.

Dantzig, G.B. 1963. *Linear Programming and Extensions*. Princeton, N.J.: Princeton University Press.

Debussche, D.M.; Godron, M.; Lepart, J.; and Romane, F. 1977. An account of the use of a transition matrix. *Agro-Ecosystems* 3:81–92.

D'Elia, C.F. 1988. The cycling of essential elements in coral reefs. pp. 195–246. In *Concepts of Ecosystem Ecology*, eds. L.R. Pomeroy and J.J. Alberts, pp. 195–246. New York: Springer-Verlag.

Diamond, J.M., and May, R.M. 1981. Island biogeography and the design of nature reserves. In *Theoretical Ecology: Principles and Applications*, 2nd ed., pp. 228–52. Oxford: Blackwell.

Dickinson, R.E. 1984. Modeling evapotranspiration for three-dimensional global climate models. Climate processes and climate sensitivity *American Geophysical Union Monographs* 29:58–72.

Dorfman, R.; Samuelson, P.A.; and Solow, R.M. 1968. *Linear Programming and Economic Analysis*. New York: McGraw-Hill.

Earth Systems Science Committee. 1988. Earth System Science: A Closer View. Washington, D.C.: NASA.

Elton, C.S. 1927. *Animal Ecology*. London: Sidwick and Jackson.

Environmental Protection Agency. 1975a. Storm water management user's manual, version II. Environmental Technology Series, EPA 67012–75–017, U.S. EPA 4, Cincinnati, Ohio.

Environmental Protection Agency. 1975b. Rates, constants, and kinetics formulations in surface water quality modeling, 2nd ed. EPA 600/3–85/040, Environmental Research Laboratory, Athens, Ga.

Finerty, J.P. 1971. Cyclic fluctuations in biological systems: a revaluation. Ph.D. Thesis, Yale University, New Haven, Conn.

Focardi, S.; Deneubourg, J.L.; and Chelazzi, G. 1989.. Theoretical analysis of rhythmical clustering in an intertidal gastropod. *Ecological Modelling* 44:177–94.

Forrester, J.W. 1961. *Industrial Dynamics*. Cambridge, Mass.: MIT Press.

Frank, P.W. 1957. Coactions in laboratory populations of two species of Daphnia. *Ecology* 38:510–19.

Gates, W.L., and Schlesinger, M.E. 1977. Numerical simulation of the January and July global climate with a two-level atmospheric model. *Journal of Atmospheric Science* 34:36–76.

Gause, G.F. 1932. Experimental studies of the struggle for existence: I. mixed populations of two species of yeast. *Journal of Experimental Biology* 9:389–402.

Gause, G.F. 1934. *The Struggle for Existence.* Baltimore: Williams and Wilkins.

Gayle, T.L. 1975. Systems modeling for understanding eutrophication in Lake Okeechobee. M.S. Thesis. University of Florida, Gainesville.

Getis, A., and Boots, B. 1978. *Models of Spatial Processes: An Approach to the Study of Point, Line and Area Patterns.* London: Cambridge University Press.

Gilpin, M.E., and Diamond, J.M. 1980. Subdivision of nature reserves and the maintenance of species diversity. *Nature* (London) 285:567–68.

Graham, L.A. 1986. HAREMS: A generalized data base manager and simulator for barrier island feral horse populations. Coop. Park Studies Unit Tech. Rep. 32.

Griffith, D.A., and R.D. MacKinnon, eds. 1981. *Dynamic Spatial Models.* Rockville Md.: Sitjhoff and Noordhoff.

Grilches, Z. 1957. Hybrid corn: an exploration in the economics of technological change. *Econometrica* 25:501–22.

Hägerstrand, T. 1953. *Innovations förlopper ur korologisk synpunkt.* Lund: Gleerup (1969). *Innovation Diffusion as a Spatial Process,* Trans. A. Pred. Chicago: University of Chicago Press.

Hägerstrand, T. 1967. On Monte Carlo simulation of diffusion. *Northwestern University Studies in Geography* 13:1–32.

Haggett, P. 1966. *Locational Analysis in Human Geography.* New York: St. Martin Press.

Haggett, P., and Chorley, R.J. 1967. Models, paradigms and the new geography. In *Models in Geography,* eds. R.J. Chorley and P. Haggett, pp. 19–51. London: Methuen.

Hall, C.A.S. 1988. An assessment of several of the historically most influential theoretical models used in ecology and of the data provided in their support. *Ecological Modelling* 43: 5–31.

Hall, C.A.S., and Day, J.W., Jr. 1977. Systems and models: terms and basic principles. In *Ecosystem Modeling in Theory and Practice,* eds. C.A.S. Hall and J.W. Day, Jr. pp. 6–36. New York: Wiley-Interscience.

Hall, F.G.; Strebel, D.E.; and Sellers, P.J. 1988. Linking knowledge among spatial and temporal scales: Vegetation, atmosphere, climate and remote sensing. *Landscape Ecology* 2:3–22.

Hardin, G. 1969. The competitive exclusion principle. *Science* 131:1292–97.

Harper, J.L. 1977. *Population Biology of Plants.* New York: Academic Press.

Harper, J.L., and McNaughton, J.H. 1962. The comparative biology of closely related species living in the same area: II. interference between individuals in pure and mixed populations of Papaver. *New Phytologist* 61:175–88.

Harwell, M.A., and Weinstein, D.A. 1982. Modelling the effects of air pollution on forest ecosystems. Ecosystems Research Center Report 6. Cornell University, Ithaca, NY.

Henderson, J.M. 1958. *The Efficiency of the Coal Industry: An Application of Linear Programming.* Cambridge, Mass.: Harvard University Press.

Hett, J. 1971. Land use changes in east Tennessee and a simulation model which describes these changes for 3 counties. ORNL–IBP–71–8. Oak Ridge National Laboratory, Oak Ridge, Tenn.

Hoover, E.M. 1975. *An Introduction to Regional Economics.* New York: Knopf.

Hopkinson, C.S., and Day, J.W. 1980. Modeling the relationship between development and storm water and nutrient runoff. *Environmental Management* 4:315–24.

Horn, H.S. 1976. Markovian properties of forest succession. In *Ecology and Evolution of Communities,* eds. M.L. Cody, J.M. Diamond pp. 196–211. Cambridge, Mass.: Harvard University Press.

Horton, R.E. 1945. Erosional development of streams and their drainage basins: hydrophysical approach to quantitative morphology. *Bulletin of the Geological Society of America* 56:275–370.

Hudson, J. 1969. A location theory for rural settlement. *Annals of the Association of American Geographers* 59:365–81.

Huff, J. 1984. Distance-decay models of residential search. In *Spatial Statistics and Models*, eds. G.L. Gaile and C.J. Wilmott, Boston: Kluwer Academic.

Huffaker, C.B. 1958. Experimental studies on predation; dispersion factors and predator-prey oscillations. *Hilgardia* 27:343–83.

Huggert, R. 1980. *Systems Analysis in Geography.* Oxford: Clarendon Press.

Hutchinson, G.E. 1947. A note on the theory of competition between two spacial species. *Ecology* 28:319–21.

Hutchinson, G.E. 1978. *An Introduction to Population Ecology.* New Haven, Conn.: Yale University Press.

Isdard, W. 1975. *Introduction to Regional Science.* Englewood Cliffs, N.J.: Prentice-Hall.

Jeffers, J.N.R. 1978. *An Introduction to Systems Analysis: With Ecological Applications.* London: Edward Arnold.

Johnson, W.C. 1977. A mathematical model of forest succession and land use for the North Carolina Piedmont. *Bull. Torrey Bot. Soc.* 104:334–46.

Kadlec, R.H., and Hammer, D.E. 1988. Modeling nutrient behavior in wetlands. *Ecological Modelling* 40:37–66.

Kamp-Nielsen, L. 1983. Sediment-water exchange models. In *Application of Ecological Modelling in Environmental Management, Part A*, ed. S.E. Jørgensen, pp. 387–420. Amsterdam: Elsevier Scientific.

Kasahara, A., and Washington, W.M. 1967. NCAR global circulation model of the atmosphere. *Mon. Wea. Res.* 95:389–402.

Kasahara, A., and Washington, W.M. 1971. General circulation experiments with a six-layer NCAR model, including orography, cloudiness and surface temperature calculations. *Journal of Atmospheric Science* 28:657–95.

Kessell, S.R. 1975. The Glacier National Park basic resources and fire ecology model. *Bulletin of the Ecological Society of America* 56:49.

Kessell, S.R. 1977. Gradient modeling: a new approach to fire modeling and resource management. In *Ecosystem Modeling in Theory and Practice*, eds. C.A.S. Hall and J.W. Day, Jr., pp. 576–605. New York: Wiley-Interscience.

Lande, R. 1987. Extinction thresholds in demographic models of territorial populations. *American Naturalist* 130:624–35.

Levin, S.A. 1970. Community equilibria and stability, and an extension of the competitive exclusion principle. *American Naturalist* 104:413–24.

Levin, S.A. 1974. Dispersion and population interactions. *American Naturalist* 108:207–28.

Levin, S.A. 1979. Non-uniform stable solutions to reaction-diffusion equations: applications to ecological pattern formation. In *Pattern Formation by Dynamic Systems and Pattern Recognition*, ed. H. Haken, pp. 210–222. New York: Springer-Verlag.

Lindeman, R.L. 1942. The trophic-dynamic aspect of ecology. *Ecology* 23:399–418.

Lippe, E.; DeSmidt, J.T.; and Gleen-Lewin, D.C. 1985. Markov models and succession: A test from a heathland in the Netherlands. *Journal of Ecology* 73:775–91.

Littlejohn, C.B. 1977. An analysis of the role of natural wetlands in regional water management. In *Ecosystem Modeling in Theory and Practice*, eds. C.A.S. Hall and J.W. Day, Jr., pp. 451–76. New York: John Wiley and Sons.

Loehle, C. 1987. Applying artificial intelligence techniques to ecological modelling. *Ecological Modelling* 38:191–212.

Lotka, A.J. 1925. *Elements of Physical Biology.* Baltimore, Md.: Williams and Wilkins (reissued, 1956, as *Elements of Mathematical Biology.* New York: Dover.

MacArthur, R.H. 1972. *Geographical Ecology: Patterns in the Distribution of Species,*Philadelphia: Harper and Row.

Maguire, L.A., and Porter, J.W. 1977. A spatial model of growth and competition strategies in coral communities. *Ecological Modelling* 3:249–71.

Mahmood, K., and Yevjevich, V., eds. 1976. *Unsteady Flow in Open Channels.* vol. I. Fort

Collins, Colorado: Water Resources Publications.

Mansfield, E. 1961. Technical change and the rate of innovation. *Econometrica* 29:741–66.

Margalef, R. 1968. *Perspectives in Ecological Theory.* Chicago: University of Chicago Press.

Maxfield, D.W. 1972. Spatial planning of school districts. *Annals of the Association of American Geographers* 62:582–90.

May, R., and MacArthur, R.H. 1972. Niche overlap as a function of environment variability. *Proceedings of the National Academy of Science,* U.S.A. 69:1109–13.

Mollison, D. 1977. Spatial contact models for ecological and epidemic spread. *Journal of the Royal Statistical Society* B 39:283–326.

Nachman, G. 1987a. Systems analysis of acarine predator-prey interactions: I. a stochastic simulation of spatial processes. *Journal of Animal Ecology* 56:247–65.

Nachman, G. 1987b. Systems analysis of acarine predator-prey interactions: II. the role of spatial processes in system stability. *Journal of Animal Ecology* 56:267–281.

Naveh, Z., and Lieberman, A.S. 1984. *Landscape Ecology, Theory and Application.* New York: Springer-Verlag.

Newbold, J.D.; O'Neill, R.V.: Elwood, J.W.; and Van Winkle, W. 1982. Nutrient spiralling in streams: implications for nutrient limitation and invertebrate activity. *American Naturalist* 120:628–52.

Nigam, S.; Held, I.M.; and Lyons, S.W. 1986. Linear simulation of the stationary eddies in a GCM. part I: the 'mountain' model. *Journal of Atmospheric Science* 43:2944–61.

Nigam, S.; Held, I.M.; and Lyons, S.W. 1988. Linear simulation of the stationary eddies in a GCM. part II: the 'mountain' model. *Journal of Atmospheric Science* 45:1433–52.

Nisbet, R.M., and Gurney, W.S.C. 1982. *Modelling Fluctuating Populations.* New York: John Wiley.

Noy-Meir, I. 1981. Spatial effects in modelling of arid ecosystems. In *Arid-Land Ecosystems: Structure, Functioning and Management,* vol. 2, eds. D.W. Goodall, R.A. Perry, and K.M.W. Howes, pp. 411–32. Cambridge, At the University Press.

Odum, H.T. 1957. Primary production in eleven Florida springs and a marine turtle grass community. *Limnology and Oceanography* 2:85–97.

Odum, H.T. 1971. *Environment, Power and Society.* New York: Wiley Interscience.

Odum, H.T. 1983. *Systems Ecology.* New York: Wiley Interscience.

Okubo, A. 1980. *Diffusion and Ecological Problems: Mathematical Models.* New York: Springer-Verlag.

Orlob, G.T., ed. 1983. *Mathematical modeling of water quality: streams, lakes and reservoirs.* New York: Wiley.

Pacala, S.W. 1987. Neighborhood models of plant population dynamics: 3. models with spatial heterogeneity in the physical environment. *Theoretical Population Biology* 31:359–92.

Pacala, S.W., and Silander, J.A. 1985. Neighborhood models of plant population dynamics: I. single-species models of annuals. *American Naturalist* 125:385–411.

Paine, R.T. 1966. Food web complexity and species diversity. *American Naturalist* 100:65–75.

Paine, R.T., and Levin, S.A. 1981. Intertidal landscapes: disturbance and the dynamics of pattern. *Ecological Monographs* 51:145–78.

Park, T. 1962. Beetles, competition and populations. *Science* 138:1369–75.

Pastor, J.; Gardner, R.H.; Dale, V.H.; and Post, W.M. 1987. Successional changes in nitrogen availability as a potential factor contributing to spruce declines in boreal North America. *Canadian Journal of Forest Research* 17:1394–1400.

Pastor, J., and Post, W.M. 1986. Influence of climate, soil moisture, and succession on forest carbon and nitrogen cycles. *Biogeochemistry* 2:3–27.

Patil, G.P.; Pielou, E.C.; and Waters, W.E., eds. 1971. *Spatial Patterns and Statistical Distributions.* University Park: Pennsylvania State University Press.

Patten, B.C. 1971–1976. *Systems Analysis and Simulation in Ecology,* vols. 1–4. New York: Academic Press.

Patten, B.C., and Matis, J.H. 1982. The macrohydrology of Okefenokee Swamp. In *Proceedings of the international workshop on ecosystem dynamics in freshwater wetlands and shallow water bodies*, vol. 2, eds. D.O. Logofet and K.N. Lyckyanov, pp. 218–35. United Nations Environmental Program. U.S'.S.R. Academy of Sciences, Moscow.

Paynter, R. 1982. *Models of Spatial Inequality: Settlement Patterns in Historical Archeology.* New York: Academic Press.

Pearl, R., and Reed, L.J. 1920. On the rate of growth of the population of the United States since 1790 and its mathematical representation. *Proceedings of the National Academy of Sciences, U.S.A.* 6:275–88.

Pearlstine, L; McKellar, H.; and Kitchens, W. 1985. Modeling the impacts of a river diversion on bottomland forest communities in the Santee River Floodplain, South Carolina. *Ecological Modelling* 29:283–302.

Phillips, N.A. 1956. The general circulation of the atmosphere: a numerical experiment. *Quarterly Journal of the Royal Meteorological Society* 82:123–64.

Phipps, R.L. 1979. Simulation of wetland forest vegetation dynamics. *Ecological Modelling* 7:257–88.

Pickett, S.T.A., and White, P.S., eds. 1985. *The Ecology of Natural Disturbance and Patch Dynamics.* New York: Academic Press.

Pielou, E. 1977. *Mathematical Ecology.* New York: Wiley.

Potter, G.L.; Ellsaesser, H.W.: MacCracken, M.C.; and Luther, F.M. 1979. Performance of the Lawrence Livermore Laboratory zonal atmospheric model. In Report of the JOC study conference on climate models: performance, intercomparison and sensitivity studies, ed. W.L. Gates, pp. 825–71. Global Atmospheric Research Program Series No. 22.

Puccia, C.J. 1983. Qualitative models for east coast benthos. In *Analysis of Ecological Systems: State-of-the-Art in Ecological Modelling*, eds. W.K. Lavenroth, G.V. Skogerboe and M. Flug, pp. 719–24. Amsterdam: Elsevier Scientific.

Quandt, R.E. 1960. Models of transportation and optimum network construction. *Journal of Regional Science* 2:27–45.

Radford, P.J. 1972. The simulation language as an aid to ecological modelling. In *Mathematical Models in Ecology: The Twelfth Symposium of the British Ecological Society*, ed. J.N.R. Jeffers, pp. 277–95. Oxford: Blackwells.

Ravenstein, E.G. 1885. The laws of immigration. *Journal of the Royal Statistical Society* 4:165–235.

Reiners, W.A. 1988. Achievements and challenges in forest energetics. In *Concepts of Ecosystem Ecology*, eds. L.R. Pomeroy and J.L. Alberts, pp. 75–114. New York: Springer-Verlag.

Ripley, B.D. 1981. *Spatial Statistics.* New York: Wiley.

Risser, P.G.; Karr, J.R.; and Forman, R.T.T. 1984. *Landscape Ecology: Directions and Approaches.* Special Publication No. 2. Illinois Natural History Survey, Champaign.

Saarenmaa, H.; Stone, N.D.; Folse, L.J.; Packard, J.M.; Grant, W.E.; Makela, M.E.; and Coulson, R.N. 1988. An artificial intelligence modelling approach to simulating animal/habitat interactions. *Ecological Modelling* 44:125–41.

Schlesinger, M.E., and Zhao, Z.-C. 1989. Seasonal climatic changes induced by doubled CO_2 as simulated by the OSU atmospheric GCM/mixed-layer ocean model. *Journal of Climate* 2:463–99.

Schoener, T.W. 1974. Competition and the form of habitat shift. *Theoretical Population Biology* 6:265–307.

Schumm, S. 1956. Evolution of drainage systems and slopes in badlands at Perth Amboy, New Jersey. *Bulletin of the Geological Society of America* 67:597–646.

Selkirk, K.E. 1982. *Pattern and Place: An Introduction to the Mathematics of Geography.* Cambridge: Cambridge University Press.

Sellers, P.J.; Mintz, Y.; Sud, Y.C.; and Dalcher, A. 1986. A simple biosphere model (SiB) for use within general circulation models. *Journal of Atmospheric Science* 43:505–31.

Shannon, C.E. 1948. The mathematical theory of communication. *Bell System Technical*

Journal. 27:379–423; 623–56.

Shoemaker, C.A. 1977. Mathematical construction of ecological models. In *Ecosystem Modeling in Theory and Practice,* eds. C.A.S. Hall and J.W. Day, Jr., pp. 75–114. New York: Wiley Interscience.

Show, I.T., Jr. 1979. Plankton community and physical environment simulation for the Gulf of Mexico region. In *Proceedings of the 1979 Summer Computer Simulation Conference,* pp. 432–39. San Diego: Society for Computer Simulation.

Shugart, H.H. 1984. *A Theory of Forest Dynamics: An Investigation of the Ecological Implications of Several Computer Models of Forest Succession.* New York: Springer-Verlag.

Shugar, H.H. West, D.C. 1980. Forest succession models. *BioScience* 30:308–13.

Simmons, A.J., Bengtsson, L. 1984. Atmospheric general circulation models: their design and use for climate studies. In *The Global Climate,* ed. J.T. Houghton, pp. 37–62. Cambridge: At the University Press.

Sklar, F.H. 1983. Water budget, benthological characterization, and simulation of aquatic material flows in a Louisiana freshwater swamp. Ph.D. Thesis. Louisiana State University, Baton Rouge.

Sklar, F.H., and Costanza, R. 1986. A spatial simulation of ecosystem succession in a Louisiana Coastal landscape. In *Proceedings of the 1986 Summer Computer Simulation Conference,* eds. R. Crosbie and P. Luker, pp. 467–72. San Diego, California. Society for Computer Simulation.

Sklar, F.H.; Costanza, R.; and Day, J.W., Jr. 1985. Dynamic spatial simulation modeling of coastal wetland habitat succession. *Ecological Modeling.* 29:261–81.

Sklar, F H.; Costanza, R.; and Day, J.W., Jr. The conceptualization phase of wetland model development. In *Symposium on Modeling Freshwater Wetlands,* ed. B. Patten. USSR/SCOPE Proceedings, in press.

Sklar, F.H.; Costanza, R.; Day, J.W.; and Conner, W.H. 1982. Dynamic simulation of aquatic material flows in an impounded swamp habitat in the Barataria Basin, La. In *Analysis of Ecological Systems: State-of-the-Art in Ecological Modeling,* ed. W.K. Lauenroth, pp. 741–50. Amsterdam: Elsevier Scientific.

Sklar, F.H.; White, M.H.; and Costanza, R. *The Coastal Ecological Landscape Spatial Simulation (CELSS) Model: Structure and Results for the Atchafalaya/Terrebonne Study Area.* U.S. Fish and Wildlife Service, Slidell, La., in press.

Slobodkin, L.B. 1961. *Growth and Regulation of Animal Populations.* New York: Holt, Rinehart, and Winston.

Slobodkin, L.B. 1968. How to be a predator. *American Zoologist* 8:43–51.

Sokal, R.R. 1979. Testing statistical significance of geographical variation patterns. *Systematic Zoologist* 28:227–232.

Sokal, R.R.; and Wartenberg, D.E. 1981. Space and population s ructure. In *Dynamic Spatial Models,* eds. D.A. Griffith and R.D. MacKinnon. Netherlands: Sijthoff & Noordhoff, Alphen aan den Rijin.

Solomon, A.M. 1986. Transient response of forests to CO_2-induced climate change: simulation modeling experiments in eastern North America. *Oecologia* 68:567–79.

Sonis, M. 1981. Space and time in the geography of aging. In *Dynamic Spatial Models,* eds. D.A. Griffith and R.D. MacKinnon. Netherlands: Sijthoff & Noordhoff, Alphen aan den Rijin.

Sprunt, B. 1972. Digital simulation of drainage basin development. In *Spatial Analysis in Geomorphology,* ed. R.J. Chorley, pp. 371–89. New York: Harper and Row.

Steel, R.G.D., and Torrie, J.H. 1960. *Principles and Procedures of Statistics.* New York: McGraw-Hill.

Strahler, A.M. 1980. Systems theory in physical geography. *Physical Geography* 1:1–27.

Swartzman, G.L., and Kaluzny, S.P. 1987. *Ecological Simulation Primer.* New York: Macmillan.

Swartzman, G.L., and Zaret, T.M. 1983. Modeling fish species introduction and prey extermination: the invasion of *Cichla ocellaris* to Gatun Lake, Panama. In *Analysis of Ecological Systems: State-of-the-Art in Ecological Modelling,* eds. W.K. Lavenroth, G.V. Skogerboe, and M. Flug, pp. 361–71. New York: Elsevier Scientific.

Tansley, A.G. 1935. The use and abuse of vegetational concepts and terms. *Ecology* 16:614–24.

Taylor, A.H. 1988. Characteristic properties of models for the vertical distribution of phytoplankton under stratification. *Ecological Modelling* 40:175–99.

Thomas, R.W. 1977. An interpretation of the journey-to-work on Merseyside using entropy-maximizing methods. *Environment and Planning* A9:817–934.

Thomas, R.W., Huggett, R.J. 1980. *Modelling in Geography: A Mathematical Approach.* London: Harper and Row.

Tilman, D. 1982. *Resource Competition and Community Structure.* Monographs in Population Biology 17. Princeton, N.J. Princeton University Press.

Tornqvist, G. 1967. *Growth of TV Ownership in Sweden, 1956–1965.* Uppsala, Sweden: Uppsala University Press.

Turner, M.G. 1987. Spatial simulation of landscape changes in Georgia: a comparison of 3 transition models. *Landscape Ecology* 1:29–36.

Turner, M.G. 1988. A spatial simulation model of land use changes in a piedmont county Georgia. *Applied Matematical Computation* 27:39–51.

Turner, M.G., Costanza, R., Sklar, F.H. 1989. Methods to compare spatial patterns for landscape modeling and analysis. *Ecological Modelling* 48:1–18.

Ulnaowicz, R.E. 1986. *Growth and Development: Ecosystems Phenomenology.* New York: Springer-Verlag.

Usher, M.B. 1981. Modelling ecological succession, with particular reference to Markovian models. *Vegetatio* 46:11–18.

Van Hulst, R. 1979. On the dynamics of vegetation: Markov chains as models of succession. *Vegetatio* 40:3–14.

Volterra, V. 1926. Varizioni e fluttuazioni del numero d'individui in specie animali conviventi. *Mem. R. Acad. Naz. dei Lincei* (ser. 6), 2:31–113.

Wang, F.C., Wei, J.S., Amft, J.A. 1985. Computer simulation of Western Terrebonne parish marsh hydrology and hydrodynamics. Technical Report., Louisiana Sea Grant Publication, Baton Rouge.

Washington, W.M., Dickinson, R.; Ramanathan, V.; Mayer, T.; Williamson, D.; Williamson, G.; and Wolsi, R. 1979. Preliminary atmospheric simulation with the third-generation NCAR circulation model: January and July. In *Report of the JOC Study Conference on Climate Models: Performance, Intercomparison, and Sensitivity Studies,* ed. W.L. Gates, pp. 95–138. GARP publication series No. 22, Washington, D.C.

Washington, W.M., Meehl, G.A. 1984. Seasonal cycle experiment on the climate sensitivity due to a doubling of CO_2 with an atmospheric general circulation model coupled to a simple mixed-layer ocean model. *Journal of Geophysical Research* 89:9475–9503.

Washington, W.M., and Williamson, D.L. 1977. A description of the NCAR global circulation models. In *Methods in Computational Physics,* vol. 2, *General Circulation Models of the Atmosphere,* ed. J. Change, pp. 111–72. New York: Academic Press.

Weins, J.A. 1976. Population responses to patchy environments. *Annual Review of Ecological Systems* 7:81–120.

Werner, C. 1985. *Spatial Transportation Modeling.* Beverly Hills, Calif.: Sage Publications.

White, G.C.; Adams, L.W., and Bookhout, T.A. 1978. Simulation model of tritium kinetics in a freshwater marsh. *Health Physics* 34:45–54.

Whittaker, R.H. 1975. *Communities and Ecosystems.* New York: Macmillan.

Whittaker, R.H.; Levin, S.A.; and Root, R.B. 1973. Niche, habitat and ecotope. *American Naturalist* 107:321–38.

Wilcox, B.A. 1980. Insular ecology and conservation. In *Conservation Biology: An Evolutionary-Ecological Perspective*, eds. M.E. Soulé and B.A. Wilcox, pp. 95–118. Sunderland, Mass.: Sinauer.

Wilkie, D.E., and Finn, J.T. 1988. A spatial model of land use and forest regeneration in the ituri forest of northwestern Zaire. *Ecological Modelling* 41:307–23.

Williams, J.; Barry, R.G.; and Washington, W.M. 1974. Simulation of the atmospheric circulation using the NCAR global circulation model with ice age boundary conditions. *Journal of Applied Meterology* 11:305–17.

Willmott, C.J. 1984. On the evaluation of model performance in physical geography. In *Spatial Statistics and Models*, eds. G.L. Gaile and C.J. Willmott, pp. 443–60. Boston: Kluwer Academic.

Wilson, A.G. 1981. The evolution of urban spatial structure: The evolution of theory. In *European Progress in Spatial Analysis*, ed. R.J. Bennett, pp. 201–35. London: Pion Limited.

Wilson, M.F.; Henderson-Sellers, A.; Dickinson, R.E.; and Kennedy, P.J. 1987. Sensitivity of the biosphere-atmosphere transfer scheme (BATS) to the inclusion of variable soil characteristics. *Journal of Climate and Applied Meteorology* 26:341–62.

Woldenberg, M.J. 1972. The average hexagon in spatial hierarchies. In *Spatial Analysis in Geomorphology*, ed. R.J. Chorley, pp. 323–52. New York: Harper and Row.

Yeates, M.H. 1963. Hinterland delimitation: a distance-minimizing approach. *Professional Geography* 15:7–10.

Zedler, J., and Zedler, P. 1969. Association of species and their relationship to microtopography within old fields. *Ecology* 50:432–42.

11. Pattern, Process, and Predictability: The Use of Neutral Models for Landscape Analysis[1]

Robert H. Gardner and Robert V. O'Neill

11.1 Introduction

Understanding the relationships between pattern and process at landscape scales was the focus of some of the earliest works in ecology (e.g., Cowles 1899; Cooper 1923). Despite the wealth of empirical and conceptual investigations that have been carried out since these early studies, the problem of predicting ecological processes at broad scales remains largely unresolved. This lack of resolution is due, in part, to the complexity of the problem and an intellectual tradition that has assumed that detailed measurements of fine-scale processes are necessary to predict broad-scale patterns.

An approach that evaluates the usefulness of fine-scale detail in explaining broad-scale patterns is an important step in developing useful and reliable models. The use of a neutral model (sensu Caswell 1976) accomplishes this objective by eliminating the mechanism(s) in question while retaining other relevant biological and physical details and then comparing predictions with empirical information. If the predictions adequately fit the data, then, by Occam's razor, there is no reason to postulate additional complexity (Hoffman and Nitecki 1987). However, if the

[1]This research was funded by the Ecological Research Division, Office of Health and Environmental Research, U.S. Department of Energy, under contract DE–AC05–84OR21400 with Martin Marietta Energy Systems, Inc. Publication No. 3540, Environmental Sciences Division, Oak Ridge National Laboratory.

model and data diverge, one may need to postulate additional mechanisms to explain the phenomena of interest.

This chapter reviews the application of neutral models for landscape analysis and the procedures for developing, testing, and applying neutral models to landscape-scale problems.

11.2 Theory of a Neutral Landscape Model

Many philosophical and semantic difficulties arise with the use of the term *neutral model* (the interested reader should see Slobodkin [1987] and Hoffman and Nitecki [1987]). However, the value of the neutral model approach was demonstrated long before the term was ever coined. For instance, Grieg-Smith (1952, 1964) suggested that vegetation should show a distribution pattern similar to the Poisson if individual plants are randomly distributed with no pattern or autocorrelation. Thus, the Poisson serves as a neutral model to test for a statistically significant departure from random.

We recognize that landscape pattern and their associated processes are complex and that all models are simplifications of reality, but models will continue to provide a useful means for summarizing information and predicting future behaviors. If we recognize that no model is "perfect" then we also realize that no neutral model will be "perfectly" neutral to all factors of interest. The intent here is to show how the development of a simple neutral model allows us to investigate the importance of a particular process (or set of processes) by first excluding this process from the model and testing the adequacy (or lack of fit) of the results against data.

Therefore, the key components for development of neutral models are (1) a clear statement of the problem, (2) the definition of a simple model that allows the importance of each variable to be examined, (3) a comparison of the model predictions with available data and observations, and (4) an objective measure of the adequacy of the results.

For example, suppose that the process of urban development affects both the total area and spatial arrangement of suitable habitat sites for an endangered animal. Further suppose that factors associated with urban development (e.g., noise and pollution) adversely affect the daily activities of this organism. The question (step 1) to be asked is this: "How does the amount and arrangement of suitable habitat affect the abundance of the organism independent of other urban effects?" The model (step 2) need not be concerned with other variables associated with urban development; that is, it will be neutral to variables that do not directly affect habitat pattern. The results (step 3) of the model are then compared with observed species patterns and, if the results are satisfactory, abundance levels (step 4) can be predicted from pattern alone. If the results are not satisfactory, other variables *must be included in the model in order to predict changes in species abundances*. This approach will ultimately result in the simplest model that can adequately explain the observed phenomena and is, therefore, conceptually similar to stepwise regression methods (Kleinbaum et al. 1988).

11.3 Neutral Models of Landscape Pattern

A recent review of landscape models by Baker (1989) notes that insufficient knowledge of the factors that may affect landscapes has hampered the development of useful models for relating pattern and process. Although a number of successful studies were reviewed by Baker (e.g., Franklin and Forman 1987; Turner 1987; Browder et al. 1985; Wilkie and Finn 1988), the independent approaches that were utilized have not resulted in a coherent method for landscape studies. We submit that an approach developed from percolation theory is useful for the systematic study of pattern and process at landscape scales.

11.3.1 Random Maps Developed from Percolation Theory

Percolation theory has been developed to deal with the effects of randomly varying the connections between spatially distributed systems. The term *percolation* was first used in 1957 by the mathematician J. M. Hammersley for models of spatial systems (Zallen 1983). Methods developed from percolation theory have been useful as neutral models for landscape pattern (Gardner et al. 1987; O'Neill et al. 1988; Gardner et al., in press). The change in pattern as a function of map size (Gardner et al. 1987) and scale (Turner et al. 1989b) has been quantified and the dynamics of species dispersal (Gardner et al., in press) and habitat utilization (O'Neill et al. 1989) simulated with simple percolation models. The interaction between landscape pattern and disturbance (Turner et al. 1989a) has also been examined with spatial models based on percolation theory.

A map can be created with methods derived from percolation theory by specifying the number, m, of columns and rows in the map and randomly selecting the habitat type of the m^2 sites with a probability of p (Fig. 11.1a–c). For sufficiently large maps (or square lattices in percolation jargon) an average of pm^2 sites will be designated as the habitat type of interest (e.g., a forested site), while $(1 - p)m^2$ sites will be designated as other (e.g., nonforested) sites. When the length, l, of a single site is specified (e.g., 100 m for a 1-ha resolution), the total area of the map may can be calculated as $(lm)^2$. The scale of the map is defined by these two parameters: l is the resolution or grain size and $(lm)^2$ is the area or extent of the map.

There are two basic types of percolation processes that can be simulated on random square lattices: *bond percolation*, which defines connectivity by individually defining the four possible "bonds" between nearest-neighbor sites, and *site percolation*, which defines connectance when sites share common edges (Zallen 1983). A set of connected bonds or sites is defined as a cluster or patch (Stauffer 1985). For both bond and site percolation, the number, size, and shape of the randomly generated patches are similar from map to map but change as a function of p. Rapid changes in the size and shape of patches occur near the critical probability, p_c, when the largest patch just manages to extend from one edge of the map to the other. The value of p_c for extremely large maps has been empirically determined to be 0.5928 for site percolation and analytically defined as 0.5 for bond percolation (Stauffer 1985). The relationship between patch size and bound-

R.H. Gardner and R.V. O'Neill

RANDOM   CONTAGION

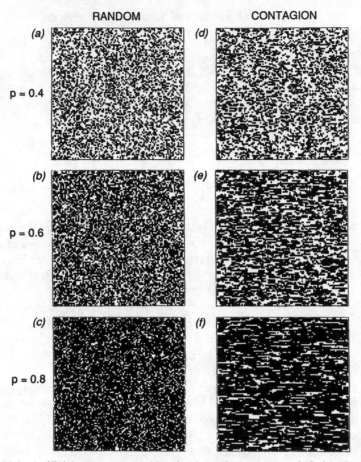

Figure 11.1. Artificial maps generated by simple random processes (a,b,c) and with contagion (Q_{11} = 0.7) between similar sites (d,e,f). The probability, p, for a and d = 0.4, b and e = 0.6, and c and f = 0.8.

ary shape, as measured by the fractal dimension, is also affected by p (Stauffer 1985): boundaries are short and straight when $p < p_c$ but longer and more convoluted when $p \geq p_c$.

Percolation models can be developed to include connection to next nearest neighbors (percolation jargon defines the "nearest neighbors" of a square lattice as the four adjacent sites with edges that touch the site in question, and next nearest neighbors as the four diagonal sites whose corners touch the site in question). The value of p_c for site percolation when next nearest neighbors are included is (1 − p_c) = 1 − 0.5928 = 0.4072. Other lattice shapes have also been used in percolation theory. Triangular lattices show p_c of 0.3473 for bond percolation and 0.5 for site percolation, while hexagonal lattices show 0.6527 and 0.7 for bond and site percolation, respectively.

Three examples of simple random maps with values of p ranging from 0.4 to 0.8 are illustrated in Fig. 11.1a–c. Figure 1a shows a map with $p = 0.4$ that has 1120 patches, with the largest patch composed of 50 sites (0.5% of the map). Figure 11.1b shows a map with $p = 0.6$, which is slightly above the critical probability for site percolation of 0.5928. The total number of patches in Fig. 11.1b is 284, and the largest patch occupies 45% of the available sites and connects opposite edges of the map. When $p = 0.8$ (Fig. 11.1c) the total number of patches is reduced to 28 and a single patch occupies nearly 80% of the grid, effectively connecting all but 37 map sites.

When actual vegetation patterns (e.g., forested landscapes) are compared to random maps, the number, size, shape, and arrangement of habitat patches and their boundaries are similar when the fraction of the map occupied by forest is greater than 0.6 (Gardner et al. 1987; Gardner et al., in press). When the fraction of the map occupied by forest is below 0.6, then actual landscape patterns show significant departures from random, suggesting that factors such as topography, land use history, human development, and disturbance can be expected to produce "unique" results that are landscape specific. Thus, a neutral model allows one to distinguish the expected from the unique.

11.3.2 Generating Maps with Contagion

The frequency distribution of patch sizes observed for actual landscapes shows greater contagion between sites than is observed for randomly generated landscapes (Gardner et al., 1987; Gardner et al., in press). The forces that produce this contagion and the processes affected by it are a principal interest of landscape ecology. A method that generates artificial landscapes with specific levels of contagion would be useful as a neutral model that produces properties of "clumped" systems independent of the forces that generate this clumping. For instance, how is the pattern of habitat patches and their edges affected by contagion and how does this contagion, in turn, affect the probability and extent of disturbance effects?

Simple maps with contagion can be generated by forming an adjacency matrix, Q, with elements Q_{ij}, which are the probabilities that habitat type i will be adjacent to a site of habitat type j. If we consider only two habitat types (i.e., $i = 1$ and $j = 2$), then Q will be a 2×2 matrix with the rows of the Q matrix summing to 1.0. Because the element Q_{11} represents the probability that a site of habitat type 1 is found adjacent to a site of the same habitat type, Q_{11} is the degree of contagion that can be measured from the frequency of site adjacency of actual landscape maps. For a completely random landscape, $Q_{11} = p$ and Q_{12} will equal $(1 - p)$. If the landscape is more dispersed than is expected of a random landscape, then $Q_{11} < p$. If the habitat is contagiously distributed, then $Q_{11} > p$. Given values of Q_{11} and p, a stable landscape distribution will be produced (O'Neill et al., in press) if Q_{22} is equal to $(1 - 2p + pQ_{11})/(1 - p)$.

The effect of contagion on the number of patches and their edges is illustrated in Fig. 11.1. Three maps are illustrated in Fig. 11.1, with levels of contagion (Q_{11}) set at 0.7 and p values of 0.4 (Fig. 11.1d), 0.6 (Fig. 11.1e) and 0.8 (Fig. 11.1f). The

Table 11.1. The number of Patches and Total Amount of Edge on Simple Random Maps
and Maps with Contagion Between Sites

P	Q_{11}	Realized p	Number of Patches	Total Edge
Simple random maps				
0.4	—	0.3947	1120	9634
0.6	—	0.6046	284	9744
0.8	—	0.7953	28	6838
Maps with contagion between sites				
0.4	0.7	0.4094	682	8548
0.6	0.7	0.5857	111	7378
0.8	0.7	0.7884	6	4762

p is the probability level used to randomly select sites on the map and Q_{11} is the parameter controlling
the degree of contagion among sites. Realized p is the fraction of sites actually obtained for each
randomly generated map. See text for an explanation of the contagion algorithm.

effect of the contagion among similar sites is to reduce the number of patches by
approximately 50% and decrease the total amount of edge by 11.2% at p of 0.4
to 30.2% at p of 0.8 (Table 11.1).

A graph of the probability that a single habitat patch will connect opposite edges
of a map as a function of p is a step function (Fig. 11.2a). If p is slightly less than
0.5928, the probability of finding a percolating patch is 0.0. If p is even slightly
above this value, the map will "percolate" with a probability equal to 1.0. However, a perfect step function will only occur when habitats are randomly scattered
across an infinitely large landscape. Because real landscapes are finite in size and
the distribution of habitats tends to show some degree of contagion, the expected
probability curve as a function of sample size (i.e., spatial extent of the map) and
contagion can be generated from our neutral models for landscape pattern.

Results of a series of simulation experiments of finite size for simple random
maps (100 rows by 100 columns) without and with contagion are illustrated in
Figs. 11.2a and 2b, respectively. For each combination of p and Q_{11}, twenty
landscape maps were generated and the probability of finding a percolating patch
estimated. The dashed line in Fig. 11.2a represents the probability of percolation
on a finite (100 × 100) landscape in the absence of contagion. Instead of a step
function, the probability curve has a sigmoid shape with a probability of 0.6 at p
= 0.5928. Figure 11.2a shows that a p of approximately 0.65 is needed to produce
percolating patches on 100% of the finite landscapes.

The other curves on Fig. 11.2a show results for relatively high levels of contagion, $0.8 < Q_{11} < 0.99$. Levels of contagion above 0.9 produce very few, large,
complex patches that are so tightly packed that the largest patches often fail to
reach across the map. At $Q_{11} = 0.97$, the largest average patch contains 5904 sites.
This is more than half of the total of 10,000 sites on the 100 × 100 map and
approximately twice the size of the largest patch on a random map (Table 2).

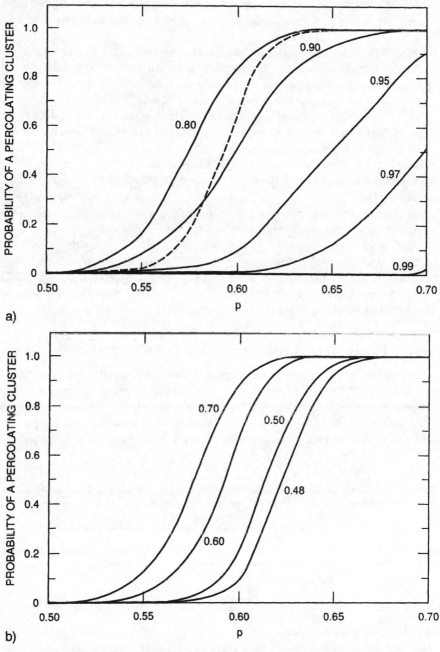

Figure 11.2. Probability of finding a percolating patch on a finite (100 × 100) map with p cells occupied by the resource of interest and at various levels of contagion. *Contagion*, Q_{11}, is defined as the probability that adjacent cells are occupied. The dashed line represents $Q_{11} = p_c = 0.5928$. Figure 11.2a considers $0.8 < Q_{11} < 0.99$ and Fig. 11.2b shows values of $0.48 < Q_{11} < 0.7$.

When Q_{11} = 0.97, the largest patch is also the most complex, as indicated by a fractal dimension of 1.25, and the map has only 7 patches as opposed to 462 patches when Q_{11} = 0.45.

Figure 2b shows the results for lower levels of contagion, $0.48 < Q_{11} < 0.7$. At values of contagion less than 0.6, the curves again move to the right. Under these levels of contagion, the patches on the map are numerous, small, and simple in shape so that none of them percolates. For example, at Q_{11} = 0.45, the largest patch (Table 11.2) contains 2128 sites and has a simple shape with a fractal dimension of 1.09. Table 11.2 shows that one can expect about 460 separate patches on the map. Under these circumstances, p_{c50}, the value of p that results in 50% of the maps having a percolating cluster, must be relatively large.

Contagion levels between 0.6 and 0.85 cause the values of p_{c50} to decline (Fig. 11.3). Table 11.2 shows that, for Q_{11} = 0.7, the largest patch is 3268 sites, slightly larger than the random map and with virtually the same shape complexity. This larger patch size tends to percolate at smaller values of p.

The simple algorithm used to simulate the effects of contagion provides a general description of pattern change with nonrandom associations occurring between similar habitat sites. A neutral model for the effects of contagion will allow the significance of this effect to be identified and, therefore, indicate the importance of processes that produce contagion.

11.4 Relating Pattern and Process with Neutral Models

Because species respond to environmental conditions at specific spatial and temporal scales (Milne et al. 1989; Swihart et al. 1988), an organism-centered view (sensu Wiens 1985) is essential for relating changes in landscape pattern to processes that affect the distribution and abundance of species. Previous applications of models based on percolation theory have shown that the spatial patterning

Table 11.2. Landscape Properties Estimated from Twenty Randomly Generated Maps

		Largest Patch		
P	Q_{11}	Patch Size	Fractal Dimension	Number of Patches
0.6353	0.45	2128	1.09	462
0.5928	0.5928	2976	1.11	296
0.5941	0.7	3268	1.13	204
0.6887	0.97	5904	1.25	7

Map dimensions were 100 rows by 100 columns with values of p and Q_{11} that resulted in 0.5 probability of observing a percolating patch. p is the probability level used to randomly select sites on the map, and Q_{11} is the parameter controlling the degree of contagion among sites. See text for an explanation of the contagion algorithm. The fractal dimension for the largest patch is estimated by dividing the logarithm of patch size by the logarithm of ¼ of the patch perimeter. The complexity of patch shape increases as the fractal dimension increases.

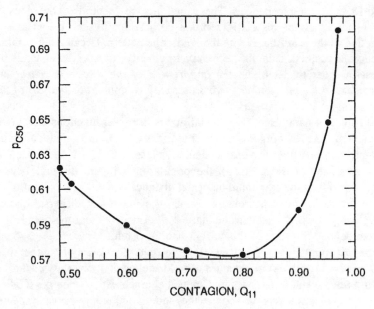

Figure 11.3. Values of landscape occupancy, p_{c50}, that result in a 50% probability of a percolating patch at various levels of contagion, Q_{11}. Contagion is defined as the prob ability of adjacent cells being occupied.

of resources constrains the potential movement of populations across the land-scape (O'Neill et al. 1988), with a sparse or patchy distribution of resources requiring organisms to operate at broader spatial scales.

The movement of populations across a heterogeneous landscape can be simu-lated by two rules derived from percolation theory: *bond* percolation rules simulate random independent movement to individual nearest-neighbor sites while, *site* percolation "rules" simulate random simultaneous movement to all nearest-neigh-bor sites (see von Niessen and Blumen 1988 for further description of bond and site percolation). When the probability of movement to new sites, i, is equal to 1.0, the movement patterns generated by bond or site percolation are equivalent and all sites within a patch will be colonized by the dispersing population. Differences between bond and site percolation become more apparent as the value of i decreases. Because habitat patches are highly fragmented when $p < p_c$, and because movement rules based on bond or site percolation restrict movement to a single patch, these dispersal rules will not allow the spread of a population across the map when $p < p_c$.

We have (Gardner et al., in press) simulated the spread of a population with dispersal generated by (1) selecting a random angle (0° to 360°); (2) selecting a random distance from either the uniform, triangular, or exponential distribution; and (3) converting the angles and distances to appropriate map coordinates. The three types of continuous frequency distributions all produce circular patterns but with different central tendencies. These methods are "neutral" to the many attributes of individual species that determine both fecundity and dispersal and the interaction of these attributes with the landscape pattern. Because a neutral model approach is adopted, it is not necessary to specify the species or landscape of interest in order to determine *the importance of the dispersal characteristics independent of the many other factors that may confound the analysis of pattern and process at landscape scales.*

The choice of parameters for the continuous dispersal functions affect both the shape and distance that organisms are able to move from the parent population. If the maximum possible dispersal distance is less than ½ *l* (recall that *l* is the grain-size of the landscape map), the population will not disperse beyond its current site. When the maximum dispersal distance equals *l*, the pattern of movement is similar to bond percolation. When the maximum dispersal distance is greater than *l*, then the habitat boundary is "permeable" and the population can disperse beyond the edge of the current habitat patch.

Replicate simulations (N = 10) were performed on a series of random maps (*m* = 100 and *l* = 1) with different values of *p*. Model simulations showed that results were not sensitive to differences in dispersal characteristics when *p* > 0.6 because above *p_c* all landscapes are well connected, and no matter how far each species can disperse per generation, all populations will be able to reach a large fraction of the available habitat. Below the critical threshold the habitat patches are highly fragmented, and large differences in species abundances and habitat utilization can be produced by small changes in maximum dispersal distance. Species capable of moving greater distances are, therefore, not affected by heterogeneity on landscapes below the grain-size set by this dispersal parameter.

Percolation studies of diffusion in two dimensions have shown that movement is affected by the size and shape of patches (Gefen and Aharony 1983), with sudden changes in the dynamics occurring near the critical threshold, *p_c* (Stauffer 1985). These changes occur because the structure of the system shifts from diffusive flow in a disconnected system (below *p_c*) to convective flow in a connected network (above *p_c*, see Ohtsuki and Keyes 1986). Percolation theory defines the "backbone" of a patch as the critical subset of sites through which there is flow from edge to edge (Stauffer 1985). The subset of sites that forms the backbone can be identified by labeling sites whose removal will disrupt movement through the patch. The concept of a backbone is of interest to landscape studies because it suggests that disturbance effects may be site dependent. For instance, removal of sites adjacent to the backbone may not affect the movement of organisms or the spread of disturbance, but the removal of a single site from the backbone itself can effectively disconnect the patch. Although this concept has direct application to the study of corridors and metapopulation dynamics, it has not yet been exploited as a neutral landscape model.

11.5 Other Forms of Neutral Models

Other theoretical approaches allow a variety of neutral models to test specific aspects of spatial dynamics. For instance, epidemiology models predict the spread of disease through a population and can be modified to deal with disturbance spread on the landscape. Diffusion-reaction and patch dynamics models deal with biological mechanisms that can cause pattern on landscapes. Thunen models can be adapted from economic theory to relate energy costs of travel to spatial patterns.

11.5.1 Epidemiology Models as Neutral Models of Disturbance

Spatial epidemiology models (Bailey 1965; Mollison 1977; Kuulasmaa 1982; Cardy 1983; Faddy 1986) follow the spread of a disease in two dimensions. The analogy with the spread of a disturbance, such as a forest fire or insect pest, is obvious. The basic model can be combined with percolation theory to introduce spatial considerations of disturbance spread (Smythe and Wierman 1978; Cardy and Grassberger 1985). It has been demonstrated that critical thresholds exist for epidemiological processes (Grassberger 1983; Kuulasmaa and Zachary 1984; Mollison 1986) and that simple epidemics die out at a critical threshold of occupancy on a landscape (Cox and Durrett 1988).

The basic epidemiology model (Bailey 1975) considers a population composed of three groups: X susceptibles, Y infected, and Z immune individuals. In the context of a fire disturbance, Z represents the acreage of unburned forest, Y represents the forest actually burning, and Z represents burned forest that is no longer susceptible to burning. The model takes the form:

$$dX/dt = - iqXY$$
$$dY/dt = iqXY - bY$$
$$dZ/dt = bY \tag{11.1}$$

Intensity, i, represents the probability of a disturbance spreading to an adjacent site. The rate of disturbance extinction, b, represents the inverse of the length of time a site burns, and contagion, q, represents the probability that a susceptible site will be found adjacent to a disturbance.

One of the most useful features of Eq. 11.1 is the ability to predict the total extent of a single disturbance. The disturbance continues to spread only as long as $d/dt > 0$. When $dY/dt = 0$, the spread of the disturbance is stopped and begins to decrease. At this threshold point,

$$dY/dt = 0 = iqXY - bY \tag{11.2}$$

Equation 11.2 implies a threshold value of $X = k = b/iq$ that defines the turning point in the disturbance spread. If the initial value of $X(0) = pN$ is less than or equal to k, the disturbance does not spread. If the initial number of susceptibles is greater than the threshold value, i.e., $X(0) = k + m$, then the total extent of the disturbance will be $2m$. There will be m sites disturbed up to the point that $dY/dt = 0$ and an equal number of sites disturbed until $Y(t) = 0$.

Then, working from $X(0) = k + m = pN$ and $k = b/iq$, the model predicts the total extent of the disturbance as

$$2m = 2pN - 2b/iq \tag{11.3}$$

Inspection of Eq. 11.3 reveals that $2m$ may be larger than the total number of susceptible sites, pN. For example, if i is greater than $2b/qpN$, the theory predicts that all susceptible sites will be disturbed. Any further increase in intensity may increase the rate at which the disturbance spreads but will have little effect on the total extent of the disturbances.

Contagion, q, is a simple way to represent landscape pattern (see section 11.3.2). The effects of contagion begin to become important at $p_c = 0.5928$ and are increasingly important as p becomes smaller. At small values of p, the landscape with contagion contains patches that are significantly larger than the purely random landscape. High levels of contagion tend to increase the extent of a single disturbance, particularly when landscape occupancy is small and disturbance intensity is large. Low levels of contagion tend to increase the probability that a disturbance will become endemic on a landscape.

Used as a neutral model, the epidemiology model can predict the size and duration of disturbances in the absence of specific spatial features such as topography or corridors. Natural landscapes will ordinarily contain far more spatial structure than can be represented by the contagion parameter, q. Therefore, comparison of model predictions with field data on a specific disturbance could be used to establish the effect of specific landscape features on disturbance spread and duration.

11.5.2 Diffusion-Reaction as a Mechanism for Pattern Formation

Diffusion-reaction theory attempts to describe the mechanisms that produce spatial patterns (Levin 1976; 1978). The distribution of a population $U(r,t)$ across spatial coordinates r at time t is given by (Levin 1978)

$$\delta U/\delta t = f(U, V) + \nabla (D \nabla U) \tag{11.4}$$

where V is a vector of environmental parameters and D is a matrix of diffusion or migration coefficients. The symbol ∇ indicates a spatial gradient of environmental conditions.

The mechanisms that generate spatial patterns involve, for example, spatial heterogeneity in the environment combined with minimal migration rates resulting in locally asynchronous dynamics. Interesting patterns are generated when the population takes on multiple steady states based on vary small changes in environmental conditions. As the population migrates, it can pass through bifurcation points resulting in distinct boundaries in space, i.e., distinct patterns.

Pattern can be due to differences in the mobility of interacting species. For example, in the predator (H) – prey (P) systems (Levin 1978):

$$\delta P/\delta t = P(a - eP - bH) + D_1 \nabla^2 P$$
$$\delta H/\delta t = H(-d + cP - gH) + D_2 \nabla^2 H \tag{11.5}$$

where a, b, c, d, e, and g are parameters describing changes in populations number due to birth, death, and predation. Eq. 11.5 produces spatial pattern if the movement rate, D_2, of the second population is sufficiently large compared to the movement rate, D_1, of the first population.

Diffusion-reaction dynamics offer a potential mechanism by which spatial pattern results from the movement and interaction of organisms. Therefore, the theory holds potential to explain how ecological interactions and processes can result in the spatial patterns we observe on the landscape.

Used as a neutral model, the diffusion-reaction equations ignore many landscape features that may determine pattern. The equations deal with those aspects of pattern which result from organism interactions. Therefore, comparison of predictions with field data on pattern could be used to test for the influence of other landscape features in determining the observed pattern.

11.5.3 Patch Dynamics Theory

Patch dynamics theory (Levin and Paine 1974; Paine and Levin 1981) considers the landscape as containing populations of bare patches, representing disturbance gaps in vegetation. The theory explains pattern by the interaction of disturbance rate, $B(t)$, and recolonization rate, $D(t)$. The fraction, $M(t)$, of the landscape that is composed of patches is obtained from (Paine and Levin 1981):

$$dM/dt = B(t) - D(t) \qquad (11.6)$$

Patches or gaps undergo "death processes" as they are recolonized by succession following disturbance and "birth processes" as new disturbed areas are created.

A further development of the theory considers the age distribution of the patches such that (Paine and Levin 1981):

$$M(t) = N(t, a) \, da$$
$$N(t, a) = B(t - a) \, s(t - a) \qquad (11.7)$$

where s is survival at age a of a gap created at time $t - a$. Paine and Levin (1981) show that this theory is able to project patterns of intertidal patches in mussel beds to within 5% of the observed values.

The patch dynamics model focuses on the aspects of spatial pattern resulting from disturbance and recovery. The model considers the landscape as a homogeneous, two-dimensional plane. Therefore, the model serves as a neutral model for testing the effects of features, such as substrate or topography, on landscape pattern.

11.5.4 Thunen Location Models from Economic Theory

The Thunen model (Hall 1966; Samuelson 1983) attempts to explain how economic activity is spread across the landscape. In the ecological context, the important observation is that an organism must move across the landscape in search of food, and this movement incurs a metabolic cost. A minimum territory size is required for a sufficiently large food supply to support an organism. A maximum territory size results from the increasing metabolic cost of travel (Jones and Krummel 1985) in search of food:

Equations can be developed for metabolic energy, E_m, and available food energy, E_f (Jones and Krummel 1985):

$$E_m = m + t(v)$$
$$E_f = a\ v^2\ v \tag{11.8}$$

where v is the radius of the territory, m is basal metabolism, $t(v)$ is the energy cost of searching the territory, f is the available food per unit area, and a is a shape parameter equal to 2.249 for a hexagonal territory. Equations 11.8 can be set equal to each other, and the resulting quadratic equation can be solved for v to determine the minimum and maximum territory sizes.

Equations 11.8 can be used to predict territory size in the absence of any spatial pattern in the food resource. As a neutral model, the predictions serve as tests of the importance of pattern, such as patchiness, on the size of individual territories. The equations can also be modified to consider the contagious distribution of food and test the importance of other landscape features such as natural or man-made barriers.

11.6 Aggregation Errors in Landscape Studies

Landscape data seldom exist in a "pure" form; that is, information is usually aggregated in both time and space to produce images of landscape pattern. The errors produced by the process of aggregation will affect the accuracy and precision of results and make the estimation of model parameters difficult. The a priori quantification of these errors is, therefore, of interest for the development of all models. Specification of the changes in mean and variance as spatial resolution changes is critical for determining the scales that result in minimum errors of analysis.

Satellite imagery from Landsat has a grain of 30 m × 30 m; imagery from Advanced Very High Resolution Radiometer (AVHRR) has a grain of 1 km × 1 km. As grain size becomes more coarse, information is always lost. If you increase extent (without changing grain), information is always gained. But even specifying the grain and extent is insufficient unless you are specific about how the change is made. It is possible to manipulate this data in a manner similar to the neutral model approach to quantify the expected changes in information with changes in scale.

We have studied the effect of changes in resolution by beginning with a fine-grained data set and performing progressively coarser aggregations of the data (Turner et al. 1989b). The United States Geological Survey digital land use maps, or Land Use Data (LUDA) data (Fegas et al. 1983), provides seven major land cover categories with a resolution of 4 hectares. Successive coarser-grained data sets were produced by aggregating adjacent sites into a single unit. The process was repeated through nineteen levels of aggregation, from 2 × 2 (4 sites) to 50 × 50 (2500 sites).

Three rules were established for aggregating the data. The first rule, S_1, defines the aggregated by the land cover type of the majority of the fine-scaled sites. If two land covers occur in equal proportions, the assignment of the aggregate land cover type was decided at random.

The second rule, S_2, attempts to preserve information about infrequent land cover types by assuming that the aggregate site belongs to a class i if even a single grid point is class i. Therefore, an aggregate can belong to several classes, and the information that an aggregate contains a particular land use at a finer grain is not lost. The probability of encountering a particular land use is calculated by counting the aggregations assigned to each land use type, then dividing by the number of aggregates times the number of land uses actually observed.

A third aggregation rule, S_3, attempts to preserve all possible combinations of patterns. Thus, all unique fine-scale combinations of land cover are assigned to the aggregate cover types. For instance, in aggregating four grid cells with two cover types, the possible combinations of aggregates would be (1) all type 1, (2) all type 2, (3) two of each type, (4) one of type 1 and three of type 2, and (5) three of type 1 and one of type 2. The five pattern combinations are then uniquely assigned to the aggregate. With seven land cover categories, a large set of pattern types is possible. Each aggregate is assigned to one of the pattern types, and the probabilities are calculated by dividing the number in each pattern type by the number of aggregates times the number of unique patterns actually observed.

We compared the aggregation rules by examining how landscape diversity changed with a change of grain. Landscape diversity was defined as

$$H = -1/m \; \Sigma_i \; P_i \; \ln(P_i) \tag{11.9}$$

where P_i is the proportion of the landscape in land cover type i and m is the total number of land cover types. The value H was recalculated for each level of aggregation and the values of H were then regressed against the logarithm of grain size. The regressions were repeated for each aggregation rule.

All of the regressions (Table 3) show a linear relationship between diversity and the logarithm of grain size. The R^2 are greater than 90%, and all regressions are significant at $p < 0.0001$. A striking aspect of Table 3 is that the three aggregation rules given quite different relationships between diversity and grain size. The majority rule, S_1, indicates that diversity decreases slowly with increases in grain (mean slope is −0.055). This occurs because cover types with small P_i get lost in the aggregation process, indicating a lower diversity. The pattern rule, S_3, also indicates a decrease in diversity with increasing grain, but the rate of decrease is much larger (mean is −0.77). As grain size increases, there are larger numbers of potential patterns, but very few of the potential patterns are realized in the LUDA data. Since the number of possibilities increases much faster than the realized cases, the diversity index (Eq. 11.9) decreases rapidly. Notice that the landscapes with the slowest decrease in diversity according to S_1 tend to lose diversity with S_3 faster.

Table 11.3. Changes in Parameter H (see Equation 8 in text) with Progressive Increases in Grain Size on Seven LUDA Landscape Scenes

Landscape Scene	Aggregation Rule		
	S_1	S_2	S_3
Goodland, KS	−0.047 (0.98)	0.094 (0.90)	−0.755 (0.96)
Natchez, MS	−0.055 (0.96)	0.106 (0.98)	−0.820 (0.98)
Knoxville, TN	−0.068 (0.98)	0.126 (0.99)	−0.661 (0.97)
Greenville, SC	−0.036 (0.94)	0.086 (0.98)	−0.829 (0.98)
Waycross, GA	−0.034 (0.92)	0.082 (0.92)	−0.848 (0.98)
Athens, GA	−0.090 (0.94)	0.125 (0.99)	−0.708 (0.96)
Mean	−0.055	0.103	−0.770

S_1, S_2, and S_3 refer to the rules for aggregating fine-scale information into coarser scales (see text for definition of the rules). Values are the slopes of the regression of H against the logarithm of the grain size in hectares. The number in parentheses is the R^2 of the regression.

It is intuitive that information is lost with changing resolution and, therefore, the aggregates appear to be less diverse, and there is a good correlation between the results for S_1 and S_2. It is surprising, therefore, that diversity increases with grain size under rule S_2. Because S_2 insures that every aggregate contains every land cover type, maximum diversity values will be obtained for larger grain sizes.

There are differences in slopes among the landscapes according to all of the aggregation rules. This difference is evident even though most of the landscapes are in the southeastern United States and would tend to be similar to each other. The use of alternative algorithms for aggregating data shows that it is not safe to simply extrapolate diversity values to different grain sizes or landscapes.

11.7 Summary

The complex interactions among history, topography, climate, and biota make it difficult to determine cause and effect relationships at landscape scales. Neutral models are useful for identifying the importance of fine-scale data in explaining broad-scale phenomena. The development of a neutral model requires (1) a clear statement of the problem, (2) the definition of a simple model that allows the importance of each variable to be examined, (3) a comparison of the model predictions with available data and observations, and (4) an objective measure of the adequacy of the results.

Percolation theory has been extensively used as a means for generating neutral models of landscape pattern, of disturbance propagation, for characterizing the differences in life-history characteristics on species distribution and abundance, and for comparing results against landscape data. Models developed from percolation theory show that critical thresholds in landscape pattern exist and slight changes in pattern can produce sudden shifts in broad-scale phenomena. Analysis of neutral models can also indicate those characteristics of landscapes and biota which are most interesting for studying these critical phenomena.

Diffusion-reaction theory, patch dynamics theory, epidemiology theory, and Thunen models adapted from economic theory might all serve as neutral models for predicting spatial patterns from biotic interactions. Neutral models can also be used to empirically define the quantitative effects that occur with changes in the resolution and scale of data.

The neutral modeling approach will ultimately result in the simplest model that can adequately explain the observed phenomena. The complex patterns that are produced from current neutral models indicate that this approach is useful for identifying key landscape variables and for the rigorous analysis of landscape data.

References

Bailey, N.T.J. 1965. The simulation of stochastic epidemics in two dimensions. Fifth Berkeley Symposium, vol. 4, pp. 237–57.

Bailey, N.T.J. 1975. *The Mathematical Theory of Infectious Diseases and its Applications*. New York: Hafner Press.

Baker, W.L. 1989. A review of models of landscape change. *Landscape Ecology* 2:111–33.

Browder, J.A.; Bartley, H.A.; and Davis, K.S. 1985. A probabilistic model of the relationship between marshland-water interface and marsh disintegration. *Ecological Modelling* 29:245–60.

Caswell, H. 1976. Community structure: a neutral model analysis. *Ecological Monographs* 46:327–54.

Cardy, J.L. 1983. Field theoretic formulation of an epidemic process with immunization. *Journal of Physics* A 16:L709–12.

Cardy, J.L., and Grassberger, P. 1985. Epidemic models and percolation. *Journal of Physics* A 18:L267–71.

Cooper, W.S. 1923. The recent ecological history of Glacier Bay Alaska: II. the present vegetation cycle. *Ecology* 4:223–46.

Cowles, H.C. 1899. The ecological relations of the vegetation on the sand dunes of Lake Michigan: I. geographical relations of the dune floras. *Botanical Gazette* 27:95–117, 167–202, 281–308, 361–91.

Cox, J.T., and Durrett, R. 1988. Limit theorems for the spread of epidemics and forest fires. *Stochastic Processes and Their Applications* 30:171–91.

Faddy, M.J. 1986. A note on the behavior of deterministic spatial epidemics. *Mathematical Bioscience* 80:19–22.

Fegas, R.G.; Claire, R.W.; Guptill, S.C.; Andersons, C.A.; and Hallam, K.E. 1983. Land use and land cover digital data. Geological Survey Circular 895–E, U.S. Geological Survey, Reston, Va.

Franklin, J.F., and Forman, R.T.T. 1987. Creating landscape patterns by forest cutting: ecological consequences and principles. *Landscape Ecology* 1:5–18.

Gardner, R.H.; Milne, B.T.; Turner, M.G.; and O'Neill, R.V. 1987. Neutral models for the analysis of broad-scale landscape pattern. *Landscape Ecology* 1:19–28.

Gardner, R.H.; Turner, M.G.; O'Neill, R.V.; and Lavorel, S. in press. Simulation of the scale-dependent effects of landscape boundaries on species persistence and dispersal. *Proceedings of the ESA Symposium: The Role of Landscape Boundaries in the Management and Restoration of Changing Environments*. Toronto, 1989.

Gefen, Y., and Aharony, A. 1983. Anomalous diffusion on percolating clusters. *Physical Review Letters* 50:77–80.

Grassberger, P. 1983. On the critical behavior of the general epidemic process and dynamical percolation. *Mathematical Bioscience* 63:157–72.

Grieg-Smith, P. 1952. The use of random and contiguous quadrats in the study of the structure of plant communities. *Annals of Botany* 16:293–316.

Grieg-Smith, P. 1964. *Quantitative Plant Ecology*. 2nd ed. London: Butterworths.

Hall, P., ed. 1966. *Von Thunen's isolated state: an English edition of Der Isolerte Staat by Johann Heinrich von Thunen*. Trans. Carla M. Wartenberg. Oxford: Pergamon.

Hoffman, A., and Nitecki, M.H. 1987. Introduction: neutral models as a biological research strategy. In *Neutral Models in Biology*, eds. M.H. Nitecki and A. Hoffman, pp. 3–8. New York: Oxford University Press.

Jones, D.W., and Krummel, J.R. 1985. The location theory of animal populations: the case of a spatially uniform food distributions. *American Naturalist* 126:392–404.

Kleinbaum, D.G.; Kupper, L.L.; and Muller, K.E. 1988. *Applied Regression Analysis and Other Multivariable Methods*. Boston: PWS-Kent.

Kuulasmaa, K. 1982. The spatial general epidemic and locally dependent random graphs. *Journal of Applied Probability* 19:745–58.

Kuulasmaa, K., and Zachary, S. 1984. On spatial general epidemics and bond percolation processes. *Journal of Applied Probability* 21:911–14.

Levin, S.A. 1976. Spatial patterning and the structure of ecological communities. *Lectures on Mathematics in the Life Sciences* 8:1–35.

Levin, S.A. 1978. Pattern formation in ecological communities. *Spatial Pattern in Plankton Communities*, ed. J.H. Steele, pp. 433–65. New York: Plenum.

Levin, S.A., and Paine, R.T. 1974. Disturbance, patch formation, and community structure. *Proceedings of the National Academy of Science* 71:2744–47.

Milne, B.T.; Johnston, K.; and Forman, R.T.T. 1989. Scale-dependent proximity of wildlife habitat in a spatially-neutral Bayesian model. *Landscape Ecology* 2:101–10.

Mollison, D. 1977. Spatial contact models for ecological and epidemic spread. *Journal of the Royal Statistical Society* B 39:283–326.

Mollison, D. 1986. Modelling biological invasions: chance, explanation, prediction. *Philosophical Transactions of the Royal Society of London* B 314:675–93.

Ohtsuki, T., and Keyes, T. 1986. Biased percolation: forest fires with wind. *Journal of Physics* A 19:L281–87.

O'Neill, R.V.; Gardner, R.H.; Milne, B.T.; Turner, M.G.; and Jackson, B. Heterogeneity and spatial hierarchies. In *Ecological Heterogeneity*, eds. S.T.A. Pickett and J. Kolasa. New York: Springer-Verlag, in press.

O'Neill, R.V.; Milne, B.T.; Turner, M.G.; and Gardner, R.H. 1988. Resource utilization scales and landscape pattern. *Landscape Ecology* 2:63–69.

Paine, R.T., and Levin, S.A. 1981. Intertidal landscapes: disturbance and the dynamics of patches. *Ecological Monographs* 51:145–78.

Samuelson, P.A. 1983. Thunen at two hundred. *Journal of Economic Literature* 21:1468–88.

Slobodkin, L.B. 1987. How to be objective in community studies. In *Neutral Models in Biology*, eds. M.H. Nitecki and A. Hoffman, pp. 93–108. New York: Oxford University Press.

Smythe, R.T., and Wierman, J.C. 1978. First-passage percolation on the square lattice. *Lecture Notes in Mathematics*, vol. 671. New York: Springer-Verlag.

Stauffer, D. 1985. *Introduction to Percolation Theory*. London: Taylor and Francis.

Swihart, R.K.; Slade, N.A.; and Bergstrom, B.J. 1988. Relating body size to the rate of home range use in mammals. *Ecology* 69:393–99.

Turner, M.G. 1987. Spatial simulation of landscape changes in Georgia: a comparison of 3 transition models. *Landscape Ecology* 1:29–36.

Turner, M.G.; Gardner, R.H.; Dale, V.H.; and O'Neill, R.V. 1989a. Predicting the spread of disturbance across heterogeneous landscapes. *Oikos* 55:121–29.

Turner, M.G.; O'Neill, R.V.; Gardner, R.H.; and Milne, B.T. 1989b. Effects of changing spatial scale on the analysis of landscape pattern. *Landscape Ecology* 3:153–62.

von Niessen, W., and Blumen, A. 1988. Dynamic simulation of forest fires. *Canadian Journal of Forest Research* 18:805–12.

Wiens, J.A. 1985. Vertebrate responses to environmental patchiness in arid and semiarid ecosystems. In *The Ecology of Natural Disturbance and Patch Dynamics*, eds. S.T.A. Pickett and P.S. White, pp. 169–93. New York: Academic Press.

Wilkie, D.S., and Finn, J.T. 1988. A spatial model of land use and forest regeneration in the Ituri forest of northeastern Zair. *Ecological Modelling* 41:307–23.

Zallen, R. 1983. *The Physics of Amorphous Solids*. New York: John Wiley and Sons.

12. Models of Forested and Agricultural Landscapes: Integrating Economics

Peter J. Parks

12.1 Introduction

Economic conditions can often influence the development of landscape pattern over space and time. For example, increases in nonforest area through deforestation of tropical landscapes have been linked to economic conditions and government policies (Repetto and Gillis 1988; Richards and Tucker 1988). Most landscape models lack explicit economic detail. This detail is needed to evaluate changes in landscape pattern due to changing economic conditions (e.g., changes in relative prices of commodities produced using land). For landscape models to be predictive in a human-dominated landscape, the influence of economic forces must be recognized (Turner 1987).

Landscapes can also influence economic development patterns over space and time. Landscape attributes (e.g., physiography) frequently grant comparative economic advantages to specific landscapes. The emergence of the Southeast and Northwest as dominant wood-producing regions in the United States is one example of this; the importance of the Midwest in crop production is another. Models of land use in forest and agricultural economies often resolve the landscape into classes based on the land's suitability for producing goods. The location, size, or shape of landscape elements is not explicitly considered. As a result, the usefulness of most economic land models for analyzing ecological phenomena linked to spatial landscape pattern (e.g., population dynamics and biological diversity) is limited. To analyze such phenomena, economic models of land must be integrated with spatial landscape models.

Joining these two perspectives for modeling land can increase the range of applications for both. At a minimum, alternative landscape futures under hypothetical economic conditions can be simulated. In addition, the implications of economic land use can be linked more closely to spatial landscape pattern. More ambitious integrated applications could involve planning landscapes to achieve social goals that have been quantified in economic terms, and quantifying economic trade-offs between alternative landscape patterns. This chapter provides an overview of economic approaches for modeling land. It also includes a case study that integrates a landscape model (Turner 1988) with an economic model of land use trends in Georgia.

12.2 Economic Models of Forested and Agricultural Landscapes: A Brief Overview

Economic models of forested or agricultural landscapes depend on the land's ability to produce commodities such as crops or wood products. Randall and Castle (1985) summarize pioneering work done in this field by von Thunen and Ricardo and frame modern research within the context of these important early studies. Von Thunen analyzes the location of agricultural activity, while Ricardo emphasizes the effects of land quality (e.g., soil fertility). A simple von Thunen model posits a city (where agricultural products are sold) in the midst of a featureless plain of uniform land quality surrounded by wilderness. Landscapes are series of concentric rings around the city; crops that provide higher net economic benefit occupy areas closer to the city. Net economic benefits consist of crop revenues minus production and transportation costs, and these benefits will decrease as distance from the city increases. Wilderness begins at the distance where net economic benefit reaches zero. More complex landscapes (e.g., noncontiguous patches) arise when land quality is considered (as in Ricardo's model) or when other assumptions in the model are relaxed.

Economic land use models can be classified into three categories based on modeling strategy and desired results (Healy 1986; Parks and Alig 1988). These categories include inventory/descriptive models, engineering/optimization models, and statistical/econometric models. Each category is helpful in landscape analyses. The best method to use with a particular landscape analysis will depend on the goals and constraints specific to the study.

12.2.1 Inventory/Descriptive Models

Most inventory/descriptive models do not employ explicit economic theory. With this strategy land use categories are defined, and land inventory data or expert opinions determine the exchange of acres among these categories. When this exchange is empirically determined, systems of differential equations or stochastic processes can be used to summarize changes in category areas. Often, resource or environmental phenomena depend on the exchange of land among specific cover types in the landscape.

For example, Houghton et al. (1985) examine flux of carbon from tropical, temperate, and boreal ecosystems into the atmosphere in 1980 due to changes in land use. Shifting cultivation, a systematic change in the forested and agricultural components of the tropical landscape, is particularly important in determining rates of flux. Melillo et al. (1988) examine land use changes in the Soviet Union between 1850 and 1980, and the net release of CO_2 to the atmosphere resulting from these changes. The authors analyze the correlation between fluctuating CO_2 levels and changes in the landscape such as forest clearing and harvesting.

These models offer only an implicit way to integrate economic considerations into a landscape analysis. Compared to other explicit approaches, they require the least amount of data, since no explicit theory must be empirically supported. This is a tremendous advantage in studies at large spatial (e.g., national and global) or long temporal (e.g., centuries) scales. Inventory/descriptive land use models require only data on land use trends. The interpretation of change between land inventories differs among inventory/descriptive studies. When differential equations are used, (e.g., Johnson and Sharpe 1976; Johnson 1977), the proportions of acres changing use are interpreted as deterministic rates. In contrast, if the rates of exchange are interpreted as probabilities of changing use (e.g., Burnham 1973; Van Loock et al. 1973), the evolution of the landscape loosely follows a Markov stochastic process. Further assumptions regarding landscape evolution are required for it to be a true Markov process (Hillier and Lieberman 1980). The deterministic or stochastic interpretations can be developed from the same data and will on average produce identical extrapolated trends (Shugart et al. 1973).

Although inventory/descriptive models are relatively easy to construct, they are not predictive. Unmodified historical trends are usually extrapolated, but anticipated changes from historical trends have also been projected using systematically obtained expert opinion (e.g., Wall 1981). Any extrapolation of past trends into the future is subjective. Projecting hypothetical land use scenarios with inventory/descriptive models usually involves altering the rates of exchange of area among land use categories (e.g., Melillo et al. 1988). Because economic considerations underlying land use trends are implicit, detailed sensitivity analyses of landscapes to specific economic influences (e.g., changes in product prices) cannot be made.

12.2.2 Engineering/Optimization Models

Engineering/optimization models solve for the landscape patterns that best achieve quantifiable goals and meet constraints. The planned landscape consists of an allocation of available land classes to feasible land use choices. Spatial features (e.g., location, size, or shape) of land classes are usually not made explicit. Constraints (e.g., available land by class) are explicitly incorporated by limiting the feasible set of landscape choices. Applications of these methods to agriculture are given in Hazell and Norton (1986) and to natural resources in Dykstra (1984).

Engineering/optimization models are used to plan publicly owned forests in the United States within the National Forest System (e.g., United States Department of Agriculture [USDA] Forest Service 1987a, 1987b). The approach is applied to optimize the mix of multiple forest benefits in national forests. The optimal land allocation maximizes sustained net social benefit from forest resources (e.g., timber, wildlife, and recreation), while meeting constraints such as sustained environmental quality and total available land. The USDA Soil Conservation Service (1982a, 1982b, 1987) uses this approach to evaluate national soil and water conservation policies. The objective is to minimize the cost of meeting national and international demands for food, and is constrained by available resources (e.g., land, soils, and water). Spatial aspects (e.g., number, size, shape, and location of landscape components in each class) are typically addressed using exogenous constraints in the Forest Service and the Soil Conservation Service studies.

The engineering/optimization approach can be useful in determining how to plan a landscape to accomplish specific goals. These goals or constraints need not be in economic terms, providing that the relationship between them and landscape choices can be quantified. Crucial steps in implementing this type of model include identifying goals and constraints, identifying a set of possible landscape choices, and quantifying the contributions of choices to goals and constraints. Data required for an engineering/optimization study include a land classification, sequences of activities, and estimates of the consequences of activities. The level of resolution (e.g., spatial or temporal) required for any of these data depends on the goals of the simulation study, as well as on the objectives and constraints in the problem statement. These models require engineering data that reflect technical efficiencies. Such data may be in the form of a production function that translates the usage of inputs into outputs from the landscape. The emphasis is on assembling data to determine optimal landscapes.

Simulations using these models take the form of changing model parameters or specifying different sets of objectives and constraints. Projections in the USDA Soil Conservation Service Study (1987) for 1990, 2000, and 2030 are made by adjusting coefficients and constraints to reflect anticipated future conditions. These impact the optimal landscape, which consists of crops and cropping practices. In Canada the impact of acid rain on agricultural production (Ludlow and Smit 1986), as well as the effect of climate change on land production potential (Brklacich and Smit 1986), have been simulated by changes in technical coefficients.

12.2.3 Statistical/Econometric Models

Statistical/econometric models of land use describe relationships between changes in landscapes and various economic, social, policy, or environmental variables. Behavioral models are used to identify important variables whose empirical impact is quantified using econometric or statistical procedures. Unlike the engineering/optimization approach, which describes how landscapes should be planned to accomplish goals, this approach estimates how landscapes empirically reflect goals.

For example, Alig (1986) used von Thunen's assumptions to identify economic variables that influence landscapes. Alig quantifies the regional influences of land product prices, rural population, and per capita income on regional land use trends in the southeastern United States. The USDA Forest Service projects future land use by utilizing econometric models, including Alig's (1986). These projections are used to assess future resource availability and to help design national forest programs and policies (e.g., Haynes 1988a, 1988b). As with most statistical/econometric models, spatial features other than regional location (e.g., piedmont and southeastern United States) of land allocations are not considered.

These models offer the most efficient way to integrate explicit economic considerations into existing landscape models. The Georgia case study demonstrates the convenience of replacing an inventory/descriptive model with an econometrically estimated one. The economic model serves as a component within the landscape model in the Georgia study. In contrast, the engineering/optimization approach requires landscape models to serve as components of economic optimization models.

Data required to estimate statistical/econometric models include economic variables (e.g., prices, costs, and production technologies) and land use variables (e.g., inventory plots in different land use categories). Although these models require more data than inventory/descriptive models, they are predictive. Because landscapes and economic variables are linked, landscape response to changes in economic conditions can easily be evaluated.

Simulated landscapes require economic variables. The values of these variables can be hypothetical, actual, or extrapolated and will always be associated with errors in predicting landscapes. These errors will increase as values for economic variables deviate from the mean of the data used to estimate the model. Often, models are constructed to incorporate variables that are influenced by policy or management so that changes in these factors and their effects can be simulated.

12.3 Land Use Dynamics in Georgia: A Case Study

This section develops a case study that integrates a statistical/econometric model of land use in the Georgia piedmont with a spatial landscape model for a piedmont county (Turner 1988). The combined model quantifies the link between economic variables and spatial pattern of five land uses. The landscape is characterized using three spatial pattern descriptors for each use, and landscape response to changes in economic conditions is via these descriptors.

12.3.1 Methods

12.3.1.1 Study Area and Data Collection

Georgia, located in the southeastern United States, has three distinct physiographic regions: the mountains (1,470,310 ha), the piedmont (4,606,139 ha), and the coastal plain (8,971,206 ha). The landscape in Georgia in 1982 was 65% forest,

18% cropland, 7% pasture and range, and 10% in other uses (including urban). Of the forest land 7% was publicly owned (national forest, other federal, state, county, and municipal); 25% was owned or leased by forest industry; and 68% was owned by nonindustrial private owners (Sheffield and Knight 1984). The piedmont and coastal plain are used in the economic model because land use in these regions is more prone to change.

Land use data included all nonforest and nonindustrial private forest USDA Forest Service Inventory plots for the 1972 and 1982 state inventories (Knight and McClure 1974; Sheffield and Knight 1984). Frequencies of plot transitions among the five Turner (1988) land uses are shown in Table 12.1. These uses include urban, cropland, transition (e.g., idle cropland), pasture, and forest. Economic data for agricultural returns are calculated from the Census of Agriculture. Data on timber removals are from USDA Forest Service inventories.

12.3.1.2 Model Development

The spatial model was developed (Turner 1988) and applied (e.g., Turner 1987) to evaluate changes in the pattern of five land uses (urban, crop, abandoned cropland, pasture, and forest). Land use data were obtained from aerial photographs from 1942, 1955, and 1980 and digitized into a matrix based on a 1-ha grid cell. Transitions of cells from one land use to another were dependent on transitional probabilities and neighborhood (spatial) influences. These influences are simulated by calculating a transition index based on transitional probabilities and the land uses present on adjacent cells. Probabilities for land use transition among uses are calculated using aggregate land use data from the USDA Forest Service

Table 12.1. Transition Frequencies for USDA Forest Service Inventory Plots, 1972–1982, and Transition Probabilities Used in Turner (1987) for Oglethorpe County, Georgia

	Urban	Crop	Transition	Pasture	Forest
	Plot Transition Frequencies (1972–1982)				
Urban	0.989	0.002	0.002	0.005	0.002
Crop	0.016	0.916	0.032	0.035	0.001
Transition	0.063	0.305	0.327	0.105	0.200
Pasture	0.020	0.226	0.046	0.665	0.043
Forest	0.016	0.033	0.002	0.019	0.930
	Turner Transition Matrix (1955–1980)				
	Urban	Crop	Transition	Pasture	Forest
Urban	1.00				
Crop		.430	.570		
Transition	.008		.212		.780
Pasture				1.00	
Forest					1.00

and United States (USBC) Census of Agriculture (Table 1). This case study builds on Turner's (1987) application of a spatial model by (1) introducing transition probabilities derived from plot-level land use changes, and (2) introducing an explicit relationship between these changes and economic conditions.

The econometric/statistical model used here ties land use transitions directly to economic data. Nonindustrial owners are presumed to change use according to expected benefits provided by land use alternatives. When these are sufficiently high, the owner changes land use. The value of benefits for each owner is unobservable, as are the critical levels of expected benefits that motivate changes in use. However, the data on observed land use choices reveal information about whether the critical level has been reached. From the analyst's point of view the land use choice is stochastic. Given a probability distribution for the unobservable components of expected benefits (e.g., the normal distribution) and data for observable land use attributes (e.g., net economic benefits), the probability of observing any given configuration of land use data and observable attributes can be specified. This function describes the statistical likelihood of observing the data, and it is optimized to obtain maximum likelihood estimates for model parameters (Judge et al. 1982, 1985). The land use decisions of individual forest land owners have been examined with these methods by Binkley (1981), Schuster (1983), Boyd (1984), and Royer (1987).

In contrast, the same rationale can be applied to investigate regional land use trends, which are presumed to arise from aggregated individual decisions about land use. In this context, a model estimates the proportion of owners (e.g., Romm et al. 1987) or acres in a region that fall into specific land use classes. Linear regression relates proportions to socioeconomic characteristics of owners or regions. In a linear proportions model, the proportion is used directly as a dependent variable.

This case study applies a linear proportions model to aggregate transitions of plots in each county. The dependent variables are specific to each county and 1972 land use. Potentially, this allows an econometric estimation of county plot transitions for each row of the plot transition matrix in Table 1. Let P_{ijk} describe the proportion of plots in county k changing from use i to use j (i, j correspond to urban, cropland, transition, pasture, and forest) between 1972 and 1982. Depending on empirical land use changes between 1972 and 1982, each county could provide as many as twenty-five observed proportions P_{ijk}. Empirically, each piedmont or coastal plain county provides fewer than this limit, since many transitions (e.g., urban to forest) did not occur for individual counties in the sample (Table 12.1).

The form of the linear proportions model regression equation is

$$P_{ijk} = X_k\beta_{ij} + \varepsilon_{ijk} \tag{12.1}$$

where the subscripts i, j, and k correspond to 1972 land use, 1982 land use, and county, respectively. With the linear proportions approach, predicted values must be constrained to fall within the [0,1] range. Since this imposes restrictions on the error term ε_{ijk}, other methods have been employed (e.g., logistic regression, Romm et al. 1987) to algebraically transform proportions.

Independent variables for each county (X_k) include average net returns per acre (in millions of 1972 dollars) for crops and livestock and removals (e.g., harvests) per acre for pine (in thousands of cubic feet). Acronyms used for these variables are CROPINC, STCKINC, and PINECFAC, respectively. Dummy variables are used to indicate county location within the region. For example, UNIT1 equals one when the county is in the southeast survey unit, and equals zero otherwise. UNIT2 and UNIT3 indicate counties in the southwest and central survey units, respectively. The coefficients β_{ij} are estimated using ordinary least squares (SAS Institute, 1982) and are shown in Table 12.2.

Table 12.2. Ordinary Least Squares Estimates[a] of Linear Proportions Model Parameters, Piedmont and Coastal Plain Counties, Georgia

Independent Variable	Dependent Variable (County Plot Transition Frequency)			
	Forest-Crop	Forest-Forest	Forest-Pasture	Forest-Transition
Intercept	−0.00846	0.969[f]	0.0221[e]	0.00110
	(0.0123)	(0.0166)	(0.00916)	(0.00295)
CROPINC	0.151[e]	−0.166	0.0571	0.00678
	(0.0754)	(0.101)	(0.0559)	(0.376)
STICKINC	−0.0253	0.0580	−0.0392	0.00157
	(0.0621)	(0.0836)	(0.0460)	(0.0148)
PINECFAC	0.199[f]	−0.244[f]	0.0313	0.0123
	(0.0494)	(0.0666)	(0.0367)	(0.0118)
UNIT1	0.0432[f]	−0.0259	−0.0107	0.000515
	(0.0129)	(0.0174)	(0.00959)	(0.00309)
UNIT2	0.0807[f]	−0.0797[f]	0.00356	0.00981[f]
	(0.0139)	(0.0187)	(0.0103)	(0.00333)
UNIT3	0.0127	−0.000999	−0.00390	−0.00222
	(0.0111)	(0.0149)	(0.00821)	(0.00265)
Model Summary Statistics				
R^2	0.371	0.284	0.0504	0.18
Dep Mean[b]	0.034	0.929	0.0193	0.00270
CV[c]	117.	5.80	154.	355.
$F(6,113)$[d]	10.5[f]	7.10[f]	1.02	3.86[f]

[a] Mean (standard deviation); n = 118 counties. Counties with incomplete census data omitted.
[b] Dependent variable mean.
[c] Coefficient of variation.
[d] Joint F-test for all parameters equal to zero.
[e] Denotes an estimate that is significantly different from zero for $P < 0.05$.
[f] Denotes an estimate that is significantly different from zero for $P < 0.01$.

12.3.1.3 Simulations

Simulations are made using estimated transition proportions in place of empirical ones in Table 12.1. This requires replacing the rows of Table 12.1 with hypothetical values. In this example, only proportions for land that was forest in 1972 are

Table 12.3. Actual 1980 and Simulated 1990 Land Area Proportions,[a] Number of Patches,[a] and Patch Size[a] by Land Use, Oglethorpe County, Georgia (n = 6)

| | Proportion of Land Area | | | |
| | 1980 | 1990 | 1990 | 1990 |
Land Use	Actual	Extrapolated[b]	Removals[c]	No Loss[d]
Urban	0.02 (0.02)	0.04 (0.02)	0.04 (0.02)	0.03 (0.02)
Cropland	0.15 (0.08)	0.25 (0.06)	0.26 (0.06)	0.23 (0.06)
Transition	0.29 (0.07)	0.10 (0.02)	0.10 (0.02)	0.10 (0.02)
Pasture	0.01 (0.01)	0.05 (0.01)	0.05 (0.01)	0.04 (0.01)
Forest	0.54 (0.05)	0.57 (0.05)	0.55 (0.05)	0.60 (0.05)

| | Number of Patches | | | |
| | 1980 | 1990 | 1990 | 1990 |
Land Use	Actual	Extrapolated	Removals	No Loss
Urban	1.83 (2.56)	34.7 (21.8)	34.2 (22.6)	26.5 (13.1)
Cropland	25.8 (9.74)	14.8 (8.16)	13.0 (8.10)	18.7 (9.58)
Transition	38.0 (14.8)	23.2 (5.19)	22.2 (4.26)	28.2 (7.52)
Pasture	3.17 (2.40)	23.5 (19.6)	22.2 (18.0)	34.2 (25.5)
Forest	27.5 (9.75)	11.2 (4.88)	10.7 (5.72)	18.8 (9.02)

| | Mean Patch Size | | | |
| | 1980 | 1990 | 1990 | 1990 |
Land Use	Actual	Extrapolated	Removals	No Loss
Urban	4.13 (6.13)	7.96 (14.9)	5.81 (9.35)	5.19 (8.64)
Cropland	12.4 (3.80)	46.5 (32.8)	24.5 (7.83)	19.5 (6.18)
Transition	22.7 (23.3)	9.49 (2.84)	5.4 (9.28)	14.4 (9.69)
Pasture	3.33 (2.09)	12.6 (14.8)	7.82 (6.48)	5.13 (5.34)
Forest	46.8 (17.4)	124. (54.6)	67.7 (24.9)	60.7 (26.5)

[a] Mean (standard deviation); n = 6 sample plots.
[b] Using unmodified 1972–1982 Forest Service plot transitions.
[c] Econometrically estimated forest transitions, assuming volume removals are doubled.
[d] No transitions from forest land.

Table 12.4. Hypothetical Land Use Transition Frequencies for Doubled-Removals and No-Net-Loss Scenarios, Oglethorpe County, Georgia

	Urban	Crop	Trans.	Pasture	Forest
Transition Frequencies, High Removals Scenario					
Urban	0.989	0.002	0.002	0.005	0.002
Crop	0.016	0.916	0.032	0.035	0.001
Transition	0.063	0.305	0.327	0.105	0.200
Pasture	0.020	0.226	0.046	0.665	0.043
Forest	0.016	0.059	0.004	0.024	0.897
Transition Frequencies, No-Net-Loss Scenario					
	Urban	Crop	Trans.	Pasture	Forest
Urban	0.989	0.002	0.002	0.005	0.002
Crop	0.016	0.916	0.032	0.035	0.001
Transition	0.063	0.305	0.327	0.105	0.200
Pasture	0.020	0.226	0.046	0.665	0.043
Forest	0	0	0	0	1.00

estimated. This allows the last row of the matrix in Table 12.1 to become sensitive to changes in county economic conditions, X_k. Each county provides observations for five possible transitions (forest-urban, forest-cropland, forest-transition, forest-urban, and forest-forest). Since these must sum to one, only four transitions need to be estimated; forest-urban transitions are obtained by subtraction.

Three simulations are presented in Table 12.3. The first extrapolates 1972–82 transitions; the second and third employ hypothetical matrices (Table 12.4) to simulate the effect of doubled-removals and no-net-loss of forest land. The doubled-removals scenario uses transitions estimated by doubling pine volume removals and maintaining other variables at their mean values. This increases the proportion of land changing from forest to cropland. In contrast, the no-net loss scenario allows no land to leave the forest category. These matrices are used as input to Turner's (1988) landscape model to calculate spatial effects (e.g., number and size of patches) on the landscape.

12.3.2 Results and Discussion

12.3.2.1 Econometric/Statistical Model

The estimated parameters are provided in Table 12.2. None of the models explains more than 37% of the variation in proportions, although the joint F-test suggests that three of the regressions are significant ($P < 0.01$). The forest-crop and forest-forest regressions perform best, using either R^2 (0.371, 0.284, respectively) or coefficient of variation (117%, 5.80%, respectively) as criteria. Increases in pine harvests increase the proportion of land leaving forests to crops and consequently decrease the proportion that remains in forest. Returns to crops also increase the proportion of land changing from forest to crops. Further, the results suggest that counties in the coastal plain (i.e., southeast and southwest survey units) are more likely to have land that changes from forests to cropland. The effect on the landscape of a doubling in county pine removals (i.e., PINECFAC) was analyzed using Turner's (1988) model.

12.3.2.2 Landscape Model

Transitions made during the decade 1972–1982 to extrapolate from 1980 suggest a landscape in which area in transition land declines sharply and area in all other uses increases (Table 12.3). This is also true in the doubled-removals and no-net-loss simulations. Relative to the extrapolated trend, doubling forest removals leaves less land in forest and more in agriculture; constraining no-net-loss of forest land has the opposite effect. None of the proportions other than transition land is greatly different from its 1980 level.

Simulated urban land patches are larger and more numerous in each of the three simulations (Table 12.3). On the other hand, cropland appears to consolidate into fewer, larger patches. Transition land patches grow smaller and decrease in number to contribute to the growth of the other landscape categories. In contrast to cropland, pasture land patches grow both in number and size. Forest land patches decrease in number but increase in size, leading to an increase in land area proportion. Doubling pine volume removals leads to fewer, larger patches than the no-net-loss scenario; however, the proportion of land in forest is largest with no-net-loss.

12.4 Summary

Economic conditions can often influence the development of landscape pattern over space and time. For landscape models to be predictive in a human-dominated landscape, the influence of economic forces must be recognized. Conversely, landscapes can influence economic development patterns over space and time. However, the usefulness of most economic models for analyzing ecological phenomena linked to spatial landscape pattern is limited. Integrating landscape ecological and economic approaches to modeling land can increase the range of applications for both.

This chapter presents an overview of three approaches to economic land modeling and describes opportunities for integration in models of forested and agricultural landscapes. The three approaches include inventory/descriptive, engineering/optimization, and statistical/econometric approaches. Each category is useful in landscape analyses.

The best method to use with a particular analysis will depend on the goals and constraints specific to the study. Inventory/descriptive models require the least amount of data since no explicit theory must be supported. Engineering/optimization models solve for landscape patterns that best achieve quantifiable goals and meet constraints. Statistical/econometric models measure how landscapes empirically reflect goals and are easily incorporated into existing landscape models. The Georgia case study demonstrates the convenience of replacing an inventory/descriptive model with an econometrically estimated one. The economic model serves as a component in a landscape model. In contrast, the engineering/optimization approach may require landscape models to serve as components of economic optimization models.

Integrating the two approaches in the Georgia case study allowed the impact of changes in economic conditions on landscape to be simulated. For example, doubling county timber volume removals increased the rate of transition from forest to crops in the coastal plain and piedmont (Table 12.4). In Oglethorpe County, this may contribute to fewer, larger forest land patches and smaller, more numerous cropland patches (Table 12.3).

Research is needed to evaluate alternative methods of integrating landscape ecological and economic techniques. At a minimum, the effect of economics on landscape development and vice versa can be quantified and evaluated. More ambitious integrated applications could involve the planning of landscapes to achieve quantifiable social goals and to evaluate trade-offs between alternative landscape patterns.

Acknowledgments

I thank Susan Chadwick, Brent Fewell, Bob Gardner, Wendy Hudson, Joe Stevens, Monica Turner, and two anonymous reviewers for their valuable comments. This work was supported in part by the School of Forestry and Environmental Studies, Duke University.

References

Alig, R.J. 1986. Econometric analysis of the factors influencing forest area trends in the Southeast. *Forest Science* 32(1):119–34.

Binkley, C.S. 1981. Timber supply from private nonindustrial forests: a microeconomic analysis of landowner behavior. Yale University School of Forestry and Environmental Studies Bulletin 92.

Boyd, R. 1984. Government support of nonindustrial production: the case of private forests. *Southern Economic Journal* 51(1):89–107.

Brklacich, M., and Smit, B. 1986. Effects of climatic change on agricultural land resource potential. In *Perspectives in Land Modeling*, eds. R. Gelinas, D. Bond, and B. Smit. Workshop proceedings. Toronto, Ontario, Nov. 17–20, 1986. Montreal: Polyscience Publications.

Burnham, B.O. 1973. Markov intertemporal land use simulation model. *Southern Journal of Agricultural Economics* 5(1):253–58.

Dykstra, D.P. 1984. *Mathematical Programming for Natural Resources Management.* New York: McGraw-Hill.

Haynes, R.W. 1988a. An analysis of the timber situation in the United States: 1989–2040, part I: the current resource and use situation. Draft. Technical Document Supporting the 1989 RPA Assessment.

Haynes, R.W. 1988b. An analysis of the timber situation in the United States: 1989–2040, part II: the future resource situation. Draft. Technical Document Supporting the 1989 RPA Assessment.

Hazell, P.B., and Norton, R.D. 1986. *Mathematical Programming for Economic Analysis in Agriculture.* New York: MacMillan.

Healy, R.G. 1986. *Competition for Land in the American South.* Washington, D.C.: Conservation Foundation.

Hillier, F.S., and Lieberman, G.J. 1980. *Introduction to Operations Research.* 3rd ed. San Francisco: Holden-Day.

Houghton, R.A.; Boone, R.D.; Melillo, J.M.; Palm, C.A.; Woodwell, G.M.; Myers, N.; Moore, B., III; and Skole, D.L. 1985. Net flux of carbon dioxide from tropical forests in 1980. *Nature* 316(6029):617–20.

Johnson, W.C. 1977. A mathematical model of forest succession and land use for the North Carolina Piedmont. *Bulletin of the Torrey Botany Society* 104(4):334–46.

Johnson, W.C., and Sharpe, D.M. 1976. An analysis of forest dynamics in the northern Georgia Piedmont. *Forest Science* 22(3):307–22.

Judge, G.G.; Griffiths, W.E.; Hill, R.C.; Lutkepohl, H.; and Lee, T.C. 1985. *The Theory and Practice of Econometrics.* New York: John Wiley and Sons.

Judge, G.G.; Hill, R.C.; Griffiths, W.E.; Lutkepohl, H.; and Lee, T.C. 1982. *Introduction to the Theory and Practice of Econometrics.* New York: John Wiley and Sons.

Knight, H.A., and McLure, J.P. 1974. Georgia's timber, 1972. USDA Forest Service Resource Bulletin SE–27. Asheville, N.C.

Ludlow, L., and Smit, B. 1986. Acid rain and agricultural production in Ontario. In *Perspectives in Land Modeling,* eds. R. Gelinas, D. Bond, and B. Smit. Workshop proceedings. Toronto, Ontario, Nov. 17–20, 1986. Montreal: Polyscience Publications.

Melillo, J.M.; Fruci, J.R.; and Houghton, R.A. 1988. Land-use change in the Soviet Union between 1850 and 1980: Causes of a net release of CO_2 to the atmosphere. *Tellus* (1988) 40B:116–28.

Parks, P.J., and Alig, R.J. 1988. Land base models for forest resource supply analysis: a critical review. *Canadian Journal of Forest Research* 18:965–73.

Randall, A., and Castle, E.N. 1985. Land resources and land markets, In *Handbook of Natural Resource and Energy Economics,* vol. 2, ed. A.C. Kneese and J.L. Sweeney, pp. 571–620. New York: North Holland.

Repetto, R., and Gillis, M, eds. 1988. *Public Policies and the Misuse of Forest Resources.* New York: Cambridge University Press.

Richards, J.F., and Tucker, R.P., eds. 1988. *World Deforestation in the Twentieth Century.* Durham, N.C.: Duke University Press.

Romm, J.; Tuazon, R.; and Washburn, C. 1987. Relating forestry investment to the characteristics of nonindustrial private forestland owners in northern California. *Forest Science* 33(1):197–209.

Royer, J.P. 1987. Determinants of reforestation behavior among southern landowners. *Forest Science* 33(3):654–67.

SAS Institute. 1982. SAS User's Guide: Statistics. SAS Institute, Inc. Cary, N.C.

Schuster, E.G. 1983. Evaluating nonindustrial private landowners for forestry assistance programs: a logistic regression approach. USDA Forest Service Research Paper INT–320. Ogden, Utah.

Sheffield, R.M., Knight, H.A. 1984. Georgia's forests. USDA Forest Service Resource Bulletin SE–73. Asheville, N.C.

Shugart, H.H.; Crow, T.R.; and Hett, J.M. 1973. Forest succession models: a rationale and methodology for modeling forest succession over large regions. *Forest Science* 19:203–12.

Turner, M.G. 1987. Spatial simulation of landscape changes in Georgia: a comparison of 3 transition models. *Landscape Ecology* 1(1):29–36.

Turner, M.G. 1988. A spatial simulation model of land use changes in a piedmont county in Georgia. *Applied Mathematics and Computation* 27:39–51.

United States Department of Agriculture (USDA) Forest Service. 1987a. Land and resource management plan, 1986–2000. Nantahela and Pisgah National Forests. USDA Forest Service, Southern Region.

United States Department of Agriculture (USDA) Forest Service. 1987b. Final environmental impact statement. Land and resource management plan, 1986–2000. Nantahela and Pisgah National Forests. USDA Forest Service, Southern Region.

United States Department of Agriculture (USDA) Soil Conservation Service. 1982a. 1980 Appraisal part I. Soil, water, and related resources in the United States: status, condition, trends. U.S. Department of Agriculture.

United States Department of Agriculture (USDA) Soil Conservation Service. 1982b. 1980 Appraisal part II. Soil, water, and related resources in the United States: Analysis of trends. U.S. Department of Agriculture.

United States Department of Agriculture (USDA) Soil Conservation Service. 1987. The second RCA Appraisal. Soil, water, and related resources on nonfederal land in the United States: analysis of condition and trends. Review Draft. U.S. Department of Agriculture.

Van Loock, H.J.; Hafley, W.L.; and King, R.A. 1973. Estimation of agriculture-forestry transition matrices from aerial photographs. *Southern Journal of Agricultural Economics* 5(2):147–53.

Wall, B.R. 1981. Trends in commercial timberland area in the United States by states and ownership, 1952–77, with projections to 2030. USDA Forest Service General Technical Report WO–31. USDA Forest Service Washington Office, Washington, D.C.

13. Modeling Landscape Disturbance

Monica G. Turner and Virginia H. Dale

13.1 Landscape Heterogeneity and Disturbance

Almost all landscapes are affected by natural or anthropogenic disturbances (e.g., Mooney and Godron 1983; Risser et al. 1984; Forman and Godron 1986; Pickett and White 1985; Turner 1987). The spatial spread of disturbance (e.g., fire, pests, pathogens, exotic species, and wind throw) across a landscape is an important ecological process that is influenced by landscape pattern (e.g., Turner 1987). Disturbance can be defined generally as "any relatively discrete event in time that disrupts ecosystem, community, or population structure and changes resources, substrate availability, or the physical environment" (White and Pickett 1985). Although the general role of ecological disturbance has received considerable attention (e.g., White 1979; Barrett and Rosenberg 1981; Paine and Levin 1981; Mooney and Godron 1983; Menges and Loucks 1984; Sousa 1984; Pickett and White 1985; Rykiel 1985), only recently have studies attempted to predict the relationship between landscape pattern and disturbance (Romme and Knight 1982; Knight 1987; Odum et al. 1987; Remillard et al. 1987; Turner and Bratton 1987; Franklin and Forman 1987; Foster 1988; Rykiel et al. 1988; Baker 1989a; Turner et al. 1989).

Disturbances operate in a heterogeneous manner in the landscape. Gradients of disturbance frequency and severity are often controlled by physical or vegetational features. For example, a study of the disturbance history of old-growth forests in New England between 1905 and 1985 found that site susceptibility to frequent natural disturbances (e.g., windstorms, lightning, pathogens, and fire) was controlled by slope position and aspect (Foster 1988). Plant survival and recovery following the eruption of Mount St. Helens depended on (1) the nature of the volcanic disturbance the site experienced and (2) the aspect, time of snowmelt, and ongoing disturbances such as erosion (Adams et al. 1987). Some insect outbreaks are dependent on external conditions and tend to track environmental gradients in time or space (Berryman 1987).

Disturbances may spread from a local epicenter to cover large areas, and thus disturbance propagation may be enhanced or retarded by landscape heterogeneity. In forests of the Pacific Northwest, increased landscape heterogeneity due to "checkerboard" clear-cutting patterns enhances the susceptibility of old-growth forest to catastrophic windthrow (Franklin and Forman 1987). On a barrier island, the unusually close proximity of different habitats in the landscape appears to enhance the disturbance effects that result from introduced ungulate grazers in mature maritime forest (Turner and Bratton 1987). Landscape heterogeneity may also retard the spread of disturbance. In some coniferous forests, heterogeneity in the spatial patterns of forest by age class tends to retard the spread of fires (e.g., Givnish 1981). Other examples of disturbance types that may be impeded by landscape heterogeneity include pest outbreaks and erosional problems in agricultural landscapes, in which disturbance is generally enhanced by homogeneity.

Modeling and analytic methods are needed to better understand the relationship between landscape patterns and disturbance regimes. Disturbances operate at many scales simultaneously and play an important role in promoting the coexistence of species in the landscape (e.g., Watt 1947; Loucks 1970; Wiens 1977). Results from models that simulate disturbance and population persistence indicate that populations are stabilized at broad scales by spatial extent, heterogeneity, and dispersal (DeAngelis and Waterhouse 1987). However, the development of models that can project the spatial and temporal patterns of ecological communities across whole landscapes remains challenging.

In this chapter, we distinguish two views of landscape disturbance and examine the quantitative methods that are applied in each. First, we briefly consider disturbance effects that are spatially heterogeneous on the landscape (e.g., the patchy effects of a storm) and distinguish between modeling approaches that are spatially aggregated and those that are spatially explicit. Second, we consider the spatial spread of a disturbance (e.g., pests, pathogens, or fire) across a landscape. Two modeling approaches that we have developed, a probabilistic disturbance spread model and a coupled species model, are presented in detail. Finally, we discuss the new insights that have emerged from the use of these approaches, compare their applicability to different questions, and suggest future directions and research needs.

13.2 Models of the Heterogeneous Effects of Disturbance

Models of the heterogeneous effects of disturbance must explicitly recognize the variable nature of the landscape. There are at least two approaches to representing this heterogeneity. First, information on landscape variability can be spatially aggregated in the models. In the aggregated approach, the landscape frequently is divided into homogeneous categories that may be based upon vegetation type, temperature or precipitation regime, elevation, land use, or time since disturbance. A second approach is a spatially explicit model of the map of landscape heterogeneity in which the geographic location and change in the attributes of each site are recorded and followed. The advantage of the spatially explicit approach is that the juxtaposition of sites and cover types can be considered.

13.2.1 Spatially Aggregated Approach

Markov transition models are examples of the spatially aggregated approach to modeling the effects of disturbances on heterogeneous landscapes. These models are stochastic in the sense that the probability of transition between two states is given an explicit value. For example, the probability of a disturbance resetting one community type to another type is assigned. The transition probability usually relates to a particular time period, such as the successional change from one vegetation type to another for a particular time since disturbance (Anderson 1966, Horn 1975). The Markov model can be applied at a variety of spatial scales. Applications have ranged from a spatial unit of less than a hectare (e.g., Austin 1980; Austin and Belbin 1981) to a few hundred hectares (e.g., Williams et al. 1969; Hobbs 1983; Lippe et al. 1985). Multiyear sequences of remote sensing data can be used to provide estimates of transition frequencies between ecological states, and these estimates can serve as inputs for Markov models (Hall et al. 1988). Hall et al. (1988) applied this method to examine disturbance frequencies in wilderness and managed areas in northern Minnesota boreal forests. The steady-state solution of the model (e.g., percentage of the landscape in different states) is determined by the frequency of disturbance.

Baker (1989a) used a Markov chain approach to examine potential long-term trends in the vegetation mosaic in the Boundary Waters Canoe Area, Minnesota, based on fire-year maps published by Heinselman (1973). The model was developed for heuristic purposes and assumed that the fire regime that occurred during the 141-year study period (1727–1868) remained constant into the future. Transitions occurred between forest age classes. Baker used the model to compare the final distribution of age classes at five spatial scales and to evaluate the presence of a stable mosaic and the minimum area needed for a stable mosaic to be observed. Similarity between the simulated final distributions of age classes and the steady-state final distribution was generally low across all spatial scales.

Tabular models have also been used to track spatial changes following disturbances. These models predict temporal changes in state variables for particular regions and ecosystems within each region. Such models are particularly useful when massive amounts of data are available and must be organized. A global environmental problem to which tabular models have been applied is estimating annual changes in terrestrial ecosystems following various patterns of land use change (Moore et al. 1981; Houghton et al. 1983, 1987; Detwiler and Hall 1988).

These models track the yearly changes of carbon in terrestrial vegetation, soils, and products removed from the land (e.g., lumber). The rates of buildup and decay of carbon can vary with geographic region, ecosystem type, and prior land use (the disturbance in these models). Time lags are generally associated with the growth and decay of the net flux of carbon. For each ecosystem, the model describes the response of vegetation and soils to different kinds of disturbances. Although these responses are aggregated by ecosystem type, if the spatial distribution of ecosystem types is known, then the spatial arrangement of response curves can be mapped.

Patch dynamics models of intertidal communities represent recovery from disturbance in a general manner (Levin and Paine 1974; Paine and Levin 1981). In these models, the landscape is considered to contain populations of patches representing disturbance gaps; in the absence of the disturbance, the landscape would ultimately be dominated by a single cover type. Disturbance patches may vary in size, and the age-size distributions of the patches are modeled over time based on the interaction between the disturbance rate (patch birth rate) and the recolonization rate (patch death rate). This theory projects patterns of intertidal patches in mussel beds to within 5% of observed values (Paine and Levin 1981). This approach may be generally applicable to disturbances that create gaps or patches in a variety of landscapes. DeAngelis and Waterhouse (1987) suggest that these models of "transient dynamics" may have practical applications in describing the dynamics of pest species over short periods of time when there is no basis for inferring the existence of an equilibrium point.

Spatially aggregated models of disturbance are useful for several reasons. Transition models are relatively simple and have been used extensively. From a pragmatic viewpoint, the results of matrix models may not differ significantly from continuous models for some purposes (Baker 1989b). However, spatially aggregated models of disturbance have several limitations. Disturbance events are often not spatially independent but are influenced by the surrounding environment. A Markov or tabular model does not incorporate these neighborhood effects. The frequency of disturbance may also not be constant through time but may vary, so dynamic transition probabilities might be required. Finally, as with all transition probability models, it is difficult to relate the probabilities to ecological mechanisms or processes. Baker (1989b) provides a detailed review of these approaches, their relative merits, and potential modifications to the model structures.

13.2.2 Spatially Explicit Models

Models that have spatial representation of heterogeneous landscape units and/or the processes that change following disturbance are referred to as spatially explicit models. There is great variation as to the detail with which the processes and spatial resolution are modeled.

The transient dynamics of disturbance patches have been examined in the presence of environmental gradients that control species reestablishment (Green 1989). Green (1989) simulated the patterns associated with fire, seed dispersal, and the distributions of plants and resources by using approaches derived from cellular automata. A cellular automata model of a landscape consists of a fixed array in which each cell represents an area of the land surface and is assigned a state (Green 1983, Green et al. 1985). This grid cell approach is amenable to many sources of data, such as satellite imagery or raster data in a GIS, and enables processes that involve movement through space, such as fire, to be modeled (Green 1983). Fahrig has also simulated disturbances on a landscape represented in grid cell format, focusing particularly on the magnitude and recurrence interval of disturbance (see Chapter 17, this volume).

Green (1989) used a square grid (50 × 50) to simulate a forested landscape that was susceptible to fire. Based on the disturbance rules that were developed, the cells were assumed to represent an area ranging from 10 m to 100 m in diameter. Each cell was assumed to contain a single plant (at the smallest spatial scale) or a stand (at the larger scales). Fires were ignited at random locations and were elliptical in shape, with the frequency and size having Poisson and negative exponential distributions, respectively. Simulated fires created nonuniform patches into which species could disperse, creating a mosaic of different vegetation types that varied depending on whether dispersal was influenced by an environmental gradient. The actual spread of fire was not simulated, but the spatial patterns of fire and species dispersal were represented.

Another class of models that can be used to simulate landscape disturbance includes the individual-based forest succession models (Botkin et al. 1972; Shugart and West 1977). This approach assumes that small landscape units are unstable with respect to vegetation cover. These models incorporate ecological processes and track the birth, growth, and death of all trees in a small plot (≤1 ha). Tree growth is influenced by light availability, soil moisture, temperature, and competition. Although the spatial location of each tree within a given plot is not recorded, the spatial relationships between plots can be analyzed. Regional patterns of forest disturbance effects have been projected by extrapolating results from model plots within a homogenous area to a larger area and by considering a variety of homogeneous subunits (Shugart and Noble 1981; Dale and Gardner 1987). The effects of disturbance can be modeled by considering forest development with and without disturbance or with a variety of disturbance types. For example, Doyle (1981) modeled the effects of hurricane frequency and severity on the lower montane rain forest of Puerto Rico and confirmed the importance of hurricanes in maintaining a diverse composition and structure of the forest. By modeling the effects of fire, wind throws, insect outbreaks, and logging on Pacific Northwest forests, Dale et al. (1986) were able to compare the long-term effects on forest development of these disturbances.

A series of individual-based forest stand models can be arrayed in a grid to simulate the dynamics of a larger forested landscape. This approach was taken by Smith and Urban (1988) to simulate forest changes through time as a spatially interactive unit rather than as a series of independent plots. Smith and Urban did not study disturbance processes per se (all grid cells were initialized with no trees), but this approach is clearly amenable to studying gap-creating disturbances in a spatially-explicit manner.

Spatially-explicit models of disturbance effects and patch dynamics are useful for several reasons (DeAngelis and Waterhouse 1987). First, the effects of the size, shape, and arrangement of disturbance patches on species responses can be directly addressed. Second, the local dynamics of patches can be studied empirically in the field, and thus patch-oriented models may generate hypotheses that can be experimentally tested. Finally, the creation and extinction of patches at small scales can also be extended to larger areas and temporal scales.

13.3 Models of the Spread of Disturbance

The approaches we have reviewed thus far have dealt with the spatial heterogeneity created by disturbances but have not considered the spread or propagation of a disturbance across a heterogeneous landscape. Understanding disturbance spread is important both from a practical standpoint (e.g., how far and how fast might a pest species be expected to spread?) and to predict more generally the dynamics of the vegetation mosaic in disturbed landscapes.

13.3.1 Gradient Models

One of the first spatially explicit disturbance spread models was the gradient model developed by Kessell (1976, 1977, 1979). The objective of the model is to predict the daily spread (i.e., distance traveled) of a fire front rather than to predict the spatial patterns that result from a disturbance. The model uses six environmental gradients (elevation, topographic moisture, time since last burn, primary succession, drainage, and alpine wind-snow exposure) and four categories of disturbance (intensity of last burn, slide disturbances, hydric disturbance, and influences of heavy winter grazing by ungulates) to predict the daily spread of fire in Glacier National Park. Linkages are achieved between a terrestrial resource inventory system, gradient models of the vegetation and fuel, a micrometeorological model, and a mechanistic fire behavior model (Rothermel 1972). Gradient analysis is used to estimate the vegetation and fuels present throughout a large area, and the fire behavior model is then run in each spatial cell. Postfire succession is simulated deterministically based on habitat types (Kessell 1979) or life history traits of the species (Cattelino et al. 1979).

13.3.2 Epidemiological Models

A class of models that may be applicable to the spatial spread of a disturbance is derived from the theory of epidemiology (Bailey 1975; see also Gardner and O'Neill, Chapter 11). Epidemiology, which deals with the spread of a disease from individual to individual through a population, predicts factors such as the rate of spread and the proportion of the population that is affected by the disease. Models are available that trace the spread of an epidemic through space (e.g., Bartlett 1955; Radcliffe 1973). However, spatial pattern is not included in these models; each site is considered to be equally susceptible to the disease. In other words, the theory considers the spread of a single disturbance initiation through a homogeneous population (or landscape) in which it is possible for a single individual (or site) to infect all others. Thus, in its general form, the "general epidemic" theory can be considered to be a neutral model with respect to the possible effects of spatial heterogeneity in the landscape on the spread of a disturbance. Gardner and O'Neill (Chapter 11, this volume) describe neutral landscape models in general and present the mathematics of epidemiology as adapted to landscape studies.

13.3.3 A Probabilistic Model of the Spread of Disturbance

Ecological disturbance regimes can be described by a variety of characteristics, including spatial distribution, frequency, return interval, rotation period, predictability, area, intensity, severity, and synergism (e.g., White and Pickett 1985; Rykiel 1985). These parameters can be used to simulate landscape disturbance, and simulation experiments that use ranges of values for several of these characteristics can offer some general insights into the interactions between landscape pattern and disturbance spread. The spatial movement or spread of disturbance is amenable to study by using simple probabilistic models.

13.3.3.1 Model Description

One approach to developing simple models of disturbance spread is to adapt the methods of percolation theory (Stauffer 1985) to landscape studies (Gardner et al. 1987). Consider a landscape that contains only two types of habitat: habitat that *is* susceptible to a particular disturbance (e.g., pine forests susceptible to bark beetle infestations) and habitat that *is not* susceptible to the disturbance (e.g., hardwood forests and grasslands). The spatial arrangement of the disturbance-susceptible habitat can be randomly generated on a landscape map at probability p, and the propagation of disturbances that spread within the susceptible habitat may then be studied. The spread of disturbance in these random landscapes can be considered as a neutral model (see Turner et al. 1989).

For our simple model, we have incorporated three disturbance characteristics: frequency, intensity, and severity. Each characteristic is represented as a probability that can range from zero to one. Disturbance frequency (f) is defined as the probability that a new disturbance will be initiated in a grid cell of susceptible habitat during the time period represented by the simulation (e.g., the probability of lightning striking a hectare of pine forest during a particular storm event or time period). Disturbance intensity (i) is defined as the probability that the disturbance, once initiated, will spread to adjacent sites of the same habitat. Disturbance severity (h) is the probability of the site being damaged or altered by the disturbances such that it cannot be "revisited" by the disturbance during the simulation. When severity is equal to one, all disturbed sites are altered (e.g., vegetation in all disturbed cells is completely consumed by a pest); if severity is less than one, then some sites are not altered and can be disturbed again (e.g., although a pest passes through a site, sufficient vegetation remains such that the pest can revisit the site).

Thus, the general model simulates the spread and effects of disturbance as a function of four parameters: (1) the proportion of the landscape occupied by the disturbance-prone cover type, (2) disturbance intensity, (3) disturbance frequency, and (4) disturbance severity. The model is conceptualized in terms of discrete units of space (grid cells) and discrete units of time (time steps), in which we assume that a disturbance can move a distance of one grid cell per time step. In the general model, units of space and time are not specified. To simulate particular types of disturbances, parameters can be estimated for particular spatial units (e.g., m² or ha) and time steps (e.g., day, week, or year).

13.3.3.2 Simulation Algorithm

Our simulation program is written in FORTRAN 77 to run on VAX computers. Each of the four parameters must be assigned a value at the beginning of the simulation. First, two-dimensional $m \times m$ landscape maps are randomly generated to represent the disturbance-susceptible habitat at different values of p. This is done by generating random numbers between 0 and 1, then filling each grid cell in the map based on whether the random number is greater or less than p. For large maps (we used 100×100 arrays), the probability p, can be estimated as the proportion of a landscape occupied by a disturbance-susceptible habitat. For example, if we wish to simulate a landscape that is 65% pine forest, the p value for generating the random distribution of that forest would be 0.65. The remainder of the landscape $(1 - p)$ is considered to be unsuitable for the simulated disturbance. Thus, the landscape maps are composed of 10,000 cells, each of which is randomly designated as vulnerable or resistant to a disturbance.

Next, the pattern of the landscape is analyzed before a disturbance is simulated. The chapters in Section 2 of this volume present a variety of methods to quantify landscape pattern; any of these could be incorporated into a disturbance model. We have included the following measures: the number of habitat clusters, the size and shape (measured by the fractal dimension) of the largest cluster, and the amount of inner and outer edges in the landscape.

After the initial landscape is generated and its pattern quantified, disturbance is simulated as follows. Grid cells on the map are randomly disturbed at a given frequency without replacement until exactly fpm^2 disturbances are initiated. (This is because the number of grid cells of susceptible habitat is given by pm^2, and fpm^2 therefore gives the number of discrete disturbances that occur.) Sites that are disturbed can be altered based on the severity, h, such that they cannot be disturbed again during the simulation. Disturbances then spread randomly with an intensity, i, to nearest-neighbor cells of susceptible habitat (diagonal movement is not allowed). Of course, the disturbance cannot spread if none of the adjacent cells is of susceptible habitat. The process is continued in discrete time steps until the disturbance can no longer spread (i.e., it "goes out" or consumes all available habitat). The pattern of the disturbed landscape is then analyzed, and the number, sizes, and shapes of clusters of the disturbed habitat and of the remaining undisturbed habitat are summarized.

Any combinations of the four parameters p, f, i, and h can be simulated. By selecting a variety of plausible values for each parameter, simulation experiments (sensu Fahrig, Chapter 17, this volume) can be conducted with the model. We used a factorial design in which all possible combinations of a set of parameter values were run, and each particular combination was replicated ten times. Because this model is probabilistic, the replicate runs (Monte Carlo approach) allow the predictions to be summarized and evaluated statistically.

13.3.3.3. Simulation Experiments

We present a few illustrative simulation experiments (see Turner et al. 1989 for additional information). In one set of simulations, disturbance frequency was simulated at three levels, $f = 0.01$, 0.10, and 0.50 (representing low, moderate, and high levels). For each frequency, ten replicates were simulated for paired combinations of low, moderate, and high disturbance intensity ($i = 0.25$, 0.50, 0.75) and low and high p ($p = 0.4$ and 0.8). The severity of the disturbance was fixed at 1.0, (i.e., once disturbed, a site could not be disturbed again during the simulation). The proportion of habitat affected by disturbance was influenced by disturbance frequency but varied qualitatively for landscapes above and below the critical threshold, p_c, when the susceptible habitat could extend across the landscape (Fig. 13.1). This threshold, p_c, is approximately 0.6 for large, square random maps and is empirically derived from percolation theory. When $p = 0.4$, an increase in disturbance frequency causes a substantial increase in the proportion of habitat affected, even when the probability of spread is low (Fig. 13.1). When $p = 0.8$, increasing frequency increases the amount of habitat affected only when probability of spread is low. If probability of spread is sufficiently high (e.g., $i = 0.75$), more than 90% of the habitat can be affected by disturbance of very low frequency.

In another set of simulations, disturbance frequency was fixed at $f = 0.01$, and the interaction between disturbance intensity and severity was examined in fragmented ($p = 0.4$) and connected ($p = 0.8$) landscapes. Disturbance severity was simulated at four levels ($h = 0.25$, 0.5, 0.75, and 1.0), and disturbance intensity was simulated at three levels ($i = 0.25$, 0.5, and 0.75). Disturbance intensity, i, was more important than severity, h, in determining the proportion of the susceptible habitat that was disturbed during the simulation (Fig. 13.2). However, for a given value of i, increased disturbance severity resulted in a greater percentage of the habitat being disturbed. The effect was not always proportional to the increase in h. For example, when $p = 0.4$ and $i = 0.75$, an increase in h from 0.75 to 1.0 resulted in a twofold increase in the percent of available habitat affected by the disturbance (Fig. 13.2). In general, as disturbance severity decreased below 0.75, the reductions in the percent of habitat disturbed tended to become smaller.

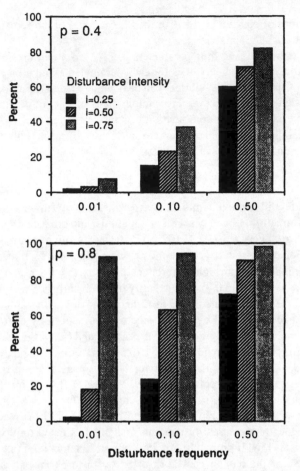

Figure 13.1. Mean ($n = 10$) percent of habitat disturbed as a function of both disturbance intensity and disturbance frequency for two initial probabilities (p) of occurrence of susceptible habitat (adapted from Turner et al. 1989).

13.3.3.4. New Insights from the Probabilistic Model

Simulation experiments in random landscapes suggest the existence of a threshold level of habitat connectivity that influences the spread of disturbance and its effects on the landscape. This insight may be of importance in resolving the question of whether disturbances are enhanced or inhibited by landscape heterogeneity because disturbance spread and effects depend on the interaction of several parameters. The distribution and spatial arrangement of the susceptible habitats help explain the differences. Habitats occupying less than the threshold, p_c, tend to be fragmented, with numerous small patches and low connectivity (Gardner et al. 1987). Disturbance propagation is constrained by this fragmented spatial pat-

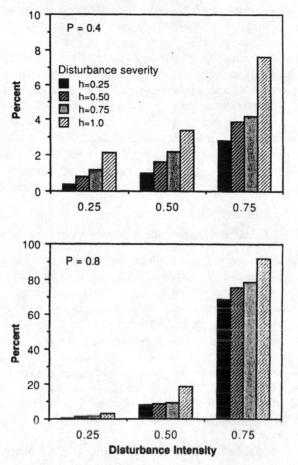

Figure 13.2. Mean ($n = 10$) percent of habitat disturbed as a function of disturbance intensity, i, and disturbance severity, h, for two initial probabilities of occurrence of susceptible habitat. Disturbance frequency, f, was held constant at 0.01.

tern, and the number and sizes of patches are not substantially affected by the probability of spread. Habitats occupying more than p_c tend to be highly connected, forming continuous patches (Gardner et al. 1987). Disturbance can spread through the landscape even when frequency is relatively low.

Results from the simulation experiments may have implications for landscape management and conservation. For example, the persistence of less dominant habitats ($p \leq p_c$) that are susceptible to disturbance appears to depend upon disturbance frequency (f). Thus, the long-term viability of remnant forest stands or other dispersed patches of rare communities (e.g., cedar barrens and granite outcrops) in a landscape may therefore depend on the number of disturbances rather than on their probability of spread. A locally intense disturbance may eliminate a single habitat patch but have little effect on the persistence of that habitat in the landscape. In contrast, many disturbances over a large region could reduce or eliminate the habitat.

Habitats that are common may be easily fragmented by disturbances of only low to moderate intensity. Intermediate levels of disturbance intensity and frequency created greater patchiness in landscapes that were dominated by a disturbance-prone habitat. The interaction among p, i, and f may thus affect the landscape in ways that are counterintuitive. Large tracts of forest, for example, may be fragmented by disturbances of relatively low frequency (e.g., Franklin and Forman 1987). Structural changes associated with this fragmentation, such as the increased number of edges, may influence the susceptibility of the landscape to other disturbances (e.g., wind throw) or cause changes in the distribution and abundance of species.

13.3.3.5. Use and Interpretation of the Probabilistic Model

Key strengths of the probabilistic modeling approach are its generality and the ability to produce testable hypotheses. The general model can be applied to numerous types of disturbance that spread spatially across heterogeneous landscapes. The expected ranges of parameter values that will give certain types of disturbance spread and effects can be predicted. Importantly, the hypotheses generated by the model (e.g., the existence of thresholds in connectivity) can be tested quantitatively by using data for real landscapes and actual disturbance regimes. However, the present simulations must be interpreted and evaluated at the scale of the whole landscape, rather than on a site-by-site basis. Because the model is probabilistic, it is not designed to predict whether a particular hectare or plot will be disturbed; rather, the objective of the model is to predict disturbance patterns over a large area.

Appropriate spatial and temporal scales must be selected for particular landscapes and disturbance regimes. Discrete units of space and time must be specified explicitly to parameterize the model for any real landscape and disturbance. For example, to simulate the spread of pine bark beetles through a forested landscape, decisions must be made regarding the time step (e.g., weekly or monthly) and the size of a cell (e.g., 1 ha or 10 ha). The parameters must be estimated in the appropriate units (e.g., the probability of spreading 10 ha within one week).

The general model could be modified as follows to simulate a particular disturbance regime and to reflect local processes more accurately. First, habitats are often differentially susceptible to disturbance (for example, susceptiblity to fire increases with age in some coniferous forests) rather than displaying the Boolean "either-or" susceptibility that we have represented. Classes of susceptibility could be introduced, although they would increase the complexity of the model. For example, forests could be differentiated by age classes and assigned unique probabilities of being disturbed. Gradients of variability in edaphic conditions, such as elevation or moisture, might also be used to influence susceptibility to disturbance. Second, disturbances frequently tend to move directionally (e.g., fire spreading with the prevailing winds). The tendency of a disturbance to spread in a particular direction could be incorporated by adding a bias factor, B, representing a probability of moving up, down, left, or right. In the random case, movement in all directions is equally probable, i.e., $B = 0.25$ (directional bias has been used by Gardner et al. [1989] to simulate the movement of organisms through a landscape). Third, ecological systems recover from disturbance within some duration of time. To simulate the long-term dynamics of a particular disturbance regime, a probability of recovery could be introduced, allowing sites to be disturbed again after a specified amount of time. Long-term recovery could also be simulated by using a transition matrix that assigned grid cells to new successional classes based on the elapsed time since the last disturbance.

13.3.4 Coupled Species Models

The spread of a pest or pathogen across a landscape represents a class of disturbance that involves the interaction of two or more species. One species causes the landscape disturbance by killing or reducing the growth or reproductive capacity of the other species. Such species frequently operate on different spatial and temporal scales. For example, the life cycle of phytophagous insects is generally on the order of days or weeks; yet outbreaks of these herbivores can drastically affect forest dynamics, which generally occur on the order of decades or longer. Models of the spread of such disturbances must take into account the different spatial and temporal dynamics of the two species. Coupled species models that operate at different spatial or temporal scales offer a means to consider the different scales as well as the different environmental factors affecting each species (Fig. 13.3).

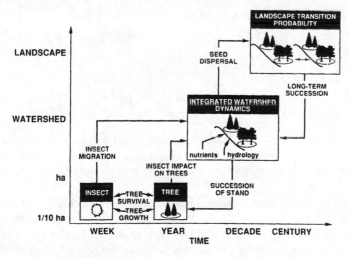

Figure 13.3. A diagram of coupled models that can address insect-tree interactions at a variety of spatial and temporal scales.

13.3.4.1 Model Description

Coupled models for tree and insect dynamics (Dale et al. 1990) have been developed by using Leslie matrix growth models (Leslie 1945, 1948; Usher 1966). The advantage of the Leslie model approach is that the dynamics of each of the life stages of the two species can be examined. Because the time scales of development for the trees and insects differ so greatly, it is necessary to use an approach that accommodates these differences. Two time scales were used simultaneously to model the tree-insect interactions: the trees were simulated with yearly time steps, while the insects were simulated with two-day time steps. The yearly time step for the tree corresponds to the annual growth rings of each tree. The two-day time step for the insects permits tracking instar development yet does not cause the simulation to be too slow (a daily time step would approximately double the simulation time).

The spatial dynamics of the coupled model are represented by implementing the tree and insect models for a series of patches or cells. Each cell can be separately parameterized to represent a spatial gradient of temperature or elevation in a linear or matrix framework. The cells interact by immigration and emigration of tree seeds and insect eggs and instars.

In the absence of insect predation, the forest patch is assumed to tend toward a stable steady state with a characteristic age structure. For simplicity, the tree population is described by the age-class vector $X(t) = (X_1, X_2, ..., X_5)$, where X_i is the number of trees in age class i and the age classes (or life stages) are one, two to five, six to thirty-five, thirty-six to sixty-five, and greater than sixty-six years old for a particular t (time in years). For each cell, the transitions of firs from one life stage to the next can be described by a Leslie matrix equation:

$$X(t + 1) = PX(t) + I_f, \qquad (13.1)$$

or

$$
\begin{bmatrix} X_1(t + 1) \\ X_2(t + 1) \\ X_3(t + 1) \\ X_4(t + 1) \\ X_5(t + 1) \end{bmatrix} =
\begin{bmatrix}
0 & 0 & 0 & f_4 & f_5 \\
p_{21} & p_{22} & 0 & 0 & 0 \\
0 & p_{32} & p_{33} & 0 & 0 \\
0 & 0 & p_{43} & p_{44} & 0 \\
0 & 0 & 0 & p_{54} & p_{55}
\end{bmatrix}
\begin{bmatrix} X_1(t) \\ X_2(t) \\ X_3(t) \\ X_4(t) \\ X_5(t) \end{bmatrix} +
\begin{bmatrix} I_s \\ 0 \\ 0 \\ 0 \\ 0 \end{bmatrix} \qquad (13.2)
$$

where $p_{i+1,i}$ is the fraction of trees transferring from one life stage to the next stage per year, p_{ii} is the fraction remaining in age class i (where $p_{ii} + p_{i+1,i} \leq 1$), I_s is input of seeds from other plots, and f_i is the number of seeds produced per mature tree in age class i.

In actual simulations, this condensed life-stage model (Eq. 13.2) is expanded into an annual age class model. To make such a simulation, the p_{ij}'s are factored into annual probabilities in a manner similar to that of Harrison and Michie (1985). The complete Leslie matrix is 66×66 in size, with identical transition probabilities within each age class. Thus a seedling that sprouts in model-year one does not reach maturity until model-year thirty-six, unlike a traditional Leslie matrix, in which individuals could reach maturity by year four.

The insect population is modeled for each spatial cell under the assumption that the insect can easily disperse within the 0.1-ha cell. The insect population is divided into five life stages (Y_i): egg, first instar (crawler), second instar, third instar, and adult. The transition from one life stage to the next can be described by a Leslie matrix equation:

$$\mathbf{Y}(d + 1) = \mathbf{Q}\mathbf{Y}(d) + \mathbf{I_a}, \tag{13.3}$$

or

$$\begin{bmatrix} Y_1(d + 1) \\ Y_2(d + 1) \\ Y_3(d + 1) \\ Y_4(d + 1) \\ Y_5(d + 1) \end{bmatrix} = \begin{bmatrix} q_{11} - E_{c1} & 0 & 0 & 0 & g_5 \\ q_{21} & q_{22} - E_{c2} & 0 & 0 & 0 \\ 0 & q_{32} & q_{33} & 0 & 0 \\ 0 & 0 & q_{43} & q_{44} & 0 \\ 0 & 0 & 0 & q_{54} & q_{55} \end{bmatrix} \begin{bmatrix} Y_1(d) \\ Y_2(d) \\ Y_3(d) \\ Y_4(d) \\ Y_5(d) \end{bmatrix} + \begin{bmatrix} I_{c1} \\ I_{c2} \\ 0 \\ 0 \\ 0 \end{bmatrix} \tag{13.4}$$

where I_{c1} and I_{c2} are the external inputs of passively transported eggs and the first instars, respectively, to a particular plot (c) from other cells (eggs dispersed per time step); $q_{j+1,j}$ and q_{jj} are the fraction of insect in life stage j that are transferred to life stage $j+1$ or that remain in life stage j, respectively, per time step (two days); E_{ci} is the fractional loss due to emigration from cell c; and g_5 is the number of eggs produced per mature adult. The ratio $q_{j+1,j}{:}q_{jj}$ indicates how rapidly insects move through their life stages. Because $(q_{j+1,j} + q_{jj})$ represents bidaily survivorship of insects starting the day in life stage j, the quantity

$$m_j = 1 - (q_{j + 1,j} + q_{jj}) \tag{13.5}$$

represents bidaily mortality in that life stage. The model includes parameters for the average fecundity of adult insects and the export of insect eggs by passive wind dispersal.

The progression through life stages of the insects is highly dependent on temperature (Amman 1968; Greenback 1970). We approximated the mean development periods of the egg ($D_{T,e}$) and nymph ($D_{T,n}$) as:

$$D_{T,e} = 5.0 + 100 \exp(-0.10(T-4)) \tag{13.6}$$

$$D_{T,n} = 10.0 + 50 \exp(-0.25(T-4)) \tag{13.7}$$

where T is temperature in °C and $D_{T,e}$ and $D_{T,n}$ are measured in days. The term *nymphs* refers to all three instar stages so that $D_{T,n}$ measures the development time from the end of the egg stage to the beginning of the adult stage. No development is allowed to occur for values of T less than four.

The ratio $q_{21}{:}q_{11}$, or the rate of development of eggs, is expressed in terms of T through

$$\frac{q_{21}}{q_{11}} = \frac{1}{D_{T,e}} \tag{13.8}$$

Mean daily temperature of a cell at time t is represented by a sine function:

$$T(d) = T_{ave} + T_{amp} * \sin \left[\frac{2\pi}{365} (d\text{-DAY}) + \frac{\pi}{2} \right] \qquad (13.9)$$

where T_{ave} is the average temperature, T_{amp} is the amplitude, and DAY is the time of the year with the maximum temperature. The actual temperature of a cell is $T(d)$ plus a factor that accounts for the variation of that cell temperature from the mean as a result of elevation.

Parasite-host interactions. The negative effects of the insect on the survival of trees (M_t) can be expressed as a cumulative effect, defined as the "effective insect days per tree" for year t, or

$$M_t = \sum_{d=1}^{183} \frac{[Y_2(d)+Y_3(d)+Y_4(d)+Y_5(d)] * f(T) \, W_d}{a_4 X_4(t) + a_5 X_5(t)} \qquad (13.10)$$

where $f(T)$ is 1 if the temperature exceeds 4°C but is 0 otherwise; W_d is two days; and a_i is the fraction of trees in each age class that are expected to be susceptible to insects $(a_i \leq 1)$. Note that only trees of age classes four and five can be infected. The number of trees dying is calculated by fraction of trees transferring from one life stage to the next for susceptible age classes by $(1 + \alpha M_t)$, where α is the aphid mortality constant that determines how many of the susceptible trees are actually infected, e.g., $p_{55} = p_{55,0}/(1 + \alpha M_t)$.

The insects are assumed to be limited by the availability of host trees and by density-dependent regulation. When susceptible trees are removed as a result of natural or insect-induced mortality, a proportional number of insects are removed. Because $a_4 X_4(t)$ and $a_5 X_5(t)$ are the numbers of infected trees at year t and $1.0 - p_{44} - p_{54}$ and $1.0 - p_{55}$ are the mortalities in each of these stages, the fraction (f_r) of the first instar removed as a result of tree mortality is

$$f_r = a_4(1.0 - p_{44} - p_{54}) + a_5(1 - p_{55}) \qquad (13.11)$$

Only the first instar is removed because the adjustment occurs on January 1, when the first instar is the only insect life stage that is living.

The carrying capacity of insects is implemented by letting the daily average fecundity of adults, g_5 (eggs per two days per insect), be a density-dependent function of the number of insects present:

$$g_5 = [g_{5,0}/(1 + kY_2 + kY_3 + kY_4)] \qquad (13.12)$$

For example, to determine the value of k that results in a total carrying capacity of 50,000 we solve Eq. 13.12 for k using the equilibrium value

$$Y_2 + Y_3 + Y_4 + Y_5 = 50,000 \qquad (13.13)$$

to get

$$k = 0.000002\left[\frac{g_{5,0}\,q_{21}\,q_{32}\,q_{43}\,q_{54}}{(1-q_{11})(1-q_{22})(1-q_{33})(1-q_{44})(1-q_{55})} -1.0\right] \quad (13.14)$$

The value k, above, is appropriate for one tree. To calculate an effective value of k (termed k') for the whole cell, we take roughly the number of trees older than sixty-six years (X_5) and the number of trees between thirty-six and sixty-five years old (X_4) that exist in the cell at the given time, and we assume that the younger trees have half the carrying capacity of the older trees. Then $k' = k/(X_5+0.5X_4)$.

The tree-insect model is used to simulate long-term effects of the insect infestation on spatial patterns of the forests. Accomplishment of this objective requires the simultaneous modeling of many patches in different climate regimes and the dispersal of insects and tree seeds between the patches. Emigration of eggs and first instars from a cell to the adjacent cell are calculated by assuming that the number of emigrants is a linear function of the number of eggs and first instars in the patch. For example, the emigration of eggs from patch c is $E_{c1}Y_1(d)$, where E_{c1} is the dispersal coefficient (eggs dispersed per two days per insect). The number of immigrants into patch c is computed by summing the number of emigrants from the neighboring patches and multiplying by the probability that an emigrant from one patch becomes an immigrant to another. Thus,

$$I_{c1} = V_{c,c-1}E_{c-1,1}Y_{c-1,1}(d) + V_{c,c+1}E_{c+1,1}Y_{c+1,1}(d) \quad (13.15)$$

where $V_{c,c-1}$ is the probability of an insect reaching patch c after leaving patch $c-1$, and the subscript c added to $Y_i(d)$ indicates the cell being referenced. Tree seed dispersal between patches is currently ignored; that is, $I_s = 0$ in Eq. 13.2. The model is flexible enough to simulate any desired configuration of a large number of patches.

13.3.4.2 Simulation Algorithm of the Coupled Species Model

The coupled tree-insect model was developed in Turbo-Pascal to run on a personal computer. A version of the model was also written in FORTRAN 77 to run on a Cray supercomputer so that numerous interacting patches could be simulated. Applying the model to specific landscapes requires obtaining data on the interaction of interest, including species transitions between age classes; inputs of seeds, eggs and instars from adjacent cells; effects of temperature on egg and nymph development or survival; insect fecundity; and temperature average and amplitude for the cells. The model projects the numbers of insects and trees by size class for each cell.

13.3.4.3 Simulation Experiment

The coupled tree-insect Leslie matrices have been applied to interactions between the exotic balsam woolly aphid (BWA) (*Adelges picae* Ratz.) and the endemic

Fraser fir (*Abies fraseri* [Purch] Poir.) in the southern Appalachians (Dale et al. 1990). The effects of BWA on the small scale of a forest stand are deterministic because BWAs kill all mature fir trees in about three to nine years after initial infestation (Amman and Speers 1965). Prediction of the broad-scale effects of BWA in a forest region requires an assessment of (1) the population dynamics of the BWA and fir, (2) the prevailing physical conditions that affect the spread of aphids, and (3) the subsequent pattern of fir mortality. The coupled models can project patterns of fir mortality and recovery on an elevation gradient using site-specific environmental conditions and ecological interactions.

The patchiness of BWA outbreaks in the fir forest (Eagar 1984) suggests that a relatively small spatial scale is appropriate for describing dynamics both within and between cells. Here, the cell size of the model is set at 0.1 ha, but it can be adjusted easily. Choice of the cell size influences the parameter value for the proportion of propagules dispersed between cells.

The simulated time from infestation of a stand to adult tree mortality and the simulated equilibrium population sizes are comparable to observed values (Table 13.1). The model projects the growth of BWA populations within each cell and the subsequent demise of both the fir and the BWA as BWAs kill their host organisms. When mature BWAs in a cell are sufficient in number, eggs and crawlers are dispersed to adjacent cells. After the BWAs kill all the mature trees within a cell, the BWAs cannot survive in that cell. The absence of BWA allows any remaining fir seedlings and saplings to grow into mature trees. If BWAs are in adjacent cells when the surviving trees mature, then BWAs can reinvade and kill the adult trees. These patterns of fir and BWA dynamics are similar to those

Table 13.1. Model Results Compared to Measured Values for Data that Were Not Used in the Development of the Model Parameters

Characteristic	Model Result[a]	Measured Result	
		Value	Reference
Years after initial BWA infestation during which fir died	7–10	3–9	Amman and Speers (1965)
Equilibrium fir population number per 0.1 ha:			
Seedlings	1984	1991	Witter and
Trees older than 6 years	67	80[b]	Ragenovich (1986)

[a] Run under temperature conditions typical for 1600 m as calculated from regression equations in Shanks (1954) by using a monthly temperature record.
[b] Trees taller than 244 cm were measured and are assumed here to be older than 6 years.

observed in the southern Appalachian spruce-fir forests (Eagar 1984). The inclusion of long-distance BWA dispersal in the model would undoubtedly increase the likelihood of infestation of a cell.

Model simulations were run for a series of six cells intended to represent a transect on an elevation gradient from 1600 to 2000 m. For all cases the BWA eggs and the first instars were allowed to disperse directly between adjacent cells. The simulation experiment was used to examine two hypotheses about the observed pattern of fir mortality, which occurs first at lower and middle elevations (1300 to 1700 m) (Eagar 1978). The hypotheses are (1) the fir mortality pattern results from dispersal of BWA between mountains versus (2) the pattern results from temperature limitations on aphid development. The first hypothesis maintains that BWAs are more likely to be dispersed from a mountain top to trees at lower elevations on the adjacent mountain because BWAs are transported by winds that generally increase in velocity as they rise toward the summit and produce eddies on the leeward size in the vicinity of the lower elevation range of the fir (Eagar 1984). Wind velocities are too high for successful BWA deposition on the windward side, but they are mild enough on the leeward side. The gradual movement of BWAs up the slope is explained by dispersal of BWAs between adjacent stands. The second hypothesis is that BWAs are dispersed to all locations at the same time, but temperature acts to slow down the development rate of the BWA and the subsequent mortality of the fir at high elevations. The simulations project the effect on fir survival (1) when the BWA is introduced to the lowest elevation cell, with all cells having identical temperature conditions, and (2) when the BWA is introduced to all cells, with the cells having a temperature gradient. In the second case, the temperature of the cells decreased linearly as elevation increased, falling from 1.2°C above the mean January temperature for the lowest elevation cell to 0.8°C below the mean for the highest elevation cell.

The tests of the hypotheses that dispersal and temperature cause fir mortality to occur first on the lower slopes suggest that temperature is a major cause of the observed pattern (Fig. 13.4). In both tests, mature firs died earlier in the lowest elevation plot and later in the highest cell. The difference between the time of fir death in the low elevation plot and that in the high elevation plot was greater in tests of the temperature hypothesis. The final size of the simulated fir population in the highest elevation cell did not depend on whether the BWAs were introduced to the lowest cell or to cells at all elevations on the temperature gradient. The simulations suggest that both BWA dispersal and temperature constraints on BWA development influence the long-term pattern of fir mortality and recovery but that temperature has the more pronounced effect when only between-plot dispersal is included.

13.3.4.4 New Insights from the Coupled Species Model

Considering causes of fir distribution on a broad spatial scale requires reevaluation of the significance of physical and biological factors in the formation of the resulting patterns. In general, physical gradients are thought to be the major

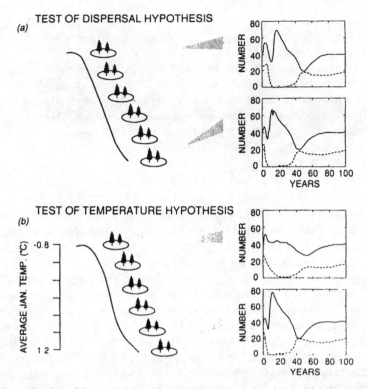

Figure 13.4. Number of fir trees by age class predicted over time for the highest and lowest elevation cells when (a) BWAs were introduced only to the lowest elevation plot and there were no temperature differences between the plots, and (b) BWAs were introduced to all plots and the plots were on a temperature gradient. The age classes depicted are trees between 5 and 35 years old (–) and trees older than 36 years (---).

determinants of pattern on the large scale, whereas biological factors may be more important at a smaller scale (Meentenmeyer and Box 1987). Whittaker (1956) has demonstrated that the general vegetation pattern in the Great Smoky Mountains is due to the physical gradients of moisture and topography (a complex gradient combining the effects of changes in precipitation, temperature, and insolation). This study builds upon the work of Whittaker by examining how a biotic disturbance can combine with physical gradients to affect vegetation pattern.

The model results show that patterns of fir mortality and spread of the BWA are strongly influenced by the relative density of fir and the prevailing temperature. Other computer simulation studies have concentrated more on the rate of dispersal have shown that persistence of both the predator and prey increased with (1) the prey's ability to disperse, (2) the predator's decreased ability to disperse (as long as the predators dispersed at some minimum level to keep ahead of exhausting their food supply), and (3) the total number of spatial cells (Caswell 1978; Hastings 1977). For the BWA-fir interaction there are a large number of cells possible over the extent of the southern Appalachians. The environmental diversity of the southern Appalachians could be simulated with the Leslie model by simulating temperature gradients of longitudinal transects.

The use of coupled tree and insect models that operate at different temporal scales has been effective in elucidating major determinants of Fraser fir distributions in the southern Appalachians. Analysis of broad-scale patterns should consider life-history characteristics of the major species as well as prevailing environmental conditions. The potential importance of small-scale interactions to broad-scale patterns has been demonstrated with the Leslie model. The model suggests that temperature can act at a small spatial and temporal scale to slow down the rate of development or to kill the BWA and thus indirectly affect fir distribution. The broad-scale implications of those results would be that temperature can act at a large resolution to limit the distribution of BWAs and thus affect the pattern of survival of fir. Therefore, predicted climate changes that would result from increases in atmospheric concentrations of CO_2 could have an indirect effect on the distribution of fir by altering the ranges where BWAs can survive.

13.4 Discussion

Each of the modeling approaches used to simulate the heterogeneous effects of disturbance or the spatial spread of disturbance across a landscape has advantages, limitations, and appropriate uses (Table 13.2). Spatially aggregated models of disturbance effects are applicable at many spatial and temporal scales and can be developed for a variety of landscape and disturbance types (Table 13.2). However, these models do not incorporate mechanisms, feedbacks, or spatial neighborhood effects. Furthermore, these models frequently imply the existence of an equilibrium state that may not exist. Spatially explicit models of disturbance effects may include more process representations and provide a means to follow the temporal changes in disturbance-created gaps or patches. These models may simulate the influence of environmental gradients on disturbance processes or the change in the size and age of disturbance patches.

Disturbance spread models are also spatially explicit, but they incorporate the spatial movement of a particular (or general) disturbance (Table 13.2). These models are particularly useful for exploring the effect of particular landscape patterns on the rate and extent of disturbance propagation. Numerical simulations are generally required in spread models because most approaches do not have an analytical solution. Models derived from epidemiology can be solved analytically, but only if the explicit pattern in the landscape is ignored. New approaches that use probabilistic models or coupled species models offer a means to address disturbances that occur at various scales.

The limitations of landscape disturbance models indicate that the relationship between landscape heterogeneity and disturbance requires additional research. The existing diversity of modeling approaches demonstrates that disturbance spread and effects can be modeled in a variety of ways. The most appropriate approach for a particular situation depends on the type of disturbance, degree of landscape heterogeneity, availability of data, and the desire to include mechanisms

Table 13.2. Summary of Approaches to Modeling Landscape Disturbances

Approach	Applications	Advantages	Limitations
		Heterogeneous Effects of Disturbance	
Spatially Aggregated Models			
Markov chain	Probabilistic disturbance events and subsequent species transitions.	Applicable at multiple scales and for many types of disturbances, organisms, and ecosystems.	Steady-state solution; no feedbacks or spatial neighborhood effects; many landscape changes not truly Markovian.
Tabular	Deterministic net changes in state variables in a region	Applicable at many scales and has been used globally; hierarchical data structure allows user to select appropriate level of detail.	No mechanisms, feedbacks, or neighborhood effects.
Patch dynamics	Disturbances that create gaps in homogeneous systems	Follows patch sizes and ages; general, applicable to many types of disturbance; has analytical solution; remote sensing and GIS data could be used.	Limited mechanistic detail (empirically determined patch birth and death rates); no interaction between patches.
Spatially Explicit Models			
Forest succession	Forest responses along environmental gradients	Processes represented; can be extrapolated to broader scales; already implemented with many types of disturbance.	Requires detailed information about species and sites to parameterize; lack of interaction between cells.

Cellular automata	General patch-creating disturbances	Has interaction between cells; includes disturbance, recovery, and environmental gradients; can be scaled for individual organisms or plots; includes size and shape of disturbance patches; can be used with remote sensing and GIS data.	Lack of mechanism (patch size determined empirically).

Spread of Disturbance

Gradient modeling	Fire spread; disturbances that vary along environmental gradient.	Uses mechanistic model, recovery and spatial data; interaction between cells; remote sensing GIS data.	Requires much detailed data in advance; patch shapes not addressed.
Epidemiology	Spread through homogeneous landscapes	Analytical solution; predicts velocity of spread.	Assumes homogeneous landscape, difficult to adapt to patterned landscape; steady-state solution.
Probabilistic spread models	Spread of generic disturbance through patterned landscape	Can be scaled for individual organisms or plots; addresses patterned landscapes; can be run using grid-cell based data (e.g., satellite imagery, GIS data); processes can be incorporated.	Lack of mechanism and feedbacks; does not include differential susceptibility.
Coupled species models	Spread and effects of pests and pathogens; any interacting species.	Includes mechanisms and feedbacks and environmental effects; links species at different spatial and temporal scales; applicable to patterned landscapes.	Requires detailed information to parameterize model.

or feedbacks. The complexity of disturbance effects remains difficult to predict, although progress is being made in determining the factors that control disturbance effects. Recent advances with geographic information systems (GIS) and satellite imagery will undoubtedly enhance the development of spatial models of disturbance because measures of landscape pattern are used both to parameterize models and as comparison against model predictions.

Predicting the spread of disturbance through landscapes and the resultant spatial pattern is of considerable current interest. It is likely that the pattern of the landscape prior to disturbance will be important at some times and not at others. For example, landscape pattern may influence the spread of disturbance when the disturbance occurs at low or moderate intensity but not at high intensity. The role of disturbance in creating landscape patterns is likely to receive greater attention regarding projected changes in the global environment. Altered disturbance regimes may be the proximal cause of large-scale changes in the environment (Clark 1988; Franklin et al. 1990), and new communities that become established may differ significantly from the undisturbed communities (Cwynar 1987; Dunwiddie 1986).

13.5 Summary

Natural and anthropogenic disturbances are common to many landscapes. The effects of disturbances frequently vary spatially across a landscape, creating a heterogeneous vegetation mosaic. Some disturbances propagate spatially and may be enhanced or retarded by the pattern of the landscape. Modeling and analytic methods are being developed to better understand and to predict the relationship between landscape patterns and disturbance regimes.

Two categories of landscape disturbance models are examined. First, we review models in which the heterogeneous effects of landscape disturbances are simulated. The models are further classified as "spatially aggregated," in which homogenous subunits of the landscape are treated together, and as "spatially explicit," in which the geographic location and change in attributes of each site are included. Spatially aggregated models (e.g., Markov chain and tabular models) are useful for addressing net disturbance effects and changes in the proportion of a land area that remains in a particular disturbance category. These models can also be used to explore the potential of a steady-state mosaic and can easily address the effects of altering the frequency of disturbance. However, the spatially aggregated models are unable to incorporate neighborhood effects. Spatially explicit models of disturbance effects do include the neighborhood effects but have differing levels of mechanism. The dynamics of disturbance-created gaps can be modeled, and the implication of changing the frequency, size, and responses to disturbance can be simulated.

The second class of models that we consider simulate the spread of disturbance. The gradient modeling approach (Kessell 1979) was one of the first spread models, and it linked several different submodels. Epidemiology theory has been used extensively to study the spread of diseases through populations, and this theory can be modified to simulate landscape disturbances. However, epidemiology theory has difficulty incorporating the explicit effects of landscape pattern. A simple probabilistic spread model can be used to examine the frequency, intensity, and severity of disturbances in patterned landscapes and to generate hypotheses that can be tested with a variety of different disturbance regimes. However, to simulate specific disturbances or landscapes, a link to empirically measurable parameters or mechanisms must be incorporated. Coupled species models can be used to predict the spatial spread of a disturbance in which one species feeds on another and the interaction of the distribution of the host and the life history attributes of both species constrain the disturbance spread.

Each model category is suited to particular types of disturbances and spatial patterns. The mathematical construct of each model and its application determine the advantages and limitations of each approach. A summary of the approaches to modeling landscape disturbances indicates that more research is needed to elucidate the relationship between landscape heterogeneity and disturbance.

Acknowledgments

Critical comments on this manuscript from D. L. DeAngelis, R. H. Gardner, and J. B. Hyman are appreciated. This research was funded by the Oak Ridge National Laboratory (ORNL) Seed Money Program; the National Science Foundation's Ecosystem Studies Program under Interagency Agreement BSR–8315185; and the Ecological Research Division, Office of Health and Environmental Research, U.S. Department of Energy, under Contract No. DE–AC05–84OR21400 with Martin Marietta Energy Systems, Inc. Publication No. 3541 of the Environmental Sciences Division, ORNL.

References

Adams, A.B.; Dale, V.H.; Kruckeberg, A.R.; and Smith, E. 1987. Plant survival, growth form and regeneration following the May 18, 1980, eruption of Mount St. Helens, Washington. *Northwest Science* 61:160–70.

Amman, G.D. 1968. Effects of temperature and humidity on development and hatching of eggs of *Adelges piceae*. *Annals of the Entomological Society of America* 61:1606–11.

Amman, G.D., and Speers, C.F. 1965. Balsam woolly aphid in the southern Appalachians. *Journal of Forestry* 63:18–20.

Anderson, M.C. 1966. Ecological groupings of plants. *Nature* 212:54–56.

Austin, M.B. 1980. An exploratory analysis of grassland dynamics: An example of a lawn succession. *Vegetatio* 43:87–94.

Austin, M.B., and Belbin, L. 1981. An analysis of succession along an environmental gradient using data from a lawn. *Vegetatio* 46:19–30.

Bailey, N.T.J. 1975. *The Mathematical Theory of Infectious Diseases and Its Applications*. New York: Hafner Press.

Baker, W.L. 1989a. Landscape ecology and nature reserve design in the Boundary Waters Canoe Area, Minnesota. *Ecology* 70:23–25.

Baker, W.L. 1989b. A review of models of landscape change. *Landscape Ecology* 2:111–33.

Barrett, G.W., and Rosenberg, R., eds. 1981. *Stress Effects on Natural Ecosystems*. New York: John Wiley and Sons.

Bartlett, M.S. 1955. *Stochastic Processes*. Cambridge: At the University Press.

Berryman, A.A. 1987. The theory and classification of outbreaks. In *Insect Outbreaks*, eds. P. Barbosa and J.C. Schultz, pp. 3–30. New York: Academic Press.

Botkin, D.B.; Janak, J.F.; and Wallis, J.R. 1972. Some ecological consequences of a computer model of forest growth. *Journal of Ecology*. 60:849–72.

Caswell, H. 1978. Predator mediated coexistence: a nonequilibrium model. *American Naturalist* 112:127–54.

Cattelino, P.J.; Noble, I.R.; Slatyer, R.O.; and Kessell, S.R. 1979. Predicting the multiple pathways of plant succession. *Environmental Management* 3:41–50.

Clark, J.S. 1988. Effect of climate change on fire regimes in northwestern Minnesota. *Nature* 334:233–35.

Cwynar, L.C. 1987. Fire and the forest history of the North Cascade Range. *Ecology* 68:791–802.

Dale, V., and Gardner, R. 1987. Assessing regional impacts of growth declines using a forest succession model. *Journal of Environmental Management* 24:83–93.

Dale, V.H.; Gardner, R.H.; DeAngelis, D.L.; Eagar, C.C.; and Webb, J.W. Broad-scale effects of the balsam woolly aphid on the southern Appalachian spruce-fir forests. Submitted to *Canadian Journal of Forest Research*, in review.

Dale, V.; Hemstrom, M.; and Franklin, J. 1986. The long term effects of disturbances on forest succession on the Olympic Peninsula. *Canadian Journal of Forest Research* 16:56–57.

DeAngelis, D.L., and Waterhouse, J.C. 1987. Equilibrium and nonequilibrium concepts in ecological models. *Ecological Monographs*. 57:1–21.

Detwiler, R.P., and Hall, C.A.S. 1988. Tropical forests and the global carbon cycle. *Science* 239:42–47.

Doyle, T.W. 1981. The role of disturbance in the gap dynamics of a montane rain forest: an application of a tropical forest succession model. In *Forest Succession: Concepts and Application*, eds. D.C. West, H.H. Shugart, and D.B. Botkin, pp. 56–73. New York: Springer-Verlag.

Dunwiddie, P.W. 1986. A 6000-year record of forest history on Mount Rainier, Washington. *Ecology* 67:58–68.

Eagar, C.C. 1978. Distribution and characteristics of balsam woolly aphid infestations in the Great Smoky Mountains. Master's thesis, University of Tennessee, Knoxville.

Eagar, C.C. 1984. Review of the biology and ecology of the balsam woolly aphid in southern Appalachian spruce-fir forests. In *The Southern Appalachian Spruce-Fir Ecosystem: Its Biology and Threats*, ed. P.S. White, pp. 36–50. U.S. National Park Service Resource Manage Report SER–71.

Forman, R.T.T., and Godron, M. 1986. *Landscape Ecology*. New York: John Wiley and Sons.

Foster, D.R. 1988. Disturbance history, community organization and vegetation dynamics of the old-growth Pisgah Forest, southwestern New Hampshire, USA. *Journal of Ecology* 76:135–51.

Franklin, J.F., and Forman, R.T.T. 1987. Creating landscape patterns by forest cutting: ecological consequences and principles. *Landscape Ecology* 1:5–18.

Franklin, J.F.; Swanson, F.J.; Harmon, M.E.; Perry, D.A.; Spies, T.A.; Dale, V.H.; McKee, A.; Ferrell, W.K.; Gregory, S.V.; Lattin, J.D.; Schowalter, T.D.; and Larson, D. 1990. Effects of global climatic change on forests in northwestern North America. In *The Consequences of the Greenhouse Effect for Biological Diversity*. New Haven, Conn.: Yale University Press.

Gardner, R.H.; Milne, B.T.; Turner, M.G.; and O'Neill, R.V. 1987. Neutral models for the analysis of broad-scale landscape pattern. *Landscape Ecology* 1:19–28.

Gardner, R.H.; O'Neill, R.V.; Turner, M.G.; and Dale, V.H. 1989. Quantifying scale-dependent effects of animal movement with simple percolation models. *Landscape Ecology* 3/4:217–28.

Givnish, T.J. 1981. Serotiny, geography, and fire in the Pine Barrens of New Jersey. *Evolution* 35:101–23.

Green, D.G. 1983. Shapes of simulated fires in discrete fuels. *Ecological Modelling* 20:21–32.

Green, D.G. 1989. Simulated effects of fire, dispersal and spatial pattern on competition within forest mosaics. *Vegetatio* 82:139–53.

Green, D.G.; House, A.P.N.; and House, S.M. 1985. Simulating spatial patterns in forest ecosystems. *Mathematics and Computers in Simulation* 27:191–98.

Greenback, D.O. 1970. Climate and ecology of the balsam woolly aphid. *Canadian Entomologist* 102:546–78.

Hall, F.G.; Strebel, D.E.; and Sellers, P.J. 1988. Linking knowledge among spatial and temporal scales: Vegetation, atmosphere, climate and remote sensing. *Landscape Ecology* 2:3–22.

Harrison, T.P., and Michie, B.R. 1985. A generalized approach to the use of matrix growth models. *Forest Science* 31:850–56.

Hastings, A. 1977. Spatial heterogeneity and the stability of predator-prey systems. *Theoretical Population Biology* 12:37–48.

Heinselman, M.L. 1973. Fire in the virgin forests of the Boundary Waters Canoe Area, Minnesota. *Quaternary Research* 3:329–82.

Hobbs, R.J. 1983. Markov models in the study of post-fire succession in heathland communities. *Vegetatio* 56:17–30.

Horn, H.S. 1975. Markovian processes of forest succession. In *Ecology and Evolution of Communities*, eds. M.L. Cody and J.M. Diamond, pp. 196–211. Cambridge, Mass.: Belknap Press.

Houghton, R.A.; Boone, R.D.; Fruci, J.R.; Hobbie, J.E.; Melillo, J.M.; Palm, C.A.; Peterson, B.J.; Shaver, G.R.; and Woodwell, G.M. 1987. The flux of carbon from terrestrial ecosystems to the atmosphere in 1980 due to changes in and use: geographic distribution of the global flux. *Tellus* 39B:122–39.

Houghton, R.A.; Hobbie, J.E.; Melillo, J.M.; Moore, B.; Peterson, B.J.; Shaver, G.R.; and Woodwell, G.M. 1983. Changes in the carbon content of terrestrial biota and soils between 1860 and 1980: net release of CO_2 to the atmosphere. *Ecological Monographs* 53:235–62.

Kessell, S.R. 1976. Gradient modeling: a new approach to fire modeling and wilderness resource management. *Environmental Management* 1:39–48.

Kessell, S.R. 1977. Gradient modeling: a new approach to fire modeling and resource management. In *Ecosystem Modeling in Theory and Practice: An Introduction with Case Histories*, eds. C.A.S. Hall and J.W. Day, Jr., pp. 575–605. New York: Wiley.

Kessell, S.R. 1979. *Gradient Modeling, Resource and Fire Management*. New York: Springer-Verlag.

Knight, D.H. 1987. Parasites, lightning, and the vegetation mosaic in wilderness landscapes. *Landscape Heterogeneity and Disturbance*, ed. M.G. Turner, pp. 59–63. New York: Springer-Verlag.

Leslie, P.H. 1945. On the use of matrices in certain population mathematics. *Biometrika* 33:183–212.

Leslie, P.H. 1948. Some further notes on the use of matrices in population mathematics. *Biometrika* 35:213–45.

Levin, S.A., and Paine, R.T. 1974. Disturbance, patch formation, and community structure. *Proceedings of the National Academy of Sciences* (USA) 71:2744–47.

Lippe, E.; DeSmidt, J.T.; and Glenn-Lewin, D.C. 1985. Markov models and succession: a test from a heathland in the Netherlands. *Journal of Ecology* 73:775–91.

Loucks, O.L. 1970. Evolution of diversity, efficiency, and community stability. *American Zoologist* 10:17–25.

Meentenmeyer, V., and Box, E.O. 1987. Scale effects in landscape studies. In *Landscape*

Heterogeneity and Disturbance, ed. M.G. Turner, pp. 15–34. New York: Springer-Verlag.

Menges, E.S., and Loucks, O.L. 1984. Modeling a disease-caused patch disturbance: oak wilt in the midwestern United States. *Ecology* 65:487–98.

Mooney, H.A., and Godron, M. eds. 1983. *Disturbance and Ecosystems*. New York: Springer-Verlag.

Moore, B.; Boone, R.D.; Hobbie, J.E.; Houghton, R.A.; Melillo, J.M.; Peterson, B.J.; Shaver, G.R.; Vorosmarty, C.J.; and Woodwell, G.M. 1981. A simple model for analysis of the role of terrestrial ecosystems in the global carbon budget. In *Carbon Cycle Modeling*, SCOPE 16, ed. B. Bolin, pp. 365–85. New York: John Wiley and Sons.

Odum, W.E.; Smith, T.J., III; and Dolan, R. 1987. Suppression of natural disturbance: Long-term ecological change on the Outer Banks of North Carolina. In *Landscape Heterogeneity and Disturbance*, ed. M.G. Turner, pp. 123–36. New York: Springer-Verlag.

Paine, R.T., and Levin, S.A. 1981. Intertidal landscapes: disturbances and the dynamics of pattern. *Ecological Monographs* 51:145–78.

Pickett, S.T.A., and White, P.S., eds. 1985. *The Ecology of Natural Disturbance and Patch Dynamics*. New York: Academic Press.

Radcliff, J. 1973. The initial geographical spread of host-vector and carrier-borne epidemics. *Journal of Applied Probability* 10:703–17.

Remillard, M.M.; Gruendling, G.K.; and Bogucki, D.J. 1987. Disturbance by beaver (*Castor canadensis* Kuhl) and increased landscape heterogeneity. In *Landscape Heterogeneity and Disturbance*, ed. M.G. Turner, pp. 103–22. New York: Springer-Verlag.

Risser, P.G.; Karr, J.R.; and Forman, R.T.T. 1984. *Landscape ecology: directions and approaches*. Special Publication No. 2., Illinois Natural History Survey, Champaign.

Romme, W.H., and Knight, D.H. 1982. Landscape diversity: the concept applied to Yellowstone Park. *BioScience* 32:664–70.

Rothermel, R.C. 1972. A mathematical model for predicting fire spread in wildland fuels. USDA Forest Service Research Paper INT–115.

Rykiel, E.J., Jr. 1985. Towards a definition of ecological disturbance. *Australian Journal of Ecology* 10:361–65.

Rykiel, E.J.; Coulson, R.N.; Sharpe, P.J.H.; Allen, T.F.H.; and Flamm, R.O. 1988. Disturbance propagation by bark beetles as an episodic landscape phenomenon. *Landscape Ecology* 1:129–39.

Shanks, R.E. 1954. Climates of the Great Smoky Mountains. *Ecology* 35:354–61.

Shugart, H.H., and Noble, I.R. 1981. A computer model of succession and fire response of the high-altitude *Eucalyptus* forest of the Brindabella Range, Australian Capital Territory. *Australian Journal of Ecology* 6:149–64.

Shugart, H.H., and West, D.C. 1977. Development of an Appalachian deciduous forest succession model and its application to assessment of the impact of the chestnut blight. *Journal of Environmental Management* 5:161–79.

Smith, T.M., and Urban, D.L. 1988. Scale and resolution of forest structural pattern. *Vegetatio* 74:143–50.

Sousa, W.P. 1984. The role of disturbance in natural communities. *Annual Review of Ecology and Systematics* 15:353–91.

Stauffer, D. 1985. *Introduction to Percolation Theory*. London: Taylor and Francis.

Turner, M.G., ed. 1987. *Landscape Heterogeneity and Disturbance*. New York: Springer-Verlag.

Turner, M.G., and Bratton, S.P. 1987. Fire, grazing and the landscape heterogeneity of a Georgia barrier island. In *Landscape Heterogeneity and Disturbance*, ed. M.G. Turner, pp. 85–101. New York: Springer-Verlag.

Turner, M.G.; Gardner, R.H.; Dale, V.H.; and O'Neill, R.V. 1989. Predicting the spread of disturbance across heterogeneous landscapes. *Oikos* 55:121–29.

Usher, M.B. 1966. A matrix approach to the management of renewable resources with special reference to trees. *Journal of Applied Ecology* 3:333–67.

Watt, A.S. 1947. Pattern and process in the plant community. *Journal of Ecology* 35:1–12.

White, P.S. 1979. Pattern, process, and natural disturbance in vegetation. *Botanical Review* 45:229–99.

White, P.S., and Pickett, S.T.A. 1985. Natural disturbance and patch dynamics: an introduction. In *The Ecology of Natural Disturbance and Patch Dynamics*, eds. S.T.A. Pickett and P.S. White, pp. 3–13. New York: Academic Press.

Whittaker, R.H. 1956. The vegetation of the Great Smoky Mountains. *Ecological Monographs* 26:1–80.

Weins, J.A. 1977. On competition and variable environments. *American Scientist* 65:590–97.

Williams, W.T.; Lance, G.N.; Webb, L.J.; Tracey, J.G.; and Dale, M.B. 1969. Studies in the numerical analysis of complex rainforest communities: III. The analysis of successional data. *Journal of Ecology* 57:515–36.

Witter, J.A., and Ragenovich, I.R. 1986. Regeneration of Fraser fir at Mt. Mitchell, North Carolina after depredations by the balsam woolly adelgid. *Forest Science* 32:585–94.

14. A Spatially Explicit Model of Vegetation-Habitat Interactions on Barrier Islands

Edward B. Rastetter

14.1 Introduction

Few landscapes have stronger or more dynamic interactions among their biotic and abiotic components than do barrier islands and barrier spits.[1] These barriers of sand and gravel enclose shallow lagoons along many coasts and are continuously being sculpted by wind and waves (Rosen 1979; Fisher and Simpson 1979). Vegetation traps sand and reduces erosion, thereby helping to stabilize and shape the topography of the barriers. However, the types and location of the vegetation are themselves highly dependent upon topography. Herbaceous and woody species tend to grow on the back sides of dunes, where they are away from water-logged saline soils (Jones and Etherington 1971) and protected from strong, salt-laden winds off the ocean (Boyce 1954; Oosting 1945). Grasses tolerant of drier, more exposed conditions tend to dominate the dune crests. Other more halophilic and hydrophilic grasses dominate the marshes that fringe the lagoon.

Soil water plays a significant role in this interaction between vegetation and topography. The depth of the water table, and hence the availability of groundwater to plants, is highly correlated with both the width and height of the barrier (Bolyard et al. 1979). Groundwater salinity will depend upon the susceptibility of

---

[1]Godfrey (1976) uses *barrier beach* as an inclusive term for both barrier islands and spits. I will use *barrier island* to mean both islands and spits to avoid confusion over the meaning of *beach*.

a location to tidal inundation and is therefore also controlled by topography. Both the availability of water and the salinity of the soil are important factors determining the zonation of vegetation on barrier islands (Schneider 1984; Ranwell 1972; Onyekwelu 1972; Jones and Etherington 1971).

One of the most intriguing aspects of barrier islands is that many of the geomorphologic and vegetation processes occur on the same time scale. This is rare for terrestrial landscapes. In most terrestrial systems, the geomorphology generally serves as a very slowly changing restraint to which the vegetation must respond. Any reciprocal effect of the vegetation on the geomorphology is generally negligible on ecological time scales. On barrier islands, however, there is a two-way interaction that occurs on a time scale relevant to many ecological questions. Thus the zones of vegetation are continuously shifting in response to the changing topography, and the topography is continuously being shaped by the vegetation.

This interaction between vegetation and geomorphologic processes links the entire barrier-island landscape. Through its influence on topography, vegetation at one location is able to influence conditions elsewhere on the barrier. For example, sand trapped by vegetation at one location will not be available for deposition downwind. Various zones of vegetation, therefore, can interact with one another through their influence on geomorphology. Because of these interactions, the barrier-island landscape behaves as a system rather than merely a sequence of independent communities along a topographic gradient.

My purpose in writing this chapter is, first, to illustrate some of the dynamic spatial interactions that I believe define the landscape concept and distinguish it from the concepts of ecosystem and community, and, second, to demonstrate how our *perception* of the relationships among species (e.g., competition, mutualism, commensalism, etc.) changes when these spatial interactions are considered. Barrier islands are ideal systems for my purpose because of the tight two-way interactions between vegetation and geomorphologic processes, because of the spatial nature of these interactions, and because of the time scale at which these interactions occur. I have therefore developed a simple model of a barrier-island landscape that simulates the interactions among four hypothetical plant species and uses very simple representations of groundwater and geomorphologic processes. Given the appropriate data and measured at the appropriate scale, the model could easily be modified to incorporate a larger species list and more rigorous treatments of groundwater and geomorphology. The added complexity of such a model, however, would not contribute significantly to the purely heuristic purposes of this chapter and would only obscure the points I wish to make. The modeling procedure itself is presented in some detail in hopes that others will extend its application beyond these heuristic purposes and that the concepts will prove useful for modeling other landscapes.

14.2 Model Structure

ISLAND is a model designed to simulate annual changes in vegetation, geomorphology, water table depth, and average groundwater salinity on a cross-sectional transect, running from ocean to lagoon, on a barrier island. The model is composed of three submodels: (1) a vegetation submodel that simulates the development of a four-species plant community consisting of two grasses, an annual, and a perennial shrub; (2) a geomorphology submodel that simulates the redistribution of sand by water and wind; and (3) a groundwater submodel that simulates the average depth of the water table and the salinity of the groundwater. Information calculated in each of the submodels is reevaluated and made available to the other submodels after each time iteration. The model works on a one-year time step and acts on 5-m sections along the transect. It is assumed that each 5-m section is homogeneous with respect to vegetation, height above sea level, water table depth, and groundwater salinity.

14.2.1 Vegetation Submodel

Most vegetation stand models are designed to simulate the development of individual trees on small forest plots of about the size of a dominant canopy tree (e.g., FORTNITE, Aber and Melillo 1982; FORET, Shugart and West 1977; JABOWA, Botkin et al. 1972; see Shugart 1984 for review). High-speed supercomputers have allowed some researchers to adapt these individual-based models to examine spatial interactions in forested landscapes (e.g., Smith and Urban 1988). However, barrier island vegetation is composed largely of grasses that are often found in densities of thousands of shoots per square meter. To simulate every shoot on a meter-wide transect running across even a small barrier island by using an approach similar to the forest stand approach would be prohibitively time consuming. Therefore, a technique had to be devised to reduce computational demand.

The algorithm used in the ISLAND model is based upon a probabilistic depiction of the development of each individual plant through a discrete series of life stages or states (e.g., seed, new sprout, and mature shoot). As the plant grows, it progresses from one state to another with a certain transition probability. This probability is calculated from the environmental conditions where the plant is growing. Thus, for example, the probability that a seed will sprout, die, or remain dormant might depend upon the salinity of the soil, the availability of water, and several other environmental factors.

The dependence of the plant on specific environmental factors will, of course, change as it progresses through various life stages. A newly sprouted plant, for example, might have a higher probability of dying due to desiccation than a mature plant because the mature plant has a better root system for acquiring water. The transition probabilities will also differ among species. Some species will have lower probabilities of dying on exposed areas like the dune crests, and others will have competitive advantages in sheltered areas behind the dunes.

Such a network of state transitions where a system (in this case a plant) jumps probabilistically from one state to another is known mathematically as a Markov process. Markov processes have been widely used in ecology. For example, Horn (1975) and Usher (1979) used Markov models to examine species replacement during succession. Patch mosaics have also been examined through the use of

Markov models (Weinstein and Shugart 1983; Shugart et al. 1973; Waggoner and Stephens 1970). Turner (1987) used a Markov approach to examine land use changes in Georgia. Markov processes were also used by Olson et al. (1985) to describe uptake dynamics for multiple resources by individual plants. The widely used "Leslie matrix" approach to population modeling is based on Markovian principles (Leslie 1945, 1948; Caswell 1983). The approach developed below is virtually identical to a stochastic Leslie matrix approach in which the elements of the projection matrix vary with time in response to changes in both the biotic and abiotic environment.

In the model, it is assumed that the probability of shifting from one life state to another is only a function of the present state of the plant and the current conditions of the environment (i.e., plant development is a first-order Markov process). The major simplifying assumption of the model is that the transition probabilities for all individuals of a given species, in a given state, under the same environmental conditions, will be equal. It is also assumed that the environmental conditions are homogeneous across each 5-m segment along the length of the transect. Under these assumptions, the probability, $P\{n{:}m\}_{ij}$, that exactly n out of total of m individuals on a transect segment in life state i will shift to state j during a particular time interval can be calculated from a binomial distribution:

$$P\{n{:}m\}_{ij} = \binom{m}{n}(P_{ij})^n(1 - P_{ij})^{m-n} \tag{14.1}$$

where $\binom{m}{n} = m!/n!(m - n)!$ and P_{ij} is the probability that any one individual on the segment shifts from state i to state j during that time interval. For large values of m (> 20), the distribution of $P\{n{:}m\}_{ij}$ can be approximated by a normal distribution with mean, $m P_{ij}$, and variance, $m P_{ij} (1 - P_{ij})$.

The number of individuals making shifts from state i to various other states during a particular time interval was calculated sequentially. Thus, the number of individuals making the first transition (e.g., from state i to j) was calculated by selecting a random number from the binomial distribution (Eq. 14.1) or its normal approximation. Once the number of individuals (N) making the first transition had been determined, the value of m was decremented by N, and the number of individuals undergoing transitions to other states was calculated. The binomial distribution (Eq. 14.1.) can also be used to calculate these transitions; however, the individual probabilities (P_{ij}) have to be corrected to account for previously evaluated transitions. That is, once it has been determined that a plant does not make a transition to a particular state (e.g., from i to j), that will increase the probability that it will make the transition to other states (e.g., from i to k). This increase can be accounted for by calculating a conditional probability for the remaining transitions. The conditional probability, P_{ik*}, that a plant will make the transition from state i to state k given that it has not made any of the previously calculated transitions can be calculated as follows:

$$P_{ik*} = P_{ik}/(1 - \sum P_{ij}) \tag{14.2}$$

where the summation is done for the (unconditional) probabilities of all previously calculated transitions.

For example, consider 100 seeds that, under a particular set of environmental conditions, have unconditional probabilities of sprouting, dying, or remaining dormant of 0.2, 0.5, and 0.3, respectively. Because these are the only possible transitions the seeds can make, the probabilities sum to 1. Using 0.2 for P_{ij} in Eq. 14.1, the probability of any number of these seeds sprouting can be calculated. On average, 20 of the 100 seeds will sprout, but, because of the stochastic nature of the model, perhaps 25 of them actually make the transition. The remaining 75 can either die or remain dormant. The conditional probability that any one of the remaining 75 seeds will die is $0.625 = 0.5/(1 - 0.2)$. Thus, one would expect about 47 of the 75 seeds to die, but, again because of the stochastic nature of the model, perhaps only 45 actually die. The conditional probability that the remaining 30 will remain dormant is $1 = 0.3/[1 - (0.2 + 0.5)]$. Thus all 30 of the remaining seeds remain dormant, which, of course, is the only remaining choice, so P_{ik*} must equal 1.

Four hypothetical plant species were developed to illustrate the modeling procedure. Two were grasses, one was an annual, and one was a perennial:

Grasses. No distinction was made between grass shoots that sprouted from seeds and those that arose vegetatively. The grasses, therefore, had only two life stages: unsprouted propagules (i.e., seeds or vegetative buds) and adult plants (Fig. 14.1a). Both the seeds and vegetative propagules could either die or sprout into adult plants. Adult grasses could either remain viable adults or die.

Annual. The annual also had two life stages: seeds and adults (Fig. 14.1b). The seeds could either die or sprout. The adult died after one year regardless.

Perennial. The perennial progressed through three life stages: seed, juvenile, and adult (Fig. 14.1c). The seeds could either die or sprout into the juvenile. The juvenile could either die or mature into an adult. The adults could either die or remain viable adults.

Clearly, more complex life histories could be developed if warranted by the characteristics of the particular species being simulated (e.g., including dormant and active or reproductive and nonreproductive stages). However, the current level of complexity is sufficient for the heuristic purposes of this chapter and to illustrate the modeling approach.

The transition probabilities were calculated based on (1) the depth of the water table (as an index of soil moisture), (2) the salinity of the groundwater, (3) exposure to salt-laden winds off the ocean, and (4) competition with other plants on the same transect segment. For each of the species, functions were constructed relating the probability of survival to each of the four factors (Fig. 14.2; Tables 14.1, 14.2). These functions took on values between 0 and 1 and were used as multipliers on a basic survival probability.

The soil-moisture functions were bell shaped (Fig. 14.2) and were calculated as follows:

$$B_i = \exp\{-(w_i[T(x) - T_i])^2\} \qquad (14.3)$$

where B_i is the multiplier, $T(x)$ is the depth of the water table below the ground

GRASSES:

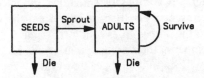

ANNUAL:

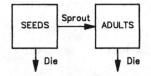

PERENNIAL:

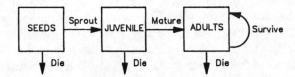

Figure 14.1. Life-stage development schemes for simulated species. Individual plants progressed through life stages on a yearly time step as a first-order Markov process with the probabilities for each transition (i.e., sprouting, maturing, surviving, dying) set by environmental conditions.

surface, T_i is the optimum depth of the water table for that species, and w_i is a parameter controlling the tolerance limits of that species.

The salinity and wind exposure functions were sigmoid (Fig. 14.2) and calculated as follows:

$$S_i = 1/(1 + \exp\{r_i[U_i - U]\}) \qquad (14.4)$$

where S_i is the multiplier, U is either the salinity or the height of the wind shadow (see below), U_i is the value of U where the multiplier is 0.5, and r_i is a parameter controlling the slope of dS_i/dU. The larger the absolute value of r_i, the steeper the slope. If r_i is positive, S_i is an increasing function of U; if r_i is negative, S_i is a decreasing function of U.

The competition function was handled in a way analogous to that used in Lotka-Volterra models. For any particular set of environmental conditions, each species could attain a certain carrying capacity in the absence of competition. In a Lotka-Volterra model, this carrying capacity is explicitly represented by the parameter K, which is either held constant or is made a function of the abiotic properties of the environment. In the ISLAND model, the carrying capacity is

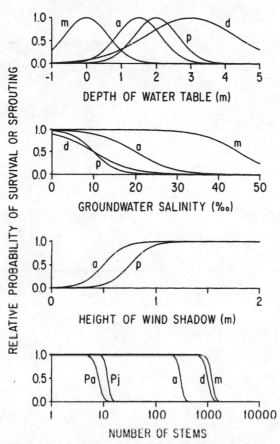

Figure 14.2. Relative probabilities of sprouting or surviving for four simulated species based on four environmental conditions. These relative probabilities were used as multipliers on a basic (fixed) probability to calculate the actual probability for the transitions shown in Fig. 14.1. Graphs show the effect of soil moisture (as indicated by depth of water table), groundwater salinity, sheltering from oceanic winds (as indicated by the height of the wind shadow), and crowding. The relative probability of sprouting was equal to the relative probability of adult survival for the grasses. The relative probability of sprouting was equal to the relative probability of survival of juveniles (i.e., probability of maturing) for the perennial. Relative survival probabilities for juvenile and adult perennials were equal for the first three factors but differed for crowding. m = marsh grass, d = dune grass, a = annual, p = perennial, p_j = juvenile perennial, p_a = adult perennial.

represented implicitly by a decrease in the probability of survival of individual plants as their population density grows (Fig. 14.2). The presence of other species on the same transect segment would decrement this carrying capacity by an amount proportional to their abundance. Thus the proportionality constants, α_{ij}, represent the equivalent number of individuals of species i that a single individual of species j would displace from the carrying capacity of species i (Table 14.2).

Table 14.1. Parameters Describing Survival Probability Functions

Effect of Water Table Depth:

Species	T_i^b	W_i^d
Marsh grass	0.0	1.25
Dune grass	3.0	0.29
Annual	1.5	1.25
Perennial	2.0	1.25

Effect of Groundwater Salinity:

Species	U_i^c	r_i^d
Marsh grass	45	−0.22
Dune grass	10	−0.22
Annual	20	−0.22
Perennial	10	−0.37

Effect of Wind Shadow Height:

Species	U_i	r_i
Annual	0.50	10
Perennial	0.75	10

Effect of Species Density:

Species	U_i	r_i
Marsh grass	1200	−0.01
Dune grass	1000	−0.01
Annual	300	−0.04
Perennial		
Juvenile	12	−1.10
Adult	8	−1.10

[a] Shown in Figure 2 and used in Equations 14.3 and 14.4.
[b] T_i = optimal depth of the water table for each species (used in Eq. 3).
[c] U_i = value of groundwater salinity, wind shadow height, or species density where survival for each species was ½ of the survival under optimal conditions (used in Eq. 4).
[d] w_i and r_i = parameters determining the range of tolerance for each species.

Table 14.2. Competition Coefficients, α_{ij}

Species	(1)	(2)	(3)	(4)	(5)
Marsh grass (1)	1.0	1.0	3.3	400.0	400.0
Dune grass (2)	1.0	1.0	20.0	400.0	400.0
Annual (3)	0.3	0.4	1.0	24.0	48.0
Juvenile Perennial (4)	0.0	0.0	0.0	1.0	2.0
Adult Perennial (5)	0.0	0.0	0.0	0.1	1.0

Values indicate the number of individuals of the species along the left side of the table that are displaced by a single individual of the species along the top of the table.

Therefore, they are directly analogous to the competition coefficients in a Lotka-Volterra model. To account for individuals displaced by competition, the operand for the competition function (Fig. 14.2) was the equivalent density of individuals of species i (N_{i*}), rather than the actual density of individuals of species i (N_i):

$$N_{i*} = \Sigma \alpha_{ij} N_j \qquad (14.5)$$

where the summation is for all j.

Data were not available to parameterize the model adequately for a particular set of species. Therefore, parameters were selected to yield reasonable stem densities in various vegetation zones on the barrier. The grasses had stem densities on the order of 1000 stems per square meter, which is consistent with data reported by Bertness and Ellison (1987) for *Spartina alterniflora* (smooth cord grass), *S. patens* (meadow cord grass), *Distichlis spicata* (spike grass), and *Juncus gerardi* (black rush). The annual and perennial had stem densities on the order of 100 and 10 stems per square meter, respectively, and are based on casual observations on species like *Solidago spp.* (goldenrod), *Rosa rugosa* (beach rose), and *Rhus radicans* (poison ivy). The parameters describing the magnitude of the effect of moisture, salinity, wind, and competition on the four species are purely hypothetical and should not be interpreted as representing the tolerances of any particular species for these factors.

Seeds for all four species were assumed to be available at all locations throughout the simulation at a predetermined constant abundance. This eliminated the need to calculate the number of seeds recruited from elsewhere on the transect and the number of seeds that remained dormant the previous year. Moisture, salinity, wind, and competition were assumed to affect the successful germination of seeds in the same way as they affected survival of the adults.

The first grass species was designed to simulate a marsh grass and therefore had a high probability of sprouting and surviving in saturated, highly saline soils and was not affected by exposure to ocean winds. One thousand marsh grass seeds per square meter were available to sprout at all locations at all times. The probability that any individual seed sprouted was $P_{12} = 0.3W_m S_m O_m C_m$, where W_m, S_m, O_m, and C_m are the functions describing the effects on the marsh grass of water table depth, groundwater salinity, strength of oceanic winds, and competition, respectively (Fig. 14.2). The probability of survival of a mature marsh grass individual was $P_{22} = 0.98W_m S_m O_m C_m$.

The second grass was designated as a dune species tolerant of low water tables and unaffected by ocean winds but intolerant of saline groundwater. One thousand dune grass seeds per square meter were available to sprout at all locations at all times. The probability that any individual seed sprouted was $P_{12} = 0.5W_d S_d O_d C_d$, and the probability of survival of an individual dune grass was $P_{22} = 0.95W_d S_d O_d C_d$.

The annual was intermediate to the two grasses in its tolerance of both salt and water table depth but was intolerant of ocean winds. Two hundred seeds per square meter of the annual were available to sprout at all locations at all times. The probability that any individual seed sprouted was $P_{12} = 0.4W_aS_aO_aC_a$, and the probability of survival of an annual shoot was $P_{22} = 0.0$.

The perennial had similar tolerance to water table depth and ocean winds as the annual but was less tolerant of salt. Fifty seeds per square meters of the perennial were available to sprout at all locations at all times. The probability that any individual seed sprouted into a juvenile was $P_{12} = 0.03W_pS_pO_pC_{pj}$. Juvenile perennials either matured or died. The probability that they matured was $P_{23} = 0.9W_pS_pO_pC_{pj}$. The survival probability of an adult was $P_{33} = 0.99W_pS_pO_pC_{pa}$.

14.2.2 Groundwater Submodel

The purpose of the groundwater submodel was to predict the average depth of the water table below the soil surface and the average salinity of the groundwater. The objective was not to make predictions of the changes in these characteristics of the groundwater in response to individual storm events or tidal cycles but rather to predict the average annual characteristics of the groundwater to serve as indices to which the vegetation submodel could respond. It was assumed, therefore, that the groundwater was in dynamic equilibrium with infiltration due to rainfall and tidal inundation.   Assuming that all flow is along the axis of the transect and following Darcy's Law (Darcy 1856, p. 638), the rate of change of the height (h) of the water table above the base of the aquifer can be calculated as follows:

$$dh/dt = \{I(x) - k\, h\, \alpha^2/\alpha x^2\}\, \Delta x\, \Delta y \tag{14.6}$$

where $I(x)$ is the net rate of water infiltration (m/year) from the surface at position x along the transect, k is the hydraulic conductivity of the soil, Δx is the length of a transect segment (5 m), and Δy is the width of the transect (1 m). To solve this equation, two assumptions were made: dynamic equilibrium (i.e., $dh/dt = 0$) and constant hydraulic conductivity (i.e., k equal at all locations). Integrating the equation numerically over the width of the barrier island and forcing the water table to equal sea level at both sides of the barrier island, the height of the water table at all locations along the transect can be calculated.

The physical interpretation of the parameter defining the hydraulic conductivity, k, is difficult. The hydraulic conductivity is not likely to be constant either with depth in the soil or with location along the transect. Thus the value of k represents an average conductivity for the whole barrier. It was assigned a value of 316 m/year, which is at the low end of the range for sandy soils (10^{-2} to 10^{-3} cm/sec; Hillel 1980, p. 101). For finer-grained soils, the conductivity can be several orders of magnitude slower (10^{-4} to 10^{-7} cm/sec for clays). The assignment of such a high conductivity, therefore, resulted in unrealistic results in the marsh where soils are of a much finer texture. In particular, the high conductivity allowed rapid infiltration of saltwater due to tidal inundation and resulted in groundwater salinities in the marsh higher than would be observed.

A physical interpretation of the depth of the aquifer base parameter, D, is even more difficult. The freshwater lens under a barrier island is perched upon saltwater, not upon an impermeable aquifer base. Thus the aquifer base will tend to move up and down with hydrostatic pressure and is not likely to be flat. The value of this parameter, therefore, was selected to match the shape of the water table and does not correspond to what would actually be measured on a barrier island. That is, the parameter D has little physical significance and must be viewed purely as a calibrating parameter. It was assigned a value of –0.9 m to yield a water table height of about 0.5 m in the center of a 160-m-wide barrier with 1 m of annual rainfall, which is consistent with data reported by Bolyard et al. (1979).

The shape of the resulting water table is, of course, also dependent upon the rate of infiltration along the transect; that is, it will depend upon the form of the function $I(x)$ in Eq. 14.6. If $I(x)$ were constant, the shape of the water table would be parabolic. However, to incorporate the effect of tidal inundation into the model, $I(x)$ had the form:

$$I(x) = I_R + I_T[Z(x)] \tag{14.7}$$

where I_R is the rainfall rate (= 1 m/year) and is assumed to be constant across the barrier, and $I_T[Z(x)]$ is the rate of infiltration due to tidal inundation (m/year) and is a function of the height, Z, of the barrier above sea level at any location, x. $I_T[Z(x)]$ was approximated as

$$I_R[Z(x)] = 2 \arccos \{2 Z(x)/Z_* - 1\}/\pi \tag{14.8}$$

where Z_* is the high-water mark (= 2 m). This equation only applied for $Z(x)$ below the high-water mark. Elsewhere $I_R[Z(x)]$ was set to zero. $Z(x)$, and therefore $I(x)$, changed through time, as will be discussed below.

Salt entered the water table at a rate of $I_s(x) = 35 I_R[Z(x)] \Delta x \Delta y$ (g/year). Salt balances were calculated for each transect segment, starting with the segment with the highest water table and moving outward. The net change in salt content of the groundwater for any segment can be approximated by

$$dC_0/dt = I_s(x) + C_1 F_1 - C_0 F_0 \tag{14.9}$$

where C_0 is the concentration of salt in the segment (g/m³ of water), C_1 is the concentration of salt in the upstream segment, F_1 is the rate of water flow (m³/year) into the segment from upstream (= $k h \Delta y \Delta h/\Delta x$), and F_0 is the rate of water flow out of the segment downstream. Assuming equilibrium, $dC_0/dt = 0$, the concentrations could be calculated sequentially starting at the high point on the water table.

On the landward side of the barrier dunes, the calculated water table often intercepted the surface of the barrier. At the locations where the water table was higher than the barrier surface, the infiltration rate was decremented by an amount equivalent to the excess volume, and the water table was recalculated. The infiltration of salt was decreased by an amount equal to the concentration of salt in the groundwater at the location times the excess volume. If the water table still intercepted the surface water after the second calculation, it was simply lowered to the surface level without further iterations of the submodel.

14.2.3 Geomorphology Submodel

The major purposes of the geomorphology submodel were (1) to determine the erosion and deposition of the sand by the wind, (2) to calculate the deposition and redistribution of sand on the beach and berm by waves, and (3) to calculate siltation in the marsh. Only aeolian transport of sand along the axis of the transect was considered in the model. Thus the net longshore aeolian transport was assumed to be zero. Longshore transport of sand by waves, however, was used as a forcing function for the simulations described below. It was included in the model as a net deposition or removal of sand on the beach and berm in addition to any removal or deposition due to wind. The calculated siltation in the marsh is in addition to any sand deposited due to aeolian transport across the barrier. It is therefore assumed to be either biogenically derived in situ or transported into the marsh from outside the transect.

To calculate the erosion of sand by wind, the extent of the wind shadows downwind of dunes was calculated first. The wind shadows were assumed to have a parabolic profile, extending horizontally from the crest of the dune and gradually decreasing in height downwind. The equation used to determine the height (H) of the wind shadow above sea level at location x was

$$H(x) = Z(x_*) - c\,(x - x_*)^2 \qquad (14.10)$$

where x_* is the location where the wind last contacted the barrier surface, $Z(x_*)$ is the height of the barrier at x_*, and c is a curvature parameter determining the length of the shadow.

This equation was applied sequentially starting at either the seaward (for ocean winds) or landward end of the transect. x_* was set equal to the location of the shore and x iterated downwind. If any transect segment had a height greater than that of the wind shadow [i.e., $Z(x) > H(x)$], x_* was reset to that location. Only those transect segments where the wind shadow contacted the surface (i.e., those locations where x_* had to be reset) were exposed to wind erosion. Segments within an ocean wind shadow could be eroded by winds off the land and vice versa. The value of c for both ocean and land winds was set at 0.0004 m/m². The height of the wind shadows on the landward side of the dunes (i.e., due to ocean winds) was also used in the vegetation submodel to calculate survival probabilities (see above).

The net changes in height of the transect segments were calculated sequentially starting at the seaward end of the transect for the ocean winds and on the lagoon end of the transect for winds off the land. For those transect segments outside wind shadows, the rate of erosion decreased as the number of stems on the segment increased. The erosion rate was calculated as

$$E = E_0/(1 + \exp\{0.1N - 10\}) \qquad (14.11)$$

where E_0 is the maximum erosion rate, $N = N_m + N_d + 3\,N_a + 50\,N_p$; and N_m, N_d, N_a, and N_p are the number of marsh grass, dune grass, annual, and perennial (both age classes) shoots, respectively. Thus, a single stem of the annual was assumed to be three times as effective as a single stem of either of the grasses at protecting against wind erosion, and a single stem of the perennial was assumed to be fifty times as effective. In addition, because aeolian erosion is inhibited in wet soils (Rosen 1979), the wind could not erode the surface below the height of the water table.

Values of E_0 were set at 0.4 m/year for ocean winds and 0.3 m/year for land winds to simulate a condition where the ocean winds were stronger than the land winds. These values are consistent with observations by Rosen (1979), who reported vertical changes in topography of up to 10 cm in four months. His data, however, were collected in the summer, when winds are weaker. Consequently, these values may be low for annual estimates of sand movement.

Sand eroded from a transect segment was available for deposition downwind. To do this, a "sand load" variable was defined. At the upwind end of the transect, the sand load was set to zero. Sand eroded from each segment was added sequentially to the sand load. Ten percent of the net accumulation of sand from upwind was redeposited on each segment and therefore lost from the sand load. This value yielded geomorphologic dynamics consistent with Rosen's (1979) observations.

The dynamics of the beach and berm are controlled largely by factors not represented in the model, such as longshore transport of sand. The deposition or removal of sand on the beach was consequently one of the forcing functions for the model. This net deposition was added to the total volume of sand between 2 m below sea level and 1.5 m above sea level for all locations < 1.5 m in height on the ocean side of the barrier. This sand was then redistributed to form a berm 1.5 m above sea level that extends as far seaward as the volume of sand would allow.

Siltation in the marsh occurred only at locations that were below sea level and where the marsh grass was present. The rate of siltation was proportional to the depth at the location $[0 - Z(x)]$ and increased asymptotically with the number of marsh grass stems present:

$$\Delta Z(x) - 0.5 \ [0 - Z(x)] \ N_m(x)/[200 + N_m(x)] \tag{14.12}$$

where $N_m(x)$ is the number of marsh grass stems at location x.

The final step in the geomorphology submodel was to ensure that the barrier profile was not too steep. The maximum stable slope angle for loose, coarse particles is about 37° (Longwell et al. 1969). Periodically submerged slopes will have stable slope angle well below this, and slopes with cohesive components like fine roots and clays can maintain slightly higher angles. The barrier profile was corrected so that the difference in height between any two adjacent transect segments was no >2.5 m (26°) for those segments below the high-water mark (Z_*) and no > 4 m (38°) for those segments above the high-water mark. This was done iteratively. Starting at the seaward end of the transect, the height of each transect segment was compared to the height of the next landward segment. If the difference was greater than allowable, enough sand was moved from the higher segment to the lower segment so that the slope was equal to the critical slope (i.e., a difference in height of 2.5 or 4 m). If the profile had to be relaxed during the first pass through the transect, the procedure was repeated, starting again at the seaward end of the transect. This process was repeated until no further relaxation was required.

14.3 Simulations[2]

Three 100-year simulations were run with the model described above (Figs. 14.3, 14.4). Simulation 1 was used as a base line from which to compare the other two simulations. Simulations 2 and 3 were run to examine the responses of the landscape to changes in one of the major driving variables and to an internal disruption of the landscape system, respectively. In simulation 2, the rate of sand deposition on the ocean end of the transect was altered, and in simulation 3 a key species was removed from the landscape. These three simulations illustrate the interdependence of the four plant species and the nature of their interaction, as well as the role that the abiotic environment plays in that interaction.

During the first twenty years for all three simulations, the seaward face of the barriers underwent a period of net erosion at a rate of 10 m^3 of sand per meter of shoreline per year. This caused the seaward side of the barriers to migrate land-ward at a rate of about 1.75 m/year. For simulation 1, sand was then deposited on the seaward side of the barrier for the duration of the simulation at a rate of 4 m^3 of sand per meter of shoreline per year (solid line in Figs. 14.3, 14.4). For simulation 2, a massive, one-time deposition of 300 m^3 of sand per meter of shoreline was added to the seaward side of the barrier during year twenty; then sand was added at a rate of 4 m^3 m^{-1} year^{-1} for the duration of the simulation (dashed line, Fig. 14.3). For simulation 3, the net deposition sequence was identical to that in simulation 1, but all the dune grass was killed in year twenty and was not allowed to recolonize for the duration of the simulation (dashed line, Fig. 14.4).

Initially, all three barriers had a broad, well-developed dune about 6 m high (Fig. 14.5a). Because of the high hydraulic conductivity of the barrier, the water table domed to a height of only about 0.5 m above sea level. Saltwater intruded a short distance inward on the ocean side of the barrier, with the groundwater being fresh about 10 m in from the beach. On the lagoon side, however, where the topography was not as steep and the zone of tidal inundation was considerably larger, saltwater intruded much further in toward the center of the island. As

---

[2]The model described in the previous sections was coded in PASCAL (Borland International Inc., Turbo Pascal [TM] version 4.0) for IBM (TM) compatible personal computers. On a 20 MHz, 80386 personal computer (Dell Computer Corporation system 310) with a 80387 math-coprocessor, all three 100-year simulations described in this section can be compiled and run, back to back, in 3 minutes, 15 seconds.

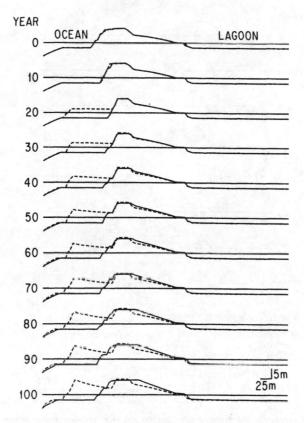

Figure 14.3. Time series of barrier profiles. Solid line shows the resulting profile for simulation 1, and the dashed line shows the resulting profile for simulation 2. The conditions under which the two simulations were run were identical except that a large berm was deposited on the seaward side of the barrier during year 20 in simulation 2.

mentioned above, this intrusion of saltwater on the lagoon side of the island is an artifact of a simplification in the groundwater model. A lower hydraulic conductivity in the marsh and a slower infiltration rate of saltwater would allow the freshwater lens to extend offshore.

A dynamic equilibrium was established between depositional and erosional processes on the dune and the survival of the dune grass. The dune grass protected the dune from erosion and allowed the dune to grow. However, as the dune grew, its crest became further removed from the water table. Once the crest of the dune was > 5 m above the water table, the abundance of dune grass became depressed (Fig. 14.5a). This left the crest exposed to erosion, and the dune could grow no higher. However, the sides of the dune were covered with abundant grass and continued to grow until they, too, reached the maximum height, leaving the dune with a flattened top (Fig. 14.5a). This may be an unreasonable result since dunes

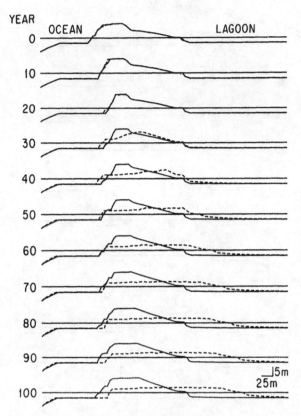

Figure 14.4. Time series of barrier profiles. Solid line shows the resulting profile for simulation 1, and the dashed line shows the resulting profile for simulation 3. The conditions under which the two simulations were run were identical except that all of the dune grass was killed during year 20 and not allowed to recolonize for the duration of simulation 2.

grow in the absence of vegetation as long as there is a source of sand. However, sand behind a stabilized barrier (e.g., a wall or the stabilized bank on the foredune) tends to accumulate until it reaches the height of the barrier and grows no further. With the additional stabilization of the sparse cover of grasses on the dune top, this flattened profile may not be as artificial as it first appears.

The patterns of vegetation were a response to a complex gradient of sheltering from wind, soil moisture, and salinity. Only the dune grass could grow on the dry, exposed dune. Immediately behind the dune, where conditions were relatively moist and sheltered, the annual and perennial excluded the dune grass. As conditions became wetter and saltier further from the dune, the annual increased in abundance and the perennial died out. Closer to the marsh, soils became even more waterlogged and saline, and the annual died out and was replaced by the marsh grass.

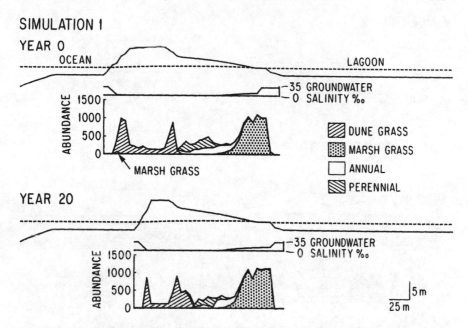

Figure 14.5. Barrier profile, water table, groundwater salinity, and zonation of vegetation for simulation 1 during years 0 and 20. Abundance of plants is given in stems per square meter. The abundance of the annual and perennial are amplifed 3 times and 50 times to reflect their sand-holding capacity relative to the grasses (see text). Thus the cumulative abundance at a location is indicative of how resistant that location is to erosion due to plant cover.

After twenty years of erosion, the beach face had migrated landward about 35 m (Fig. 14.5b). The dune became narrower, and the flats behind the dune accreted from 0 to 0.5 m of sand, with the majority of the deposition being near the dune. The marsh was virtually unchanged. The patterns of vegetation were similar to those at year 0 but more compressed. The zone of dune grass was narrower, reflecting the narrower dune. However, because the flats immediately behind the dune had accreted 0.5 m, they were drier. This allowed the dune grass to invade the higher reaches of the back-dune flats, thereby compressing the annual-perennial community toward the marsh.

The seaward edge of the barrier in simulation 1 migrated about 20 m seaward during the subsequent eighty years of deposition (0.25 m/year average). The dune broadened and the back-dune flats continued to accrete sand. The patterns of vegetation remained similar to the original patterns, except the range of the dune grass broadened with the dune, and the annual-perennial community was compressed more toward the marsh.

The large berm deposited in year twenty of simulation 2 was rapidly colonized by both the marsh and dune grasses. Ten years after the deposition event, the berm was completely covered with grasses and had begun to accrete sand on its seaward

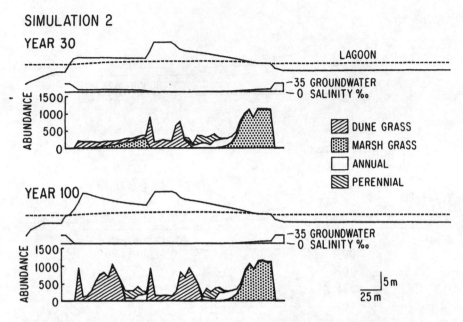

Figure 14.6. Barrier profile, water table, groundwater salinity, and zonation of vegetation for simulation 2 during years 30 and 100. A large berm was deposited on the seaward side of the barrier during year 20. Abundance of plants is given in stems per square meter. The abundances of the annual and perennial are amplified 3 times and 50 times to reflect their sand-holding capacity relative to the grasses (see text). Thus the cumulative abundance at a location is indicative of how resistant that location is to erosion due to plant cover.

edge (Fig. 14.6a). The dune, back-dune, and marsh communities remained the same as ten years earlier. Eighty years after the deposition event, the vegetation and geomorphology of the barrier, from the original dune landward, remained virtually unchanged (Fig. 14.6b). A second dune, however, had formed on the seaward end of the berm. This dune was also stabilized by the dune grass, and the area between the new and the original dunes was colonized by the annual and perennial.

After the dune grass was killed in simulation 3, the dune migrated landward (Fig. 14.4). There are two reasons for this. First, winds off the ocean were stronger and could erode more sand than the winds off the land (0.4 versus 0.3 m of sand per year maximum). Second, as long as a dune persisted, the annual and perennial could still grow on its landward side, thereby protecting that side of the dune from erosion (Fig. 14.7a). However, fifty years after the dune grass became extinct, the dune had blown away, the annual and perennial had died out, and only the marsh grass persisted on the barrier. Again, a dynamic equilibrium was established between depositional and erosional processes and the survival of the grass. However, because it was the hydrophilic marsh grass rather than the drought-tolerant dune grass stabilizing the barrier, the barrier height stabilized at only about 2 m

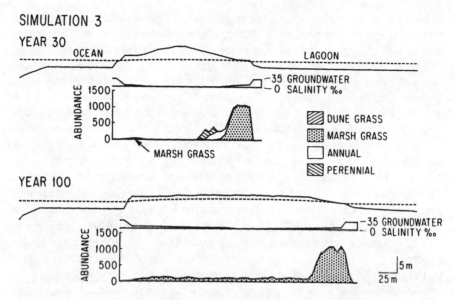

Figure 14.7. Barrier profile, water table, groundwater salinity, and zonation of vegetation for simulation 3 during years 30 and 100. All dune grass was killed during year 20 and not allowed to recolonize. Abundance of plants is given in stems per square meter. The abundances of the annual and perennial are amplifed 3 times and 50 times to reflect their sand-holding capacity relative to the grasses (see text). Thus the cumulative abundance at a location is indicative of how resistant that location is to erosion due to plant cover.

above sea level. Eighty years after the dune grass extinction, the barrier was little more than a sand bar with a sparse covering of marsh grass (Fig. 14.7b). Without the stabilization of the dune, sand was transported landward and deposited in the lagoon. By the end of the simulation, the barrier had encroached about 115 m into the lagoon.

14.4 Discussion

14.4.1 Landscape Dynamics

The characteristic that distinguishes landscapes from other units of ecological organization is the spatial interaction among the landscape elements. This interaction ties the landscape together as a system, with feedbacks and mutual dependencies among the various elements on the landscape. Thus a landscape is not merely a large, heterogeneous plot of land; the landscape elements must also exert some ecologically meaningful influence upon one another. Without this interaction, even a very heterogeneous terrain would be only a collection of independent components, each of which could be fully understood without reference to the rest of the landscape.

The model presented here illustrates the type of interaction that is characteristic of landscapes. In the simulations, the vegetation segregated into zones based upon their tolerance for ocean winds, waterlogging, and soil-water salinity. The variation in these factors across the barrier jointly characterized a complex environmental gradient to which the community had to adjust. This complex gradient changed very little with time in simulation 1, where the barrier was near equilibrium. The dynamic nature of the landscape interactions, therefore, was not obvious. On the other hand, the dynamic response of the barrier landscape to a major perturbation is clear in simulation 2. The perturbation was the deposition of a large berm in year twenty. The new berm was quickly colonized by grasses that protected it from erosion. Sand accumulated, and a new dune formed on the seaward edge of the berm. Once formed, the dune was stabilized by the dune grass. This new dune provided shelter from ocean winds, allowing the annual and perennial to colonize the area between it and the original dune. Thus, through a chain of interactions between biotic and abiotic processes, the characteristic zonation of vegetation was established on the berm. Through these interactions, the vegetation itself helped establish the complex gradient of environmental factors that regulate the zonation of the plant community.

Simulation 3 illustrates the response of a landscape to the removal of one of its key components—in this case, the dune grass. When the dune grass was removed, the dune could not persist. Without the shelter of the dune, the annual and perennial died out. The landward migration of the barrier itself was also greatly accelerated. Eventually the barrier would have merged with the mainland, and the marsh would have been lost as well. Thus the dune grass acts as a "keystone" species on this landscape in a sense analogous to that of Paine (1980); without the dune grass, the barrier community could not survive.

Until recently, these types of dynamic interactions among landscape elements have been largely overlooked in ecosystem research. Of course, there are exceptions (e.g., regeneration waves in balsam fir forests; Sprugel and Bormann 1981), but the view of spatial heterogeneity in ecosystems has generally been far more static and noninteractive. For example, many researchers (e.g., Clements 1936, Whittaker 1967) have recognized the patterns of vegetation along environmental gradients (i.e., ecoclines), but only local interactions among the plants along these gradients were needed to explain the zonation of vegetation. On a landscape, on the other hand, organisms influence one another over large distances, not just locally. For example, the dune grass not only changed its local environment (i.e., on the same transect segment) but, by stabilizing the dune, also regulated the environment everywhere else along the transect.

These types of landscape interaction are mediated by large-scale processes like sand movement (e.g., the model presented here), animal feeding and nesting patterns (e.g., Pastor et al. 1988), the spread of fire or pests (e.g., M.G. Turner and Dole Chapter 13), or the movement of soil-water nutrients (e.g., Shaver et al., Chapter 5). One of the problems in the development of an ecology of landscapes has been the recognition of these large-scale processes as integral and dynamic parts of the landscape system. Earlier researchers (e.g., Clements 1936) tended to classify many of these large-scale processes as extrinsic to the ecosystem, even as disturbances on the ecosystem. Fire, for example, was viewed as a disturbance that prevented an ecosystem from reaching its climax rather than an integral part of that climax. From a landscape perspective, these large-scale processes must be conceptually incorporated into the system because they tie the system together.

14.4.2 Species Interactions

The landscape perspective also changes the perception of interactions among species. For example, viewed from a local perspective (i.e., from an individual transect segment), the relationship between the dune grass and the perennial is clearly detrimental to the grass but has no affect on the perennial. That is, the perennial can competitively exclude the grass, although its own growth is not hindered by the presence of the grass. From the perspective of the whole landscape, however, the form of the interaction changes completely. As mentioned above, the perennial is unable to survive on the barrier without the dune grass because the grass stabilizes the dune, which in turn provides shelter for the perennial. The dune grass, however, could survive perfectly well without the perennial. Thus, by simply changing the scale at which the interaction is viewed, the relationship changed from indifferent to obligatory for the perennial and from detrimental to indifferent for the grass.

Relative position of the dune grass and the perennial also plays a part in their interaction. The survival of the perennial depends only upon dune grass growing seaward of it (i.e., growing upwind for ocean winds). Its relationship to landward dune grasses is altogether different. Consider, for example, the relationship between the perennials growing between the two dunes in simulation 2 and the dune grass on the landward dune. The landward dune cannot shelter these perennials from ocean winds; so the perennials do not depend on the dune grass to stabilize this dune. The relationship between the two patches of vegetation, however, is not neutral. As long as the dune grass can stabilize the dune and can keep it high and in the wind, the landward spread of the perennial will be impeded. Therefore, if the dune grass is growing landward of the perennial, the relationship becomes detrimental to the perennial but remains neutral for the dune grass. This is a complete reversal of the relationship as viewed from a local perspective.

One of the challenges to ecological modeling imposed by this spatial perspective in landscape ecology is the development of models that can handle the added complexity of species interactions. Pielou (1981) strongly criticized earlier ecological models for the "unnaturalness of the assumptions" about the interactions among organisms. She lists two defects in these assumptions that are germane to the present discussion. First, in earlier models it was assumed "that environmental conditions do not change with time." This criticism is not altogether justified. However, it is true that the importance of the organisms in bringing about these changes has not been adequately represented. This is especially true for large-scale changes that can mediate the interactions among organisms that are well separated on the landscape. Second, in earlier models it was assumed "that the space occupied by the interacting populations is homogeneous and sufficiently small for all the contained individuals to interact with one another in the same manner." I have already discussed the importance of spatial interactions among organisms on a landscape, how the large-scale processes that mediate these interactions must be incorporated into the system, and how the spatial orientation of organisms can affect the nature of the relationship among them.

14.4.3 Model Application

Before the model presented here can be applied to a real barrier landscape, several modifications must be made. The most obvious of these is to tabulate a list of real species and to define each of their niches with respect to the important environmental factors (Fig. 14.2). The Markov chain representation of organism development is by no means critical to the modeling approach. However it is a computationally efficient way of simulating populations with high stem densities (e.g., marshes and grasslands). A less aggregated approach would be more appropriate for less dense stands (e.g., a FORET approach for forests), or a hybrid could be developed for mixed systems.

Water table depth and groundwater salinity were used in the model as indices of water availability and soil salinity. These may be inadequate indices in drier soils like those found on dunes. For such soils, a more complex model able to predict the abundance of water held by capillary action above the water table might be more appropriate. The assumptions of constant hydraulic conductivity and equal infiltration rates across the barrier are also unrealistic for most systems. For example, the permeability of sandy soils on the dunes is clearly much higher than the permeability of the fine-textured soils in the marsh. The vertical structure of the soils is also an important determinant of soil water movement. For example, on barrier islands the dune line often migrates landward to cover old marsh, thus leaving a relatively impermeable layer beneath the sandy dunes. This added complexity could be handled by a finite-element model (e.g., Pinder and Gray 1977), but the added complexity may make such an approach impractical.

The movement of sand is difficult to simulate. It depends upon the material from which the sand is derived, the moisture of the soil, the plant cover, the slope, the strength of the winds, and several other factors. It is difficult to apply a model based upon first principles because time scales of interest here are much longer than those at which the actual processes that move sand act. Therefore, the best approach may be an empirical model calibrated at the appropriate time and space scales. To account for all the factors that influence sand movement, considerable data will be required to develop this model. The system may also need to be manipulated experimentally to unravel the interactions among various factors.

The modeling approach presented here could easily be expanded to include two horizontal dimensions instead of just one. This added dimension is an important aspect of barrier landscapes. The dunes are rarely continuous along the length of these barriers unless they have been artificially stabilized. There are usually breaks along the dune line caused by waves or high winds during storms. Sand is easily eroded from these breaks and is often blown longshore and deposited elsewhere.

Thus the longshore component of barrier-island dynamics can be important and could easily be incorporated into the model.

14.5 Conclusion

To accommodate a landscapes perspective, ecological models will have to incorporate the large-scale processes that tie landscapes together. This was accomplished in the model described here by linking a fine-scale vegetation model to large-scale models of sand movement and soil-water processes. The vegetation model was applied to each of the transect segments, with each segment treated as if it were spatially homogeneous. The two large-scale models tied the landscape together. This general approach should apply to most landscapes. Of course, the nature of the specific models used will change according to the questions being addressed. For example, instead of a vegetation-community model that simulates the interactions among plant populations, an ecosystem model that simulates the processing of nutrients might be required. Similarly, instead of simulating sand movement and soil-water salinity, the large-scale models might simulate seed dispersal and the groundwater transport of nitrate.

Landscapes are an important, but as yet poorly understood, class of ecosystem. Awareness of the importance of spatial heterogeneity and spatial interaction is a significant first step in developing our understanding of landscapes. The techniques presented here will help quantify that understanding and lead to more realistic simulations of landscape dynamics.

14.6 Summary

A model for simulating landscape dynamics on barrier islands was developed to illustrate some general principles of landscape ecology. The model is composed of three major subprograms simulating (1) the water table depth and groundwater salinity, (2) the movement of sand by wind and waves, and (3) the growth of four hypothetical plant species. Groundwater was modeled by using Darcy's Law and assuming dynamic equilibrium at each time step. Simulation of sand erosion was dependent on the amount of protective vegetation at a location and sheltering from wind by dunes. Deposition of sand at a location was dependent upon the amount of sand available due to upwind erosion. The plant community simulation was based on a Markov chain representation of the life-stage development of the organisms. At each time step, the probability that an organism progresses from one life-stage to another was reevaluated based on current environmental conditions.

Three simulations were run by using the model. The first simulation demonstrated the behavior of a barrier island during a slow erosional phase followed by a depositional phase and was used as a point of reference to which the other two simulations were compared. In simulation 2, a large berm was deposited on the seaward side of the barrier. Vegetation that colonized this berm subsequently underwent a succession that resulted in the development of a typical zonation of the dune and back-dune communities. The abiotic process of dune formation was an integral and dynamic part of this succession, not an extrinsic factor to which the succession responded. In the third simulation, one of the species, a dune grass capable of stabilizing the dunes, was made to go extinct. Without the grass, the dune was eroded away and the back-dune community could not survive.

Landscapes, like barrier islands, are not merely regions of heterogeneous terrain. Landscapes are collections of spatially interactive elements. This interaction is mediated by large-scale processes such as the movement of sand across barrier islands. To understand the dynamics of landscapes, therefore, it is necessary to incorporate these large-scale processes as integral and dynamic parts of the system, not as extrinsic controlling factors.

The landscape perspective also changes the way interactions among organisms must be viewed. From the perspective of a small plot, two species might be seen as competitors. When separated on a landscape, however, these same two species may perform functions, like dune stabilization, that are essential to one another's survival. Thus the proximity and orientation of two organisms on a landscape can change the perceived nature of their interaction.

Acknowledgments

I would like to thank Becky Schneider, who inspired this model through her master's thesis work, and Hank Shugart for the opportunity and motivation to start the work. Many thanks also to Jerry Melillo, George Kling, Mike Ryan, Anne Giblin, Gary Banta, Knute Nadelhoffer, Gus Shaver, and Bruce Peterson for reviewing the early manuscript and to Liz Griffin and Pam Clapp for their editorial help. Many thanks also to Liz Griffin for help in translating segments of Darcy's masterpiece.

References

Aber, J.D., and Melillo, J.M. 1982. FORTNITE: A computer model of organic matter and nitrogen dynamics in forest ecosystems. University of Wisconsin Research Bulletin R3130.

Bertness, M.D., and Ellison, A.M. 1987. Determinants of pattern in a New England salt marsh plant community. *Ecological Monographs* 57:129–47.

Bolyard, T.H.; Hornberger, G.M.; Dolan, R.; and Hayden, B.P. 1979. Freshwater reserves of mid-Atlantic coast barrier islands. *Environmental Geology* 3:1–11.

Botkin, D.B.; Janak, J.F.; and Wallis, J.R. 1972. Some ecological consequences of a computer model of forest growth. *Journal of Ecology* 60:849–73.

Boyce, S.G. 1954. The salt spray community. *Ecological Monographs* 24:29–67.

Caswell, H. 1983. Phenotypic plasticity in life-history traits: demographic effects and evolutionary consequences. *American Zoologist* 23:35–46.

Clements, F.E. 1936. Nature and structure of the climax. *Journal of Ecology* 24:252–84.

Darcy, H. 1856. *Les Fontaines Publique de la Ville de Dijon: Exposition et Application des Princepes a Suive et des Formules a Employer dans les Questions de Distribution d'Eau par un Appendice Relatif aux Fournitures d'Eau de Plusieurs Villes au Filtrage des Eaux et a la Fabrication des Tuyaux de Fonte, de Plomb, de Tole et de Bitume*. Paris: Dalmont.

Fisher, J.J., and Simpson, E.J. 1979. Washover and tidal sedimentation rates as environmental factors in development of a transgressive barrier shoreline. In *Barrier Islands: From the Gulf of St. Lawrence to the Gulf of Mexico*, ed. S.P. Leatherman, pp. 127–48. New York: Academic Press.

Godfrey, P.J. 1976. Barrier beaches of the East Coast. *Oceanus* 19:27–40.

Hillel, D. 1980. *Introduction to Soil Physics.* New York: Academic Press.

Horn, H.S. 1975. Markovian properties of forest succession. In *Ecology and Evolution of Communities*, eds. M. L. Cody and J. M. Diamond, pp. 196–211. Cambridge: Harvard University Press.

Jones, R., and Etherington, J.R. 1971. Comparative studies of plant growth and distribution in relation to waterlogging: IV. Growth of dune and dune slack plants. *Journal of Ecology* 59:793–801.

Leslie, P.H. 1945. The use of matrices in certain population mathematics. *Biometrika* 33:183–212.

Leslie, P.H. 1948. Some further notes on the use of matrices in population mathematics. *Biometrika* 35:213–45.

Longwell, C.R.; Flint, R.F.; and Sanders, J.E. 1969. *Physical Geology.* New York: John Wiley and Sons.

Olson, R.L.; Sharpe, P.J.H.; and Wu, H. 1985. Whole-plant modelling: a continuous-time Markov (CTM) approach. *Ecological Modelling* 29:171–87.

Onyekwelu, S.S. 1972. The vegetation of dune slacks at Newborough Warren: I. Ordination of the vegetation. *Journal of Ecology* 60:887–98.

Oosting, H.J. 1945. Tolerance to salt spray of plants of coastal dunes. *Ecology* 26:85–89.

Paine, R.T. 1980. Food webs: linkage, interaction strength and community infrastructure. *Journal of Animal Ecology* 49:667–85.

Pastor, J.; Naiman, R.J.; Dewey, B.; and McInnes, P. 1988. Moose, microbes, and the boreal forest. *BioScience* 38:770–77.

Pielou, E.C. 1981. The usefulness of ecological models: a stock-taking. *Quarterly Review of Biology* 56:17–31.

Pinder, G.F., and Gray, W.G. 1977. *Finite Element Simulation in Surface and Subsurface Hydrology.* New York: Academic Press.

Ranwell, D.S. 1972. *Ecology of Salt Marshes and Sand Dunes.* London: Chapman and Hill.

Rosen, P.S. 1979. Aeolian dynamics of a barrier island system. In *Barrier Islands: From the Gulf of St. Lawrence to the Gulf of Mexico*, ed. S. P. Leatherman, pp. 81–98. New York: Academic Press.

Schneider, R.L. 1984. The relationship of infrequent oceanic flooding to groundwater salinity, topography and coastal vegetation. Master's thesis, Department of Environmental Sciences, University of Virginia, Charlottesville.

Shaver, G.R.; Nadelhoffer, K.J., and Giblin, A.E. Biochemical diversity and element transport in a heterogeneous landscape, the North Slope of Alaska. Chapter 5, this volume.

Shugart, H.H. 1984. *A Theory of Forest Dynamics: The Ecological Implications of Forest Succession Models.* New York: Springer-Verlag.

Shugart, H.H.; Crow, T.R.; and Hett, J.M. 1973. Forest succession models: a rationale and methodology for modeling forest succession over large regions. *Forest Science* 19:203–12.

Shugart, H.H., and West, D.C. 1977. Development of an Appalachian deciduous forest succession model and its application to assessment of the impact of the chestnut blight. *Journal of Environmental Management* 5:161–79.

Smith, T.M., and Urban, D.L. 1988. Scale and resolution of forest structural pattern. *Vegetatio* 74:143–50.

Sprugel, D.G., and Bormann, F.H. 1981. Natural disturbance and steady state in high-altitude balsam fir forests. *Science* 211:390–93.

Turner, M.G. 1987. Spatial simulation of landscape changes in Georgia: a comparison of 3 transition models. *Landscape Ecology* 1:29–36.

Turner, M.G., and Dale, V.H. Modeling landscape disturbance. Chapter 13, this volume.

Usher, M.B. 1979. Markovian approaches to ecological succession. *Journal of Animal Ecology* 48:413–26.

Waggoner, P.E., and Stephens, G.R. 1970. Transition probabilities for a forest. *Nature* 225:1160–61.

Whittaker, R.H. 1967. Gradient analysis of vegetation. *Biological Reviews* 42:207–64.

Wienstein, D.A., and Shugart, H.H. 1983. Ecological modeling of landscape dynamics. In *Disturbance and Ecosystems*, eds. H. Mooney and M. Gordon, pp. 29–45. New York: Springer-Verlag.

15. A Spatial-Temporal Model of Nitrogen Dynamics in a Deciduous Forest Watershed[1]

Steven M. Bartell and Antoinette L. Brenkert

15.1 Introduction

Describing and explaining pattern in relation to underlying processes are funda-mental to science. Landscape ecologists attempt to understand and quantify how variously scaled physical, chemical, and biological processes determine ecological patterns measured in time and space on landscapes. Mathematical models have become useful tools for exploring the implications of differently scaled environ-mental processes on observed landscape patterns (e.g., Jarvis and McNaughton 1986; Kesner and Meentemeyer 1989). This modeling effort complements the necessary and corresponding development of indices, metrics and models used to describe landscape patterns (Krummel et al. 1987; Gardner et al. 1987; O'Neill et al. 1988; Turner et al. 1989).

Relating pattern to process requires consideration of scales in space and time (Platt and Denman 1975). For example, in nutrient budgets for watersheds, the concentrations and fluxes of nutrients are typically expressed in units of kilograms per hectare and kilograms per hectare per year (e.g., Johnson and Van Hook 1989). However, none of the sampling methods actually measures at these scales. This

[1]Research sponsored by the Office of Health and Environmental Research, U.S. Department of Energy, under contract DE–AC05–84OR21400 with Martin Marietta Energy Systems, Inc.

methodological limitation may be important because patterns measured at the
scale of landscapes might not always be the simple sum, statistical average, or
linear interpolation of smaller-scale population, community, and ecosystem dynamics. To meaningfully relate pattern to process, the relevant ecological structure
and correct scaling and formulation of processes will be required. These issues of
scale are not particular to landscape ecology. Considerations of scale remain a
focus of theoretical and methodological developments in related disciplines, for
example, groundwater hydrology (Cushman 1984; 1986; Dagan 1986).

Systematically varying the spatial-temporal scales of a landscape model may
help identify the scale dependence of particular landscape processes. The primary
purpose of work reported here was to examine the implications of heterogeneities
in initial conditions and process rates on simulated patterns of total nitrogen
concentrations in the Walker Branch watershed. This objective was addressed with
the use of a linear, four-component model of N cycling (Cole and Rapp 1981).
This model was implemented for each element (pixel) of a square grid imposed
conceptually on the watershed. Measured heterogeneities in vegetation type and
topography were represented in the model as different initial N concentrations in
soil and vegetation, different rates of N cycling within each pixel, and different
rates of N flux among adjacent pixels. These different values were taken from the
N budgets reported for the different vegetation types (Johnson and Van Hook
1989). The model was used to explore the implications of current N dynamics on
future nitrogen disposition within this watershed.

A second objective was to examine the dependence of model results on temporal scale. Previous analyses of models of organic chemical transport (Bartell et
al. 1983) and food web dynamics (Bartell et al. 1986, 1988) have shown changes
in the sensitivities of model parameters during the time course of simulation. In
the present study, we speculated that the spatial heterogeneities in the watershed
model would overshadow temporal changes in parameter sensitivities. Alternatively, the spatial model might also show time-dependent changes in sensitivity,
as noted in analysis of the nonspatial models (e.g., Bartell et al. 1983, 1986, 1988).

The explicit criteria for the selected scale in watershed studies are seldom stated
(e.g., Vitousek and Reiners 1975; Vitousek et al. 1979; Johnson and Van Hook
1989). Traditional studies define the watershed itself as the unit of study and
measure nutrient concentrations in stream waters at the weir (Likens et al. 1970).
It remains difficult to objectively define a priori the appropriate spatial scale for
the study of watershed nutrient dynamics. In many instances, the appropriate scale
may simply not be known. However, given spatial heterogeneity in nutrient pool
sizes and the rates of processes that determine nutrient dynamics within and
through watersheds, it is possible to imagine an experiment in which a manipulation (e.g., fertilization) imposed at one site would produce measurable changes at
the weir, whereas the same manipulation administered elsewhere in the watershed
might not produce measurable changes at a weir. The final modeling exercise
explored the effects of different spatial resolution on modeled concentrations of N.
To examine the implications of varying spatial scale, the same watershed area was
modeled using fewer, larger grid elements.

15.2 Site Description

The Walker Branch watershed, located on the U.S. Department of Energy Oak Ridge Reservation, has been the subject of diverse and intensive ecological investigation for nearly three decades. Comprehensive site description appears in Johnson and Van Hook (1989). Briefly, the watershed vegetation comprises pines (*Pinus virginiana, P. echinata, P. taeda*), oak-hickory (mixed upland species), and chestnut oak (*Quercus prinus*) on ridge tops and hill-slope forest dominated by tulip poplar (*Liriodendron tulipifera*) (Grigal and Goldstein 1977). Watershed soils are mainly Ultisols, with soils of the Fullerton series occupying ridge tops and upper slopes and Bodine soils located on the intermediate and lower slopes (Peters et al. 1970).

An approximately 11.6-ha square area (Fig. 15.1) was outlined from a 1:24,000 map of the 48-ha watershed. This area was selected because it included ridge-top, hill-slope, and riparian vegetation, and a section of Walker Branch. The topography of the area, determined from a 10-ft elevation contour map, was also characteristic of the watershed. Use of this subsection of the watershed for modeling economized the computations and permitted detailed numerical analysis of the model behavior.

15.3 The Watershed-Specific Model

Local nutrient dynamics are determined in space and time by one set of processes that cycle or retain nutrients and another set of processes that transport them. The level of detail used in formulating these processing depends on the purpose of the model. As stated, this modeling effort had several objectives: (1) to examine the

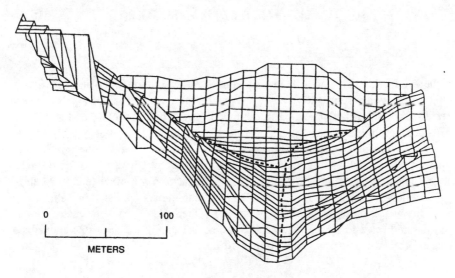

0   100

METERS

Figure 15.1. Three-dimensional projection of topography for watershed subsection.

short- and long-term implications of estimated total N concentration, cycling rates, and transport rates on total N dynamics in the watershed; (2) to explore the effects of spatial heterogeneities in topography and vegetation type on simulated N dynamics; and (3) to examine the implications of spatially aggregating these heterogeneities on model results.

Watershed nutrient budgets are commonly accounted for on an annual time scale for the average hectare (e.g., Johnson and Van Hook 1989). The units can be conveniently transformed into rate constants with the units of $1/y$. This scaling convenience suggested application of the linear model (Fig. 15.2), even though current ecological understanding of nutrient dynamics recognizes the importance of nonlinear processes. These nonlinear dynamics and additional ecological details, not explicit in the model, are aggregated in the structure and parameter estimates of the linear model. The following difference equations describe the transport, cycling, and accumulation of total N for a single pixel:

$$C1(i,j)_{t+1} = C1(i,j)_t + I_p + f(m,n) - f(i,j) \tag{15.1}$$
$$+ r12(i,j)\ C2(i,j)_t - (r31(i,j) + r41(i,j))\ C1(i,j)_t$$

$$C2(i,j)_{t+1} = C2(i,j)_t + I_c + r23(i,j)\ C3(i,j)_t \tag{15.2}$$
$$+ r24(i,j)\ C4(i,j)_t - (r12(i,j) + r32(i,j)$$
$$+ r42(i,j))\ C2(i,j)_t + g(m,n) - g(i,j)$$

$$C3(i,j)_{t+1} = C3(i,j)_t + r31(i,j)\ C1(i,j)_t + r32(i,j)\ C2(i,j)_t \tag{15.3}$$
$$- r23(i,j)\ C3(i,j)_t$$

$$C4(i,j)_{t+1} = C4(i,j)_t + r41(i,j)\ C1(i,j)_t + r42(i,j)\ C2(i,j)_t \tag{15.4}$$
$$- r24(i,j)\ C4(i,j)_t ,$$

where $C1$, $C2$, $C3$, and $C4$ are the concentrations of nitrogen in soil, litter, understory vegetation, and overstory vegetation, respectively. The (i,j)'s indices refer to the site (pixel) location in the model grid. The functions f and g in Eqs. 1 and 2 determine fluxes between pixels of soil N $(C1)$ and litter N $(C2)$ due to elevation differences in the watershed. I_p and I_c are the input of nitrogen to each pixel from wet deposition and crown wash, respectively. The $r(ij)$'s are the rate constants that determine flow among components within a pixel (e.g, $r12$ determines flow from $C2$ to $C1$). Values for model parameters are listed in Table 15.1.

The overall modeling strategy was to implement Eqs. 1 to 4 for each element in the grid (e.g., Fig. 15.3). At the smallest scale (i.e., $32 \times 32 = 1{,}024$ elements), each pixel was approximately 10 m $\times$ 10 m.

Table 15.1. Parameter Values for Vegetation-Specific Model of N Dynamics in Walker Branch Watershed[a]

Model Values	Vegetation Type		
	Oak-Hickory	Liriodendron	Nonforest
Initial N Concentration			
C1, soil	4500	7300	5630
C2, forest floor	334	187	278
C3, understory	4	14	17
C4, overstory	399	267	0
N Inputs			
Precipitation	13	13	13
Crown wash	3.2	2.3	0
Rate constants			
r01	0.05	0.05	0.05
r12	0.10	0.10	0.10
r23	0.12	0.15	0.13
r24	0.12	0.12	0
r31	0.00015	0.00040	0.00030
r32	0.002	0.017	0.006
r33	0	0	0
r41	0.015	0.009	0
r42	0.20	0.34	0
r44	0	0	0

[a] Initial concentrations and external inputs have units of kilograms per hectare per year; the rate constants, $r(i,j)$'s, are in units of 1/y. The i,j indices identify flow from component j to i.

15.3.1. Vegetation

The pixel model presented new problems in the estimation of initial conditions and model parameters. With 1,024 pixels, even this simple model structure required the estimation of more than 4,000 initial conditions and 7,000 model parameters. Initial conditions consisted of concentrations of N in the soil, litter, understory vegetation, and overstory vegetation. These values varied across the subsection of the watershed in relation to the vegetation type dominant at each pixel (Johnson and Van Hook 1989). The data were insufficient to identify a unique set of N concentrations for each pixel. Rather, the variability reported for the N concentrations for each vegetation type was used to estimate standard deviations for as-

KERNEL NUTRIENT MODEL

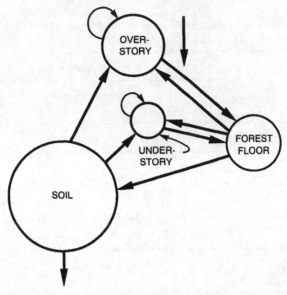

Figure 15.2. Four-component N cycling model implemented for each grid element (pixel) in a spatial model of the watershed subsection.

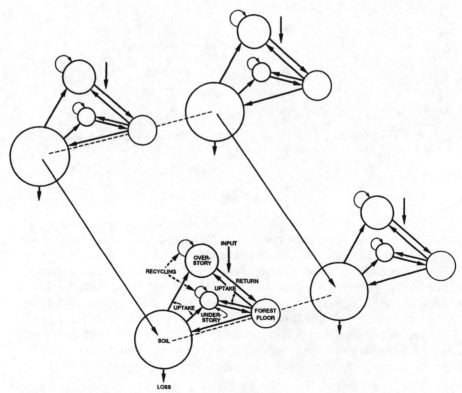

Figure 15.3. Kernel Nutrient Model: Illustration of functional connections among adjacent pixels in the model.

sumed normal distributions. Values of initial N concentrations in soil, litter, and both understory and overstory vegetation were selected for each pixel from these distributions with the use of a stratified, random procedure (Iman and Conover 1981). The set of pixels that represent each vegetation type collectively exhibits the mean values for total N concentrations reported for the watershed.

Values for each vegetation-specific rate constant were selected with the same stratified random procedure from distributions of estimated N flux among the model components for each vegetation type (Johnson and Van Hook 1989). The mean rates of N flux for the basic model for each of the vegetation types are listed in Table 15.1.

As a result of this modeling approach, each pixel represents spatially the vegetation-specific initial N concentrations and rate constants for each watershed vegetation type.

15.3.2 Topography

In the pixel model, N was assumed to move through the soil solution and to move with downslope litter transport. As an initial simplification, watershed hydrology and litter flux were not modeled directly. Rather, pixel-specific elevation data were digitized from a 10-ft contour map of the watershed and used to calculate the steepness of slope from each pixel to all adjacent pixels. The greater the relative difference in surrounding elevation values, the greater was the potential yearly downslope N movement to that pixel. The normalized slopes were used to estimate rate constants (1/y) by calibrating total downslope N losses to reported watershed annual N losses (Johnson and Van Hook 1989) and then apportioning the total annual loss among pixels according to the normalized pixel-to-pixel slopes.

15.4 The Watershed-Neutral Model

Heterogeneity is scale dependent. Once the sampling scale is defined, heterogeneity at smaller scales cannot be measured; at these scales, the system will appear homogeneous. Models that systematically include or ignore heterogeneity at different scales can be used as neutral models (Caswell 1976; Gardner et al. 1987) to produce results for evaluating the effects of adding heterogeneity to the watershed-specific model. To compare results with those of the detailed N model, a neutral model identical to the watershed-specific model in structure (i.e., Eqs. 15.1 to 15.4) was constructed. However, the neutral model's initial conditions were the grid average N concentrations for each component (*C1* to *C4*). The internal cycling parameters represented the grid average rates, and the topography was defined by the average slope measured across all grid elevations. The simulations performed with the watershed-specific model were also performed with the neutral model.

At some broader scale, landscape heterogeneities will necessarily be aggregated in models developed to study landscape phenomena. A relevant example might be to interface a spatial model with remote sensing. The integrative scale of the sensor will define the necessary scale of resolution in the corresponding spatial model.

Integration of general circulation models (GCMs) with landscape process models to address regional and global issues will also require identification of appropriate spatial-temporal scales, in which averaging may span much larger scales than watersheds. These reasons suggest examining the implications of such averaging and further justify the construction of neutral models. System theory further argues that at some scale the linear, average model should be useful for relating pattern and process (Patten 1975). Thus, model neutrality is also scale dependent, and neutral models may likely have additional heuristic value.

15.5 Simulations and Model Solution

The watershed-specific pixel model and the neutral model were use to perform one series of simulations to examine the time behavior of watershed N at the finest spatial scale (e.g., 32 × 32 = 1,024 pixels). A second series addressed the implications of modeling N dynamics for a fixed period in time but used successively less spatial detail. Fewer, larger pixels were used to represent the section of watershed. In scaling terminology (Allen and Starr 1982), the spatial extent was held constant while the grain was increased.

In each simulation, the internal dynamics of each pixel was calculated, as well as the spatial flux of soil and litter N across the entire grid. Reversing the order of this scheme for solving the model equations did not alter the results. Therefore, the solution algorithm was accepted as providing minimal numerical artifact; differences in model results were in the range of round-off errors. Of course, the solution of sets of coupled partial differential equations or their difference equation analogs remains an active area of mathematical and computational research. The subject is meaningful addressed elsewhere (e.g., Wang and Anderson 1982), and we only offer caution regarding the development and solution of these models.

15.5.1 Constant Space and Variable Time

Both the neutral model and the model incorporating Walker Branch heterogeneities were used to simulate 50 y of annual values of N concentrations in soil, litter, understory vegetation, and overstory vegetation for each of the 1,024 interconnected pixels. At the end of each model year, the simulated concentration of N in each of the four component types was recorded for each of the 1,024 pixels. Annual N dynamics were calculated using 0.1–y time steps.

15.5.2 Constant Time and Variable Spatial Scale

To explore the implications of spatial aggregation, the 32 × 32 pixel model was successively aggregated into fewer, larger pixels (e.g., a 16 × 16 grid, then 8 × 8, 4 × 4, and 2 × 2). This aggregation was accomplished by averaging watershed elevations over increasing pixel size and appropriately weighting the initial N concentrations and the vegetation-specific internal cycling parameters. The same spatial aggregations were performed by using the neutral watershed model. Simulations of fifty years of watershed N dynamics were performed with each spatial aggregation and each model.

15.5.3 Patterns of N Concentrations

At each point in time, the model calculates 1,024 concentrations in each of four model components. These concentrations can be mapped (e.g., Fig. 15.4). In addition, cumulative density function (cdf) can be calculated for modeled N concentrations for the entire grid at selected points in time from the simulations. Comparisons of the median value and the relative variability in the model results can be readily obtained by inspecting the cdf's.

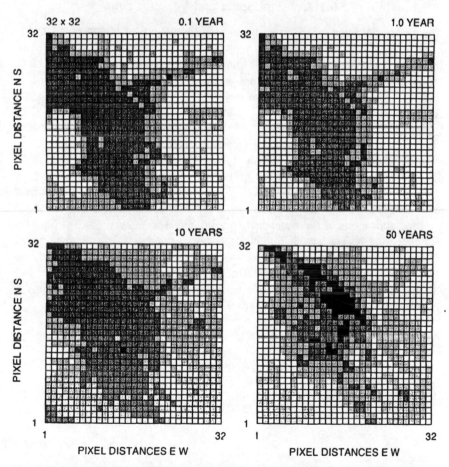

Figure 15.4. Modeled concentrations of soil N in space and time for the 32 × 32 element (1,024 pixels) watershed-specific model. Concentrations increase with darker shading.

Though not a focus of the present study, the spatial nature of the model output presents the opportunity for further sampling and characterizing the model results. For example, a watershed average can be calculated by using all pixel values. Alternatively, the simulated N concentrations can be further subsampled to explore the implications of alternative sampling designs using different plot sizes and transects in estimating the modeled patterns of N concentrations in soil, litter, understory plants, or trees.

15.6 Results

A selection of model results is presented, with particular focus on soil N concentrations.

15.6.1 Constant Space and Variable Time

The watershed-specific model produced a spatial pattern of N concentrations in each of the model components through time.[1] The temporal change in spatial location of higher soil N illustrates the net downslope movement (Fig. 15.4). Ridge-top soils (see Fig. 15.1 for topographical reference) lose N to downslope and riparian soils over the course of the 50-y simulation as the net result of the model.

Spatial heterogeneity in topography and vegetation type in the watershed-specific model resulted in increased variability in simulated concentrations of N accumulated as overstory vegetation (Fig. 15.5). The flattening of the cdf for this model through time compared to the neutral model cdf illustrates the comparative increase in variability. Both models predicted a similar median increase in overstory N during the 50-y simulations.

The implications of the watershed-specific model were even more striking for simulated soil N (Fig. 15.6). Both models demonstrated a decrease in median soil N through time; however, the trend was less clearly defined by the watershed-specific model. The vegetation-specific internal cycling parameters, coupled with the topographical soil and litter movement, produced more variable patterns of N concentrations through time, as indicated by the comparatively flat shape of its cdf's. The orderly progression of decreased soil N concentrations indicated by the neutral model was not as evident in the watershed-specific model.

15.6.2 Sensitivity Analysis

The comparatively greater heterogeneity in N concentrations produced by the

---

[1]The results are from the 32 × 32 element grid model. However, the graphical algorithm uses the value for each element to create a node and then shades between nodes to generate a grid that shows one less row and column (i.e., 31 × 31). This software limitation does not appreciably change the modeled N pattern.

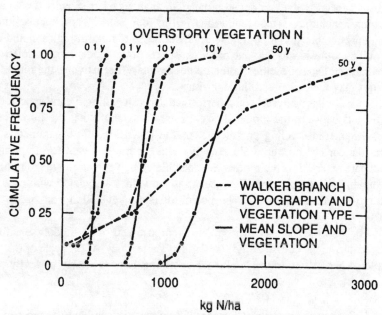

Figure 15 5   Cumulative frequency distributions for overstory N concentrations produced by the 1,024-pixel watershed specific model (dashed line) and the corresponding neutral model (solid line)

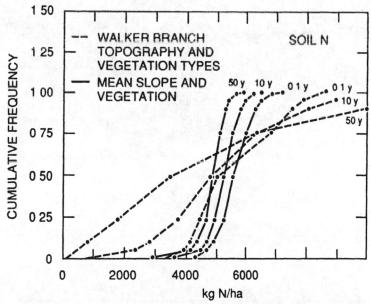

Figure 15 6   Cumulative frequency distributions for soil N concentrations produced by the 1,024-pixel watershed-specific model (dashed line) and the corresponding neutral model (solid line)

watershed-specific pixel model stimulated an analysis of its sensitivities to initial conditions, including N concentrations in the four model components, internal cycling parameters, and topographically determined N flux through soil and litter The null hypothesis was that one of these aspects of the model largely determined the calculated heterogeneities in the model results Alternatively, the model sensitivities may have changed through time

The sensitivity analyses were performed by calculating Pearson correlation coefficients between the modeled N concentrations and either the initial conditions, the internal cycling parameters, or the downslope flow parameters The higher the correlation, the more sensitive was the model result to the particular model parameter The sensitivities were calculated by vegetation type to minimize confounding effects due to differences in mean values associated with each vegetation type and because of the absence of overstory vegetation from many of the pixels

Results for the oak-hickory pixels demonstrated that model sensitivities changed during the course of the simulation (Table 15 2) At the scale of 0 1 y, the calculated N concentrations in all four components were nearly perfectly corre-

Table 15 2 Pearson Correlations Between N Concentrations in Model Components and Model Inputs and Parameter Values for the Oak-Hickory Vegetation Specific Model

	Model component			
Parameter	Soil	Forest Floor	Understory	Overstory
	After 0 1 y			
C's	0 99	0 99	0 99	0 99
	After 1 0 y			
C's	0 90	0 83	0 82	0 95
TP	−0 30	−0 30		
r31			0 45	
	After 10 0 y			
TP	−0 65	−0 59	−0 39	−0 55
r24				−0 28
r31			0 67	
r41				0 37
	After 50 0 y			
TP	−0 61	−0 63	−0 60	−0 61

Only values of the Pearson correlation coefficient >0 3 are shown The C's are the component-specific initial concentrations of nitrogen, TP is the total rate of loss of nitrogen from each pixel, and the $r(i,j)$'s are the rate constants that determine the flow N between model components within a pixel

lated with the initial N concentrations. After 1 model year, correlations between intital concentrations in litter and understory vegetation decreased to approximately 0.8. In other words, initial N concentrations still accounted for approximately 64% of the variation in these model components. The rate of N uptake from the soil by the understory vegetation increased in importance at this scale. Negative correlations between the topographic flow parameters and N content in soil and litter were small, but measurable.

On a ten-year scale, the importance of the initial concentrations was not as evident. Model results were increasingly sensitive to a set of flow parameters and internal N cycling parameters (Table 15.2). Finally, at the scale of half-centuries, N concentrations were explained primarily by the topographically determined loss parameters. Initial N concentrations and the internal cycling parameters were no longer predictive of N concentrations in model components.

15.6.3 Constant Time and Variable Spatial Scale

Cumulative frequency distributions were calculated for the 10-y model results for successive spatial aggregations of the watershed-specific model and the neutral model. As expected, spatial aggregation resulted in less variability in the predicted N concentrations in model components (e.g., Fig. 15.7). Given the methods for

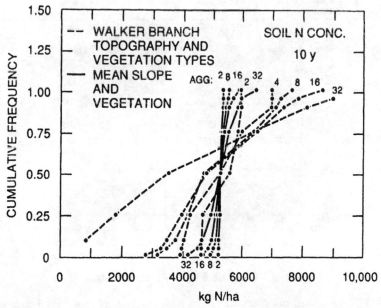

Figure 15.7. Cumulative frequency distributions for successive spatial aggregations of the watershed-specific and neutral models calculated from the results of 10-y simulations.

aggregating, the median soil concentration for the neutral model should remain the same, as seen in the model cdf.

At all levels of aggregation, the watershed-specific model demonstrated greater variability in predicted N concentrations in space. An important associated result was that the median values differed in relation to the aggregation, thereby demonstrating the scale dependence of measuring and estimating watershed nutrient budgets. This dependence is consistent with previously reported scale dependence of measured vegetation patterns (e.g., Getis and Franklin 1987), given the sensitivity of nutrient dynamics to vegetation-specific nutrient cycling characteristics (i.e., Table 15.2).

The effects of aggregation on spatial pattern are further illustrated in Fig. 15.8. These results were produced by the 8 × 8 pixel aggregation of the watershed-specific model: that is, the spatial scale is 16 times that of the 32 × 32 representation (Fig. 15.4). The pattern of soil N through time demonstrates the same general depletion of upper slope soils and accumulation in riparian regions. However, the location of highest N concentrations differs markedly from the 32 × 32 model results. The different pattern is explained partially by aggregating the watershed topography (Fig. 15.9). While the 8 × 8 aggregation retained the same overall three-dimensional character of the watershed, some distortion occurred. This distortion influenced the detailed downslope movement of soil N, especially as

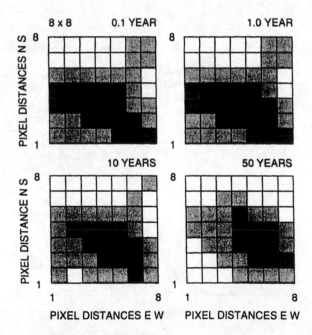

Figure 15.8. Simulated soil N concentrations resulting from the 8 × 8 pixel aggregation using the watershed specific model.

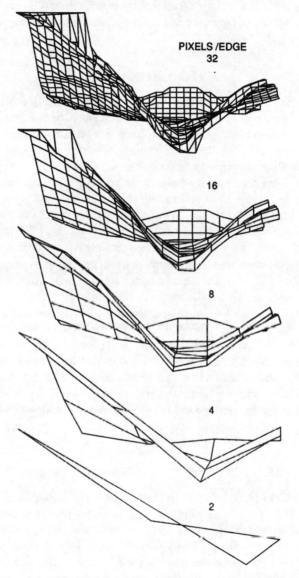

PIXELS /EDGE
32

16

8

4

2

Figure 15.9. Illustration of spatial aggregation on the three-dimensional projection of the watershed subsection topography.

time increases in the simulation. Recall that the sensitivity results emphasized the increasing importance of the flow parameters with increasing time for the 32 × 32 model (Table 15.2).

15.7 Discussion

The results of this modeling effort underline the importance of scale considerations in the design of models and experiments that relate pattern to process. Indeed, the results of the current study are open to criticism regarding some arbitrariness in scaling. In space, the smallest pixel element was defined simply by the convenience of dividing the watershed subsection so it could be easily aggregated by successively halving the number of pixels per grid edge. The spatial grain at the finest scale was ~ 10 m × 10 m, still rather large given the scale of most sampling methods used in watershed studies. While the modeling exercises clearly demonstrated a scale dependence of model results in space (e.g., Fig. 15.7) and time (e.g., Table 15.2), the study did not produce a new scaling rule. Rather, the important implication of the results was that, for any nutrient cycling phenomenon, one can identify a spatial-temporal scale that provides for minimal variance in measurement and, presumably, for maximum predictive ability.

The temporal scale of the model was determined by the annual units used in reporting the nutrient budget (Johnson and Van Hook 1989). To solve the model equations, 0.1–y time steps were used to increase model accuracy at the annual scale. Of course, this temporal scale was too coarse to measure the effects of precipitation events and their associated effects on nutrient transport and distribution. The challenge remains (1) to identify sampling scales that minimize errors in describing landscape patterns and (2) to determine how disparately scaled but influential phenomena naturally interact, as well as how they can be usefully incorporated into models.

15.7.1 Pixels and Individual-Based Models

The spatial model of watershed N dynamics can be classified as an individual-based model (IBM). Each pixel in the spatial model relates directly to an individual forest plot described by its dominant vegetation type, N concentrations, internal cycling rates, and topography. In theory, pixel-specific parameters could be developed from measurements made at their "locations" (recognizing the scale dependence) in the watershed. Similarly, model predictions can be evaluated for each location. Thus, IBMs are ecologically attractive partially because they are scaled to tangible ecological entities, usually individual organisms. IBM constructs also permit the integration of life history traits, behavior, and growth dynamics, not easily combined in more traditional population models (e.g., Leslie models, biomass models) or differential equations in numbers or mass. For these and other reasons, IBMs promise much for the future of ecology (Huston et al. 1989).

However, deriving parameter values for the watershed-specific pixel model pointed to a potential problem in developing IBMs. Even the simple four-component pixel model produced a problem in parameter estimation. It was readily apparent that sufficient data required to define the unique parameters for the 1,024 individual pixels would never become available. Instead, variances specific to the internal cycling rates within each vegetation type were used to define distributions from which individual pixel values were selected at random. As a result, the collection of pixels for each vegetation type would be expected to exhibit the average N dynamics characteristic of the vegetation type. However, it remained a matter of chance that any model pixel accurately represented a specific location in the watershed. Future refinements in parameter estimation may benefit from application of methods in spatial statistics (e.g., kriging, cokriging, cospectral analysis) that have found use in the hydrosciences (Delhomme 1978; Warrick et al. 1977). Semivariograms describing scales of correlations among watershed attributes might assist in estimation of model parameters, retaining the correlation structure that confers "individuality" at the scale of the pixels.

Therefore, one must ascertain what confers "individuality" on the basic structural components of the IBMs. This issue assumes importance in the expectation that population-level phenomena will emerge as a property of interacting model individuals, thereby making IBMs powerful tools for exploring population dynamics by using individual based information (Huston et al. 1989). If the necessary data for individual parameter estimation are lacking, "individuality" might result from correlations among parameter values for individual model units determined from sampling the population. For example, an individual with a high feeding rate might also possess a high rate of basal metabolism. If the correlations can be demonstrated, parameters for individuals can be derived independently, while retaining the collective correlations (Iman and Conover 1981). Some degree of empirical individuality will result. If the nature of the underlying mathematical description and the development of parameters are not considered, the IBMs may merely be highly disaggregated population models. For example, if the IBM mathematical construct applies directly to population-level descriptions (e.g., logistic equation), there should be little surprise in observing population level behavior as an "emergent property" of IBMS.

15.7.2 The Pixel Model and Sampling Design

This modeling approach can also be used to evaluate field sampling procedures in relation to the scale and underlying cause of heterogeneity in nutrient concentrations. The number and location of sample plots (e.g., random, stratified random, nested) or the dimension and orientation of sample transects can be easily imposed on the gridded model. Location, size, geometry, and number of sample transects, quadrats, etc., can be evaluated by using a completely defined landscape. Sample statistics can be compared with the "true" values calculated from the entire grid.

With appropriate modification or wholesale replacement of the pixel model, the sampling of other spatial ecological phenomena can be evaluated, for example, spatial-temporal vegetation dynamics in relation to environmental gradients, successional processes, or disturbance. The impacts of heterogeneity in factors affecting these phenomena on sampling can be quantified through simulation. Relationships between pattern and process that result from such model studies might then be tested on real landscapes.

15.8 Summary

A spatial model of nitrogen transport and cycling was constructed to explore the implications of heterogeneities in vegetation type and topography on N accumulation in a forested watershed. A finite difference model was implemented for a conceptual grid imposed on a section of watershed that contained ridge-top, hill-slope, and riparian forest. Annual values of nitrogen (kilograms per hectare) in soil, litter, and understory and overstory vegetation were modeled for each grid element. Simulated soil and litter N moved among adjacent elements in relation to differences in elevation. Vegetation-specific model parameters were derived from N budgets estimated for the watershed. Results from a neutral model that ignored watershed heterogeneities and simulated average N dynamics in the watershed were compared to results from the detailed model. These comparisons demonstrated the potential errors in describing N dynamics that can result from ignoring spatial heterogeneity in watershed characteristics.

Sensitivity analyses demonstrated the time dependence of model processes in determining modeled N dynamics. At annual time scales, initial N concentrations carried the most predictive power. Internal cycling parameters were most important at scales of 10 y. The long-term N dynamics (> 50 y) were determined mainly by the topography-related losses. These analyses may be useful in designing nutrient cycling models directed at different time scales. The results might also be used to design field experiments or monitoring plans for studying different components of nutrient dynamics.

Representing the same watershed area by fewer, larger grid elements changed the results for the watershed-specific model. These changes resulted in part from the distortion of the watershed topography as spatial resolution was lost through aggregation. Further explorations of the effects of changing spatial and temporal scales on model results are needed to identify appropriate scales for quantifying nutrient dynamics in a landscape context.

References

Allen, T.F.H., and Starr, T.B. 1982. *Hierarchy: Perspectives for Ecological Complexity.* Chicago: University of Chicago Press.
Bartell, S.M.; Breck, J.E.; Gardner, R.H.; and Brenkert, A.L. 1986. Individual parameter perturbation and error analysis of fish bioenergetics models. *Canadian Journal of Fisheries and Aquatic Sciences* 43:160–68.

Bartell, S.M.; Brenkert, A.L.; O'Neill, R.V.; and Gardner, R.H. 1988. Temporal variation in regulation of production in a pelagic food web model. In *Complex Interactions in Lake Communities,* ed. S.R. Carpenter, pp. 101–18. New York: Springer-Verlag.

Bartell, S.M.; Gardner, R.H.; O'Neill, R.V.; and Giddings, J.M. 1983. Error analysis of predicted fate of anthracene in a simulated pond. *Environmental Toxicology* 2:19–28.

Caswell, H. 1976. Community structure: a neutral model analysis. *Ecological Monographs* 46:577–97.

Cole, D.W., and Rapp, M. 1981. Elemental cycling in forest ecosystems. In *Dynamic Properties of Forest Ecosystems,* ed. D.E. Reichle, pp. 341–409. Cambridge: At the University Press.

Cushman, J.H. 1984. On unifying the concepts of scale, instrumentation, and stochastics in the development of multiphase transport theory. *Water Resources Research* 20:1668–76.

Cushman, J.H. 1986. On measurement, scale, and scaling. *Water Resources Research* 22:129–34.

Dagan, G. 1986. Statistical theory of groundwater flow and transport: Pore to laboratory, laboratory to formation, and formation to regional scale. *Water Resources Research* 22:120S–134S.

Delhomme, J.P. 1978. Kriging in the hydrosciences. *Advances in Water Resources* 1:251–66.

Gardner, R.H.; Milne, B.T.; Turner, M.G.; and O'Neill, R.V. 1987. Neutral models for the analysis of broad-scale landscape pattern. *Landscape Ecology* 1:19–28.

Getis, A., and Franklin. J. 1987. Second-order neighborhood analysis of mapped point patterns. *Ecology* 68:473–77.

Grigal, D.F., and Goldstein, R.A. 1977. An integrated ordination-classification analysis of an intensively sampled oak-hickory forest. *Journal of Ecology* 59:481–92.

Huston, M.H.; DeAngelis, D.L.; and Post, W.M. 1989. New computer models unify ecological theory. *BioScience* 38:682–91.

Iman, R.L., and Conover, W.J. 1981. Small sample sensitivity analysis techniques for computer models with an application to risk assessment. *Communications in Statistics* A9(17):1749–1842.

Jarvis, P.G., and McNaughton, K.G. 1986. Stomatal control of transpiration: scaling up from leaf to region. *Advances in Ecological Reserach* 15:1–49.

Johnson, D.W., and Van Hook, R.I., eds. 1989. *Analysis of Biogeochemical Cycling Processes in Walker Branch Watershed.* New York: Springer-Verlag.

Kesner, B.T., and Meetenmeyer, V. 1989. A regional analysis of total nitrogen on an agricultural landscape. *Landscape Ecology* 2:151–63.

Krummel, J.R.; Gardner, R.H.; Sugihara, G.; O'Neill, R.V.; and Coleman, P.R. 1987.. Landscape patterns in a disturbed environment. *Oikos* 48:321–24.

Likens, G.E.; Borman, F.H.; Johnson, N.M.; Fisher, D.W.; and Pierce, R.S. 1970. Effects of forest cutting and herbicide treatment on nutrient budgets in the Hubbard Brook watershed ecosystem. *Ecological Monographs* 40:23–47.

O'Neill, R.V.; Krummel, J.; Gardner, R.H.; Sugihara, G.; Jackson, B.; DeAngelis, D.L.; Milne, B.; Turner, M.B.; Zygmutt, B.; Christensen, S.; Graham, R.; and Dale, V.H. 1988. Indices of landscape pattern. *Landscape Ecology* 1:153–62.

Patten, B.C. 1975. Ecosystem linearization: an evolutionary design problem. In *Ecosystem Analysis and Prediction,* ed. S.A. Levin, pp. 182–201. Philadelphia: Society for Industrial and Applied Mathematics.

Peters, L.N.; Grigal, D.F., Curlin, J.W.; and Selvidge, W.J. 1970. Walker Branch watershed project: chemical, physical, and morphological properties of the soils of Walker Branch watershed. ORNL/TM-2968. Oak Ridge National Laboratory, Oak Ridge, Tennessee.

Platt, T., and Denman, K. L. 1975. Spectral analysis in ecology. *Annual Review of Ecology and Systematics* 6:189–210.

Turner, M.G.; Costanza, R.; and Sklar, F.H. 1989. Methods to evaluate the performance of spatial simulation models. *Ecological Modelling* 47:18–28.

Vitousek, P.M.; Gosz, J.R.; Grier, C.C.; Meljllo, J.M.; Reiners, W.A.; and Todd, R.L. 1979. Nitrate losses from disturbed ecosystems. *Science* 204:469–74.

Vitousek, P.M., and Reiners, W.A. 1975. Ecosystem succession and nutrient retention: a hypothesis. *BioScience* 25:376–81.

Wang, H.F., and M.P. Anderson. 1982. *Introduction to Groundwater Modeling.* Finite difference and finite element methods. San Francisco: W.H. Freeman.

Warrick, A.W.; Mullen, G.J.; and Nielsen, D.R. 1977. Scaling field-measured soil hydraulic properties using a similar media concept. *Water Resources Research* 13:355–62.

16. Landscape Dynamics Models

Gray Merriam, Kringen Henein, and Kari Stuart-Smith

16.1 Introduction

Landscapes are never static; their elements are in constant temporal and spatial flux. Seasonality is a common cause of temporal landscape dynamics. The annual turnover of plants, wild and crop, affects the quality of the environment by changing resource availability over time. In north temperate areas, winter decisively punctuates these changes. Spatial dynamics on several time scales, some reversible and some directional, arise from cultural management of the landscape. In agricultural areas, crop rotation shifts plant species (e.g., hay alternates with corn), resulting in habitat changes for local animal populations.

The animals have their own inherent spatial and temporal dynamics. For example, the reproductive hiatus of winter moves a north temperate population of white-footed mice (*Peromyscus leucopus*) into spring with only reproductively mature individuals, and numbers may be very low because of overwinter mortality (Middleton and Merriam 1981). As the species restocks during its reproductive period, populations undergo demographic changes: numbers grow and age and sex ratios shift over time, as may behavioral and genetic quality.

Many conceptual models of species in static landscapes have been developed (e.g.,Buechner 1987; Stamps et al. 1987; Trewhella and Harris 1988), but they are not discussed here. This chapter specifically addresses landscape dynamics models. These models simulate doubly dynamic systems in which elements of changing landscape patterns interact in space and time with demographically and spa-

Table 16.1. Summary of Model Applicability to Landscape Characteristics and Demographic/Behavioral Variables

Landscape	Demography/Behavior	Example	Text Discussion
			Applicability
Corridor quality changes Addition or removal Quality management Seasonal vegetation changes Interaction of patches/corridors with land use change/rotation	Interpatch dispersal Metapopulation demography Resource patch utilization Corridor mortality effects Trait group gene flow	Single species (e.g., woodland mice *Peromyscus leucopus* or chipmunks *Tamias striatus* in farmland)	Corridor
Matrix/patch quality changes Addition or removal Quality management Seasonal vegetation changes Land use change/rotation Successional change	Diffusive dispersal Interpatch Intrapopulation Population/metapopulation demography Diffusion mortality effects Demic structure/gene flow	Single species (e.g., woodland mice in farmland or extensive forest)	Noncorridor

Habitat patches few and isolated by
  matrix
Habitat patches distinct and
  relatively homogeneous
Habitat patches shift annually
	Demography of patch population
	  foci
	Low-success/high-vagility dispersal
	  pool
	Constrained habitat choice behavior
	Quasi-epidemiological demography
		Highly dispersive single species
		(e.g., cabbage white butterfly
		Pieris rapae in cabbage
		patches scattered in farmland)
			Pool

Model			Example[a]
Pool/corridor hybrid	Much matrix penetrable or habitat with overlaid corridor network Matrix considered pool divisible into spatial or habitat patches Differential probability of use of movement routes Within patches Between landscape elements Seasonal With landscape change	Applicable as for corridor model plus Patch or nonpatch populations and combinations Diffusive and corridor dispersal Programmable movement behavior for age/sex groups and individuals	Single species (e.g., woodland mice in farmland or extensive forest)
Sequential habitat	Successional habitat, patchy or extensive Natural succession Predictable alteration No interpatch analysis	Species assemblage dynamics driven by habitat changes Programmable pairwise competition No single-species demography	Multispecies (e.g., avifauna in successional forest)
3-D mosaic	3-dimensional distribution of penetrable patches within 3-D structural mosaic Multiple-patch scales Temporally variable matrix and patches Species-specific or shared patches	Species assemblage dynamics driven by behavior/ecology of species interacting with habitat No single-species demography	Multispecies (e.g., carnivorous bats in dry tropical forest canopy)

[a] Includes examples of use for model types discussed.

tially dynamic populations. The main question considered by this class of model is this: How do organisms match their dynamics to the fluctuations of resources they require for survival? The inverse is equally interesting: What is the spatio-temporal array of resources made available by the dynamic landscape that could supply the requirements for an organism's survival? Table 16.1 summarizes the various types of landscape dynamics models that we discuss and their utility in landscape ecology.

Species-landscape interactions can occur at many different temporal and spatial scales. Examples of rapid changes in unmanaged landscapes include wildfires, flooding, disease, and defoliation (Pickett and White 1985), as well as some types of cultural management (e.g., crop rotation). The effects of climatic change, growth and shifting of human populations, and forest fragmentation and depletion caused by technological trends in landscape management occur more slowly. Spatially, the relationship between species and landscape can be approached on scales ranging from the home range and habitat choice of an individual (Morris 1987) up to the geographic range of species or ecological communities (Seagle 1986).

Animal dispersal is the conceptual link between processes and spatial distributions. In a patchy landscape, this dispersal can be modeled in terms of individuals moving between patches with differential, patch-specific rates of mortality, natality, and resource availability. Successful dispersal is not simply a function of landscape structure, however. It also is based on the behavior of the animal. Connectivity (Merriam 1984) refers to the potential for species' access to landscape elements and includes both physical and behavioral criteria. The elements of connectivity include corridor connections, patch and corridor quality, and patch and corridor spatial position. These elements can influence the size, stability, and persistence of patch-based populations (Lefkovitch and Fahrig 1985; Seno 1988; Henein and Merriam, 1990; Stuart-Smith and Merriam, in preparation).

Modeling the relationship of population performance to the spatial distribution of populations in the landscape goes back at least to Reddingius and den Boer (1970), who modeled beetle dispersal to show the importance of exchange among patches in a heterogeneous landscape. Roff (1974) extended this work with an analytical model that demonstrates how dispersal reduces the fluctuations of local populations in such patchy areas. Even in relatively stable environments, where these fluctuations are small and empty sites may be unavailable, the selective value of an exchange of individuals among populations has been demonstrated with the use of models (Hamilton and May 1977; Comins et al. 1980).

Recent work has extended this recognition of the importance of patchiness by using models designed to investigate metapopulations. A metapopulation (Levins 1970) refers to a population of several subpopulations in scattered habitat patches separated from each other by nonhabitat. These subpopulations may be interconnected to varying degrees by individual movement. Metapopulation models have been applied to a wide range of organisms, including beetles, aphids, newts, butterflies, small mammals, and birds (e.g., den Boer 1970; Addicott 1978; Gill 1978; Merriam 1984; Ehrlich and Murphy 1987; Kareiva 1987; Urban et al. 1988).

However, these models have dealt with the long-term dynamics of a metapopulation in an unchanging set of habitat patches. The focus in this chapter is either on models that have been designed to assess the interaction of the organisms and their demographic dynamics with the landscape and its spatiotemporal fluctuations or on models with the potential to incorporate landscape changes.

We categorize these models according to their utility for studying particular aspects of landscape ecology and give some examples illustrating their use. Categories include (1) corridor models in which connectivity is the key element affecting demographics (Fahrig and Merriam 1985; Lefkovitch and Fahrig 1985; Seno 1988; Henein and Merriam, 1990); noncorridor models in which dispersal takes place on a broader front and corridors are not a factor (Urban and Shugart 1986; Verboom et al., in prep.); pool models in which dispersal may be indirect through a pool of potential immigrants (Fahrig and Paloheimo 1988); and mixed pool/corridor models for species that use a landscape matrix as a subdivisible pool with an overlay of directed corridor movement (DeAngelis et al. 1977; Merriam 1988; Stuart-Smith and Merriam, in prep.; Merriam and Gauvin, unpub.). Other types of dynamic landscape models discussed are those for species assemblages (Dueser et al., 1986; Seagle 1986) and a proposed three-dimensional model using behavioral parameters to relate a community of species to the geometric structure of the landscape. Program listings are available on a disk exchange basis for any unpublished models presented here.

16.2 Corridor Models

The effects of various levels of connectivity on populations that disperse along corridors can be detected by simulating the population dynamics of the species in connected and unconnected patches. Models of simple spatial distribution and dispersal offer insight into the elements determining connectivity and their importance to a metapopulation. This is basic to an understanding of the effects of landscape pattern.

Fahrig and Merriam (1985) developed a simple corridor model to answer the question, "Does population survival in *Peromyscus leucopus* depend on the degree to which the population is isolated from others?" The model followed the population dynamics of *P. leucopus* in a number of interconnected patches over time. A series of Leslie matrices was used to compute weekly population sizes based on stochastic rates of birth, death, and graduation from one age class to another. A matrix giving the location of connections in the landscape as the probability of successful travel between any two patches controlled the movement of dispersers. For various spatial arrangements of a four-patch metapopulation over 100 years, the model clearly showed that local extinctions occur more frequently for *P. leucopus* populations in isolated woodlots than in connected ones (Fig. 16.1A). An independent field study validated the model's predictions (Fahrig 1983), confirming the importance of connectivity to *P. leucopus* survival in a fragmented landscape.

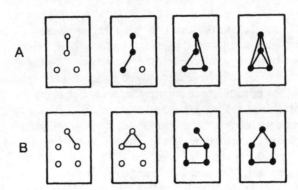

Figure 16.1. Occurrence of local extinctions over 100 years for 4-and 5-patch metapopulations with different patterns of connectivity: (A) 4-patch metapopulations and (B) 5-patch metapopulations. Hollow symbols are local extinctions.

Lefkovitch and Fahrig (1985) isolated the elements of connectivity in 'this model by expanding the metapopulations to five patches and testing various attributes of these five-patch arrangements. The probability of survival for a population depended on whether or not its patch was connected to others and increased significantly with the size of the largest geometric figure of which it was a part; that is, a patch in a group of five connected as a polygon had less chance of local extinction over a 100-year period than did a patch connected as part of a linear or triangular figure (Fig. 16.1B). Habitat type or quality can be made dynamic in this model by the assignment of patch-specific demographic parameter values to each of the four, five, or more patches on any chosen time schedule.

Seno (1988) developed a mathematical model to study the effects of a "singular patch" on the persistence of a metapopulation composed of a linear series of connected patches. All the patches had similar rates of growth and emigration, except the singular patch. Migration was possible between any two adjacent patches, with some built-in leakage to and from the system. Eigenvalues were calculated and used to quantify and compare different positions of the singular patch within the linear arrangement. By varying the position of this patch in the linear series and giving it lower or higher quality based on growth rate and emigration, Seno found that (1) the more centrally located the singular patch, the greater was its effect on population persistence; (2) a critical number of total patches existed below which persistence became dependent on the quality of the critical patch; and (3) even with a low-quality, centrally located singular patch, metapopulation persistence was still possible if a subsystem could survive on its own. These results have implications for changes to archipelago populations or other linearly arranged metapopulations.

These corridor models assume that all connections between patches contribute equally to connectivity. However, this is not true in nature (Merriam and Lanoue, 1990). Henein and Merriam (1990) have shown that corridor quality is critical to metapopulation dynamics. In their deterministic model, corridor quality was based on the survival rates of dispersers, with high-quality corridors having higher survival rates than low-quality corridors. Metapopulations with only high-quality connections between patches supported larger and more persistent populations than similar arrangements with any combination of high-and low-quality connections. Low-quality corridors, with their increased mortality, were a drain on metapopulation numbers. With corridor quality held constant, metapopulations with more patches always produced bigger populations. However, a three-patch metapopulation with high-quality corridors between patches maintained a larger population than did a four-patch metapopulation when the additional patch was connected by low-quality corridors (Fig. 16.2). The model also predicted that geometrically isolated patches connected by low-quality corridors were most likely to suffer local extinctions. With the addition of variable corridor quality, the elements of connectivity, as determined by Lefkovitch and Fahrig (1985), require modification to include the fact that large geometric figures increase the size and stability of a metapopulation only if the connections are predominantly of high quality. These results begin to address the notion that corridors may have costs as well as benefits (Simberloff and Cox 1987). Temporal changes to corridor quality in space can be incorporated into this model, and the model can also be modified to simulate differential quality of patches and spatiotemporal changes in patch quality.

Corridor models can be used to isolate variables affecting metapopulation demography (including dispersal) and to reveal metapopulation effects of habitat choice behavior and of changing resource availability. These models can be applied at the landscape scale to questions of changing vegetation type, land use, and management of corridor quality.

16.3 Noncorridor Models

Noncorridor dispersal models are useful for species whose movements are not restricted to corridors. Elements in these models include a source population, a direction, and a probability of success based at least partly on distance.

Urban and Shugart's (1986) model of avian demography in a mosaic of discrete

 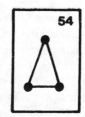 

Figure 16.2. Size of metapopulation after 10 years for varying numbers and qualities of connections (redrawn from Henein and Merriam, 1990): high-quality connection ———, low-quality connection ------, size of metapopulation after 10 years (nn).

habitat patches simulated a metapopulation for a mobile group whose demographic response to patchy landscapes is not clear (compare Opdam et al. 1985;
Lynch 1987; Urban et al. 1988). Emigration of birds without territories (floaters)
from one habitat patch to another was modeled by the use of an electrical flow
analogy to control the probability of dispersal between patches. Interpatch distance, navigability of the matrix between patches, and attraction based on site
fidelity and territory availability were the factors driving the system. The model
was based on life history attributes of the species rather than on inferred relationships such as minimum area requirements and incidence functions and thus better
represents landscape and population dynamics than do approaches based on island
biogeographic theory.

Current landscape management research in The Netherlands includes the development of several "winking patch" models to simulate metapopulations of
potentially interconnected patches in a dynamic landscape (Verboom et al., in
prep.) The landscape is modeled as a square grid of patches, either occupied or
empty, which are linked by dispersal. Two variants of the model were developed:
(1) a corridor model in which each patch supplies emigrants to its four immediate
neighbors and (2) a pool design where all patches are equally accessible. In the
first version, the colonization rate is proportional to the number of occupied
nearest neighbors, while in the second, with no distance effect, the colonization
rate is proportional to the total number of occupied patches in the grid. A more
recent version of the model falls into this noncorridor category because patches are
randomly distributed and dispersal is related to distance through a Bessel function.
Successful colonization rate here depends on the distance of other source populations from a patch. All three versions of the model assume negligible immigration
into occupied patches, but these may suffer local extinctions from catastrophes
occurring randomly in time and space in combination with stochastic demographic
changes. Survival of the metapopulation depends on the ratio between colonization and extinction of patches. Such generalized models may be adapted for
different species by calculation of appropriate colonization and extinction rates
and can be used to predict the changes in distribution of a population in a
fragmented landscape over a long period.

16.4 Pool Models

Pool models differ from the previous types by the presence of a pool of dispersing
individuals with potential access to otherwise unconnected landscape elements.
This pool reduces the effects of landscape structure, source, distance, and colonization behavior, leaving time and chance as driving variables. Since the pool is
usually large relative to patch capacity, it can dominate the system.

Fahrig and Paloheimo (1988) designed a pool model for a metapopulation of
cabbage butterflies (*Pieris rapae*). The metapopulation consisted of three sets of
patches, suitable for ovipositing, located at varying distances from each other. The
model simulated the growth of the butterfly population in each patch, using Leslie

matrices for both within- and between-patch dynamics. Seventy-four percent of the population had low host plant detection ability and dispersed long distances in random directions. These factors, combined with a system open to butterflies originating outside the study site, created a large pool of potential immigrants that did not travel directly from patch to patch but could remain between patches for several days. Because the numbers in the pool were so large, the number of individuals going between patch and pool greatly exceeded direct dispersal from one patch to another. Figure 16.3 illustrates this finding, showing one isolated patch from which individuals can only reach the pool and two proximate patches between which individuals can exchange directly. Butterflies from the pool can reach the patches only by moving to within range of some detection gradient.

This model shows that the spatial location of patches has a minimal effect on the relative size of patch populations but that these population sizes are influenced by the dispersal characteristics of the species. The pool effectively lowers the importance of direct dispersal between patches, a process that results in a similar immigration rate for all patches and reduces the effect of spatial arrangement on local population abundance. In such cases, the spatial arrangement of patches can have a relatively small influence on metapopulation dynamics. This is in contrast to the results from corridor models and generates questions concerning how exact the knowledge of an organism's dispersal behavior must be to predict metapopulation effects of habitat fragmentation. These findings support the concept that connectivity is intrinsically tied to species behavior and confirm an earlier suggestion that habitat fragmentation affects different species in different ways (Merriam 1984; Baudry and Merriam 1988). There are important implications here for conservationists and for those concerned with the design of nature reserves and the management of fragmented habitats.

16.5 Combined Pool/Corridor Models

Another type of model combines elements of both pool and corridor designs and allows a dynamical landscape pattern to affect the demography of a particular

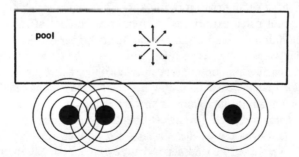

Figure 16.3. Spatial arrangement of cabbage patches with respect to each other and to a pool of cabbage-patch butterflies. Butterflies may move between intersecting units.

species. One example modeled the habitat choice and movement behavior of the white-footed mouse (*P. leucopus*) in an agricultural landscape (Merriam 1988; Stuart-Smith and Merriam, in prep.). This model allowed direct movement of mice along fencerow corridors and indirect movement through the farmland matrix, treated here as the pool. The pool was subdivided into patches of different habitat, each with different mouse demographic parameters. Movement was permitted between all adjacent pairs of patches and was therefore possible indirectly between any two points in the landscape, in addition to corridor movement between selected patches. Habitat types could be changed to simulate seasonal variations and crop rotation.

This model accommodates recent field data showing that *P. leucopus* at the northern edge of its range is not restricted to woody habitats but also is adapting to move and breed in both corn and grain fields (Wegner and Merriam, in press). As in the pool model of Fahrig and Paloheimo (1988), a large percentage of the population dispersed, creating a sizable pool of immigrants. These mice flowed across and could breed differentially in any suitable habitat patch. Cornfields, grainfields, woodlots, fencerows, and farm buildings were all treated as patches. Each patch was given a separate set of probabilities for mouse reproduction, death, emigration, and immigration to be applied weekly to mice within or passing through that patch. These probabilities can be determined empirically and may vary substantially between different habitat types. For example, fencerows showed a higher immigration rate because they were connected to more patches than were other landscape units and were therefore more likely to be used as movement corridors.

Runs of the model revealed that patch population size was related not only to the quality of the particular patch and the number of other patches to which it was connected but also to the quality of adjacent patches. For example, a hay field (an unsuitable patch), acted as a barrier to movement among surrounding fields and led to the occurrence of local extinctions in these neighboring patches. This result suggests that rotation of crops in space (a common three- to five-year practice) could have a significant influence on local population dynamics.

Another interesting result involved patch quality and the distribution of spring populations. A small number of adults in every patch, regardless of quality, at the beginning of the breeding season resulted in a low overall metapopulation over time, with several patch extinctions. An identical number of mice in the same landscape beginning the season exclusively in the higher-quality patches (woodlots or cornfields) resulted in a much larger metapopulation over the same period.

Finally, increasing mortality in fencerow corridors had a greater effect on the metapopulation than increasing mortality in fields. This finding may be attributed to the higher probability that animals encounter the corridors and subsequently move along them. In a highly connected landscape, changes in corridor quality may be important even for species not restricted to their use.

Behavioral responses to landscape changes are important in these models. Behavior modifications may occur over short time periods. For example, we know that white-footed mice in the Ottawa area have adapted to make use of grain corn

(*Zea mays*) within the last few decades, because historical records show that this crop was previously uncommon. Such a response can be driven by heterogeneity in a landscape and may ultimately influence evolution. The ability to adapt behaviorally could be a better predictor of species' long-term survival than stereotyped habitat specificity has been. Figure 16.4 shows how demographies in a landscape mosaic conform to a hierarchical system, with the landscape pattern at the top and selection of individuals at the bottom. Patch populations and metapopulation dynamics are driven from below by individual reproductive life history and constrained or shaped from above by the landscape pattern (Urban and Shugart 1988). Population features interact with spatial features through behavioral mediators such as dispersal. If the landscape pattern and species behavior interact to give low connectivity, local extinctions may accumulate despite selection on individuals.

Another *P. leucopus* model has been designed to track individuals with programmable behavioral characteristics and to predict their location and success in a changing agricultural mosaic over time. Each individual animal was programmed with reproductive life history information and a set of probabilities to control its patch-to-patch movement. The probabilities may be temporally varied among and within individuals. The landscape was modeled as a set of patches of different types (grainfields, cornfields, woodlots, fencerow corridors, etc.). Move-

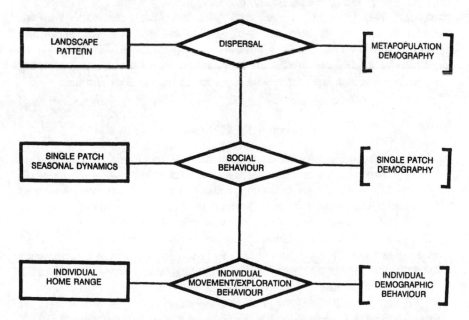

Figure 16.4. Hierarchy of spatial elements (left) and demographic elements (right) interacting through behavioral mediation. Selection of individuals acts at bottom right. Landscape pattern constrains from top left to top right by potential extinction of metapopulations.

ment was related to both the habitat patches and the behavioral characteristics of the mice. Because field measurements can be used to parameterize this model, it provides a tool for predicting the movement and survival patterns of a tagged study population in a specific landscape based on demographic and behavioral composition (Merriam and Gauvin, unpub.).

DeAngelis et al. (1977) designed a digital model to simulate seed dispersal by organisms such as small mammals or birds. Although the objective was to predict the distribution pattern of seeds from a source rather than dispersal of animal populations, the techniques employed and the behavior modeled are appropriate to this discussion. Animal movement away from a seed source was modeled stepwise on a grid system by the use of randomly selected directions modified by the animal's behavioral interactions with features in the landscape (i.e., some species are inclined to follow corridors, while others are less restricted by the landscape). The tendency for the animal to continue its direction, the influence of attractor points and repulsion points, and the boundaries between habitats all influenced the direction of movement. There are some similarities to the previous model. For example, both models are individual-based. Such models permit the integration of parameters from several levels of hierarchical scale and can cause the animals to interact more precisely with their habitat, providing insights not possible using age-or sex-class-based simulations (Huston et al. 1988).

Hybrid pool/corridor models allow individually variable parameter values for all landscape elements of interest and permit temporal dynamics of these on any time scale. This flexibility, together with the programmable reciprocity between corridor dispersal and diffusive dispersal, makes these models broadly applicable to species moving through complex mosaics. These characteristics make hybrid pool/corridor models very useful for experimenting with the effects of varying connectivity levels in landscapes with a patchy matrix, such as farmland.

16.6 Sequential Habitat Models

When a spatial distribution of landscape patches shifts over time, the composition of surviving species in that landscape may be influenced not only by the availability, distribution, and size of suitable habitat sites but also by the number of other species competing for them. Models in this class can simulate species turnover and restructuring of species assemblages.

Dueser et al. (1986) coupled a multivariate habitat prescription to a forest gap revegetation model (Shugart and West 1977) to generate a dynamic landscape model that will predict the changing array of habitat types in a forest. This model is based on the premise that multivariate habitat structure will predict species-specific habitat quality and that predictive equations can be used to project the dynamic avifaunal composition after the fashion of Anderson and Shugart (1974). Parameter values for the FORET model are readily available for many forest types and can incorporate management or disturbance influences. A two-group discriminant function is used as a classification function to couple the habitat structures to the species's requirements. The authors point out that the model does not accommodate behavioral or demographic adaptations by the species to the land-scape dynamics. However, the model could easily predict results from known behavioral modifications of habitat preference because it will forecast the dynamic availability of habitat types for the future forest. This ability would be particularly useful in fragmented rather than fully forested landscapes.

Seagle (1986) modeled the same process in a dynamic landscape with the use of a first-order Markov model. Variables included landscape size, disturbance frequency, and competition among colonizing species. A pool of 150 vertebrate species provided potential colonists, and a habitat transition matrix contained the probabilities of any patch of one cover type becoming a patch of any other cover type in a study year. Thus patch dynamics were local and relatively unpredictable. A competition coefficient was calculated for each pair of species, incorporating a competition index and a habitat overlap for each pair. The habitat range and size tolerance for each species defined its niche. The number of available patches in the landscape that met these criteria determined the carrying capacity of the landscape for that species. Carrying capacities varied through time because of the dynamic nature of the landscape. Larger landscapes showed greater habitat diver-sity and constancy, species richness increased with increasing landscape area, and species-area curves rose rapidly to a limit of 38 species with no competition. Small landscapes supported smaller animal populations, and these were more susceptible to local extinctions. Competition had little effect on the slope of the species-area curve but lowered the equilibrium number of species supported by the landscape. The probability of extinction for a species thus increased in landscapes where species were allowed to compete.

16.7 Three-Dimensional Landscape Mosaics– An Immediate Possibility

Model predictions would be useful to direct the experimental analysis of how species assemblages fit the three-dimensional habitat resource template (South-·wood 1977). Three-dimensional modeling has been used extensively in engineer-ing applications (Barsky 1984) and in architecture (Arnold 1973) but rarely in landscape ecology. Thompson et al. (1974) modeled the foraging behavior of flocking birds in trees, a three-dimensional phenomenon, by reducing the habitat to two dimensions and varying the parameters to simulate the vertical gradient.

A vector approach to the problem of movement in three-dimensional space is currently under development for the diverse bat fauna occupying the dry tropical forests of the Zambesi valley in Zimbabwe. Flight capability dictates the size of open spaces above, within, and below the forest canopy that each bat species can use (Aldridge and Rautenbach 1987). This is calculated from bat parameters such as wingspan, wing loading, aspect ratio, wing-tip shape, flight speed, and turning radius (Findlay and Wilson 1982). Whether each species can effectively navigate and forage in those openings is further constrained by the capability of their specific biosonar system, which is calculated from parameters for signal frequency, bandwidth, power, modulation, and pulse rate. The habitat mosaic is modeled as an array of open spaces in the three-dimensional structure of the forest (Fig.16.5). Bat movement and foraging success are determined by an algorithm based on flight and biosonar capabilities and stochastic elements.

Both the prey distribution and the habitat template could be given a variety of dynamics in the model, for example, if wet and dry season changes were applied to both. Such a simulation should generate hypotheses that can be tested by independent data from ecological and behavioral experiments. Predictions would also be possible about the fit of the whole array of bat species, or any subset of it, to any static or dynamic three-dimensional habitat template. The model could challenge the array of species to maintain a "best fit" to a dynamic habitat template specified by the most probable landscape scenario, given current major forces such as human populations, economics, land use practices, and climate.

By eliciting needed parameter values from planners, population geneticists, and behaviorists, models like this could improve our understanding of the pattern of species' successes and losses as fitness requirements are shuffled in space and time by the dynamics of the landscape and could ultimately help us predict which species will survive.

16.8 Looking Ahead

Understanding and predicting population changes in a heterogeneous landscape requires consideration of both demographic and spatial dynamics and the linkage of the two by dispersal. Dispersal allows the needs generated by seasonally changing demography to be met by the resource mosaic as it shifts in time and space. Landscape dynamics models can address such systems by incorporating the behavioral parameters that mediate the interactions between demography and

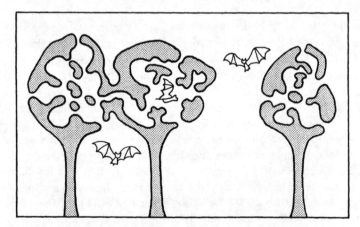

Figure 16.5. Bat movement above, below, and within a modeled three-dimensional forest tree canopy.

spatial configuration. Behavioral variability within and between individuals through time and among behavioral subsets can all clearly affect the linkage between demographic and spatial aspects.

We believe these models to be widely applicable in scale and in topic. However, their full assessment awaits the field studies needed to give improved parameter values and to test many conventional assumptions. Beyond good working models, additional field experiments will be required to answer questions arising from the models. What is the relationship between mortality risk and corridor travel? What are the elements of corridor quality? What are the elements of movement behavior, exploratory behavior and habitat choice, and how do they vary between and within individuals over time? How are metapopulations bounded, and what are the relationships between these limits and the spatial configuration of genetic trait groups? Investigations of questions relating to the integrity, interactions, and spatial configurations of metapopulations may best be directed through these models, as may similar questions about genetic trait groups and their relationships to demographic units. It is clear that changes in demography, environment, and genetics must all be considered in any landscape dynamics model that hopes for useful predictive power. It is in such models that the artificial separation of demographics from genetics could be alleviated, permitting a comprehensive approach to problems of conservation biology in heterogeneous environments.

16.9 Summary

Landscapes and populations undergo changes seasonally and over longer time periods, and species interact with landscapes at many different spatial scales. This results in a doubly dynamic system in which elements of changing landscape pattern constrain the populations they support in space and time. The populations have their own inherent temporal and spatial dynamics and must match available resources to their shifting needs. Dispersal provides the link between spatial distribution of resources and population dynamics that determines species' success over time.

Models of spatial distributions of populations that relate landscape dynamics to species' performance help uncover some of the complexities facing conservation biology. Different models are appropriate to behaviorally different organisms and may be categorized accordingly. Metapopulations that use linear landscape features as connections between habitat patches may be studied with corridor models. Species that move between habitats across nonlinear landscape elements require a noncorridor model.

Some organisms leave a habitat patch in a completely undirected pattern and move extensively before detecting new habitat patches. Pool models are appropriate for such species. Model types may be combined to provide for species with more than one type of movement behavior, and these models may also be applicable to species assemblages in dynamic landscapes. Three-dimension modeling has potential for species not restricted to movement along the ground even where habitat types cannot be aggregated by simple stratification.

A well-designed model can direct and complement field studies. Parameterized with reliable demographic, environmental, and genetic information, a model enables preliminary experimentation at the landscape scale. The results can provide testable hypotheses about short- and longer-term effects of dynamic, heterogeneous environments on populations. Field testing of these predictions allows an economic, explicit investigation of multiscale, heterogeneous systems.

References

Addicott, J.F. 1978. The population dynamics of aphids on fireweed: a comparison of local and metapopulations. *Canadian Journal of Zoology* 56:2554–64.

Aldridge, H.D.J.N., and Rautenbach, I.L. 1987. Morphology, echolocation, and resource partitioning in insectivorous bats. *Journal of Animal Ecology* 56:763–78.

Anderson, S.H., and Shugart, H.H. 1974. Habitat selection of breeding birds in an east Tennessee deciduous forest. *Ecology* 55:828–37.

Arnold, D. 1973. A computer model of housing layout: 1. sunlight and daylight analysis. *Land Use and Built Form Studies*. Working Paper No. 77, University of Cambridge Department of Architecture, Cambridge, England.

Barsky, B. 1984. A description and evaluation of various 3-D models. *Computer Graphics and Applications* (Institute of Electrical and Electronic Engineers) 4:38–52.

Baudry, J., and Merriam, G. 1988. Connectivity and connectedness: functional vs. structural patterns in landscapes. In *Connectivity in Landscape Ecology*, ed. K.F. Schreiber, pp. 23–28. Munster Geographische Arbeiten 29.

Buechner, M. 1987. A geometric model of vertebrate dispersal: tests and implications. *Ecology* 68(2):310–18.

Comins, H.N.; Hamilton, W.D.; and May, R.M. 1980. Evolutionarily stable dispersal strategies. *Journal of Theoretical Biology* 82:205–30.

DeAngelis, D.L.; Stiles, E.W.; Johnson, W.C.; Sharpe, D.M.; and Schreiber, R.K. 1977. A model for the dispersal of seeds by animals. EDFB/IBP–77/5, Oak Ridge National Laboratory, Oak Ridge, Tenn.

den Boer, P.J. 1970. On the significance of dispersal power for populations of carabid beetles (Coloeoptera, Carabidae). *Oecologia (Berlin)* 4:1–28.

Dueser, R.D.; Shugart, H.H.; and Connor, E.F. 1986. The dynamic landscape approach to habitat management. In *Wilderness and Natural Areas in Eastern North America: A Management Challenge*, eds. D.L. Kulhavy and R.N. Conner. Nacogdoches, Texas: Stephen F. Austin State University.

Ehrlich, P.R., and Murphy, D.D. 1987. Conservation lessons from long-term studies of checkerspot butterflies. *Conservation Biology* 1:122–31.

Fahrig, L. 1983. Habitat patch connectivity and population stability: A model and case study. Master's thesis, Carleton University, Ottawa.

Fahrig, L., and Merriam, G. 1985. Habitat patch connectivity and population survival. *Ecology* 66:1762–68.

Fahrig, L., and Paloheimo, J. 1988. Effect of spatial arrangement of habitat patches on local population size. *Ecology* 69(2):468–75.

Findley, J.S., and Wilson, D.E. 1982. Ecological significance of chiropteran morphology. In *Ecology of Bats*, ed. T.H. Kuntz, pp. 243–60. New York: Plenum Press.

Gill, D. 1978. The metapopulation ecology of the red-spotted newt *Notophthalmus viridescens* (Rafinesque). *Ecological Monographs* 48:145–66.

Hamilton, W.D., and May, R.M. 1977. Dispersal in stable habitats. *Nature* 269:578–81.

Henein, K.M., and Merriam, G. 1990. The elements of connectivity where corridor quality is variable. *Landscape Ecology* 4:157–70.

Huston, M.; DeAngelis, D.; and Post, W. 1988. New computer models unify ecological theory. *Bioscience* 38(10):682–91.

Kareiva, P. 1987. Habitat fragmentation and the stability of predator-prey interactions. *Nature* 326:388–90.

Lefkovitch, L.P., and Fahrig, L. 1985. Spatial characteristics of habitat patches and population survival. *Ecological Modelling* 30:297–308.

Levins, R. 1970. Extinction. In *Some Mathematical Questions in Biology. Lectures on Mathematics in the Life Sciences*, vol. 2, ed. M. Gerstenhaber, Providence, R.I.: American Mathematical Society.

Lynch, J.M. 1987. Responses of bird communities to forest fragmentation. In *Nature Conservation: The Role of Remnants of Natural Vegetation* eds. D.A. Saunders, G.W. Arnold, A.W. Burbidge, and A.J.M. Hopkins, pp. 123–40. Chipping Norton, New South Wales: Surrey Beatty and Sons Pty Ltd.

Merriam, G. 1984. Connectivity: a fundamental ecological characteristic of landscape pattern. In *Methodology in Landscape Ecological Research and Planning*, vol. 1, eds. J. Brandt and P.A. Agger, Roskilde, Denmark: University Centre.

Merriam, G. 1988. Modelling woodland species adapting to an agricultural landscape. In *Connectivity in Landscape Ecology*, ed. K.F. Schreiber, Munster Geographische Arbeiten 29.

Merriam, G., and Lanoue, A. 1990. Corridor use by small mammals: field measurements for three experimental types of *Peromyscus leucopus*. *Landscape Ecology*, 4:123–31.

Middleton, J.D., and Merriam, G. 1981. Woodland mice in a farmland mosaic. *Journal of Applied Ecology* 18:703–10.

Morris, D.W. 1987. Ecological scale and habitat use. *Ecology* 68(2):362–69.

Opdam, P.; Rijsdijk, G.; and Hustings, F. 1985. Bird communities in small woods in an agricultural landscape: effects of area and isolation. *Biological Conservation* 34:333–52.

Pickett, S.T.A., and White, P.S. 1985. *The Ecology of Natural Disturbance and Patch Dynamics*. New York: Academic Press.

Reddingius, J., and den Boer, P.J. 1970. Simulation experiments illustrating stabilization of animal numbers by spreading of risk. *Oecologia (Berlin)* 5:240–84.

Roff, D.A. 1974. Spatial heterogeneity and the persistence of populations. *Oecologia (Berlin)* 15:245–58.

Seagle, S.W. 1986. Generation of species-area curves by a model of animal habitat dynamics. In *Wildlife 2000: Modeling Habitat Relationships of Terrestrial Vertebrates*, eds. J. Verner, M. Morrison, and C.J. Ralph, pp. 281–85. Madison: University of Wisconsin Press.

Seno, H. 1988. Effect of a singular patch on population persistence in a multi-patch system. *Ecological Modelling* 43(3/4):271–85.

Shugart, H.H., and West, D.C. 1977. Development of an Appalachian deciduous forest succession model and its application to assessment of the impact of the chestnut blight. *Journal of Environmental Management* 5:161–79.

Simberloff, D., and Cox, J. 1987. Consequences and costs of conservation corridors. *Conservation Biology* 1(1):63–71.

Southwood, T.R.E. 1977. Habitat, the templet for ecological strategies? *Journal of Animal Ecology* 46:337–65.

Stamps, J.A.; Buechner, M.; and Krishnan, V.V. 1987. The effects of edge permeability and habitat geometry on emigration from patches of habitat. *American Naturalist* 129(4):534–52.

Stuart-Smith, K., and Merriam, G. Effects of level and spatial configuration of patch quality in a metapopulation model for chipmunks (in preparation).

Stuart-Smith, K., and Merriam, G. A landscape model with linear corridor movement and non linear patch moment (in preparation).

Thompson, W.A.; Vertinsky, I.; and Krebs, J.R. 1974. The survival value of flocking in birds: a simulation model. *Journal of Animal Ecology* 43:785–820.

Trewhella, W.J., and Harris, S. 1988. A simulation model of the patterns of dispersal in urban fox (*Vulpes vulpes*) populations and its application for rabies control. *Journal of Applied Ecology* 25(2):435–50.

Urban, D.L., and Shugart, H.H. 1986. Avian demography in mosaic landscapes: modeling paradigm and preliminary results. In *Wildlife 2000: Modeling Habitat Relationships of Terrestrial Vertbrates*, eds. J. Verner, M. Morrison, and C.J. Ralph, pp. 273–79. Madison: University of Wisconsin Press.

Urban, D.L.; Shugart, H.H., Jr.; DeAngelis, D.L.; and O'Neill, R.V. 1988. Forest bird demography in a landscape mosaic. ORNL/TM–10322, Oak Ridge National Laboratory, Oak Ridge, Tenn.

Verboom, J.; Lankester, K.; and Metz, H. A metapopulation model: the case of the European Badger (in preparation).

Wegner, J., and Merriam, G. Use of spatial elements in a farmland mosaic by woodland rodents. *Biological Conservation* (in press).

17. Simulation Methods for Developing General Landscape-Level Hypotheses of Single-Species Dynamics

Lenore Fahrig

17.1 Introduction

As described in the introductory chapter (Turner and Gardner, Chapter 1), landscape ecology includes a wide range of concepts. However, one of the dominant themes is the importance of spatial and spatiotemporal heterogeneity in landscape pattern on ecological processes. At the level of the single population, the main aim in studies of landscape ecology is to answer this question: Does our understanding of population dynamics depend on the spatial or spatiotemporal heterogeneity of the landscape in which the population occurs? In this chapter I discuss methods for developing general hypotheses about population dynamics within a landscape.

17.2 Why the Analytical Approach Will Not Work

A major goal of population ecology is the development of central, general hypotheses around which research can focus. Traditional examples are the competitive exclusion hypothesis and density-dependent (logistic) population growth. Such hypotheses usually have originated from simple, analytically tractable mathematical models; reviews of the origins of such models are included in May (1976).

In the present context the term *general* implies that the theory and model from which the hypotheses arise are, in principle, not restricted to one or a few species.

Assuming that they are measurable quantities, the terms used in the model are general enough that they could be measured for any of a broad range of species. The broader this range, the more general is the theory. This kind of generality should not be confused with the actual range of species for which the theory is found to be confirmed. Onstad (1988) uses a more narrow definition of generality; he restricts the generality of a theory to the range of species that actually fits the theory. However, the typical procedure in theoretical ecology is to first propose a theory that is not intrinsically restricted to a narrow range of species (i.e., a general theory). The actual range of applicability is not discovered until after much further work, particularly field studies (Fahrig 1988a).

Many of the simple cornerstone theories and models in ecology have recently undergone a marked loss of credibility, as they have been found to be lacking in their ability to describe observable nature (Hall 1988; Hastings 1988; Ulanowicz 1988). This development is generally attributed to the fact that these models do not include the most critical factors influencing population dynamics (DeAngelis 1988). However, inclusion of such factors often renders the models analytically intractable (Hall 1988).

One factor that makes models of population dynamics intractable is the effect of spatial heterogeneity (Fahrig 1988b). Because landscape ecology explicitly considers spatial heterogeneity, it follows that models of population dynamics in a landscape context, even those that include only simple dynamics, are analytically intractable. A possible alternative is the use of simulation models. Dynamic simulations have been used in landscape ecology to look at the changes in spatial distributions of landscape elements over time (e.g., Sklar and Costanza, Chapter 10; Turner 1987). Many landscape models do not deal with population dynamics, and they usually are specific to a particular study system, making generalization difficult (but see also Hyman et al., Chapter 18; and Merriam et al., Chapter 16).

The question I address in the remainder of this chapter is this: Is it possible to use computer simulation to develop general hypotheses of population dynamics in the landscape context?

17.3 The Simulation Approach: Problems and Preferences

The reason that analytical approaches are still largely preferred for development of general theory in ecology is that, if a solution can be found, the range of parameter values over which the solution is applicable is known perfectly. For example, the solution of the Lotka-Volterra competition equations leads to the simple hypothesis that, if interspecific competition is stronger than intraspecific competition, there is no stable equilibrium point with both species coexisting. If intraspecific competition is stronger, a stable coexistence is possible, but only if $1/a_{21} > K_1/K_2 > a_{12}$, where a_{ij} is a measure of the extent to which species j puts pressure on the resources of species i and K_i is the carrying capacity of the environment for species i.

Simulation models are fundamentally different from analytical models because to run the simulation one must specify values for each parameter in the model.

This means that the outcome of the simulation holds for the particular parameter values selected; it is not known to what extent the outcome may or may not hold for other parameter values.

There are two possible approaches to developing general hypotheses from simulations. The first is advocated by Onstad (1988). He suggests that we "create valid models for one or more specific cases and then attempt to generalize . . . ; models are produced as realistically as possible based on knowledge of a few specific cases; the modeler then attempts to generalize and tests the hypothesis in a variety of other situations." Although this may seem a reasonable approach, there are few, if any, examples of this procedure having been fully carried through. Many specific, detailed models have been developed, but these have not led to generalizations leading to general theories. This is because the detail in the model renders one unable to "see the forest for the trees," and any attempt to generalize is viewed as an oversimplification of the model.

The second possible approach, which I discuss for the remainder of this chapter, is to develop a simple simulation model with a small number of parameters, analogous to an analytical model. The model should be as simple as possible, while remaining faithful to a realistic, if general, picture of the system. The level of detail is greater than if the analytical approach were used because the constraint of analytical tractability has been removed. By conducting a large number of simulations in which the parameters take on different values, one can formulate relationships between the parameters (or combinations of parameters) and the output (i.e., population dynamics) of the model. These relationships are analogous to the general hypotheses that one develops from analytical models.

Although this approach is analogous to the analytical approach, it is not completely equivalent because, even though a large number of simulations is conducted, each simulation represents only one possible point in the parameter space. This space is infinitely large if any of the parameters is continuous or if the range of possible values for any of the parameters is unbounded. In this case it is not possible to cover the whole parameter space, no matter how many simulations are conducted. The descriptions of the relationships between the parameters and the model output (i.e., the general hypotheses) are always subject to the qualification · that they may not be reliable for some as yet unsimulated point(s) in parameter space. This problem is particularly important in landscape ecology, where there is a large array of possible landscape configurations in which the population may occur.

The success or failure of the simulation approach, therefore, depends on the choices of parameter values for the many simulations conducted. The simulation experiment is then analogous to an actual experiment in the sense that each simulation run will cost time and money, and one would like to gain the maximum possible amount of information for the number of runs conducted. The set of methods used to determine the best choices of values for the independent variables or factors in an experiment (parameter values in simulation runs) is termed "design of experiments." The principles (e.g., Hicks 1982) apply equally well to actual experiments and simulation experiments.

In the two examples that I describe in this chapter, I illustrate two of the most useful methods for conducting such simulation experiments. The first makes use of the standard factorial experimental design in which a few values are chosen for each parameter, spanning most or all of its range, and at least one simulation is run using each of the possible combinations of the chosen parameter values. The second method makes use of the Latin hypercube design in which Monte Carlo simulations are conducted; random values of the parameters are chosen for each of a large number of simulations. In both cases statistical techniques are then used to relate the parameters to the simulation outcomes.

The latter of these methods is similar to Monte Carlo approaches used in sensitivity analysis (Gardner et al. 1981; Downing et al. 1985) The purpose of sensitivity analysis, however, is different from the purpose of the analysis described here. In the former, the question is this: How sensitive is the model outcome to changes in the estimated value of a parameter? The question here is rather this: What is the qualitative relationship between the model outcome and the parameters over all of the possible parameter space? The general hypotheses are formed from these relationships.

This difference in purpose results in some differences in designs of the simulation experiments and in analyses of the simulation results. In the case of sensitivity analysis, it is usually assumed that there is either a single "correct" parameter value with some error in its estimation or that the parameter may take on a range of values described by a probability distribution with a central tendency (usually a normal distribution; e.g., Gardner et al. 1980; Warwick et al. 1986).

The current analysis differs from this approach because there is no a priori reason to choose any particular parameter value over any other. Each combination of parameter values is viewed as being representative of a particular combination of conditions. This is not to say that some situations are not more common in nature than others, but the goal is to understand the response of the model to all the parameters over all possible values in combination with all possible values of other parameters. This means that simulations must be conducted over the entire possible parameter space. For example, in the case of the Latin hypercube design, parameters are chosen from a uniform distribution that extends over the whole range for that parameter instead of from a normal distribution (as in the sensitivity analysis).

17.4 Two Examples

Since model development always has an intuitive component, it is not possible to present a perfectly standardized methodology. The choices that are made about the structure of the model ultimately depend on the question(s) that the model will be used to explore.

In the development of any simulation model there are four things that need to be resolved at the outset: (1) What are the model components? (i.e., what is in the boxes?); (2) What is the currency in the model (i.e., what is transferred into and out of and between boxes?); (3) What is the spatial scale? (i.e., what is the total spatial area represented in the model; if it is subdivided, what sizes are the pieces?); (4) What is the temporal scale? (i.e., what does one time step in the model represent in actual time?) In a population-level model, the components of the model are populations that may be subdivided by spatial location into local populations or further by age class into local age-classified populations. The currency, then, in a population-level model is the number of individuals moving into or out of or between the local age-classified populations. The appropriate spatial and temporal scales normally depend on the particulars of the species and spatial location that one has in mind.

From this description it is clear that the natural tendency is to develop models that are most appropriate for one or a small number of species, since different species will naturally have different appropriate population subdivisions by space and age class. It is also tempting to make the model specific for a particular spatial location since the same species may react differently to different landscapes. However, this level of specificity is counterproductive when one is attempting to use the model for development of general hypotheses. The challenge is therefore to develop a model that is flexible in terms of the number and kind of local populations, the ways in which individuals can move among local populations, and the spatial and temporal scales that the model can represent. A further challenge is to obtain this flexibility while still using parameters that can be obtained from observed data. In the following example I describe two general models, as well as the methods by which they have been used to develop general hypotheses.

17.4.1 Example 1: Population Response to Disturbance

In the first example I develop general landscape-level hypotheses of population dynamics for the case in which spatial heterogeneity occurs because of local disturbances. The model is a discrete-space and discrete-time simulation model. It is not restricted to a particular spatial scale or temporal scale. Instead, space is divided into a grid of spatial *cells*, and time is divided into time *steps*. The numbers of cells and steps are specified at the start of a simulation.

17.4.1.1 Disturbance

Disturbance frequency is a standard value calculated in field studies of ecological disturbance. However, including a parameter for disturbance frequency in a model is less straightforward than one might suppose. Since the basic spatial unit of the model is the cell, one would like to have a parameter that defines the probability that a cell is disturbed. However, one would also like to have the flexibility to vary the spatial extent of disturbances (i.e., number of cells disturbed) and the duration of disturbances (i.e., number of time steps disturbances last). The disturbance probability per cell would depend not only on the frequency of disturbance events, but also on the spatial extent of disturbances, the total area over which the disturbance frequency is calculated, and the duration of disturbances. Therefore, to standardize disturbance frequency, I use a parameter called the "mean disturbance incidence" (d). This is the mean probability (averaged over all cells and all steps) per time step of a cell being in a disturbed state. This is not the same thing as the probability of a disturbance event. For a particular value of d, the probability of a disturbance event will decrease as the sizes of disturbances increase or as their durations increase.

The size or magnitude of disturbances is determined by the parameter m, which is the number of adjacent cells disturbed by each disturbance event. The choice of which of the cell(s) adjacent to the initially disturbed cell is disturbed is random. The duration of disturbance events, in number of time steps, is specified by the parameter t. If t is 1, the disturbance lasts only one time step; if it is 2, the disturbance lasts 2 time steps, and so on. The parameters m and t are illustrated in Fig. 17.1.

At the beginning of a simulation d, m, and t are specified. The probability of a disturbance event (p) is calculated in each time unit as

$$\frac{d - df}{m(1 - df)}$$

where df is the fraction of cells remaining disturbed from the previous time period (due to disturbance duration; df is 0 if t is 1). In this way, the magnitude or the duration of disturbances can be changed without affecting the mean disturbance incidence (d).

Once the probability of a disturbance event (p) is determined, the decision of whether a particular undisturbed cell becomes disturbed is made by using a uniform random number generator that returns a value between 0 and 1. The cell is disturbed if the random number is less than or equal to the current value of p.

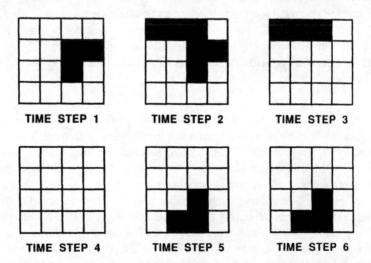

| TIME STEP 1 | TIME STEP 2 | TIME STEP 3 |
| TIME STEP 4 | TIME STEP 5 | TIME STEP 6 |

Figure 17.1. Illustration of model parameters disturbance magnitude (m) and disturbance duration (t). A small grid is shown for 6 time steps for $m = 3$ cells and $t = 2$ steps.

From these parameters for the disturbance regime, one can calculate the mean time between disturbance events at a cell and the mean distance from a disturbance to the disturbance nearest it. The mean time between disturbance events at a cell (wt) is

$$\frac{(m)\,(t)}{d} \text{ steps}$$

The mean distance from a disturbance to the disturbance nearest it (i.e., distance between closest edges assuming approximately circular disturbance patches, nn) is

$$-2x + \sum_{z=0}^{\infty} [(z + 2x)\,(1 - (d/m))^{\pi(z+x)^2}(1 - (1 - (d/m)^{\pi(z+3x)^2 - \pi(z+x)^2}))]$$

cells, where x is the mean radius of disturbed areas and z is a distance from the center of the disturbance.

17.4.1.2 Within-Cell Population Demographics

The population in each grid cell is age classified; the number of age classes is specified at the beginning of a simulation. Inputs also include age-specific survival rates and birth rates. The number of time steps between reproductive events is also an input parameter; this gives the model the flexibility to include reproductive timing strategies such as dormancy. Although there is no density-dependent adjustment of population growth within cells, there is a ceiling to the cell population size that I call the carrying capacity. The carrying capacity is expressed in units of the smallest age class to allow for age classes having different resource requirements.

17.4.1.3 Dispersal from Cells

Dispersal is set up for species for which the shape of the curve away from the point of dispersal follows a negative exponential function; this assumption appears to be appropriate for many organisms, including plants (Farah et al. 1988) and microorganisms (Lindow et al. 1988). In many cases, dispersal is modeled as a diffusion process (Fleischer et al. 1988; Kareiva and Shigasada 1983; Lande 1987). This approach assumes that the distribution of dispersers follows a Gaussian distribution centered at the starting point of the dispersal. For the current study, in which qualitative questions are being asked, the negative exponential is sufficiently close to the Gaussian so that the conclusions drawn from the simulation experiments should apply to these cases as well. The number of individuals (e.g., seeds) immigrating to a cell is the sum over all the other cells of

$$((dfo\, N_a)/2\pi)e^{-dfo\, s}$$

where dfo is the "density fall-off rate" in the negative exponential, N_a is the number of emigrants leaving the donor cell a, and s is the distance between the recipient cell and the donor cell (DeAngelis et al. 1985).

The equation represents the relative contributions of the cells by calculating the point dispersal. This means that many of the dispersers are "lost" in the calculation. The number lost is the total number that leaves cells minus the total number that disperses into cells. These individuals are allocated among the cells in proportion to the actual immigration rates into the cells. This adjustment works well except when *dfo* is very small. In this case many dispersers should actually leave the grid completely; so the adjustment produces an overestimate of the population sizes in all cells. The mean dispersal distance is then calculated as $2/dfo$.

Besides the dispersal distance, the other components of dispersal in the model are the dispersal rate or fraction of organisms dispersing per dispersal event (*dr*), the time between dispersal events (*dt*), and the survival probability of dispersing individuals (*ds*). The structure of the model is shown in Fig. 17.2.

17.4.1.4 Experimental Design

The ultimate goal of the development of this model is to predict which categories of species types (described by their life history and dispersal characteristics) are most likely to occur in various types of disturbance regimes. The most straightforward approach to this problem is to conduct a factorial simulation experiment (Hicks 1982). Several values are chosen for each parameter, spanning the relevant range for that parameter. Simulations are then conducted for every combination of each of the parameter values. Although this model is simple compared with specific models of particular species, the level of complexity of the model (required to make it realistic enough to produce believable and testable results) means that it is not possible to conduct a single set of simulations to study this problem.

In fact, the number of possible parameters in the model is large. There are four disturbance parameters, four life history parameters and four dispersal parameters; each of these twelve parameters may have different values for each of any number of different age classes. It is therefore not practical to conduct a complete set of simulations to analyze for the effects of each parameter over its complete range. In addition, there is likely to be a large number of important interactions among the effects of the parameters; a high value of parameter "a" may have a positive impact on predicted population survival if parameter "b" is at a low level but a negative impact if "b" is at a high level. For example, through the use of a different model, Fahrig (1990) showed that there is a predicted interaction between disturbance rate and dispersal rate on population survival.

Even if the number of age classes in the model were limited to two, there would be twenty-four parameters, each of which should be studied over its relevant range and in combination with all other parameters. The required number of simulations would be on the order of 5^{16} (if five levels are used for each parameter), which is not practical, since each model run takes five to thirty min. Even if this could be done, it would certainly lead to intractable results analogous to the results of field studies in which too many factors are simultaneously measured (Caswell 1988).

To deal with this problem I decided not to try to conduct the entire simulation experiment simultaneously but to begin with a subproblem; only a subset of the parameters was varied. Two stages were then required in the simulation experi-

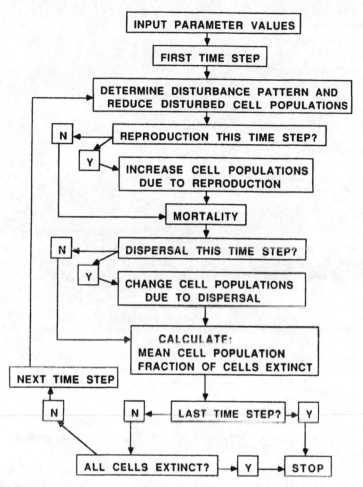

Figure 17.2. Flow diagram of the model of population dynamics in the presence of disturbance.

ment; after the first set of simulations was conducted, it became obvious that the results possibly depended on the particular values chosen for the parameters that were held constant. This realization led to two further simulation experiments.

17.4.1.5 Factorial Simulation Experiment—First Stage

The purpose of the first simulation experiment was to determine the relative effects of the four components of dispersal on population survival in the presence of disturbance. Five hundred simulations were run with the use of a factorial design for the four dispersal parameters (Table 17.1c). Each simulation ran for 150 time steps, and only the last 100 time steps were used in analyses of results. The grid size was 25 by 25 cells in all runs; all of the grid cells were initially assumed to

Table 17.1. Values of Parameters Used in the Simulation Experiment (Example 1)

Parameter	Units	Value(s)	
a. *Disturbance Regime*			
Mean incidence	Rate/cell/step	0.1	
Intensity	Fraction killed	1.0	
Magnitude	Cells	1	
Duration	Steps	1	
b. *Demographics*			
Carrying capacity	Number	500	
Reproductive frequency	Steps	1	
Number of age classes		2	
		Age 1	Age 2
Starting values	Number	10	10
Birth rates	Number per individual	0	1
Survival rates	Fraction	0.8	0.8
c. *Dispersal*			
Rate	Fraction of population	0, 0.25, 0.5, 0.75, 1	
Survival rate	Fraction of population	0, 0.25, 0.5, 0.75, 1	
Time between Dispersal events	Steps	1, 2, 3, 4, 5	
Mean distance	Cells	1, 1.3, 2, 4	

be identical, with spatiotemporal heterogeneity being imposed by the process of disturbance.

I calculated three output values for each simulation: (1) the time at which all populations died (this was 150 for all simulations in which at least one cell was still occupied at the end of the simulation), (2) the mean population size per time step in each cell, and (3) the mean fraction of extinct local (cell) populations per time step before complete extinction (a measure of local extinction probability).

The parameters of the disturbance regime and the within-cell demographics were held constant during all simulations. To set these parameter values in a meaningful range relative to the goal of the simulations, I ran preliminary simulations in which I varied the values of birth rate and disturbance incidence for the cases of no dispersal and maximum dispersal. The goal was to run the simulations under conditions such that the population would not survive without dispersal and would reach near-maximal levels with maximum dispersal. The simulation experiment could then be used to determine which components of dispersal caused the population to shift along this gradient from low to high survival rate.

The FASTCLUS procedure from SAS statistical software (1985) was used to cluster the output data into groups ranging from low success (i.e., low survival rates and population sizes) to high success. Discriminant functions analysis (procedure DISCRIM in SAS [1985]) was then used to ensure that the clusters produced a unique categorization of the data. Finally, I conducted two-way contingency tests with the procedure FREQ from SAS (1985) for the levels of each dispersal component with the clusters. I repeated this procedure with from two to eleven clusters. This allowed me to discover which dispersal components were important for determining population level for those runs in which the population survived.

The cluster analysis led to the delineation of three general categories of simulation outputs: populations were either unsuccessful, moderately successful, or highly successful. The unsuccessful populations had low survival probability. The moderately and highly successful populations had high survival probability, but the latter group had greater population abundances than the former group.

The results of the analysis are summarized in Fig. 17.3A–D. In simulations with no dispersal (i.e., dispersal rate = 0 or disperser survival rate = 0), the populations crashed due to the accumulated effects of disturbed cells that were not recolonized. Even for low dispersal rates, the population sizes jumped to much higher levels, and the mean fractions of extinct cells dropped markedly. For populations that did survive, those with relatively small population sizes had low dispersal rates, low disperser survival rates, and high values of the time between dispersal events. Groups with large population sizes had the reverse.

Dispersal distance is the only dispersal component that had no effect on population size or survival; a possible explanation for this outcome is given in the "Interpretaton of Results" section. Dispersal distance is also the only component that requires the explicit inclusion of space in the model; without it the model would not be a landscape-level model. The question of whether the landscape has an important effect on the population dynamics is the same as asking the question, Under what conditions is the explicit consideration of space a necessary component of the model for understanding population dynamics, and under what conditions is it not? In the context of the current modeling exercise, this is the same thing as asking the question, Under what conditions does the dispersal distance have an important impact on population recovery from disturbance?

17.4.1.6 Factorial Simulation Experiment—Second Stage

The first set of simulations indicated that the spatial component of dispersal, dispersal distance, was not important for population recovery from disturbance. However, it seemed likely that this result might depend on the spatial characteristics of the disturbance regime: disturbance magnitude and disturbance incidence, which controls the distance between disturbed sites.

To study the effect of disturbance magnitude on the relationship between dispersal distance and population size, I ran a factorial simulation experiment with seven values of dispersal distance, ranging from 1.0 to 4.0 cells, and five values of disturbance magnitude ranging from 1 to 9 cells. Analysis of variance (procedure GLM in SAS [1985]) showed that there was a significant positive relationship between dispersal distance and population size for disturbance magnitudes of three cells and above.

To study the effect of disturbance incidence on the relationship between dis-

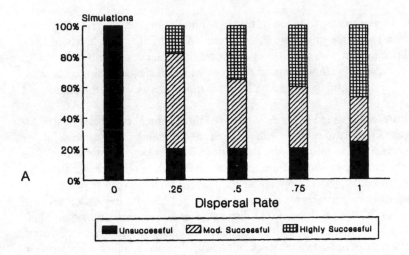

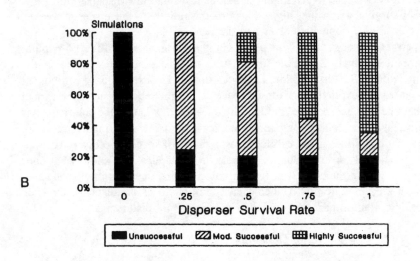

Figure 17.3. Proportions of simulated populations falling into each of three categories: unsuccessful, moderately successful, and highly successful, at various levels of four parameters. (A) Dispersal rate. (B) Disperser survival rate. (C) Time between dispersal events. (D) Dispersal distance.

persal distance and population size, I ran a factorial simulation experiment with sixteen values of dispersal distance, ranging from 1.0 to 4.0 cells, and nineteen values of disturbance incidence fro 0.05 to 0.95. Analysis of variance (procedure GLM in SAS [1985]) showed that there was a significant positive relationship between dispersal distance and population size, but only for high values of disturbance incidence.

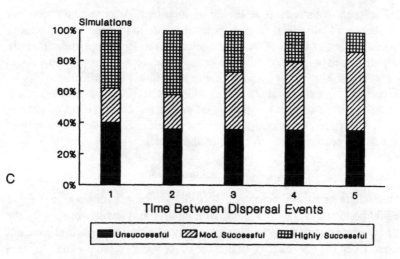

C

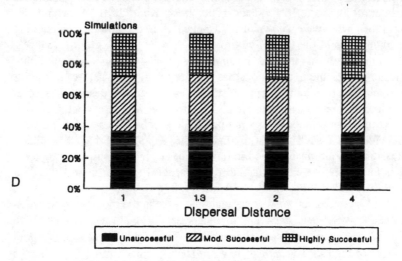

D

Figure 17.3. *Continued.*

17.4.1.7 General Hypotheses

The results of these simulation experiments can be phrased in terms of a series of landscape-level hypotheses about single-species dynamics in the presence of disturbance: (1) to ensure survival in the presence of disturbance, it is most important for a species to have a high dispersal rate and high disperser survival rate; (2) given that a population has a high survival probability, the highest regional population sizes are obtained by species that have the highest dispersal rates and disperser survival rates and the shortest time intervals between dispersal events; (3) dispersal distance is relatively unimportant for population recovery from disturbance; (4) dispersal distance becomes more important with increasing disturbance incidence and disturbance magnitude; and (5) populations that are frequently disturbed and occur in widespread habitats are unstable at low to moderate levels, so they must maintain high levels to avoid extinction.

In the present context this study serves as an illustration of a simulation experiment that uses the factorial experimental design for development of general landscape-level hypotheses about population dynamics. The following is a brief discussion of the ecological significance of the results.

17.4.1.8 Interpretation of Results

The most obvious pattern in the output is the distinction between successful and unsuccessful populations. In simulations with no dispersal (i.e., dr or ds = 0), the population crashed due to the accumulated effects of disturbed cells that were not recolonized. Even for low dispersal rates (dr or ds = 0.25), the population sizes jumped to much higher levels and the mean fractions of extinct cells dropped markedly. This result is similar to that of many others (e.g., May 1974; Reddingius and den Boer 1970; Roff 1974; Vance 1980, 1984; Kuno 1981; Agur and Deneubourg 1985) who have shown that dispersal is important for maintaining regional populations in the face of environmental variability (e.g., disturbance).

However, the result that none of the simulated populations survived with both low sizes and low extinction fractions has not been suggested before. It implies that, for populations that are frequently disturbed but occur in widespread habitats, the population must be able to maintain high levels, or it will go extinct. Note that this result (and the simulations) does not pertain to species that occur in a habitat that is not widespread, such as most fugitive species and species that occur in rare but stable habitats. Patterns observed in field studies of relationships among local population variability, species abundances, and species ranges support this idea (Gaston 1988; Glazier 1986; Grulke and Bliss 1988).

Within the successful category of populations, those in the moderately successful group had low dispersal rates (dr), low disperser survival rates (ds), and high values for the time between dispersal events (dt). The highly successful category had the reverse: high dr and ds and low dt. These results can be explained as follows. If the dispersal rate is high, the population is more evenly spread over all cells. This spreads out the effects of disturbance, leaving room in all cells (relative to their carrying capacities) so that the population can increase over the whole grid. Disperser survival rate has a slightly larger impact than dispersal rate because it incorporates two effects: (1) it enhances overall population size of the regional population and (2) it enhances the effect of high dr. The importance of disperser survival was also shown by Levin et al. (1984). The importance of time between dispersal events (dt) for distinguishing between moderate and high success has not been shown before. The explanation is straightforward: the more often organisms disperse, the faster they can recolonize empty cells.

The lack of importance of dispersal distance in the outcome of the simulations was somewhat surprising. Previous studies have shown that dispersal distance is important for fugitive species that make use of widely scattered patches of habitat (Platt 1975; Green 1983; Turin and den Boer 1988). Fahrig and Paloheimo (1988) showed that dispersal distance is important for determining the effect of spatial distribution of patches on local population dynamics. The second stage in the simulations was conducted to examine the dispersal distance result in the context of varying disturbance magnitude and incidence. There was a significant positive relationship between population size and dispersal distance for high values of both disturbance magnitude and incidence.

In both cases the results indicated a threshold in the relationship. The result for disturbance magnitude is qualitatively similar to that of simulations by Coffin and Lauenroth (in press) for a prairie grass species. The result for disturbance incidence has not been suggested before. It may be explained as follows. In the first set of simulations, the disturbance incidence (0.1) produced a grid that was composed of disturbed patches surrounded by a matrix of undisturbed habitat. In this situation, even if dispersers move only one grid cell on average, disturbed cells are easily recolonized by the neighboring cells. If the disturbance incidence is much higher, the grid consists of patches of occupied habitat surrounded by a matrix of unoccupied (disturbed) habitat. In this situation, dispersal distance would be important because if the population in a particular patch were to go extinct, it would need to be replenished from another (possibly distant) patch.

The transition point between these two cases may be related to the critical point (p_c) described in percolation theory (Stauffer 1985) and applied to the spread of disturbances across landscapes (Gardner et al. 1987). In this analysis p_c is the disturbance probability at which the disturbed cells on a grid are connected through at least one path from one side of the grid to the other. For random square grids such as the one in the present simulations, this number has been identified as 0.5928. For disturbance rates lower than this, the grid has patches of disturbance surrounded by a matrix of undisturbed habitat. Below this point dispersal distance should have little, if any, effect. This idea is supported by the result here that the population size increased with increasing *dd* only for simulations with *d* = 0.55 or greater.

17.4.2 Example 2: Population Dynamics in a Patchy Landscape

The second example is a study in which general hypotheses of population dynamics were sought for the case where habitat heterogeneity in the landscape is due to the patchy distribution of breeding sites. Details of the model are given in Fahrig (1988b) and Fahrig and Paloheimo (1988). The following is a summary.

The model is a stochastic discrete-time simulation model. Both time and space are in arbitrary units (time steps and spatial units). All parameters are expressed relative to these arbitrary units. The region is assumed to consist of discrete patches of breeding habitat surrounded by nonbreeding habitat. The explicit spatial arrangement of breeding habitat is included; distances between all pairs of patches are required to calculate dispersal rates among the patches.

There are six parameters in the model, which determine the population sizes in each breeding patch at each time step. These are (1) intrinsic population growth rate, (2) patch-carrying capacity, (3) fraction of organisms that disperse from patches in each time unit, (4) dispersal distance relative to the mean distance between patches, (5) distances from which dispersers detect new patches (detection radius) relative to the mean distance between patches, and (6) rate of immigration into the region. A seventh factor, patch detection probability, was calculated for each simulation; this value depends on the detection radius and the dispersal distance (Fahrig and Paloheimo 1988).

Dispersal from each patch is assumed to be, on average, equally likely in all directions, except when one patch is within detection range of another. Dispersal is directly proportional to population size in the patches (i.e., dispersal rate is not assumed to be density dependent). Dispersers from patches are assumed to travel to some mean distance, expressed as a fraction of the average distance among patches. It is not assumed that all dispersers move exactly the mean dispersal distance, but rather that they are spread out over space around the mean.

The distance from which dispersers can detect a new patch (e.g., visually or by chemoreception) is called the detection radius. Although this is a characteristic of the species, for computational ease I refer to the detection zone of patches. This is a circular area around the patch from within which dispersers are able to detect the patch. Patches are assumed to "attract" all those dispersers that, by chance, fall within the detection zone of the patch.

The structure of the model is shown in Fig. 17.4.

17.4.2.1 Experimental Design

The goal of the simulation experiment was to determine the relative importance of the six parameters for the mean population size and the degree to which spatial pattern of patches affects local population size. Simulation runs of ten patches for 150 time steps require about five minutes on the computer. To conduct a factorial experiment with the six parameters at a minimal number of levels (say five) would require about 1300 h of simulations. Since in this case I did not wish to break the problem down into smaller units, I decided to abandon the factorial design in favor of a Latin hypercube design (Iman and Conover 1980). This design allows one to include more factors than the factorial design, for the same number of simulations.

17.4.2.2 Simulation Experiment

Two thousand runs of the model were conducted. In each run there were ten patches and 150 time steps. A different spatial arrangement of ten patches was used for each run. The x and y coordinates of each patch were chosen by a uniform random number generator, and the distances between all pairs of patches were then calculated. The initial population size in all patches and for all runs was 100 organisms. The mean values of the six parameters were chosen at random from a uniform distribution at the beginning of each run. The ranges of these parameters are given in Table 17.2. Stochastic variation in the parameters was included, so that their actual values fluctuated randomly among the 10 patches and between the time periods. The population size in each of the patches was calculated for the 150

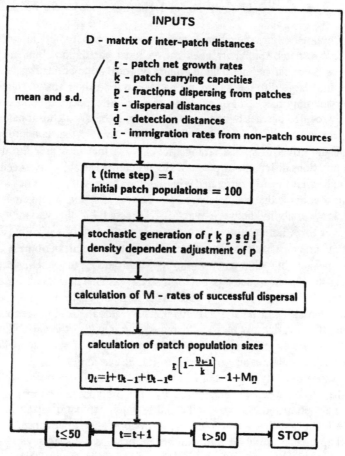

Figure 17.4. Flow diagram of the model of population dynamics in a patchy landscape (Fahrig 1988b).

Table 17.2. Upper Limits of Parameter Values Used in the Simulation Experiments (Example 2)

Parameter	Upper Limit of Mean Value
Growth rate	0.1
Carrying capacity	10000
Dispersal rate	1.0
Dispersal distance/mean interpatch distance	2.0
Dispersal radius/mean interpatch distance	0.3
Immigration rate	1000

Lower limit in all cases is 0, standard deviation/mean is 0.1 (Fahrig and Paloheimo 1988).

time steps. Only the results from the final 100 time steps were included in the analyses.

The following outlines the rationale and procedure used for obtaining a measure of the effect of patch spatial arrangement on local population abundance. If the spatial location of patches has a large impact on local population sizes, one would expect to find large differences among patches due to their spatial relationships with the remaining patches. For example, one might expect the population sizes in centrally located patches to be higher than in peripheral patches if patch spatial arrangement is important. Therefore, the type of variable that is appropriate for measuring the effect of spatial arrangement is one that measures the degree to which populations differ among patches. However, the variability between patches is likely to be significantly correlated with the variability within patches over time. Therefore, a variable that accurately reflects effects of spatial arrangement should measure the variability between patches, corrected for the variability within patches over time. For each run, the mean and coefficient of variation (CV) over the 100 time units for each patch were calculated; then the mean of these CVs was taken as a measure of within-patch variability over time for that run (*tempvar*). The CV among the ten mean patch population sizes was also calculated for each run (*patvar;* i.e., variability between patches). A significant quadratic relationship was found between ln(*tempvar*) and ln(*patvar*). The residuals from the regression are a measure of the variability between patches, corrected for the variability within patches. They are therefore a measure of the importance of patch spatial arrangement on local population size or the pure spatial variability.

The purpose of analyzing the simulation results was to determine which types of organisms (i.e., species characterized by which combinations of independent variables) are predicted to have small or large patch population sizes and a small or large effect of patch spatial arrangement on population abundance. A standard statistical approach to problems in which relationships between several independent variables and one dependent variable are sought is to use the least-squares method to build the best polynomial model relating the independent variables to the dependent variable (Box and Draper 1987). This approach is often used in uncertainty analysis, in which one attempts to measure the sensitivity of model predictions to changes in model parameters. If the relationships between the parameters and the model output are approximately linear, first-order terms will dominate the regression analysis, and the relative importance of parameters can be ranked by using correlation coefficients (Gardner et al. 1981; Gardner 1984).

In the present analysis, however, the results of such polynomial regression analyses were not interpretable; relationships between the parameters and the model output were nonlinear and complex. The resulting polynomial regression models consisted of large numbers of statistically significant terms (fifteen to twenty), most of them higher order interactions involving two or three parameters and each of them explaining only a small portion of the total variation. Therefore, although the analyses resulted in adequate empirical models, in this case they did not aid in understanding of the qualitative relationships between the independent and dependent variables. In particular, they did not result in an estimate of the relative importance of each parameter in explaining variation in the model output.

The following method was used for analyzing the simulation results. First, polynomial regression equations were calculated for each of the dependent variables (i.e., population size and spatial variability) on each of the independent variables (i.e., the six parameters plus detection probability). This provided estimates of the forms of the underlying relationships. To determine the relative importance of the independent variables, the polynomial regression equations were then used as polynomial variables in stepwise regressions of the dependent variables.

The polynomial variables for mean patch detection probability and mean dispersal rate were the most important factors determining average local population size, with partial R^2 values of 0.3890 and 0.2923, respectively. The polynomial regression equations for the parameters are plotted in Figs. 17.5A and B. The polynomial variable for mean dispersal distance was the most important factor determining the spatial variation among local populations, with a partial R^2 value of 0.1404. The equation for this variable is shown in Fig. 17.5C.

17.4.2.3 General Hypotheses

This simulation experiment led to the following hypotheses about single-species dynamics in a patchy landscape: (1) the most important determinants of mean local population size are the probability of dispersers detecting new patches (positive relationship) and the fraction of organisms dispersing from the patches (negative relationship); and (2) the main factor that determines whether a local population will be influenced by exact spatial relationships among patches is its dispersal distance (negative relationship).

In the present context this study serves as an illustration of a simulation that uses the Latin hypercube experimental design for development of general landscape-level hypotheses about population dynamics. A detailed discussion of the results and their significance is in Fahrig and Paloheimo (1988). The following is a brief summary of the points made there.

17.4.2.4 Interpretation of Results

The most important factor determining patch population size was the probability that dispersers successfully detect new patches. This result is analogous to that of Levin et al. (1984), who found that the optimal level of dispersal increases with an increase in the probability of a dispersing propagule successfully attaining a new site. The relative unimportance of dispersal distance and detection radius results from the fact that the effects of these two parameters are largely encompassed by the effect of detection probability. The importance of the dispersal rate (fraction of local population dispersing) reflects the high risk associated with dispersal; high dispersal rates generally result in lower mean patch population sizes, unless the probability of dispersers detecting new patches is very high. A

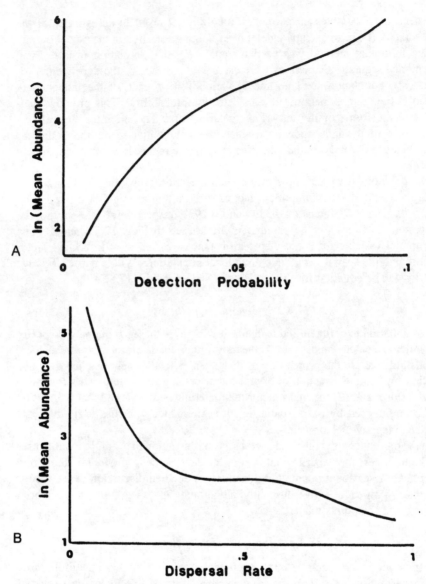

Figure 17.5. (A) Relationship between patch detection probability and log-transformed values of average patch population abundance (averaged over 10 patches). (B) Relationship between dispersal rate of organisms from patches and log-transformed values of average patch population abundance (averaged over 10 patches). (C) Relationship between dispersal distance and spatial variation among 10 patch population means due to patch spatial arrangement (Fahrig and Paloheimo 1988).

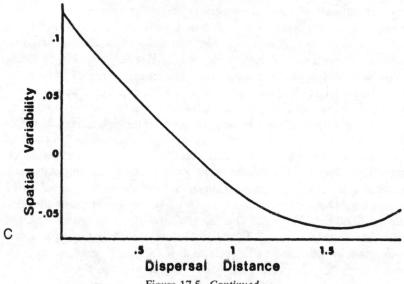

Figure 17.5. *Continued.*

negative relationship between dispersal rate and population size has also been found in studies by Lomnicki (1980) and Roff (1974).

Intrinsic growth rate and carrying capacity were predicted to have little effect on average population sizes in a patchy cnvironment; their effects were almost completely swamped by the dynamics of intcrpatch dispersal. This result suggests that between-patch processes (i.e., dispersal) will often be more important than within-patch processes (i.e., births and deaths) in determining local population size within a habitat patch.

The results also indicate that the dynamics of dispersal among patches are most important in determining the level of the effect of patch spatial pattern on local population size. The most important factor in this case is the dispersal distance. · The greater the dispersal distance, the less important is the spatial arrangement of patches in causing differcnccs among local population sizes. This is because the same number of dispersing organisms is spread over a much larger area when the dispersal distance is large. The result is that dispersal from a particular patch has a more general influence (i.e., affects more patches) for large dispersal distances than for small ones. If the dispersal distance is small, then those patches that have neighbors at close distances receive many more dispersers than those that do not. In this case the spatial relationship among patches is an important determinant of the local population size, and the spatial variation among patch sizes is high.

It is generally believed that migratory bird species disperse shorter distances from their natal site to their reproductive site than do nonmigratory birds (Whitcomb et al. 1981), mainly because the migratory phase uses up time that might otherwise be spent moving between breeding sites. If this is true, the simulation results suggest that the spatial arrangement of bird breeding sites should be a more important determinant of local abundance of migratory species than of non-migratory species. This pattern was found in a study of birds in woodlots in Maryland (Lynch and Whigham 1984). Also, for species that disperse long distances in the wind, such as small insects (e.g., aphids; Kennedy and Stroyan 1959), small plant seeds, or spores, the result indicates that the spatial pattern of potential breeding sites is unlikely to have much effect on local population abundance.

The reasons for the lack of importance of the rate of immigration from outside the study region are not obvious. One would expect that for high rates of immigration, the effects of interpatch dispersal would be swamped, and in this case the patch spatial arrangement would have less effect on local population size. However, the effect of dispersal distance far outweighed the effect of immigration rate. It is possible that for much higher immigration rates the swamping effect of immigration would be more pronounced. In this study, the upper limit to immigration rate was 1000 individuals per time unit. Relative to the ranges of the other parameters in the model (Table 1; Fahrig and Paloheimo 1988), this represents a large immigration rate. The results indicate therefore, that immigration from outside the study area is not expected to significantly swamp the effects of interpatch dispersal, except possibly at extremely high immigration rates.

17.5 The Possibility of Chaos

Chaotic dynamics are likely to throw a wrench into many types of simulation experiments, particularly those aimed at the development of general hypotheses. Kot et al. (1988) give a summary of the history and current status of the study of chaos. Chaotic behavior has been shown in some of the simplest analytically tractable nonlinear models. Also, several researchers have apparently found chaotic dynamics in real population dynamics (see Table 1 in Kot et al. 1988). Although it is not as commonly noted, chaotic behaviors are now being noticed in some simulation models of particular systems; for example, Rejmanek et al. (1987) found chaotic behavior in a simulation model of forest tent caterpillar.

The problems presented by chaotic behavior can be dealt with by limiting simulations to the parameter space in which the dynamics are not chaotic (Ulanowicz 1988). However, this solution is problematic for the purpose of developing general hypotheses, because one does not know before starting the simulations where the nonchaotic region of parameter space is.

This problem is illustrated in "Example 2: Population Dynamics in a Patchy Landscape" presented above. When the analysis was completed, 78% of the variability in population size and only 23% of the variability in spatial variability were explained by variability in the seven parameters (Fahrig and Paloheimo 1988). Some of this lack of explanatory power may be due to the stochastic nature of the simulations and because the method of postsimulation statistical analysis was not sensitive enough to adequately describe the relationships. However, it is also quite possible that some of the unexplained variability was due to chaotic behavior of the population dynamics for certain values of the parameters. If this is true, the simulation output would appear to be unrelated or only slightly related to the input parameter values, and simulations in these parts of the parameter space would add to the unexplained variability.

To find the areas of parameter space in which chaotic behavior occurs, one would need to conduct an analysis for chaotic behavior of the dynamics of each of a large number of runs then attempt to delineate the chaotic region(s) of parameter space. Parameter values in these regions would represent species types for which the general hypotheses would not be expected to hold. Given the large number of simulations required, conducting such an analysis for anything other than an extremely simple model (e.g, two parameters) would be impractical.

17.6 Summary

I argue that the constraints imposed by simple, analytically tractable models make them unlikely to be of use for the development of general hypotheses of population dynamics in the landscape context. At the other extreme, development of complex simulation models for specific systems is equally unlikely to render general hypotheses. Different situations appear to be less and less comparable as the level of detail in the simulation models increases. The problems associated with extrapolation from detailed models to general hypotheses are analogous to the problems encountered in attempts to extrapolate from detailed field studies to general hypotheses; extrapolation is viewed as oversimplification.

Between these two extremes is the approach I advocate here: (1) state the question; (2) develop the simplest possible simulation model to study the question; (3) design and conduct a simulation experiment(s), using principles of experimental design, to study the effects of each parameter over its relevant range; (4) analyze the output of the simulation experiment as one would analyze the results of an ordinary experiment; and (5) phrase the uncovered relationships between the input parameters and the output in terms of hypotheses of population dynamics.

I illustrate this approach with two examples. The first is a general model of population response to disturbance. By using a factorial experimental design, I ran simulation experiments to develop the following hypotheses about single-species dynamics in the presence of disturbance: (1) to ensure survival in the presence of disturbance, it is most important for a species to have a high dispersal rate and high disperser survival rate; (2) given that a population has a high survival probability, the highest regional population sizes are obtained by species that have the highest dispersal rates and disperser survival rates and the shortest time intervals between dispersal events; (3) dispersal distance is relatively unimportant for population recovery from disturbance; (4) dispersal distance becomes more important with increasing disturbance incidence and disturbance magnitude; and (5) populations that are frequently disturbed and occur in widespread habitats are unstable at low to moderate levels—they must maintain high levels to avoid extinction.

The second example is a general model of population dynamics in a patchy landscape (Fahrig 1988b; Fahrig and Paloheimo 1988). By using a Latin hypercube design, I ran simulation experiments that led to the following hypotheses: (1)

the most important determinants of mean local population size are the probability of dispersers detecting new patches (positive relationship) and the fraction of organisms dispersing from the patches (negative relationship) and (2) the main factor that determines whether a local population will be influenced by exact spatial relationships among patches is the organism's dispersal distance (negative relationship).

Although the method has many advantages, it also has limitations. First, there is always a trade-off between the number of parameters included in the simulation experiment and the accuracy of the results. This is a standard trade-off in real experiments as well, and it is one reason that the first step in the above process (statement of the question) is important. Secondly, one must be aware that this type of simulation is not immune to the possibility of chaotic behavior that has been described for analytical models and is being observed with increasing frequency in nature.

Acknowledgments

I am grateful to Phil Burton, Bill Lauenroth, Dean Urban, an anonymous reviewer, and the editors of this volume, Monica Turner and Bob Gardner, for their comments on the manuscript. This work was supported by NSF LTER grant BSR8702333.

References

Agur, Z., and Deneubourg, J.L. 1985. The effect of environmental disturbances on the dynamics of marine intertidal populations. *Theoretical Population Biology* 27:75–90.

Box, G.E.P., and Draper, N.R. 1987. *Empirical Model-Building and Response Surfaces.* Toronto: Wiley.

Caswell, H. 1988. Theory and models in ecology: a different perspective. *Ecological Modelling* 43:33–44.

Coffin, D.P., and Lauenroth, W.K. 1989. Disturbances and gap dynamics in a semiarid grassland: A landscape-level approach. *Landscape Ecology* 3:19–27.

DeAngelis, D.L. 1988. Strategies and difficulties of applying models to aquatic populations and food webs. *Ecological Modelling* 43:57–73.

DeAngelis, D.L.; Waterhouse, J.C.; Post, W.M.; and O'Neill, R.V. 1985. Ecological modelling and disturbance evaluation. *Ecological Modelling* 29:399–419.

Downing, D.J.; Gardner, R.H.; and Hoffman, F.O. 1985. An examination of response-surface methodologies for uncertainty analysis in assessment models. *Technometrics* 27:151–63.

Fahrig, L. 1988a. Nature of ecological theories. *Ecological Modelling* 43:129–32.

Fahrig, L. 1988b. A general model of populations in patchy habitats. *Applied Mathematics and Computations* 27:53–66.

Fahrig, L. 1990. Interacting effects of disturbance and dispersal on individual selection and population stability. *Comments on Theoretical Biology* 1:275–297.

Fahrig, L., and Paloheimo, J. 1988. Determinants of local population size in patchy habitats. *Theoretical Population Biology* 34:194–213.

Farah, K.O.; Tanaka, A.F.; and West, N.E. 1988. Autoecology and population biology of dyers woad *Isatis tinctoria. Weed Science* 36:186–93.

Fleischer, S.J.; Gaylor, M.J.; and Hue, N.V. 1988. Dispersal of *Lygus lineolaris* (Heteroptera: Miridae) adults through cotton following nursery host destruction. *Environmental Entomology* 17:533–41.

Gardner, R.H. 1984. A unified approach to sensitivity and uncertainty analysis. In *Applied Simulation and Modelling: Proceedings of the IASTED International Symposium,* ed. M.H. Hamza, pp. 155–57. Calgary: Acta Press.

Gardner, R.H.; Milne, B.T.; Turner, M.G.; and O'Neill, R.V. 1987. Neutral models for the analysis of broad-scale landscape pattern. *Landscape Ecology* 1:19–28.

Gardner, R.H.; O'Neill, R.V.; Mankin, J.B.; and Carney, J.H. 1981. A comparison of sensitivity analysis and error analysis based on a stream ecosystem model. *Ecological Modelling* 12:173–90.

Gardner, R.H.; O'Neill, R.V.; Mankin, J.B.; and Kumar, D. 1980. Comparative error analysis of six predator-prey models. *Ecology* 61:323–32.

Gaston, K.J. 1988. Patterns in the local and regional dynamics of moth populations, *Oikos* 53:49–57.

Glazier, D.S. 1986. Temporal variability of abundance and the distribution of species. *Oikos* 47:309–14.

Green, D.S. 1983. The efficacy of dispersal in relation to safe site density. *Oecologia* 56:356–58.

Grulke, N.E., and Bliss, L.C. 1988. Comparative life history characteristics of two high Arctic grasses, Northwest Territories, Canada. *Ecology* 69:484–96.

Hall, C.A.S. 1988. An assessment of several of the historically most influential theoretical models used in ecology and of the data provided in their support. *Ecological Modelling* 43:5–31.

Hastings, A. 1988. Food web theory and stability. *Ecology* 69:1665–68.

Hicks, C.R. 1982. *Fundamental Concepts in the Design of Experiments.* New York: Holt, Rinehart and Winston.

Iman, R.L., and Conover, W.J. 1980. Small sample sensitivity analysis techniques for computer models, with application to risk assessment. *Communication in Statistics: Theory and Methods* A9:1749–1842.

Kareiva, P.M., and Shigesada, N. 1983. Analyzing insect movement as a correlated random walk. *Oecologia* 56:234–38.

Kennedy, J.S., and Stroyan, H.L.G. 1959. Biology of aphids. *Annual Review of Entomology* 4:139–60.

Kot, M.; Schaffer, W.M.; Truty, G.L.; Graser, D.J.; and Olsen, L.F. 1988. Changing criteria for imposing order. *Ecological Modelling* 43:75–110.

Kuno, E. 1981. Dispersal and the persistence of populations in unstable habitats: a theoretical note. *Oecologia* 49:123–26.

Lande, R. 1987. Extinction thresholds in demographic models of territorial populations. *American Naturalists* 130:624–35.

Levin, S.A.; Cohen, D.; and Hastings, A. 1984. Dispersal strategies in patchy environments. *Theoretical Population Biology* 26:165–91.

Lindow, S.E.; Knudsen, G.R.; Seidler, R.J.; Walter, M.V.; Lambou, V.W.; Amy, P.S.; Schmedding, D.; Prince, V.; and Hern, S. 1988. Aerial dispersal and epiphytic survival of *Pseudomonas syringae* during a pretest for the release of genetically engineered strains into the environment. *Applied Environmental Microbiology* 54:1557–63.

Lomnicki, A. 1980. Regulation of population density due to individual differences and patchy environment. *Oikos* 35:185–93.

Lynch, J.F., and Whigham, D.F. 1984. Effects of fragmentation on breeding bird communities in Maryland, USA. *Biological Conservation* 28:287–324.

May, R.M. 1974. Ecosystem patterns in randomly fluctuating environments. In *Progress in Theoretical Biology,* vol. 3, eds. R. Rosen and F.M. Snell, pp. 1–50. London: Academic Press.

May, R.M., ed. 1976. *Theoretical Ecology: Principles and Applications.* London: Blackwell Scientific Publications.

Onstad, D.W. 1988. Population-dynamics theory: the roles of analytical, simulation, and supercomputer models. *Ecological Modelling* 43:111–24.

Platt, W.J. 1975. The colonization and formation of equilibrium plant species associations on badger disturbances in a tall-grass prairie. *Ecological Monographs* 45:285–305.

Reddingius, J., and den Boer, P.J. 1970. Simulation experiments illustrating stabilization of animal numbers by spreading of risk. *Oecologia* 5:240–84.

Rejmanek, M.; Smith, J.D.; and Goyer, R.A. 1987. Population dynamics of the forest tent caterpillar (*Malacosoma disstria*) in water tupelo (*Nyssa aquatica*) forest: a simulation model. *Ecological Modelling* 39:287–305.

Roff, D.A. 1974. Spatial heterogeneity and the persistence of populations. *Oecologia* 15:245–58.

SAS Institute Inc. 1985. *SAS/STAT Guide for Personal Computers,* version 6 ed. Cary, N.C.: SAS Institute Inc.

Stauffer, D. 1985. *Introduction of Percolation Theory.* London: Taylor and Francis.

Turin, H., and den Boer, P.J. 1988. Changes in the distribution of Carabid beetles in the Netherlands since 1880:II. isolation of habitats and long-term trends in the occurrence of Carabid species with different powers of dispersal (Coleoptera: Carabidae). *Biological Conservation* 44:179–200.

Turner, M.G. 1987. Spatial simulation of landscape changes in Georgia: A comparison of 3 transition models. *Landscape Ecology* 1:29–36.

Ulanowicz, R.E. 1988. On the importance of higher-level models in ecology. *Ecological Modelling* 43:45–56.

Vance, R.R. 1980. The effect of dispersal on population size in a temporally varying environment. *Theoretical Population Biology* 18:343–62.

Vance, R.R. 1984. The effect of dispersal on population stability in one-species, discrete-space population growth models. *American Naturalist* 123:230–54.

Warwick, J.J.; Asce, M.; and Cale, W.G. 1986. Effects of parameter uncertainty in stream modeling. *Journal of Environmental Engineering* 112:479–89.

Whitcomb, R.F.; Robbins, C.S.; Lynch, V.F.; Whitcomb, B.L.; Klimkiewicz, M.K.; and Bystrak, D. 1981. Effects of forest fragmentation on avifauna of the eastern deciduous forest. In *Forest Island Dynamics in Man-Dominated Landscapes,* eds. R.L. Burgess and D.M. Sharpe, pp. 125–205. New York: Springer-Verlag.

18. An Individual-Based Simulation Model of Herbivory in a Heterogeneous Landscape

Jeffrey B. Hyman, Jay B. McAninch, and Donald L. DeAngelis

18.1 Introduction

One of the potentially richest areas of ecological study, for both theoretical and applied questions, is the dynamic interaction between mammalian herbivores and the plants they consume. There is an extensive literature on the interactions between grazers and grasslands (e.g., Crawley 1983; McNaughton and Georgiadis 1986; Mc Naughton et al. 1988) The impact of herbivores on woody trees also is widely recognized. Herbivory can modify the course of succession by suppressing certain species (Ross et al. 1970; Anderson and Loucks 1979; Naiman 1988), may partially explain tree diversity and distribution in the tropics (Janzen 1970; Sterner et al. 1986; Schupp 1988), and is an important component of applied problems such as conservation of native trees (e.g., Scowcroft and Hobdy 1987) and forest management (Adams 1975).

Most past models of herbivore-plant systems considered the interaction between herbivore and plant at the population level and in a homogeneous setting (Crawley 1983), partly so the problem would be amenable to analytical analysis. In nature, however, heterogeneous resource distributions are likely to be the rule (Ford 1983). The recent expansion of two fertile areas of ecological research, individual-based modeling (Huston et al. 1988) and landscape ecology, has provided a new context for the study of herbivore-plant interactions. The modeling approach that emerges naturally from this new context is able to consider the interactions between landscape elements, herbivores, and plant species of interest (Fig. 18.1).

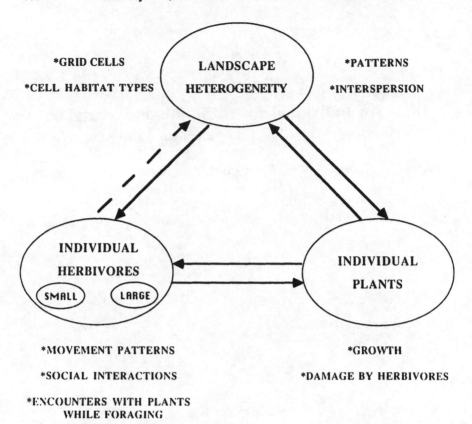

Figure 18.1. A schematic diagram of the interactions between landscape elements, individual herbivores, and individuals of a particular plant species consumed or damaged by the herbivores. We do not consider the direct effects of herbivores on landscape heterogeneity (dashed line) in this study.

This approach can be used to investigate exciting new sets of questions that could not be addressed with the earlier models, such as how landscape heterogeneity affects the space-use patterns of herbivores and ultimately the degree and distribution of plant survival. We can begin to systematically examine how differences in the characteristics of the landscape, herbivore, and plant can lead to observed variation across systems in the outcome of herbivore-plant interactions.

Our objectives in this chapter are (1) to describe the development of a spatially explicit, individual-based simulation model of herbivore-plant interactions on a heterogeneous landscape; and (2) to discuss the variety of general and specific questions that can be addressed with the model. The model considers movement patterns, foraging, and social behaviors of herbivore individuals, and growth and damage of individual plants, from a multiscale perspective. The development of the model was motivated by our desire to examine the potential role of small and large mammalian herbivores in slowing the invasion of utility rights-of-way

(ROWs) by trees. Although the model we constructed exhibits a degree of generality sufficient to examine herbivory in other systems as well, we will present the model in the context of the ROW study for the sake of clarity and effectiveness.

18.2 Applied Problem and Study System

A potential component of a management strategy for keeping tree establishment and growth under power lines in check is to take account of the activity of mammalian herbivores that naturally utilize ROWs. It has been observed that some 30 to 40% of sampled seedlings in selected ROWs show recent evidence of herbivory. Therefore, seedling damage resulting from herbivory can potentially affect the rate of invasion of ROWs by trees to a significant degree. Of course, the change in the numbers and distribution of trees from germination to sapling stage results from many interacting factors affecting growth and mortality: resource competition with other plants, allelopathic inhibition of growth, physical stress or disturbance, pathogens, and herbivory. The ability to predict the potential role of herbivory in altering the invasion process, however, and how this role depends on the characteristics of the herbivore, seedling, and landscape, has important applied and theoretical implications. The ROW system thus provides an excellent opportunity to examine some important questions concerning herbivore-plant interactions in heterogeneous landscapes.

The study system comprises selected electric transmission ROWs in the mid-Hudson River Valley of southeastern New York. These ROWs exhibit a high degree of spatial heterogeneity of the shrub and herbaceous vegetation. The heterogeneity of the larger-scale landscape surrounding a ROW depends upon past and present land use.

The general ROW study incorporates four scales of sampling units. At the largest scale, ROWs in areas separated by tens of kilometers, in the Catskill Mountains and the Hudson River lowlands, are compared. Within each area are three sites, each of which is 8000 to 10,000 m². These sites are from 3 to 5 km apart and may or may not be on the same ROW. Each site contains three 1800- to 2400-m² plots located at 50- to 80-m intervals. Plots are the basic units on which animal, vegetation, resource, and microclimate data are developed. Seedling data are gathered from six 1-m² quadrats on each plot.

In 1987 vegetation within the study ROWs was classified by applying ordination-space partitioning methods (Pielou 1984) to data on floristic dominance. Twelve patch types were identified: five were dominated by herbaceous species, four were dominated by shrubs, and three were composed of an herb-shrub mixture. Typical dominant species included little bluestem grass (*Andropogon scoparius*), hayscented fern (*Dennstaedtia punctilobula*), gray dogwood (*Cornus racemosa*), and mountain laurel (*Kalmia latifolia*). The most abundant tree seedling species found on the ROWs were red maple (*Acer rubrum*), sugar maple (*Acer saccharum*), white ash (*Fraxinus americana*), black cherry (*Prunus serotina*), and black oak (*Quercus velutina*).

18.3 Selecting the Major Scales of the System

Typically, we refer to the concept of scale as comprising both extent and grain (grain is equivalent to resolution). An adequate understanding of the relationships between pattern and process in a system is likely to be enhanced by, and indeed may require, observations at a range of scales (Risser et al. 1984; Addicott et al. 1987). In studying the interactions between plants and small and large herbivores, we must often consider organisms that operate at very different spatial and temporal scales. The growth and survival of tree seedlings may change drastically over the distance of a few meters (Collins and Good 1987; Harmon and Franklin 1989), yet a large herbivore like a deer chooses foraging habitats on the scale of hectares (Tierson et al. 1985). Similarly, vegetation dynamics usually occur on a much slower time scale than the activities of individual herbivores. When selecting the most important scales, all possible scales need to be considered and evaluated, because processes that occur at one scale can affect those that occur at a different scale. For example, the movement of deer over their home range, which occurs at a relatively large spatial scale, has implications for the survival of tree seedlings in any given segment of a ROW. Conversely, it has been argued that small-scale foraging processes can explain the habitat use of herbivores at larger scales (Senft et al. 1987). Our objective is to simultaneously capture all essential processes that contribute to seedling herbivory, even though they may occur over a wide range of scales.

In this section, we outline the rationale behind our choice of those scales which we have included in the model. Because our objective is to elucidate the relationships between animals and vegetation, we must choose scales of investigation that reflect real and biologically meaningful scales of response for the organisms considered. Thus, two questions are critical: (1) What extent, both spatial and temporal, will encompass the dynamics important to our study? (2) At what scales of resolution are differences in the environment significant to the individual; or conversely, at what scales of resolution can the environment be considered homogeneous? The answers depend on the type and size of the individual (e.g., small mammal, large mammal, or seedling), the context or process (e.g., foraging, nest site selection, or survival), and the environmental variable measured (e.g., plant cover, resource availability, or air temperature) (Addicott et al. 1987). The responses of the seedling, the small herbivore, and the large herbivore thus dictate the important scales of our system.

18.3.1 Spatial Scales

The largest spatial extent of the investigation is set by the space-use patterns of the most mobile herbivore considered, to the degree that these patterns relate to the rate of herbivory on a particular plant species. For the ROW study, this is the area used by a large group of deer during a specified time period. The identification of grain sizes is more complex. We can identify three discrete levels of spatial resolution that are of primary importance to the herbivore-vegetation interaction. At the coarsest scale, ROWs can be viewed as elongated patches, or corridors, of habitat. Such patch sizes demonstrate the megahabitat scale, which is most important when we are considering home range movements and habitat selection of large mammals. Variation in the population densities of small mammals and seedlings would be large at this scale.

Within the ROW, we can identify two additional scales of resolution: the macrohabitat and the microhabitat. Although our use of these terms corresponds conceptually to the definitions given by Morris (1987, p. 363), the grain sizes we intend these terms to represent here will likely be smaller than those considered by many small-mammal researchers.

The macrohabitat scale is employed to consider differences between areas on the order of 100 to 1000 m² in size. For small mammals, there are two potential causes of variation in landscape use at this scale. Firstly, some areas may lack a sufficient quantity or quality of resources (cover or food) required by small mammals during a particular time period, and so will remain unused during that period, whereas other areas may contain at least the minimum requirements of these resources. Differences between these unused areas and areas occupied by resident individuals (i.e., home ranges) constitute the macrohabitat scale of resolution for small mammals. Secondly, this scale of resolution is needed also to describe differences between areas that correspond to occupied territories and recently vacated territories. Both areas contain sufficient resources, but areas left vacant due to the death of a resident may remain vacant until a disperser re-occupies it (Baird and Birney 1982b). We can also observe variation in landscape use on the macrohabitat scale by large mammals (e.g., deer). This variation reflects features of topography and vegetation structure that enhance or inhibit use. For example, a large mammal may be restricted to the central area of some ROWs because of dense shrub growth along the edges of ROWs bordered by forests. Seedlings also exhibit variation on this same macrohabitat scale. For seedlings, the main response at this scale is differential recruitment and growth as a function of the distance from the edge of the ROW. In general, areas near the ROW edge receive a higher level of seed input, but a lower level of light input, than areas near the center of the ROW.

Our microhabitat scale of resolution refers to spatial variation between areas on the order of 1 to 10 m² in size. Heterogeneity at this scale has been found to critically influence the responses of small mammals to the landscape (Cockburn and Lidicker 1983). For a small mammal, consideration of this scale is necessary to predict the variation in seedling consumption that results from preferences for particular patch types, from social interactions, and from variation in resource density within the home range. Pockets of unused areas may remain within or between home ranges or territories (Ford 1983; Sherman 1984). Furthermore, areas of overlap between home ranges may receive either relatively high or low use, depending on intraspecific interactions. The sizes of home ranges themselves can be a function of the distribution of cover (Birney et al. 1976) and the density and spatial distribution of food patches, with individuals having to move greater distances in search of food in habitats that have a lower density of high-quality food (Ims 1987; Mares and Lacher 1987). Consideration of the microhabitat scale may be necessary also for studying the effects of large mammals such as deer. For a deer, differential impact on seedlings at this scale will depend on forage preferences and the reach of the individual from a stationary position (Senft et al. 1987). Seedlings, too, respond at the microhabitat scale of resolution. Differential establishment, growth, and survival of tree seedlings at this scale are likely to be a function of differences in microclimate, resource availability, and competition from shrub and herb species in different patch types (Collins and Good 1987).

Notice that we have defined the spatial scales in such a way that the responses of small and large herbivores and of tree seedlings differ only in the extent of spatial scales they cover. For example, we do not redefine patches at the microhabitat scale when we switch from vole to deer to seedling.

18.3.2 Grid Cells

We have used the grid-cell modeling approach to capture the spatial dynamics of individuals on heterogeneous landscapes. This approach has several advantages: (1) it facilitates comparison of results with theoretical models of landscapes (e.g., Murai et al. 1979; Gardner et al. 1987; Gardner and O'Neill, Chapter 11); (2) it provides a suitable method for mapping real landscapes that can be used for model parameterization, verification, and validation (e.g., Stocker and Gilbert 1977; Ford and Krumme 1979; Getty 1981a,b; Lancia et al. 1982; Browder et al. 1985; several chapters within this volume); (3) it allows application of standard statistical methods (Jumars et al. 1977; Upton and Fingleton 1985; S. J. Turner et al., Chapter 2); and (4) it allows modeling of landscape patterns, distances between patches, and animal movements in a straightforward and realistic fashion.

We have identified the major spatial scales of resolution for our study without yet specifying the exact sizes of the model grid cells used to capture the system dynamics at these scales. For general theoretical modeling studies, this choice may not be important, and the cells can be given arbitrary relative sizes. In a model that is applied to a specific field situation, however, cell size becomes an issue. The accuracy of our predictions about spatial differences in plant damage by herbivores will be greatest if the model can reflect heterogeneity at resolutions that are most relevant to the organisms involved.

In specifying grid cells in the model, only the microscale and megascale will be considered (for reasons discussed later). For each of these scales considered, the lower limit of cell size is set by the resolution of the field measurements that generated the data used for model parameterization. For example, seedling data for the ROW study are gathered from 1-m² quadrats, so this is our minimum cell size at the microscale. We may want to aggregate these data, however, so that we can use larger cells in the model to decrease computation time, but we do not want to sacrifice significant accuracy in our model predictions. One of the main issues to consider before aggregating is whether the larger cells made up of aggregated smaller cells will still be homogeneous. The cell should represent a habitat area that the organisms treat as a homogeneous unit (Smith 1986), given a specified context and environmental variable. For the microscale (and megascale), we have defined habitat types which differ floristically and probably also structurally, and we assume that they integrate environmental and ecological variation so that herbivores and seedlings respond primarily to differences between these types. Thus, the criterion of homogeneity allows us to set our cell size larger than the grain size of the field measurements (1 m²), as long as each cell still contains a single habitat type. The specific sizes of our grid cells are discussed in Section 18.4.

18.3.3 Temporal Scales

The largest temporal extent of the study must necessarily be on the order of several years to follow the growth and development of trees. The extent must be at least a year in length, because we would like to be able to assess the effects of seasonal differences in herbivore behavior and seedling availability. As with spatial scale, resolution as well as extent must be considered for temporal scale. Because we consider many varied components of the system, we must consider several temporal scales of resolution. Besides the basic daily and yearly cycles, we can identify other major temporal scales of change for our system (Table 18.1). Here we combine the scales for small and large mammals, because their temporal responses are comparable for those processes considered. The temporal dynamics of vegetation patches require elaboration. We define patch or cell types in such a way that the types change on a slow time scale relative to the dynamics of the herbivores and seedlings. Thus, we can treat the pattern of vegetation types as constant over the length of simulation. This is not to say that the attributes of structure, abundance, and quality of vegetation within a patch remain constant, only that its classification as a particular type remains fixed. Clearly, there is often a significant change in these three attributes between seasons, especially for

Table 18.1. Major Temporal Scales of Change as Considered in the Model

Context	Time Scale of Change
Herbivores	
Foraging within cells	1–30 minutes
Daily foraging activity	<24 hours
Species abundance	Season to yearly
Seedlings	
Herbivory	?
Regeneration through growth	Weekly
Regeneration through recruitment	Yearly
Shrubs and herbs	
Patch type	Decades
Patch atrributes (structure, quantity, and quality)	Seasonal or subseasonal

herbaceous vegetation. We assume that the temporal dynamics of these attributes will be characteristic for each patch type.

18.4 Model Description

In this section we consider the main details of the model and their biological justification. We have incorporated the following main concepts or factors into our model: (1) the adjustment of model resolution to deal with differently scaled components, (2) the herbivory and growth of tree seedlings, (3) the social interactions of herbivores, (4) the general movement patterns of herbivores, (5) foraging and home range dynamics, and (6) dispersal dynamics. The independent variables of primary importance are listed in Table 18.2, and these variable names are cited in the following discussion. We have chosen voles and deer as model herbivores because they are relatively common and well studied; they will be considered representative of small and large mammalian herbivores. For both voles and deer, individuals are assigned to classes on the basis of their maturity and sex, under the assumption that the major variation in the behavior patterns within the population results from differences between these classes.

Our primary dependent variables, measured after a particular length of time, are (1) the percentage of tree seedlings damaged; (2) the percentages of living seedlings that grow to given heights, as summarized in a frequency histogram of seedling size classes; and (3) the spatial pattern of living tree seedlings on the ROW. Combinations of values for the independent variables listed in Table 18.2 will produce response surfaces for these dependent variables.

In Fig. 18.2 we present a summary of the primary modules and operations of the model, including the locations of subroutine calls that access values for the variables listed in Table 18.2. Within each of these subroutines, the required value is generated from a table lookup or regression surface.

18.4.1 Scaling and Cell Types of Landscape Grids

The model must simultaneously consider processes that occur on the megahabitat, macrohabitat, and microhabitat scales. To do this, we could increase the extent of the landscape while holding the resolution (i.e., the cell size) at the microscale, at great cost in computer time. A more efficient method is to store different versions of the landscape simultaneously, each having a different spatial extent and resolution, and which together comprise the three scales mentioned.

The megascale landscape is approximately 6000 by 6000 m in extent. This scale gives a landscape large enough to provide an adequate context for most of the dynamics of a large group of deer that are directly related to ROW utilization (Tierson et al. 1985). The cells of this landscape (i.e., megacells) were chosen to be 50 by 50 m (0.25 ha) in size, based on the size of the ROW study plots (see Section 18.2) and a consideration of homogeneity of cover types. As already mentioned, there are no cell sizes specifically associated with the macrohabitat scale, because both the space use (i.e., movement and foraging) of herbivores and the variation of seedling recruitment at the macroscale can be modeled as a direct result of processes at the microscale. The size of the microscale landscape is equal to the size of a single megacell (50 by 50 m). Because seedling data are gathered

Table 18.2. Summary of Primary Independent Variables in the Model, with Their Units and Functional Arguments[a]

Variable Name	Description	Units	Functional Argument
TSTEP	Basic time step	1 – 10 minutes	
NFORAGR	Number of individuals		Species, functional class
STRUCTUR	Vegetation structure	Relative units of height and leaf area	Cell type, period
SDAMAGE	% height taken off seedling by herbivore	Dimensionless	Species, period
REGROW	Regrowth of height by seedling	cm/day	Cell type, period, cumulative SDAMAGE, period of latest herbivory
SREPULSE	Strength of scent	Relative units	Species, functional class, residence time at site, functional class of scent, site residence time of scent
LREPULSE	Longevity of scent	# TSTEPs	Same as SREPULSE
RESIDE*	Residence time on cell	# TSTEPs	Species, cell type, period, landscape
STEPSIZE	step size	Distance/TSTEP	Species, period, landscape
RETURN*	Time before revisiting a cell	# TSTEPs	Species, dispersal status, landscape
SUITRANK	Suitability ranking for cell types	Integer value	Cell type
SF*	Function converting SUITRANK into effective distance	Dimensionless	Cell type
MAXHR	Max travel distance from refuge	Straight-line distance	Species, functional class, period
PROBENC*	Probability of seedling encounter	Dimensionless	Species, cell type, seedling density, STRUCTUR, maximum seedling height
DAYLNGTH	Maximum daily foraging time	# hours	Species, functional class, period
HRTD	Threshold for dispersal readiness	# contacts w/dominants	Species, period, landscape

[a]An asterisk denotes variable or parameter values neither derived from data gathered by the right-of-way project nor available from literature sources.

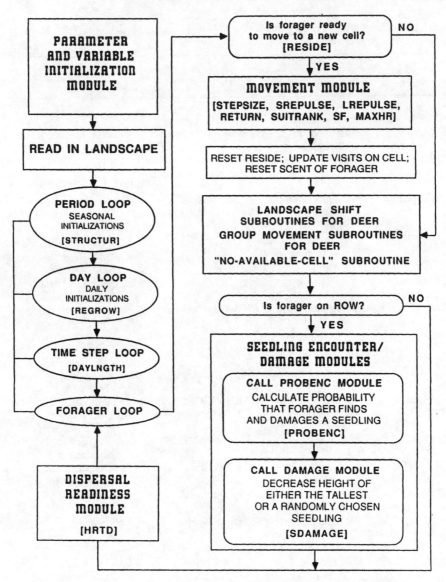

Figure 18.2. Diagrammatic summary of the main modules and computations of the simulation model. Independent variables from Table 18.2 are in brackets at the locations where they are referenced in the model. The order in which individuals are considered depends on their age, social dominance status, and residency time at the current home range site; the oldest, most dominant, longest-established residents are considered first. See Fig. 18.5 for the details of the movement module.

from 1-m² quadrats, microcell sizes were set initially at 1 m². A microscale landscape exists for every ROW megacell that we are monitoring for seedling damage.

Besides the monitored and unmonitored ROW cells, the cell types of the megascale landscape include major cover types such as forest, shrub/thicket, old field, agricultural land, urban land, and water. The cell types of the microscale landscape are based on the species composition and dominance of shrub and herbaceous vegetation, as described in Section 18.2. The interactions between herbivores and tree seedlings take place within these microcells, which may contain, at least initially, one or more seedlings. The vegetation on a microcell can serve as a source of cover or food for the herbivores, although not all cell types contain adequate cover or edible food. Seedlings also are affected by the major herb and shrub species on a cell through competition for resources, resource enhancement, or modification of the microclimate. A patch is defined as one or more contiguous cells of a particular type that are isolated from other cells of that type (see Wiens 1976). Because by definition patches are composed of cells of only a single type, the type of a patch is equivalent to the type of its constituent cells.

We define vegetation structure (STRUCTUR in Table 18.2) as a relative measure incorporating both vegetation height and leaf area. We assume that the vegetation (other than tree seedlings) on a cell retains a fixed structure within discrete time intervals (periods) but that structure may change between periods, with potentially important implications. Specifically, a change in STRUCTUR may make seedlings of a given height more (or less) apparent to herbivores, and the apparentness of the seedlings in turn will affect the degree to which they are vulnerable to herbivory. For model variables that are likely to depend on the quantity and quality of vegetation on a cell (e.g., see RESIDE in Table 18.2), the time period is used as a surrogate for these attributes, and so they are not accounted for explicitly in the model. The lengths of the periods correspond to seasons or to intervals within seasons.

18.4.2 Herbivory and Growth of Tree Seedlings

At the start of each year, seedlings are established in densities according to cell type and the distance from seed sources. The subsequent growth rates of seedlings are also functions of cell type, but the conditions for establishment are not necessarily correlated with the conditions for growth. Upon this baseline relationship between cell type and seedling establishment and growth, herbivores are allowed to change the density, the heights, and the growth rates of seedlings.

The loss of tissue as a result of herbivory may have one of several effects on seedlings besides reduced height: (1) direct mortality, (2) reduced growth and development, or (3) indirect mortality due to the effects of reduced growth. In the model, once a herbivore encounters a seedling, the seedling is reduced in height by a certain percentage of its current height, or it may be killed (see SDAMAGE in Table 18.2). Indirect mortality is not modeled at this time.

Several factors may influence the chance of encounter and the degree of subsequent damage to seedlings. Herbivory rates in ROWs have been observed to vary from 0 to 60% across different patch types, as a result of interactions not only between the herbivore and the patch type but also between the seedling and the patch type. For example, the availability of a seedling to an herbivore (as a function of vegetation structure and seedling height) may significantly influence seedling damage. Furthermore, seasonal changes in nutrition and water requirements of herbivores may change the attractiveness of seedlings to herbivores.

Three possible mechanisms for tree seedling regeneration exist, depending on the degree of damage: (1) recruitment from seeds, (2) root and stump sprouting, and (3) lateral bud sprouting. The rate of tissue growth by a seedling will often be high enough to compensate for the loss resulting from herbivory (Crawley 1983). In the model, the height of all seedlings not killed by herbivory is increased daily by an amount depending on the extent and timing of herbivory (if any), cell type (e.g., resource and competitive conditions), and period (see REGROW in Table 18.2). The model parameters relating regeneration to the degree of herbivory will be based on data from simulated herbivory (i.e., clipping) experiments.

In order to track the heights of seedlings and because different types of damage and regeneration may occur, we consider seedlings as individuals. This would not be practical if we were considering, say, grasses (Rastetter, Chapter 14). We consider tree species either separately or within small assemblages, because (1) differences in the life history and physiological characteristics of various species will likely change the relationship of seedlings to patch type, and (2) herbivores may use some tree species preferentially.

18.4.3 Interactions of Herbivores

Often, patterns of space use emerge from the behavorial interactions of individual herbivores. The two general social behaviors that must be considered are grouping and repulsion. Grouping is handled by assigning each individual herbivore to a particular group on the basis of characteristics such as relatedness and sex. The model can then track the location and behavior of both the group and its members, so that groups can be broken apart and rejoined when necessary. Repulsion, in contrast, incorporates aggression, dominance, and avoidance behaviors. It is a property of a specific interaction between individuals, or groups of individuals, with particular characteristics (e.g., maturity, sex, and relatedness). The same modeling construct can be applied to both intraspecific and interspecific repulsions.

18.4.3.1 Grouping

In some cases it is efficient, or necessary, to simulate an aggregation of individuals that share common essential characteristics. For the herbivore component, these characteristics may include similar habitat or food preferences and similar movement patterns. A common example of such an aggregation is a group of related individuals that travel together. Deer form several types of social aggregations:

associations of adult females and fawns in summer; male bachelor groups in summer; groupings with an older female, her female offspring, and their collective fawns in fall; and groups including all sexes and ages during winter (Hawkins and Klimstra 1970; Hirth 1977; Marchinton and Hirth 1984). For a group of deer moving on the megascale landscape, a group leader is identified, and the movements of all members of the group are determined by the movements of the leader. Voles have been observed to form groups of individuals that share nest sites and home ranges during fall and winter (Webster and Brooks 1981; Madison et al. 1984; Merritt 1984; Madison and McShea 1987). Individuals that share home ranges, however, may avoid foraging in the same locations (Karlsson and As 1987), and these different movement patterns may be important to model.

18.4.3.2 Repulsion

During the time that an herbivore resides on a particular cell and for a certain time after it leaves the cell, the cell bears the scent of the individual. This scent provides information about that particular individual to other herbivores that encounter the cell. The strength of the repulsion (see SREPULSE in Table 18.2) and the length of time it exerts its effect (longevity) in the absence of the individual (see LREPULSE in Table 18.2) depend on the relative positions of the interacting individuals in the social hierarchy of the population. For example, the repulsiveness of an adult female meadow vole (*Microtus pennsylvanicus*) would have high strength and longevity for other adult females but not necessarily for adult males. Residents of an area would have stronger and longer-lasting repulsion toward nonresidents than toward each other (Turner and Iverson 1973). Also, strength and longevity can vary independently; the repulsiveness of an adult male would be strong for another adult male but not necessarily long lasting. This is the basis of spatial time-sharing (Getty 1981a). Although repulsion can change as a result of factors such as forage levels and the location of the interaction site in relation to. the nest sites of the individuals (Ims 1987), for simplicity we initially assume that the repulsion strength and the longevity for a given type of interaction are constant within periods (see Madison and McShea 1987). Within a species, cell-by-cell repulsion between small mammals results in the delimitation of home ranges and territories. If the actual radius of repulsion between animals is larger than the radius of the microcell, a larger aggregate cell (say 16 m^2) is defined for use when repulsion is assessed.

Figures 18.3A and B show some possible effects that repulsion has on the use of space in the ROW by small mammals, according to our model. A 10 by 10 grid landscape was used to facilitate illustration. Sixty-seven percent of the cells are suitable for use. When individuals move independently of one another, movements overlap considerably, and cell use is locally intense but limited in space. When individuals avoid each other, as do territorial vole females, more of the landscape cells are used, but the intensity of use of any one cell decreases. We can thus modify the parameters of repulsion in the model, and the implications for the patterns of plant survival can be assessed. These parameter changes represent both

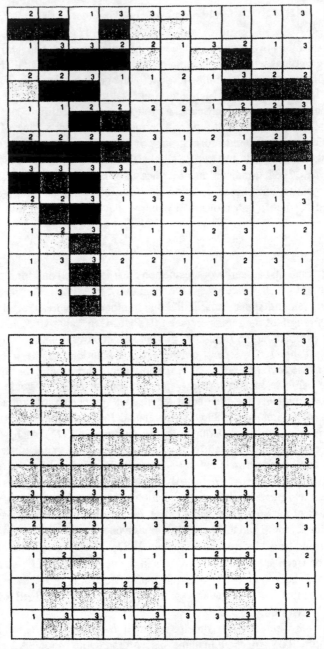

Figure 18.3. A 10 by 10 landscape showing the intensity of cell use for 8 simulated small mammals foraging for 12 steps. The mammals forage independently of one another in (A), whereas in (B) the current or previous use of a cell by a conspecific repulses the individual (i.e., SREPULSE and LREPULSE were set to high values). The darkness of shading represents the number of different individuals that have used that cell, ranging from 0-white to 4-black. The number in the corner of each cell is the cell type; cell types were randomly assigned to cells. Type 1 is unsuitable, whereas both types 2 and 3 are equally suitable. The initial nest locations were randomly chosen.

changes in species behavior across seasons (Webster and Brooks 1981; Madison and McShea 1987) and interspecies differences in social dynamics (e.g., see Stenseth et al. 1988).

18.4.4 Movement Patterns

Mammalian herbivores exhibit several types of movements. For example, Madison (1980) defines four types of movements in meadow voles: (1) residency, (2) shifting, (3) wandering, and (4) true dispersal. For the same species, Baird and Birney (1982a) distinguished three types of movement: (1) residency, (2) dispersing, and (3) moving. During the breeding season, residents tend to be reproductive females, dispersers tend to be nonreproductive males and females avoiding intraspecific encounters and exploring vacated habitat, and movers tend to be resident adult males that travel relatively long distances in the course of daily activities, possibly in search of estrus females (Baird and Birney 1982a). Deer exhibit relatively short-range movements during daily foraging and shelter seeking (Rongstad and Tester 1969; Nudds 1980; Wiles and Weeks 1986; Kufeld et al. 1988), as well as long-range movements corresponding to disperal (Hawkins et al. 1971; Kammermeyer and Marchinton 1976; Marchinton and Hirth 1984) or, in northern areas, seasonal migration (Ozoga and Gysel 1972; Verme 1973; Drolet 1976; Tierson et al. 1985). Our megascale landscape may not be large enough to include the distances involved in seasonal and dispersal movements of deer in northern areas. Thus, our analysis of deer movement patterns will be limited to home range dynamics and in some cases will be applicable to winter and summer dynamics only.

The movements of model herbivores between cells, on both megascale and microscale landscapes, are associated with a stimulus, a step size, a movement rule, and a maximum activity range. Importantly, these same model constructs are used for all species and landscape scales; only the parameter values are changed. The **stimulus** to move between cells is due to factors such as social interaction and hunger. The time between stimuli defines the residence time (RESIDE in Table 18.2) on a cell. Residence time is based on one rule or a mixture of rules that the animal uses to assess whether it should leave a patch or continue foraging there (McNair 1982, 1983; Green 1984; Stephens and Krebs 1986, Chapter 8); however, we are not concerned here with the specific rule that generates a particular residence time. For small mammals, RESIDE will likely be largely determined by the exposure of the individual to predators, which depends upon vegetation structure (Thompson 1982; Wywialowski 1987; Newman et al. 1988; Brown et al. 1988). For deer, RESIDE in microcells probably depends primarily on forage quality and quantity and environmental conditions. For simplicity, we assume that RESIDE is characteristic for a particular herbivore species and cell type, and RESIDE will be estimated from data on movement rates and daily activity budgets. Note that RESIDE can be made a function of the distance from the nest site, or of the distances to the nest sites of neighbors, to simulate cell-use intensities that decay with distance from the home range core (Ford 1983). We assume that

RESIDE does not depend on the herbivore's past foraging experience (see Ollason 1980; McNamara and Houston 1985) or on the alternative cells available.

The **step size** (STEPSIZE in Table 18.2) is the maximum distance the animal can travel, under ideal conditions, in a time that is short in relation to the minimum residence time. When a herbivore is ready to move, the available choices of forage cells are limited to those within an area of radius STEPSIZE.

The **rule** determines the next cell moved to and reflects four factors: (1) preferences for certain habitat types, (2) the current or past occupancy of a cell by other herbivores (SREPULSE and LREPULSE), (3) the minimum time allowed between subsequent visits of a given cell by the same individual (see RETURN in Table 18.2), and (4) the distances to prospective cells from the current position. As a simplification, cells can be classified as either suitable or unsuitable; unsuitable cell types are not visited. Where more information is available, cell types are ranked by their relative suitabilities (SUITRANK in Table 18.2). In these cases, visitation of a certain habitat type, unlike residence time, depends on the alternative choices available (Kufeld et al. 1988). The second factor of occupancy and repulsion is described in detail in Section 18.4.3. The variable RETURN is included because foragers often avoid cells that they have recently visited. In nature, this may be due to several factors: (1) "trapline foraging" (Thomson et al. 1982; Gill 1988), the repeated visitation of sites in a predictable sequence to allow for resource regeneration; (2) home range patrol (Thompson 1982); or (3) directionality of movement (Senft et al. 1987). The value of RETURN can be modified to simulate different strategies relating to the proportion of the total home range used in one or a series of days. Some herbivores may use several different regions of their home range over the course of one or a few days, whereas others may repeatedly use the same region for many days before moving to a new area (Sparrowe and Springer 1970; Ford 1983).

Given that one or more prospective cells are within radius STEPSIZE and satisfy the constraints of LREPULSE and RETURN, their distances from the present location are assessed. The travel distance (TD) to a prospective cell may be quite different from the straight-line Euclidean distance. For example, in many cases the TD between two cells separated by a lake is at least the path length around the lake. This level of realism is necessary in order to consider a variety of questions, especially in the context of human-modified landscapes. The TD to each of the prospective forage cells is estimated, and all cells with TD greater than STEPSIZE are discarded. Then, for each of the remaining cells we define the effective distance:

$$\text{Effective Distance} = (\text{SF}) \times (\text{suitability ranking}) \times (\text{TD}) \qquad (18.1)$$

where SF (Table 18.2) is a function of the suitability ranking. Thus, the difference in suitability between two cells is converted into a difference in effective distance. The forager is moved to the forage cell with the mimimum effective distance, and so the cell type chosen depends on the alternatives available. With this modeling construct, at any step the herbivore may choose an inferior but close cell over a superior but distant cell, a phenomenon observed in both invertebrate and vertebrate foragers (Levey et al. 1984; Murphy et al. 1984). Although estimating the TD between cells is quite computer intensive, in some cases TD may be taken as simply the straight-line distance.

The **maximum activity range** (MAXHR in Table 18.2) reflects the compressive tendency due to increased costs with increasing home range or territory size. If a home range core is respected and the activity range is limited by site tenacity, then associated movement patterns are called home range dynamics.

We assume that a mammalian herbivore has perfect knowledge of the locations and types of resources within its home range and that movements of a particular step size between suitable cells are on a short time scale in relation to residency within a cell (Karlsson and As 1987). This allows us to model movement between cells as saltatory and instantaneous, and search behavior is not included in the model.

Figure 18.4 illustrates an example of model results of deer movement on both the megascale and microscale landscapes. The deer travel as a group on the megascale landscape, but when they reach a monitored ROW cell they forage as individuals. We therefore have two sets of rules for the home range movements of deer: (1) choices of cells at the megascale, which represent the dual needs for cover and forage and which are restricted by external influences (e.g., snow depth); and (2) choices at the microscale (within the ROW), which are set by the specific patterns of foraging behavior. The length of time spent on the ROW cell depends on the abundance of microcells with food and the length of time the herbivores reside on them. Although the time a deer spends foraging within any megacell can be modeled as the result of foraging processes at the microscale (Senft et al. 1987), here megascale processes will depend on microscale movements for monitored ROW cells only.

18.4.5 Foraging and Home Range Movements

The model implementation of the mechanisms discussed in the previous section is illustrated in Fig. 18.5 for an herbivore moving to a new forage cell. We assume that the herbivore forages randomly once it has moved to a new cell; the finer details of foraging choices within cells are beyond our resolution of mechanism. While foraging, an herbivore may, with a particular probability, find and damage a seedling (see PROBENC in Table 18.2). The probability of seedling encounter is conditional, given herbivore visitation of a particular cell. The modeling of the herbivore component tells us how intensely each cell is used by herbivores of a particular species and functional class. In practice, the model can be run for different scenarios to generate spatial distributions of cell-use intensities, and total seedling damage can then be calculated based on these distributions.

A more detailed model would track both the change in vegetation biomass in each cell visited and the amount of food assimilated by the herbivore as the day progresses. We have not explicitly modeled the nontree seedling vegetation for three reasons: (1) these data are not readily available, especially for small mam-

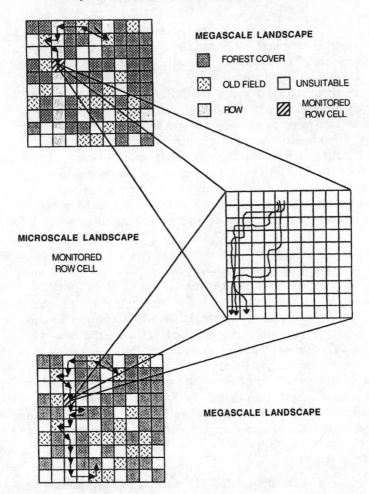

Figure 18.4. An example of movements of a single group of deer composed of three individuals. Arrows indicate the direction of travel for two foraging days. The megascale landscape has two equally suitable food cell types interspersed with cover and unsuitable cells; all cells on the microscale landscape are suitable food types. The locations of non-ROW cell types were randomly generated. On the microscale landscape, the movements of individuals are constrained toward one randomly chosen far corner. Except for the last food cell visited in a day, previously visited cells are not revisited.

mals; (2) voles and deer have very general diets; and (3) our main concern with nontree seedling vegetation is its role within the context of a patch type and how this role mediates the herbivore-seedling interaction. Explicit modeling of the energetics of the herbivores is considered to be beyond the scope of the present objectives. Parameters such as DAYLNGTH and RESIDE (Table 18.2) take into account the effect of differences in metabolic needs on foraging behavior. The features of energy expenditure and uptake can be readily included in the model, however, and we plan to consider energetics in future model uses.

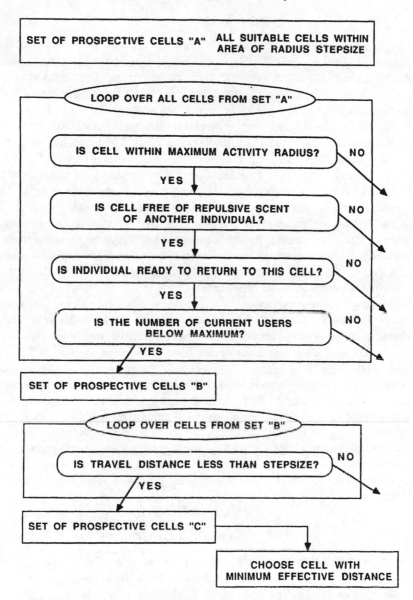

Figure 18.5. Flow diagram of the movement module, which locates the next cell for an herbivore ready to move from its current position. See Section 18.4.4 of the text.

18.4.6 Disperal Dynamics

Dispersal movements are considered only for voles on the microscale landscape. Megascale dispersal mainly affects the densities of the herbivore populations associated with the ROW, but these densities are treatment variables to be manipulated in model experiments. The main result of vole dispersal in the model will be to reorganize initial home range boundaries with the ROW patch.

Like foraging and home range movements, dispersal is associated with a stimulus, a step size, and a rule. The stimulus or readiness to disperse, like the stimulus to change cells during foraging, is evaluated at each time step (see HRTD in Table 18.2). Dispersal readiness is a function of the frequency of encounters with dominant individuals at the present site and the level of repulsion associated with these encounters. Baird and Birney (1982b) argue that for *M. pennsylvanicus*, vacant areas at the macroscale are usually filled by dispersers that are smaller and younger than surrounding residents and are thus likely to be subordinate individuals. As in foraging, the rule for movement to a new site depends on habitat preference, occupancy by others, and previous occupancy by self. The step size, as before, controls the scale of movement.

The length of time that a dispersing individual remains in a new location (i.e., makes the new cell its new home range core) will depend on the frequency and outcome of subsequent intraspecific encounters while the individual forages around this site. This conclusion is consistent with the suggestion of Baird and Birney (1982b) that potential colonizers are inhibited from settling in an area by behavioral interactions with long-term residents rather than by density per se and that dispersers settle in the first available good habitat. Actual settling in an area can thus be inhibited over many days by encounters with dominants and/or residents. Because foraging and dispersal are evaluated at the same time scale, whether habitat vacancies are filled by dispersers or by neighbors expanding their home ranges depends on the size and isolation of the area left vacant and on the food needs (as modeled by the daily foraging time) and nest-site affinities of the residents.

18.5 A Simulation Experiment

We performed a simple analysis to explore the response of the model to changes in three of the independent variables listed in Table 18.2, for both a random and a clumped landscape. The three variables chosen for analysis were PROBENC, RETURN, and RESIDE. The dependent variable used to measure model response was the degree of seedling damage. One input variable was varied at a time, with all other model variables and parameters held at their "control" values, and the change in the dependent variable was measured. These particular independent variables were chosen for analysis because they are the least likely to be known from field measurements, and because they have characteristic effects on the amount of seedling damage.

The number of seedlings in a particular cell that are damaged by herbivores depends on both the probability that an herbivore encounters the seedlings while foraging on the cell and the number of herbivore-hours of use of that cell. The total damage to the plant population over the landscape is a function of these two factors. The influence of PROBENC on plant abundance is through the first factor alone. A change in PROBENC does not directly influence cell-use intensities in our model, since we are not considering seedlings as an energy source for the herbivores. The variables RETURN and RESIDE, however, primarily influence the second factor. A decrease in RETURN lowers the minimum time between visits to a cell for an individual, resulting in decreased movement over the home range, a smaller number of cells visited, and more intense use of those cells which are visited. A decrease in RETURN is similar in effect to an increase in the resource renewal rate of other models (Ford 1983). The effects of an increase in RESIDE, the time spent on each cell, are similar to those which result from a decrease in RETURN. Unlike RETURN, however, RESIDE does not control the number of different foraging circuits made by the herbivore.

For this analysis, we considered six nonterritorial individuals of a single species on our microscale landscape (i.e., a 50- by 50-m landscape with 1-m² grid cells). Twenty percent of the cells were assigned to a single food cell type, with all other cells unsuitable for foraging but suitable for travel. The activities of the model herbivores were simulated at five-minute time steps for three hours per day, for thirty days. All forage cells began with ten seedlings, for a total initial abundance of 5000 seedlings, and it was assumed that the herbivores killed any seedlings they encountered. The input variables were changed in a direction that we hypothesized would lead to a decreased impact on the seedlings. The values of PROBENC were set relatively high so changes in the output variable (seedling damage) would be readily apparent. The step size was set at 10 m, and for our simulations there was always at least one food cell available to a given herbivore at any move.

Table 18.3 shows the response of total seedling damage to changes in the independent variables, expressed as the ratio of the percent change in output to the percent change in input. For this particular scenario, seedling damage does not appear to be very responsive to changes in the variables chosen. Altering one variable at a time, however, does not readily expose interactions among the variables, and this scenario is one of the simplest the model is capable of simulating. The number of cells visited at least once during the simulation (NCV) is also shown, and, not unexpectedly, there is a positive relationship between NCV and seedling herbivory.

There are patterns that emerge even from this simplest scenario. First, and not surprisingly, the time at which the output variable is measured has an influence on the percent response. For RETURN and RESIDE, the output response is larger after thirty days than after ten days. The response to a change in PROBENC, however, is smaller for the thirty-day run. Because of the high mortality rate on seedlings and the fact that PROBENC is modeled as an exponential function of seedling density (see Table 18.3), most seedlings that will suffer herbivory are killed within a relatively short time. After thirty days, both values of PROBENC lead to similar total mortality.

Table 18.3. Effects of Altering the Values of Selected Input Variables on the Total Number of Seedlings Killed After 10 and 30 Days (Output Variable)[a]

Input Variable or Parameter Changed	Control Input Value	Altered Input Value	NCV[b]	Output Value	Ratio of Output Change/ Input Change
Random landscape, 10 days					
CONTROL			331	1567	—
RETURN[c]	144	72	188	1378	0.24
RESIDE[d]	2	4	182	1283	0.18
PROBENC[e]	0.2	0.1	331	1154	0.53
Random landscape, 30 days					
CONTROL			331	2951	—
RETURN	144	72	188	1867	0.73
RESIDE	2	4	182	1804	0.39
PROBENC	0.2	0.1	331	2421	0.36
Clumped landscape[f], 10 days					
CONTROL			307	1553	---
RETURN	144	72	190	1398	0.20
RESIDE	2	4	193	1354	0.13
PROBENC	0.2	0.1	307	1163	0.50
Clumped landscape, 30 days					
CONTROL			307	2798	---
RETURN	144	72	190	1885	0.65
RESIDE	2	4	193	1910	0.32
PROBENC	0.2	0.1	307	2361	0.31

[a] PROBENC was changed through its exponential parameter, whereas the values of the other variables were changed directly. The percentage change in the input or output value is calculated as [altered value - control value]/control value.
[b] The number of cells with at least one visit during the simulation run.
[c] The minimum number of time steps (5 minutes) before revisitation of a cell by a particular forager.
[d] The number of 5-minute time steps that a forager resides on a cell.
[e] The parameter a in PROBENC = $[1 - \exp\{-a *SEEDING\ DENSITY\}]$ is altered.
[f] A landscape with 50 clusters of food cells, each cluster of maximum radius 4 m, with 10 food cells per cluster. Clusters are randomly located and allowed to overlap.

Second, the changes in the input variables have less effect on model response in the clumped landscape than in the random one. The reason for this difference can be seen if we compare the values of NCV across runs. At the control values of RETURN and RESIDE, foragers cover relatively large distances over the course of several days, and the activity ranges of the individuals overlap. This overlap of cell use, however, is higher in the clumped landscape than in the random landscape. Between-landscape variation in overlap is expressed in the variation of NCV values, since for given values of RESIDE and RETURN the total number of visits does not differ between landscapes. The higher overlap, and thus the lower NCV, leads to a lower seedling mortality for the clumped landscape. When RETURN is decreased or RESIDE is increased, the activity ranges shrink. Now, cell-use overlap is higher for the random landscape than the clumped landscape (on the clumped landscape, nest sites tend to be in separate clumps). The clumped landscape now shows the higher NCV, and thus the higher seedling mortality. By plotting the output values given in Table 18.3, we can see that this change in which landscape shows the higher NCV, as the control input values are altered, can explain the trend in the output change/input change ratio. Similarly for PROBENC, the lower ratio on the clumped landscape can be seen to be the result of the lower NCV (and thus higher overlap) for that landscape pattern in relation to the random pattern ($NCV_{clumped}=307$ versus $NCV_{random}=331$). When PROBENC is changed, fewer cells are affected on the clumped landscape. We do not know to what degree these differences between the two landscapes analyzed are general features of clumped versus random landscapes. But this simple analysis provides a good example of the sometimes unpredictable complexities that emerge from this model.

18.6 Discussion and Hypotheses

The model that we have described above can be used to ask many different types of questions of a variety of systems. In many cases it is useful to phrase these questions in the context of the interaction between landscape pattern (i.e., abundance and spatial distribution of patches) and ecological processes: (1) What is the effect of the pattern of patch types, at both the megascale and microscale, on the use of patches by herbivores, and ultimately on the degree and distribution of plant survival and growth over the landscape; and (2) What is the contribution of herbivore foraging to the enhancement or suppression of the spatial patterns of plants? Because the model is individual based, we can also ask questions about the mechanistic effects of herbivore species diversity on plant survival and growth.

18.6.1 Megascale Landscape Pattern and Herbivore Impact at the Microscale

In the context of the megascale landscape, the megacell is a patch of habitat connected to other such patches through herbivore movement. For small mammals, fluxes between megacells are due to dispersal, whereas deer may use several of these cells during a single day's movements within the home range (e.g., Sparrowe and Springer 1970; Drolet 1976). Because of these different scales, the implications of megascale landscape pattern differ for small and large mammals.

18.6.1.1 Small Mammals

Fluctuations in population numbers are damped when the metapopulation is spread over a number of distinct patches that are connected by dispersal (Reddingius and Den Boer 1970; Vance 1984; DeAngelis and Waterhouse 1987). In this case, disturbances that may cause population extinction are not likely to affect all patches simultaneously, and those which are affected can be recolonized from unaffected yet dispersal-connected patches. The distances between these patches in relation to the average dispersal distance of the animal have been shown to be important in the rate of recolonization of disturbed patches and thus in the long-term persistence of the metapopulation (Hanson 1977; Lefkovitch and Fahrig 1985; Fahrig and Merriam 1985). Although we could use this model to explore the effect of megascale landscape patterns on the densities of small mammals in the ROW, we consider such an enterprise outside the scope of our study because of the relatively long time scales involved in population transfer.

18.6.1.2 Large Mammals

We have already mentioned that the length of time a large mammal forages in a megacell, once it arrives, depends primarily on the foraging process and the pattern of patches at the microscale (Senft et al. 1987). Whether or not the animal visits a megacell, however, depends on the pattern of suitable patches over the megascale landscape. Recent trends in habitat analysis as applied to wildlife management and species preservation make use of data on the spatial interspersion of different habitat types (e.g., Farmer et al. 1982; Lancia et al. 1982; Matulich et al. 1982; McNay et al. 1987; Harris and Kangas 1988; Urich and Graham 1988). Lancia et al. (1982), in a study designed to estimate habitat quality for bobcats in North Carolina, used a grid-cell representation to map habitat types over a 144-km^2 area. They assigned an "interspersion index" to each cell to reflect the influence of the spatial proximity of other cell types on the quality of that cell. Thus an isolated cell, even if it is critical habitat, may be assigned a relatively low overall quality.

The same considerations of habitat juxtaposition and interspersion can be applied to the ROW study. In an area where the ROW is surrounded by forest (e.g., in the Catskill Mountains), the ROW may provide the only good foraging habitat (Loft and Menke 1984) over several kilometers, and ROW use will likely be high if nearby cover is suitable. In contrast, lowland megascale landscapes may provide more open areas due to human modification, so that the ROW may be only one of many suitable foraging sites in the area. The intensity of use of the ROW would then likely depend on its location relative to these other sites and may be relatively low. In winter, when energy is the major constraint on activity in northern areas (Parker et al. 1984), the spatial isolation of a suitable food patch away from other food and cover patches may make its use prohibitively expensive energetically. In these ways the heterogeneity and pattern of the megascale landscape can influence the intensity of deer use of a ROW segment and thus affect the impact of deer on the resident seedlings. Assuming that we have determined the intensity of deer use required to produce a particular level of seedling damage, we can then ask which megascale landscape patterns would produce this level of use, for a certain set of population (e.g., abundance, sex ratio, and age structure) and environmental (e.g., snowfall) parameters.

18.6.2 The Effects of Microscale Heterogeneity

Heterogeneity of the microscale landscape exerts its effects on both herbivores and the plants they damage. First, even if a particular plant species is distributed uniformly on the landscape, not all plant individuals will be equally vulnerable to herbivory. Because of small-scale habitat heterogeneity, herbivores will use some areas more than others. Patches with dense cover may get a lot of use, whereas those that have poor cover, that are far from the home range core, or that are between territory boundaries may get less use (Ford 1983). A preferred plant species may be hidden or sheltered by other plant species; for example, patches with dense or inedible shrubs may inhibit mammals from browsing on seedlings (McAuliffe 1986). Some plants then, by virtue of being rather isolated in poor or remote patches, or because they are growing under other plant species, may not be exposed to herbivory at all. Such isolation can lead to intrinsic stabilization of the herbivore-seedling system by making it difficult for the consumer to eliminate all of the seedlings from an area.

It is unlikely that the plants will be distributed uniformly in space, however. A particular patch type may inhibit or encourage the establishment and/or growth of a plant species, for a given level of seed input. For example, the level of resource competition our plant species must face is likely to vary across patch types. Thus, the relative abundances and spatial distribution of patch types can determine the abundance, biomass, and spatial distribution of the plant species on the landscape.

Because these effects of heterogeneity on the herbivores and plants operate simultaneously, microscale heterogeneity to some degree controls the herbivore-plant interaction. As an illustration, assume there are two patch types I and II on the landscape. Consider three cases relating the responses of herbivores and tree seedlings to these patch types: Case (1) herbivores use mainly type I, but seedling establishment and/or growth is best on II; Case (2) herbivores use mainly II, but seedlings do best on I; Case (3) herbivores and seedlings both do best on type I. In cases 1 and 2, the responses of herbivores and seedlings to patch type are inversely related, they most often occur on different patches, and high seedling escape should result. Case 3 is the most interesting, because it represents a balance between the detriment to seedling survival as a result of herbivory and the increment to survival as a result of enhanced establishment and/or growth. Such a situation was found by Schupp (1988) for *Faramea occidentalis* seedlings and their mammal predators in tree-fall gaps in Panama. In this case, the overall survival rate of seedlings may be strongly influenced by the ability of the herbivores to utilize type I patches, which may in turn depend on the relative abundances and distribution of these two patch types.

18.6.3 Microscale Processes and Tree Spatial Patterns

We have seen how landscape vegetation patterns can affect herbivore pressure and, ultimately, seedling densities. In addition, herbivore dynamics may influence the spatial pattern of trees on the landscape. Trees of different species, or of different size classes within a species, have been shown to exhibit different spatial patterns (Armesto et al. 1986). The relative contributions of various factors such as canopy gaps, thinning, soil nutrients, and seed/seedling predation to tree spatial patterns in an area are difficult to determine (Hubbell 1980; Pemadasa and Gunatilleke 1981; Becker et al. 1985; Armesto et al. 1986; Newbery et al. 1986; Leps and Kindlmann 1987; Schupp 1988). To what extent can unequal encounter or damage of tree species by herbivores contribute to interspecies differences in pattern? And does herbivory play a role in the observed changes of pattern over time within a species, beginning with established seedlings? If we consider that herbivores (1) often show preferences for certain tree species (e.g., Ross et al. 1970), (2) use the landscape patchily (e.g., McAuliffe 1986), and (3) can exert great pressure on seedling populations, then it is clear that herbivores may have a large influence on pattern under certain circumstances. If established seedlings are randomly dispersed across a patchy landscape, when can the herbivore response to this patchiness produce a clumped tree distribution? And under what conditions can they drive the pattern in the opposite direction, from clumped to random? Our simulations will provide a null model that we can use to examine these questions.

Similar questions can be applied to the ROW study also. At the time of seedling establishment, a distance-from-edge effect is sometimes observed in seedling density as a result of the limitations of seed dispersal across wide ROWs. This original macroscale pattern in seedling density becomes obscured over time. The question we can ask is this: How does herbivory affect this initial seedling pattern? Specifically, what is the level of herbivory that would be needed to suppress an initial seedling pattern of given strength? We hypothesize that realistic herbivory rates can swamp out the original pattern of germinated seedlings observed in the field.

18.6.4 The Effects of Herbivore Diversity on Plant Survival and Growth

Species diversity of mammal assemblages can be extremely variable in space and time (Anthony et al. 1981). Variation in the types and relative abundances of herbivore species using the landscape may affect plant survival and growth in ways that can be explored only with a spatially explicit, individual-based model. The simplest case is where several species have different patch type preferences. Plants growing in less-preferred patch types would thus escape herbivory from any one of the herbivore species acting alone, but all herbivore species together would cover a wider array of patch types.

Even where all herbivore species use the same patch types, different species may have different effects on a given plant. Some species may browse an encountered plant without killing it, whereas other species may kill it outright. Moreover, the size range over which a plant is most vulnerable to herbivory may be different for various species of herbivores, although there will likely be overlap between these ranges. In the ROW case, tree seedlings growing beyond a certain size may become less vulnerable to damage by small mammals like voles, which tend to gnaw or cut tree seedlings near their base, but may at the same time become more vulnerable to larger browsers like deer. These phenomena together may generally lead to two consequences when more than one species of herbivore is present. First, consider a small herbivore that exerts an early filter on plant survival by killing small seedlings. The resulting spatial pattern of the remaining plants, which are now becoming more vulnerable to larger herbivores, may depend on the behavioral strategy of the small herbivore (see Section 18.6.3). The ability of the larger herbivores to subsequently harvest those plants which remain may be affected by this spatial pattern (e.g., whether the plants are clumped or random), and therefore by the strategy of the smaller herbivore. Second, a reduction in plant growth rate due to browsing by one species may prolong the plant's availability to another species, possibly increasing the mortality rate of the plant population. Both examples illustrate that the combined effects of different herbivore species may be nonadditive (Turner 1988).

The types of herbivore species present can also have an influence on herbivory rates when patch types have different suitability rankings and individuals of different species move about the landscape at different scales (i.e., different step sizes) (O'Neill et al. 1988). As an illustration, assume that all herbivores use two food cell types I and II, but that type I is preferred to type II. Also assume (1) that at any given location on the landscape a type I cell is, on average, further away than a type II cell and (2) that the herbivore knows their distances. If the herbivore's STEPSIZE (Table 18.2) is relatively large, then it is likely that a type I cell will be within its step size, and type II cells may be consistently passed over. Thus, type II cells may go relatively unused. If by contrast the step size is relatively small, then type II cells may be the only alternative available for many moves, and type II cells will receive greater use. Depending on the relationship between the plant species of interest and the different habitat types, several herbivore species with different step sizes, operating simultaneously on the landscape, may have a greater impact on seedlings than a single species with a relatively large step size.

In the above examples, no social interaction between the individuals of different species has been assumed. Interspecific repulsion between small mammals at the microscale, however, may result in spatial segregation of species over the landscape (e.g., Iverson and Turner 1972; Glass and Slade 1980; Linzey 1984). This may have an influence on the number of plants that escape herbivory. We predict that whether repulsions between herbivores result in greater or lesser control of seedlings than if herbivores moved independently of one another will depend on landscape pattern, herbivore densities, and the characteristics of the species involved.

In general, we would like to know how plant damage on the landscape is related to the types and relative abundances of herbivore species present. Our model is well suited to explore to what extent differences in the behavioral strategies of herbivore species can influence plant population densities. We can ask questions about the effects of different types of species individually and jointly, and about the effects of different types of herbivore species assemblages. We hypothesize that a community of herbivores from a variety of species that operate on different spatial scales and/or respond to different size ranges of plants will better control plant density and growth than the same number of individuals of a single species with some average strategy.

18.6.5 Future Directions

We began this chapter by arguing that the recent expansion of the individual-based modeling approach, combined with the growth of landscape ecology as a discipline, has provided a new context for the study of herbivore-plant interactions. It is clear from the literature over the past decade that spatial heterogeneity and the reactions of individual organisms to it are increasingly important and obvious factors in ecological research. Moreover, the methodological limitations on modeling complex systems are becoming less imposing with time, allowing spatially explicit, individual-based models to become a viable modeling alternative. The general advantages of individual-based models have been reviewed by Huston et al. (1988).

Three important contributions of the model we have presented are: (1) it emphasizes a multiscale approach to ecological investigation, (2) it makes use of the type of data generated by studies that are designed with an appreciation of the potential effects of landscape heterogeneity, and (3) it can be used to explore the implications of the interactions between organism response and landscape heterogeneity for a variety of systems and problems. The major limitation of our model is that it is currently implemented in Fortran, which is neither a very efficient nor an elegant programming language for the tasks our model must perform. Other programming methods (e.g., Loehle 1987) are being explored. Regardless of implementation, however, we believe that the modeling approach we have described in this chapter has enormous potential for increasing our understanding of ecological systems in general, and herbivore-plant systems in particular.

18.7 Summary

In this chapter we describe the development of an individual-based, spatially explicit simulation model of herbivore-plant interactions on a heterogeneous landscape. The model considers movement patterns, foraging, and social behaviors of individuals of two types of mammalian herbivores, represented by voles and deer. It also considers recruitment, growth, and damage of individual tree seedlings. The model was developed to examine the potential role of mammalian herbivores in slowing the invasion of utility rights-of-way by tree seedlings; however, the modeling approach can be used to ask many different types of questions of a variety of systems. The ROW system provides an excellent opportunity to examine some general ecological questions concerning herbivore-plant interactions.

The main details of the model are presented, along with their biological justification. In studying the interaction between plants and small and large herbivores, we must often consider organisms that operate at very different spatial and temporal scales. Our approach is to use a hierarchy of landscapes of differing resolutions to allow consideration of the simultaneous impact of large and small herbivores. The model is behaviorally mechanistic, so that the dynamics of space use by model herbivores are responsive to variation in landscape characteristics.

We consider several types of questions that can be addressed with our model and emphasize the general nature of the approach. Some questions are framed in the context of the interaction between landscape pattern and ecological processes: (1) What is the effect of the pattern of patch types, at both large and small scales, on the use of patches by herbivores, and ultimately on the degree and distribution of plant survival and growth over the landscape; and (2) What is the contribution of herbivore foraging to the enhancement or suppression of the spatial patterns of plants? Because the model is individual based, we can also ask questions about the mechanistic effects of herbivore species diversity on plant survival and growth. The model described here can be used to explore the implications of the interactions between organism response and landscape heterogeneity for a variety of systems and problems.

Acknowledgments

We are grateful for useful comments and suggestions by C. D. Canham, M. A. Huston, R. V. O'Neill, J. Wiens, and an anonymous reviewer. Support for the work of J. B. M. was provided by the Institute of Ecosystem Studies, Millbrook, NY 12545, under contract No. 11010 with the Central Hudson Gas and Electric Co. This research was sponsored by the National Science Foundation's Ecosystem Studies Program under Interagency Agreement No. 40-689-78 with Martin Marietta Energy Systems, Inc., under contract DE-AC05-84OR21400 with the U.S. Department of Energy. Publication No. 3557, Environmental Sciences Division, Oak Ridge National Laboratory.

References

Adams, S.N. 1975. Sheep and cattle grazing in forests: A review. *Journal of Applied Ecology* 12:143–52
Addicott, J.F.; Aho, J.M.; Antolin, M.F.; Padilla, D.K.; Richardson, J.S.; and Soluk, D.A. 1987. Ecological neighborhoods: scaling environmental patterns. *Oikos* 49:340–46.
Anderson, R.C., and Loucks, O.L. 1979. White-tailed deer (*Odocoileus virginianus*) influence on structure and composition of *Tsuga canadensis* forests. *Journal of Applied Ecology* 16:855–61.
Anthony, R.G.; Niles, L.J.; and Spring, J.D. 1981. Small-mammal associations in forested and old field habitats--a quantitative comparison. *Ecology* 62:955–63.
Armesto, J.J.; Mitchell, J.D.; and Villagran, C. 1986. A comparison of spatial patterns of trees in some tropical and temperate forests. *Biotropica* 18(1):1–11.
Baird, D.D., and Birney, E.C. 1982a. Characteristics of dispersing meadow voles *Microtus pennsylvanicus*. *American Midland Naturalist* 107(2):262–83.
Baird, D.D., and Birney, E.C. 1982b. Pattern of colonization in *Microtus pennsylvanicus*. *Journal of Mammalogy* 63(2):290–93.
Becker, P.; Lee, L.W.; Rothman, E.D.; and Hamilton, W.D. 1985. Seed predation and the coexistence of tree species: Hubbell's models revisited. *Oikos* 44:382–90.

Birney, E.C.; Grant, W.E.; and Baird, D.D. 1976. Importance of vegetative cover to cycles of *Microtus* populations. *Ecology* 57:1043–1051.

Browder, J.A.; Bartley, H.A.; and Davis, K.S. 1985. A probabilistic model of the relationship between marshland-water interface and marsh disintegration. *Ecological Modeling* 29:245–60.

Brown, J.S.; Kotler, B.P.; Smith, R.J.; and Wirtz II, W.O. 1988. The effects of owl predation on the foraging behavior of heteromyid rodents. *Oecologia* 76:408–15.

Cockburn, A., and Lidicker, W.Z., Jr. 1983. Microhabitat heterogeneity and population ecology of an herbivorous rodent, *Microtus californicus*. *Oecologia* 59:167–77.

Collins, S.L., and Good, R.E. 1987. The seedling regeneration niche: habitat structure of tree seedlings in an oak-pine forest. *Oikos* 48:89–98.

Crawley, M.J. 1983. *Studies in Ecology,* Vol. 10, *Herbivory: The Dynamics of Animal-Plant Interactions.* Berkeley: University of California Press.

DeAngelis, D.L., and Waterhouse, J.C. 1987. Equilibrium and nonequilibrium concepts in ecological models. *Ecological Monographs* 57(1): 1–22.

Drolet, C.A. 1976. Distribution and movements of white-tailed deer in southern New Brunswick in relation to environmental factors. *Canadian Field Naturalist* 90:123–36.

Fahrig, L., and Merriam, G. 1985. Habitat patch connectivity and population survival. *Ecology* 66:1762–68.

Farmer, A.H.; Armbruster, M.J.; Terrell, J.W.; and Schroeder, R.L. 1982. Habitat models for land-use planning: assumptions and strategies for development. *Transactions of the North American Wildlife and Natural Resources Conference* 47:47–56.

Ford, R.G. 1983. Home range in a patchy environment: optimal foraging predictions. *American Zoologist* 23:315–26.

Ford, R.G., and Krumme, D.W. 1979. The analysis of space use patterns. *Journal of Theoretical Biology* 76:125–55.

Gardner, R.H.; Milne, B.T.; Turner, M.G.; and O'Neill, R.V. 1987. Neutral models for the analysis of broad-scale landscape pattern. *Landscape Ecology* 1(1):19–28.

Getty, T. 1981a. Territorial behavior of eastern chipmunks (*Timias striatus*): encounter avoidance and spatial time-sharing. *Ecology* 62:915–21.

Getty, T. 1981b. Structure and dynamics of chipmunk home range. *Journal of Mammalogy* 62(4):726–37.

Gill, F.B. 1988. Trapline foraging by hermit hummingbirds: competition for an undefended resource. *Ecology* 69:1933–42.

Glass, G.E., and Slade, N.A. 1980. The effect of *Sigmodon hispidus* on spatial and temporal activity of *Microtus ochrogaster*: evidence for competition. *Ecology* 61:358–70.

Green, R.F. 1984. Stopping rules for optimal foragers. *American Naturalist* 123:30–40.

Hansson, L. 1977. Spatial dynamics of field voles *Microtus agrestis* in heterogeneous landscapes. *Oikos* 29:539–44.

Harmon, M.E., and Franklin, J.F. 1989. Tree seedlings on logs in *Picea-Tsuga* forests of Oregon and Washington. *Ecology* 70:48–59.

Harris, L.D., and Kangas, P. 1988. Reconsideration of the habitat concept. *Transactions of the North American Wildlife and Natural Resources Conference* 53:137–44.

Hawkins, R.E., and Klimstra, W.D. 1970. A preliminary study of the social organization of the white-tailed deer. *Journal of Wildlife Management* 34:407–19.

Hawkins, R.E.; Klimstra, W.D.; and Autry, D.C. 1971. Dispersal of deer from Crab Orchard National Wildlife Refuge. *Journal of Wildlife Management* 35:216–20.

Hirth, D.H. 1977. Social behavior of white-tailed deer in relation to habitat. *Wildlife Monographs* 53. Washington, D.C.: The Wildlife Society.

Hubbell, S.P. 1980. Seed predation and the coexistence of tree species in tropical forests. *Oikos* 35:214–29.

Huston, M.; De Angelis, D.; and Post, W. 1988. New computer models unify ecological theory. *Bioscience* 38:682–91.

Ims, R.A. 1987. Responses in spatial organization and behavior to manipulations of the food resource in the vole *Clethrionomys rufocanus*. *Journal of Animal Ecology* 56:585–96.

Iverson, S.L., and Turner, B.N. 1972. Winter coexistence of *Clethrionomys gapperi* and *Microtus pennsylvanicus* in a grassland habitat. *American Midland Naturalist* 88(2):440–45.

Janzen, D.H. 1970. Herbivores and the number of tree species in tropical forests. *American Naturalist* 104:501–28.

Jumars, P.A.; Thistle, D.; and Jones, M.L. 1977. Detecting two-dimensional spatial structure in biological data. *Oecologia* 28:109–23.

Kammermeyer, K.E., and Marchinton, R.L. 1976. Notes on dispersal of male white-tailed deer. *Journal of Mammalogy* 57:776–78.

Karlsson, A.F., and As, S. 1987. The use of winter home ranges in a low density *Clethrionomys glareolus* population. *Oikos* 50:213–17.

Kufeld, R.C.; Bowden, D.C.; and Schrupp, D.L. 1988. Habitat selection and activity patterns of female mule deer in the Front Range, Colorado. *Journal of Range Management* 41(6):515–22.

Lancia, R.A.; Miller, S.D.; Adams, D.A.; and Hazel, D.W. 1982. Validating habitat quality assessment: an example. *Transactions of the North American Wildlife and Natural Resources Conference* 47:96–110.

Lefkovitch, L.P., and Fahrig, L. 1985. Spatial characteristics of habitat patches and population survival. *Ecological Modeling* 30:297–308.

Leps, J., and Kindlmann, P. 1987. Models of the development of spatial pattern of an even-age plant population over time. *Ecological Modeling* 39:45–57.

Levey, D.J.; Moermond, T.C.; and Denslow, J.S. 1984. Fruit choice in neotropical birds: The effect of distance between fruits on preference patterns. *Ecology* 65:844–50.

Linzey, A.V. 1984. Patterns of coexistence in *Synaptomys cooperi* and *Microtus pennsylvanicus*. *Ecology* 65:382–93.

Loehle, C. 1987. Applying artificial intelligence techniques to ecological modeling. *Ecological Modeling* 38:191–212.

Loft, E.R., and Menke, J.W. 1984. Deer use and habitat characteristics of transmission-line corridors in a douglas-fir forest. *Journal of Wildlife Management* 48:1311–16.

McAuliffe, J.R. 1986. Herbivore-limited establishment of a Sonoran desert tree, *Cercidium microphyllum*. *Ecology* 67:276–80.

McNair, J.N. 1982. Optimal giving-up times and the marginal value theorem. *American Naturalist* 119(4):511–29.

McNair, J.N. 1983. A class of patch-use strategies. *American Zoologist* 23:303–13.

McNamara, J.M., and Houston, A.I. 1985. Optimal foraging and learning. *Journal of Theoretical Biology* 117:231–49.

McNaughton, S.J., and Georgiadis, N.J. 1986. Ecology of African grazing and browsing mammals. *Annual Review of Ecology and Systematics* 17:39–65.

McNaughton, S.J.; Ruess, R.W.; and Seagle, S.W. 1988. Large mammals and process dynamics in African ecosystems. *Bioscience* 38:794–800.

McNay, R.S.; Page, R.E.; and Campbell, A. 1987. Application of expert-based decision models to promote integrated management of forests and deer. *Transactions of the North American Wildlife and Natural Resources Conference* 52:82–91.

Madison, D.M. 1980. Space use and social structure in meadow voles, *Microtus pennsylvanicus*. *Behavioral Ecology and Sociobiology* 7:65–71.

Madison, D.M., and McShea, W.J. 1987. Seasonal changes in reproductive tolerance, spacing, and social organization in meadow voles: a microtine model. *American Zoologist* 27:899–908.

Madison, D.M.; FitzGerald, R.W.; and McShea, W.J. 1984. Dynamics of social nesting in overwintering meadow voles (*Microtus pennsylvanicus*): possible consequences for population cycling. *Behavioral Ecology and Sociobiology* 15:9–17.

Marchinton, R.L., and Hirth, D.H. 1984. Behavior. In *White-tailed deer: Ecology and Management,* ed. L.K. Halls, pp. 129–68. Harrisburg, Pa.: Stackpole Books.

Mares, M.A., and Lacher, T.E., Jr. 1987. Social spacing in small mammals: patterns of individual variation. *American Zoologist* 27:293–306.

Matulich, S.C.; Handson, J.E.; Lines, I.; and Farmer, A. 1982. HEP as a planning tool: an application to waterfowl enhancement. *Transactions of the North American Wildlife and Natural Resources Conference* 47:111–27.

Merritt, J.R. 1984. *Winter Ecology of Small Mammals.* Special Publication of the Carnegie Museum of Natural History, No. 10. Pittsburgh, Pa.

Morris, D.W. 1987. Ecological scale and habitat use. *Ecology* 68:362–69.

Murai, M.; Thompson, W.A.; and Wellington, W.G. 1979. A simple computer model of animal spacing. *Researches on Population Ecology (Kyoto)* 20:165–78.

Murphy, D.D.; Menninger, M.S.; and Ehrlich, P.R. 1984. Nectar source distribution as a determinant of oviposition host species in *Euphydryas chalcedona. Oecologia* 62:269–71.

Naiman, R.J. 1988. Animal influences on ecosystem dynamics. *Bioscience* 38:750–52.

Newbery, D. McC.; Renshaw, E.; and Brunig, E.F. 1986. Spatial pattern of trees in kerangas forest, Sarawak. *Vegetatio* 65:77–89.

Newman, J.A.; Recer, G.M.; Zwicker, S.M.; and Caraco, T. 1988. Effects of predation hazard on foraging "constraints": Patch-use strategies in grey squirrels. *Oikos* 53:93–97.

Nudds, T.D. 1980. Forage "preference": theoretical considerations of diet selection by deer. *Journal of Wildlife Management* 44(3):735–40.

Ollason, J.G. 1980. Learning to forage-optimally? *Theoretical Population Biology* 18:44–56.

O'Neill, R.V.; Milne, B.T.; Turner, M.G.; and Gardner, R.H. 1988. Resource utilization scales and landscape pattern. *Landscape Ecology* 2:63–69.

Ozoga, J.J., and Gysel, L.W. 1972. Response of white-tailed deer to winter weather. *Journal of Wildlife Management* 36:892–96.

Parker, K.L.; Robbins, C.T.; and Hanley, T.A. 1984. Energy expenditures for locomotion by mule deer and elk. *Journal of Wildlife Management* 48:474–88.

Pemasada, M.A., and Gunatilleke, C.V.S. 1981. Pattern in a rain forest in Sri Lanka. *Journal of Ecology* 69:117–24.

Pielou, E.C. 1984. *The Interpretation of Ecological Data.* New York: John Wiley and Sons.

Reddingius, J., and Den Boer, P.J. 1970. Simulation experiments illustrating stabilization of animal numbers by spreading of risk. *Oecologia* 5(3):240–84.

Risser, P.G.; Karr, J.R.; and Forman, R.T.T. 1984. *Landscape Ecology: Directions and Approaches.* Illinois Natural History Survey Special Publication No. 2. Illinois Natural History Survey, Champaign.

Rongstad, O.J., and Tester, J.R. 1969. Movements and habitat use of white-tailed deer in Minnesota. *Journal of Wildlife Management* 33(2):366–79.

Ross, B.A.; Bray, J.R.; and Marshall, W.H. 1970. Effects of long-term deer exclusion on a *Pinus resinosa* forest in north-central Minnesota. *Ecology* 51:1088–93.

Schupp, E.W. 1988. Seed and early seedling predation in the forest understory and in treefall gaps. *Oikos* 51:71–78.

Scowcroft, P.G., and Hobdy, R. 1987. Recovery of goat-damaged vegetation in an insular tropical montane forest. *Biotropica* 19(3):208–15.

Senft, R.L.; Coughenour, M.B.; Bailey, D.W.; Rittenhouse, L.R.; Sala, O.E.; and Swift, D.M. 1987. Large herbivore foraging and ecological hierarchies. *Bioscience* 37(11):789–96.

Sherman, L.J. 1984. The effects of patch food availability on nest-site selection and movement patterns of reproductively active female meadow voles *Microtus pennsylvanicus. Holarctic Ecology* 7:294–99.

Smith, T.M. 1986. Habitat-simulation models: integrating habitat-classification and forest-simulation models. In *Wildlife 2000: Modeling Habitat Relationships of Terrestrial Vertebrates,* eds. J. Verner, M.L. Morrison, and C.J. Ralph, pp. 389–93. Madison: University of Wisconsin Press.

Sparrowe, R.D., and Springer, P.F. 1970. Seasonal activity patterns of white-tailed deer in eastern South Dakota. *Journal of Wildlife Management* 34(2):420–31.

Stenseth, N.C.; Bondrup-Nielsen, S.; and Ims, R.A. 1988. A population dynamics model for *Clethrionomys*: sexual maturation, spacing behavior and dispersal. *Oikos* 52:186–93.

Stephens, D.W., and Krebs, J.R. 1986. *Foraging Theory.* Princeton, N.J.: Princeton University Press.

Sterner, R.W.; Ribic, C.A.; and Schatz, G.E. 1986. Testing for life historical changes in spatial patterns of four tropical tree species. *Journal of Ecology* 74:621–33.

Stocker, M., and Gilbert, F.F. 1977. Vegetation and deer habitat relations in southern Ontario: Application of habitat classification to white-tailed deer. *Journal of Applied Ecology* 14:433–44.

Thompson, S.D. 1982. Spatial utilization and foraging behavior of the desert woodrat. *Neotoma lepida lepida. Journal of Mammalogy* 63(4):570–81.

Thomson, J.D.; Maddison, W.P.; and Plowright, R.C. 1982. Behavior of bumble bee pollinators of *Aralia hispida* Vent. (Araliaceae). *Oecologia* 54:326–36.

Tierson, W.C.; Mattfeld, G.F.; Sage, R.W., Jr.; and Behrend, D.F. 1985. Seasonal movements and home ranges of white-tailed deer in the Adirondacks. *Journal of Wildlife Management* 49:760–69.

Turner, B.N., and Iverson, S.L. 1973. The annual cycle of aggression in male *Microtus pennsylvanicus*, and its relation to population parameters. *Ecology* 54:967–81.

Turner, M.G. 1988. Multiple disturbances in a *Spartina alterniflora* salt marsh: are they additive? *Bulletin of the Torrey Botanical Club* 115:196–202.

Upton, G.J.G., and Fingleton, B. 1985. *Spatial Data Analysis by Example.* Vol. 1. New York: John Wiley and Sons.

Urich, D.L., and Graham, J.P. 1988. Applying national assessment data to wildlife management in Missouri. *Transactions of the North American Wildlife and Natural Resources Conference* 53:157–63.

Vance, R.R. 1984. The effect of dispersal on population stability in one-species, discrete-space population growth models. *American Naturalist* 123(2):230–54.

Verme, L.J. 1973. Movements of white-tailed deer in upper Michigan. *Journal of Wildlife Management* 37(4):545–52.

Webster, B.A., and Brooks, R.J. 1981. Social behavior of *Microtus pennsylvanicus* in relation to seasonal changes in demography. *Journal of Mammalogy* 62(4):738–51.

Wiens, J.A. 1976. Population responses to patchy environments. *Annual Review of Ecology and Systematics* 7:81–120.

Wiles, G.J., and Weeks, H.P., Jr. 1986. Movements and use patterns of white-tailed deer visiting natural licks. *Journal of Wildlife Management* 50(3):487–96.

Wywialowski, A.P. 1987. Habitat structure and predators: choices and consequences for rodent habitat specialists and generalists. *Oecologia* 72:39–45.

4. Synthesis

19. Translating Models Across Scales in the Landscape

Anthony W. King

19.1 Introduction

Ecology has accumulated a vast body of data, theory, models, methodologies, and understanding that is, albeit developed for and characteristic of relatively small scales, indispensable in the study of large-scale environmental questions involving landscapes, regions, and the earth as a whole. Application of this knowledge requires translation from the small or fine scales at which information exists to the larger or coarser scales most appropriate to the questions being asked. The disparities between these scales may be extreme, and, furthermore, larger spatial scales frequently involve increased spatial heterogeneity and longer time scales, which further complicate the "scaling problem" (Jeffers 1988). Methods for explicit translation across disparate spatial scales that consider changes in both spatial heterogeneity and temporal scale are an important part of the investigation of large-scale environmental change (Risser 1986, 1987; Rosswall et al. 1988).

This chapter addresses the problem of "scaling up" (Shugart et al. 1986), the translation or extrapolation of ecological information from small local scales to larger landscape and regional scales. The problem of scaling up is certainly not limited to landscape ecology or ecology in general; geographers and hydrologists have addressed the issue in some detail. However, given my background and the objectives of this book, my presentation will be restricted to considerations within landscape ecology. The inverse problem of "scaling down," translating information from larger to smaller scales is an important one (for example, in the prediction of landscape response to climate change predicted by climate models [Gates 1985]), but it will not be discussed here.

19.2 Perspective and Theory

19.2.1 The Landscape as an Aggregate Entity

Scaling up presupposes the existence of an aggregate landscape property that can be derived from the appropriate transformation of smaller-scale information from sites, stands, or patches. The landscape may be an element or component within a yet larger spatial extent (a region, Forman and Godron 1986) or function as a distinct unit within a larger context (as part of the global biosphere, for example [Woodmansee 1988]). The concept of the landscape as an integrated entity with aggregate properties and dynamics that are tied to the properties and dynamics of smaller-scale entities is consistent with a hierarchical view of landscapes, and it suggests ecological hierarchy theory (Webster 1979; Allen and Starr 1982; Patten 1982; Allen et al. 1984; O'Neill et al. 1986) as a framework for addressing scale translations (Urban et al. 1987; O'Neill 1988a,b; Shugart and Urban 1988).

19.2.2 Hierarchy Theory and Translating Information Between Scales

A landscape can be described as a nested hierarchy. Urban et al. (1987) defined one such nested forest hierarchy with gaps, stands, watersheds, and landscapes representing successively higher levels of organization. Levels of organization in a nested hierarchy are ordered as a monotonically increasing sequence of time and space scales (Fig. 19.1). Lower levels are characterized by smaller spatial scales (e.g., areal extent) and smaller temporal scales (e.g., high-frequency behaviors or rapid turnover times). Higher levels are characterized by larger spatial scales and larger temporal scales (e.g., low-frequency behaviors or slow turnover times) (Allen and Starr 1982; Allen et al. 1984; O'Neill et al. 1986; Urban et al. 1987).

Levels of organization in a nested hierarchy can be explicitly linked by changes in grain and extent of the observation set describing that system (Allen et al. 1984). Grain is the finest level of temporal or spatial resolution in an observation set, and it sets the lower limit on how fine a distinction can be made with that observation set. Extent is the areal expanse or the length of time over which observations with a particular grain are made, and it sets an upper, large-scale limit on the distinctions that can be made. Because changes in level of organization within a nested hierarchical system involve changes in both spatial and temporal scale (Fig. 19.1), manipulations of grain and extent are a means of translating information between the time and space scales of nested landscape hierarchies. Larger scales can be reached by increasing the extent of the observation set; smaller scales can be distinguished by making the observation set more fine grained. In particular, scaling up involves an increase in extent.

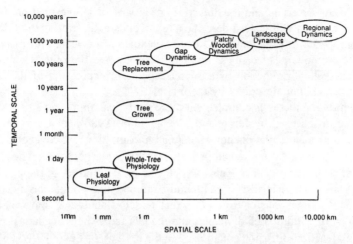

Figure 19.1. A nested hierarchical representation of carbon/biomass dynamics in a forested landscape. The ellipses represent levels of hierarchical organization. Adapted from King et al. (1990a) after Urban et al. (1987).

19.2.3 Scale, Heterogeneity, and Aggregation

Changes in scale usually involve changes in heterogeneity (Meentenmeyer and Box 1987; Levin 1989). An increase in the extent of an observation set of fixed grain reveals differences between components not seen with the smaller extent and increases the heterogeneity of the observation set. Observed with the same grain, a patch of vegetation is (virtually by definition) less heterogeneous than the landscape, and daily weather metrics (e.g., maximum temperature) are more variable when observed over a period of several months than when observed for two or three consecutive days. At the same time, a change to coarser-grained observations reduces observed heterogeneity. Differences between components are obscured by smoothing, averaging, integration, or aggregation. Because of the generally inverse relationship between grain and extent, a change from small to large scale typically involves both an increase in extent and aggregation to coarser grain (Allen et al. 1984; Meentemeyer and Box 1987). The aggregation requires the integration of the finer-grained heterogeneity of the larger extent. It is this aggregate property of the larger landscape extent that scaling up attempts to quantify. In brief, the problem of scaling up information in landscapes can be stated as a problem in correctly aggregating or integrating landscape heterogeneity (O'Neill 1988a,b; Rastetter et al. [manuscript]).

The challenges of scaling up landscape information lie in (1) correctly defining the spatial (and temporal) heterogeneity of the fine-scale information, and (2) correctly integrating or aggregating this heterogeneity. An exhaustive and definitive resolution of these challenges is not possible here, in part because the definition of *correctly* is surely dependent upon the specific nature of the in-

formation being scaled up and the objectives of a particular investigation. In the
sections that follow I will instead describe a relatively small set of methodologies
that are appropriate to situations in which the available fine-scale information is
a mathematical model, especially a simulation model, and the objectives of the
investigation include prediction of the landscape-scale expression of the local,
fine-scale phenomenon described by the model.

These simulation methods complement more analytical approaches to scaling-
up. In the approach used by Jarvis and McNaughton (1986), for example, processes describing a particular phenomenon (e.g., transpiration) are conceptualized,
and functional representations developed, at successively larger scales. The resulting series of scale-dependent models translates an understanding of the phenomenon across scale. There is also a rich literature on modeling population dynamics
in patchy or heterogeneous environments that includes analytical approaches to
the translation of within-patch dynamics and models across heterogeneous spatial
extents (e.g., reaction-diffusion equations; see the reviews of Levin [1976a,b] and
DeAngelis et al. [1986]). Similarly, the considerable literature on aggregation in
ecological systems (e.g., Cale and Odell 1979, 1980; O'Neill and Rust 1979;
Gardner et al. 1982; Cale et al. 1983; Luckyanov et al. 1983; Luckyanov 1984;
Hirata and Ulanowicz 1986; Iwasa et al. 1987, 1989) is a valuable resource in the
investigation of methods for translating models across scales in the landscape,
especially the explicit considerations of spatial aggregation (e.g., O'Neill et al.
1979; Gardner et al. 1981; Iwasa et al. 1987).

The simulation approaches I present here share many similarities with these
more analytical approaches. The strengths of the simulation methods lie in their
wide and general applicability (especially for the often complicated, cumbersome,
and analytically intractable ecosystem simulators) and in the ability to select a
method based on general considerations of simple model and landscape characteristics.

19.3 Scaling Up Ecological Models: General Methods

19.3.1 Model Grain and Extent

The scale of an ecological model can be defined in terms of grain and extent in
a manner analogous to that for observation sets. Indeed, the model and simulations
with that model can be regarded as an observation set.

Model grain is the spatial/temporal resolution of the model. At these scales the
model "sees" the system as spatially and temporally homogeneous. Any finer-
scale heterogeneity is already integrated (usually implicitly by assumptions of
homogeneity or insignificant heterogeneity) as part of system specification and
model definition (e.g., the mathematical formulation of the model, the assignment
of values to variables and parameters, the definition of observables (Rosen 1977)
or state variables, the determination of what is inside and outside the system, and
the definition of interactions and dynamics). Once the grain of the model is fixed,

faster temporal dynamics or finer spatial heterogeneity cannot be described with that model.

The temporal grain of an ecological model is usually explicit in the temporal resolution of the model. Annual models have a temporal grain of 1 year, and they cannot be used to simulate day-to-day (e.g., seasonal) dynamics. Similarly, a daily model has a fixed temporal grain and cannot be used to describe diurnal patterns. Importantly, a model's temporal grain is defined by the temporal resolution of the data used to determine the model's rate constants and not by the time step used in numerical solutions of the model's equations.

The spatial grain of ecological models is usually not so explicit. Rather, the spatial grain is usually implicit in the model type (e.g., a forest stand model or a lake ecosystem model) and in the frequently unstated assumption of an area over which the system is assumed to be spatially homogeneous. The actual spatial dimensions of these implicit areas are often not specified. There are, of course, exceptions to this generality. Shugart et al. (1974, p. 232), for example, state that the spatial scale of the TEEM forest ecosystem model is "a forest stand which is assumed to have minimal heterogeneity," and Shugart and West (1977) and Shugart (1984) discuss in some detail the importance of grain size (ca. 0.08 ha) in the gap models of forest stand dynamics. The degree to which explicit considerations of scale influence the process of system specification and model definition is often not clear in the presentation of models and probably varies considerably.

Model extent is the area or time period over which a model of fixed grain is used to simulate the phenomena under investigation. When model extent exceeds model grain (i.e., a model is used to simulate more than one site or time period), simulations normally require changes in the values assigned to model parameters and variables. These quantitative changes are expressions of spatial and temporal heterogeneity at the scale of the model's grain.

From a temporal perspective, consider that daily resolution models are commonly used to simulate dynamics over a sequence of many days to several years. For example, Shugart et al. (1974) state that their daily resolution forest-ecosystem model is designed to simulate forest response for up to two or three years. The temporal grain of the model is one day, and the temporal extent is two or three years. Simulations require time series of environmental input (e.g., temperature and light) with daily resolution for that extent.

Historically, the spatial extents of simulations with ecological models are equivalent to the models' spatial grains. Models are frequently applied in a site-specific manner to a single homogeneous spatial unit within a given study area. The ecosystem simulations of the International Biological Program and the bulk of classical population and community models are examples. It has, however, become increasingly commonplace to simulate larger contiguous spatial extents with multiple simulations from a smaller-scale, finer-grain model. Examples can be found in Duncan et al. (1967), Vollenvieder (1970), Levin (1974, 1976a,b), Wiens (1976), Curry et al. (1977), O'Neill et al. (1979), Bolla and Kutas (1984),

Shugart (1984), DeAngelis et al. (1986), King et al. (1987, 1990b), Coffin and Lauenroth (1988), Smith and Urban (1988), Running et al. (1989), Burke et al. (1990), and the references they cite.

Maximum extent is the extent over which a model can be used to simulate the system without changes in model structure. Over some time period or spatial extent, new phenomena or dynamics will appear that the model is not designed to consider. When acceptable simulation of system phenomena requires qualitative changes in model structure (e.g., the addition of time-dependent parameters, new state variables, or new functional representations), the maximum extent of the model has been exceeded. For example, a model of leaf photosynthesis that does not consider leaf-age effects can only be applied over the temporal extent in which aging effects are insignificant. Application beyond that extent would require modifications, perhaps in the form of time-dependent parameters or the addition of mechanistic functions for leaf aging.

As a spatial example, consider a forest stand model of some fixed grain that is used to simulate a collection of similar forest stands within some expanded extent. With the appropriate quantitative changes representing spatial variations, the model adequately simulates a relatively heterogeneous collection of forest stands. But, if the spatial area under consideration is expanded to encompass grassland vegetation, it is highly unlikely that the forest model can reasonably simulate the phenomena of interest for grassland sites. In this simple example, the maximum extent of the forest model is the area of the landscape occupied by forest.

19.3.2 Scaling Up by Manipulation of Model Grain and Extent

Continuing the analogy with observation sets, quantitative manipulations of model grain and extent can be used to translate models across scales in the landscape. The small-scale model (e.g., a model of a patch or stand) provides an explicit definition of the local fine-scale information at a fixed grain, and the size of the landscape in question specifies the required increase in extent. Scaling up requires a quantitative description of the spatial heterogeneity of that extent. The grain of that description is determined by and equal to the grain of the small-scale model. The description may be an explicit function of space or a statistical description (e.g., a probability distribution), either univariate or multivariate, discrete or continuous. If the description of the landscape is static, and the translation is limited to spatial scales, the spatial function is time invariant and the probability distribution is stationary. However, if the translation involves both space and time, the distribution may be nonstationary and the spatial function may involve time-dependent parameters. Integration of the fine-scale heterogeneity over the extent of the landscape yields the aggregate prediction at the landscape scale.

Here I consider four general methods that implement this approach to scaling up. For convenience, I have identified these as (1) lumping, (2) direct extrapolation, (3) extrapolation by expected value, and (4) explicit integration. Each method satisfies the required integration of landscape heterogeneity. They are distinguished by the method of integration and the model or data transformations involved in the manipulations of grain and extent. I will present the application of these methods to extrapolation across spatial scales, but with modification the methods can also be extended to temporal scales.

It should be noted that these methods are limited to the spatial extent within which all relevant phenomena are simulated by the smaller-scale model; i.e., they are limited to the maximum extent of the model. Similarly, the methods are restricted to translations within a system description and within a particular level of organization. For example, the methods are appropriate for translating stand-scale atmospheric gas exchange to landscape-scale gas exchange, but they are inappropriate (at least incomplete) for translating between whole-tree physiology and landscape-scale biomass dynamics. This latter type of scale translation is treated elsewhere (e.g., King et al. 1990a, Luxmoore et al. 1990; Huston and Smith 1987; Smith and Huston 1989). Furthermore, the methods presented here are most applicable to circumstances in which there are no strong flows or influences among the local sites described by the smaller-scale model (a later section will consider relaxation of this constraint).

19.3.2.1 Extrapolation by Lumping

In "lumping," one of the simplest methods for scaling up ecological models, the increased heterogeneity that accompanies scaling-up's increase in extent is integrated by averaging across heterogeneity in the landscape data and calculating mean values for the model arguments (i.e., the variables and parameters used to calibrate, initialize, and drive the model). Averaging increases both the grain and extent of these quantities, and the larger-scale mean values replace their smaller-scale counterparts in the model. Importantly, values within the model are altered, but model structure is unchanged. Simulation with the mean, or lumped, values is used to represent the larger-scale, landscape expression of the smaller-scale phenomena described by the original model.

There are immediate problems with this simple approach to scaling up. Lumping, as defined and used here, assumes that system properties reflected in model structure (e.g., the mathematical formulation of mechanistic processes) do not change with scale. This is equivalent to the assumption that the larger-scale system behaves like the average smaller-scale system. From a modeling perspective, this assumption only holds strictly if the equations describing the system are linear. If the equations describing the system are nonlinear, the assumption can introduce a bias error, the nature and magnitude of which is determined by the specifics of the nonlinear equations and the heterogeneity of the system (see O'Neill 1979a). The potential for error in lumping or averaging has been discussed from a variety of perspectives (e.g., O'Neill 1979a,b; Leduc and Holt 1987; Welsh et al. 1988; Rastetter et al. [manuscript]).

The basic requirements for scaling up by lumping are easily met. One only needs estimates of model arguments averaged across the extent of the landscape. But the simplicity of lumping is beguiling. Many ecological models are nonlinear, and the accompanying aggregation or scaling error can be significant. Lumping should only be used with careful and explicit consideration of this potential error. However, if the accompanying error is known and acceptable, lumping may in some circumstances provide a useful first approximation. Considerations that minimize aggregation error (e.g., Gardner et al. 1982; Iwasa et al. 1989; Rastetter et al. [manuscript]) may be used to guide the application of this method.

19.3.2.2 Extrapolation by Increasing Model Extent

Model predictions can be scaled up or extrapolated to larger scales by increasing the extent of model simulations. The smaller-scale model for a single site or patch in the landscape is used to simulate the same processes for a collection of such sites across the landscape. A standard framework for dealing with spatially distributed systems (see Section 19.3.1), each individual simulation requires appropriate quantitative changes in the model arguments (e.g., environmental input variables) that reflect the spatial heterogeneity of the landscape. The multiple simulations are then combined to represent the aggregate expression at the landscape scale. In common with lumping, quantitative input to the model is changed, but the model structure is unaltered (e.g., the number and definition of state variables or the model's mathematical functions are not changed). In contrast with lumping, the grain of the data used to quantify model arguments is not changed, only the extent. An expanded collection of data is used, but there is no averaging of that data. The limits of the extrapolation are set by the points at which an increase in spatial extent encounters new conditions or phenomena that are not simulated by the model or that require modifications in the model beyond quantitative changes in model arguments.

The key issues in extrapolation by increasing model extent are (1) the appropriate description of the larger-scale heterogeneity of the arguments of the smaller-scale model (e.g., which arguments exhibit significant variation and how are they distributed?) and (2) the appropriate combination of output from the smaller-scale model to describe the aggregate expression at the large scale (e.g., is the aggregate expression a simple summation, or is there some more complex relationship among the smaller-scale outputs?). Here I consider two general methods of extrapolation that involve increased model extent.

19.3.2.2.1 Direct Extrapolation. The heterogeneous landscape mosaic of discrete and homogeneous grid cells, patches, or other landscape elements is a common construct in landscape ecology, and it provides an obvious means of scaling up from the local model to the landscape. Given a local small-scale model with reasonable computational demands, a tractable number of discrete elements, and estimates of the spatially distributed variable for all elements, it is a relatively simple matter to estimate the landscape property by "direct extrapolation." The local small-scale model is applied to each discrete element for which the model is appropriate, and output from each simulation is scaled (multiplied) by the area of the discrete element. The individual simulations for all elements are then combined to represent the landscape. For independent elements, the combination is often a simple summation.

Spatially-distributed, georeferenced modeling that links ecosystems models with geographical information systems (e.g., Grossmann and Schaller 1986; Band

and Wood 1988; Coughlan and Running 1988; Running et al. 1989; Burke et al. 1990) and other grid-based modeling (e.g., Ziegler 1979; Sawyer and Haynes 1985; Smith and Urban 1988; Wilkie and Finn 1988; and Bartell et al. 1989, among many others) are examples of this approach. Other modeling that does not explicitly include the spatial coordinates or arrangement of individual elements but includes the area of each patch type and involves simulations for each type are even more common and are also examples of direct extrapolation (see the reviews of Levin 1976a,b and DeAngelis et al. 1986 and the specific examples of Dale and Gardner 1987 and Pastor and Post 1986, 1988). Some variation of what I have identified here as direct extrapolation is probably the most commonly employed method for applying a local small-scale model to a larger, heterogeneous spatial extent.

Direct extrapolation may be of limited use, however, when the local model is a large system of differential equations with time-consuming numerical solutions and the landscape involves a large number of cells (Band and Wood 1988). There are also questions of error introduced by representing a continuous landscape as a discrete system and error in defining the homogeneity of the discrete elements. There must be some trade-off in error and computational demands between assumptions of within-cell homogeneity and the number of individual cells simulated. Thus, while it may often be possible to extrapolate a local model by imposing a finite grid on the landscape and applying direct extrapolation, it may be more computationally efficient, and perhaps more accurate, to seek and apply alternative approaches.

19.3.2.2.2 Extrapolation by Expected Value. Local fine-scale models can be extrapolated across heterogenous areas by calculating the expected value of the model output. This approach assumes (perhaps as a working hypothesis) that the larger-scale expression of the local, finer-scale behavior is the product of the landscape area and the expected value of the model output simulating the local processes, or

$$Y = AE[f(\mathbf{x}, \mathbf{p}, \mathbf{z})] = AE[y] \qquad (19.1)$$

where Y is the larger-scale, landscape expression of local behavior, y; A is the area of the landscape; $E[]$ is the expected value operator; f is the local model; and $\mathbf{x}$, $\mathbf{p}$, and $\mathbf{z}$ are vectors of model arguments (state variables, parameters, and driving variables, respectively).

Arguments in the local model that vary spatially across the landscape are treated as random variables, and their joint probability distribution defines the spatial heterogeneity of the landscape. Output from the local model, as a function of the spatially distributed random variables, is also a random variable, and the expression of local behavior at the landscape scale is determined by the mathematical expectation of this random variable. For discrete variables,

$$Y = A \sum_{j=1}^{q} \sum_{k=1}^{r} \sum_{l=1}^{s} f(x_j, p_k, z_l) g(x_j, p_k, z_l) \qquad (19.2a)$$

and for continuous variables

$$Y = A \int \int \int f(x,p,z) g(x,p,z) \, dx \, dp \, dz \qquad (19.2b)$$

where $g(\ )$ is the joint probability function for the spatially heterogeneous arguments of the local model $f(\ )$. For simplicity, I have assumed one-dimensional vectors for **x**, **p**, and **z**, but the approach can easily be generalized.

The relationship between the landscape and the local model expressed by Eq. 19.1 is quite general but limited to those situations where y (the local small-scale behavior or phenomenon) at one site i is not a function of y at another site j (i.e., $y_i \neq f(y_j)$). Spatial correlations or complex spatial patterns in the arguments (x, p, and z) of the local model are perfectly compatible with the theory behind Eq. 19.1. These patterns may, however, produce very complex joint probability distribution functions.

Many ecological processes are independent of the concurrent processes at other sites (e.g., gas exchange between the vegetation and the atmosphere), and the extrapolation expressed by Eq. 19.1 is readily applicable. If the problem is appropriately bounded (especially the temporal scale), the abstraction can also be applied to ecological processes where between-site dependency is common (e.g., population growth). For example, the abstraction could be applied to a landscape occupied by sessile organisms during the nondispersal stage of their life cycle or to motile organisms over a sufficiently short time interval so that dispersal is small. Consequently, "extrapolation by expected value" can be applied to a large number of ecological problems and the models describing them. Models describing site-level processes with strong between-site interactions (or with feedbacks within a site [e.g., $x_i = f(y_i)$]) require a modified approach that I will not address here.

The main components of a general algorithm for extrapolation by expected value are (1) the model simulating system behavior at a local, relatively fine spatial scale; (2) the larger areal extent, the landscape, over which the local model is to be extrapolated; (3) the frequency distributions of the variables describing landscape heterogeneity; and (4) the calculation of the expected value. The local model, the landscape, and the description of landscape heterogeneity as spatially distributed random variables are, of course, specific to each application. The choice of methods for calculating the expected values is largely dependent upon the available description of the probability distributions describing landscape heterogeneity.

If the probability density functions, the $g(\ )$s, of Eqs. 19.2a and 19.2b are known (or can be estimated) explicitly, the expected value can be evaluated directly. If, for example, the spatial variables are continuous, and if the local model, $f(\ )$, and the probability function, $g(\ )$, are relatively simple, it may be possible to evaluate the integral of Eq. 19.2b analytically in closed form. Alternatively, if a closed-form solution does not exist or cannot be found, numerical quadrature techniques may be used to approximate the expected value of a model with continuous variables. Similarly, if the spatially distributed random variables are discrete, all discrete values are known, and their number is not untractably large, the summation of Eq. 19.2a can be evaluated explicitly. Finally, and if in particular the mathematical form of the probability density function is unknown, Monte Carlo simulations may be used to calculate the expected value of the local model (c.g., Shugart 1984; King et al. 1987; Prentice et al. 1989). The Monte Carlo, or sampling, approach is powerful in that it can be applied to complex simulation models and requires information only on the types of probability distributions (e.g., normal, uniform, etc.), and estimates of central tendency (e.g., means or modes), variance, (e.g., standard deviations or minima and maxima), and covariance or correlation (in multivariate models). The expected value of the Monte Carlo output distribution provides the estimate of $E[f(\mathbf{x}, \mathbf{p}, \mathbf{z})]$, and the standard deviation of that distribution can be used to calculate an unbiased confidence interval for the extrapolation's landscape prediction. This approach is often used to extend gap-scale models of forest dynamics to the landscape (Shugart 1984), and King et al. (1987, 1990b) applied this approach to estimate regional CO_2 exchange with the atmosphere by extrapolation from site-specific ecosystem models.

The principle source of error in extrapolation by expected value lies in the estimation of the probability distributions (types and moment). A joint probability distribution for several variables is more difficult to estimate than a distribution for a single independent variable and can be more difficult to implement in the calculation of the expected value (whether by Monte Carlo simulation or more direct evaluation). Furthermore, there is some potential for error in actually calculating the expected value. There is, for example, potential error in (1) using too few Monte Carlo iterations, (2) estimating a truly continuous variable with discrete values (Eq. 19.2a), and (3) numerically evaluating the expected value of a continuous value model by quadrature. However, if these sources of error are controlled (experience shows that they tend not to be very important in Monte Carlo applications [O'Neill et al. 1982; Robert H. Gardner, personal communication]), and the fundamental requirement of independent sites is satisfied, calculation of the local model's expected value is an accurate and effective method for extrapolation from the small scale to the scale of the landscape.

19.3.2.3 Extrapolation by Explicit Integration

Extrapolation from smaller to larger spatial scales may be achieved by explicitly evaluating the integral of the smaller-scale model in exact or closed form, with space as the variable of integration. Consider a two-dimensional landscape defined as a region R in the xy-plane. If the small-scale model can be defined as a function of space, i.e., a function of x and y, that is integrable over R, then the landscape-scale expression of the smaller-scale model can be estimated by the evaluation of

$$\int\int_R f(x, y)\, dA \tag{19.3}$$

where A is the area of the landscape R. The double integral of Eq. 19.3 can be evaluated with the use of iterated integrals and the Fundamental Theorem of Calculus to determine the definite integral of the original model on the interval representing the extent of the landscape.

In general terms, consider the small-scale model

$$l = m(z) \tag{19.4}$$

where l is some local quantity and z is a spatially heterogeneous variable. If z can be described as an explicit function of space $g(x, y)$ where (x, y) specifies a unique point in the landscape, then

$$l = m(g(x,y)) = f(x,y) \tag{19.5}$$

Assume a rectangular landscape bounded in the xy-plane by a and b in the x-dimension and c and d in the y-dimension. L, the aggregate landscape-scale expression of the local small-scale model, can then be predicted by

$$L = \int_c^d \int_a^b f(x, y)dx\, dy = \int_c^d \left[\int_a^b f(x, y)dx \right] dy \tag{19.6}$$

The partial integration with respect to x, $\int_a^b f(x,y)\,dx$, may be achieved by evaluation of the indefinite integral (provided that integral exists) and application of the Fundamental Theorem of Calculus to obtain a function in y, a, and b. The same procedure may then be used to integrate this function with respect to y and obtain a new model for the landscape in a, b, c, and d.

In contrast with the previous methods, "extrapolation by explicit integration" involves a change in model structure. In direct extrapolation and extrapolation by expected value, the increase in the extent of model simulation requires an increase in the extent of the data used to quantify the simulations. In lumping both the extent and grain of this data are changed. However, in each case the structure of the small-scale model is unaltered. In contrast, the indefinite integration of this method represents a transformation or rescaling of the original small-scale model. The new model now describes the larger-scale landscape behavior as a function of the spatial limits of the landscape rather than as a function of the local spatially distributed variables. The grain of the new model now coincides with the extent of the landscape, and the landscape is simulated with one evaluation of the model.

Approximating the double definite integral by numerical methods in the absence of closed-form indefinite integrals is virtually equivalent to extrapolation by increasing the extent of model simulations. There is no transformation of the original smaller-scale model structure, and the small-scale model is evaluated repeatedly with values of the independent variables that describe the spatial heterogeneity of the landscape. Traditional quadrature (e.g., Simpson's Rule or Newton-Cotes formulas) converges on direct extrapolation, and Monte Carlo quadrature converges on extrapolation by expected value using Monte Carlo simulation. These relationships emphasize the degree to which the distinctions I have made between various approaches, while useful, are somewhat arbitrary and a bit "fuzzy." Other useful categorizations could be made.

Extrapolation by explicit integration assumes that landscape heterogeneity can be described by explicit functions of space. Obviously, there is potential for error in approximating landscape heterogeneity in this way, and, consequently, there is potential for error in extrapolating from a local model to the landscape by this method. Furthermore, the method assumes that the indefinite integral of the model with respect to space exists and can be found. Complex functions describing spatial variability or a complex local model may make it difficult or impossible to find an explicit solution by indefinite integration. However, if these difficulties can be overcome, explicit integration can provide an accurate and efficient estimate for the landscape. There is no numerical approximation error in evaluating the definite integral once the indefinite integral is determined, and only one simulation with the rescaled model (the indefinite integral) is needed to estimate the landscape behavior (in contrast with the several hundred or thousand simulations involved in direct extrapolation or extrapolation by expected value).

19.4 Scaling Up Ecological Models: Simple Examples

In this section I consider four simple and familiar ecological models: (1) a model of exponential population growth, (2) a model of logistic population growth, (3) a model of competition between two species, and (4) a model of population growth with immigration and emigration. These models are assumed to represent local, small-scale, point-space phenomena in heterogeneous ecological landscapes. I will apply the general methods for scaling up from the previous section to estimate the aggregate, landscape-scale expression of the local phenomena. The scaling up will be limited to translation across spatial scales.

The models represent a sequence of model structures and assumptions that increasingly test the assumptions and requirements of the four general methods I have presented. As such, these examples are simply meant to illustrate the application of the general methods, clarify their finer points, and help define specific requirements and limitations. They are not intended and should not be viewed as preferred methods for scaling up the particular models or classes of models involved. For example, Vlad (1988) treats the analytical aggregation of populations satisfying nonlinear growth equations in a patchy environment with a much more refined approach than I present in Section 19.4.2. The reader may draw conclusions about the general application of the methods to other models but should not draw general conclusions about the models involved. To that end the reader is referred to the reviews of Levin (1976a,b) and the work of Iwasa et al. (1987, 1989).

19.4.1 Model 1: Exponential Growth

19.4.1.1 The Landscape and the Local Model

Consider a square landscape of 10^6 ha, 100 km on a side. At any point in this landscape the population growth of some species of interest (perhaps a colonizing species at the beginning of a successional sequence or an annual species at the beginning of the growing season) is described by the differential equation

$$\frac{dn}{dt} = rn \tag{19.7a}$$

and the population at time t during this period of exponential growth is given by the general solution to Eq. 19.7a:

$$n = n_0 e^{rt} \tag{19.7b}$$

where n is population size (e.g., number of individuals), n_0 is the initial population at $t = t_0$, and r is the intrinsic per capita growth rate.

The landscape is heterogeneous in r; at any point (x,y) on the landscape, r is given by

$$r(x, y) = \frac{-p_1 x + p_2 y}{p_3} + p_4 \tag{19.8}$$

where x and y are normalized Cartesian coordinates (0 to 100), and the p_i ($i = 1,2,3,4$) are parameters for the landscape (Fig. 19.2). For simplicity, I assume that the landscape is homogeneous with respect to n_0; n_0 is a constant ($n_0 = 0.0001$) across the landscape.

Within the time period under consideration, individual points or sites in the landscape are independent. A site experiences neither immigration nor emigration, and local population dynamics at a site are completely described by Eqs. 19.7a and b. N, the population of the entire landscape at time t, is the landscape-scale property of interest.

19.4.1.2 Scaling Up by Direct Extrapolation

A uniform grid of 50 × 50 square elements or cells is superimposed on the landscape. If it is assumed that variation in r across each grid cell is negligible and r at the centroid of each cell is a reasonable approximation of r for the entire cell, Eq. 19.7a can be used to model the change in the population of each cell. The change in population of any grid cell i is

$$\frac{dn_i}{dt} = r_i n_i \tag{19.9a}$$

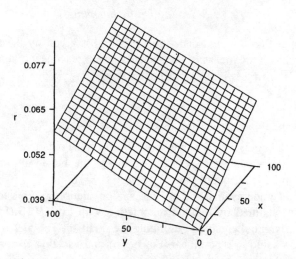

Figure 19.2. The spatial pattern of r, the intrinsic growth rate, in the local model of exponential population growth. The corresponding parameters in Eq. 19.8 (see text) describing r as an explicit function of the normalized spatial coordinates x and y are $p_1 = -1.9048$, $p_2 = -1.8962$, $p_3 = -10000$, and $p_4 = 0.0392$.

and

$$n_i = n_0 a_i e^{r_i t} \tag{19.9b}$$

where n_i is the population per grid cell, r_i is given by Eq. 19.8 evaluated at the centroid of cell i, n_0 is the initial population at the centroid of each cell, and a_i is the area of cell i.

The extremely fine-scaled, local model of Eqs. 19.7a and b, with ill or undefined spatial grain and extent, has been rescaled in the simplest of ways, by assumptions of homogeneity, to model the larger-scale grid cell (Eqs. 19.9a and b). The grain of this model is better defined (i.e., a_i, the spatial extent of the grid cell) and is coincident with the model's extent. Rather than exploring the implications and potential errors of moving from the extremely fine-scaled point model to the grid cell model, I will, instead, begin with the grid-cell model and explore the same issues by looking toward the larger scale of landscape.

Applying Eq. 19.9b to each cell and summing all grid cells on the landscape provides an estimate of the landscape's population:

$$N = \sum_{i=1}^{m} n_0 a_i e^{r_i t} \tag{19.10}$$

where m is the number of grid cells. Because of the clarity in this direct extrapolation, I will use estimates of N from Eq. 19.10 as a reference simulation for evaluating estimates from the other methods for scaling up.

19.4.1.3 Scaling Up by Lumping

Lumping assumes that the landscape-scale property is given by the small-scale model evaluated with the means of the spatially heterogeneous variables (Section 19.3.2.1). In this example, lumping estimates that

$$N = A\bar{n}_0 e^{\bar{r}t} \tag{19.11}$$

where A is the total landscape area and $\bar{r}$ (= 0.0581) is the mean value of r for the landscape (i.e., the mean of Eq. 19.8). Note that the mean for the spatially homogeneous initial local population, $\bar{n}_0$, is $n_0 = 0.0001$.

Equation 19.11 underestimates the landscape population relative to the reference simulation obtained by direct extrapolation with Eq. 19.10 (Fig. 19.3A). Although the differential equation describing the small-scale model is linear (Eq. 19.7a), the model solution is nonlinear (Eq. 19.7b), and a scaling error is expected. For this model and landscape, the lumping error is significant, exceeding 20% near the end of the simulation period (Fig. 19.3B).

This simple example nicely illustrates the error that can occur if lumping is used to translate a local small-scale model to the landscape scale. The example also illustrates another important point: since the solutions of linear differential equations are often nonlinear, one cannot safely assume that because a local ecological model involves linear differential equations, the landscape-scale expression of that model can be accurately estimated by using lumping to scale up.

19.4.1.4 Extrapolation by Expected Value

If the landscape's heterogeneity is described not by the spatially explicit function of Eq. 19.8 or the collection of resulting grid-cell values but instead by the probability distribution of r (Fig. 19.4), the landscape population can be estimated by

$$N = \sum_{j=1}^{q} n_0 e^{r_j t} a_j \tag{19.12}$$

where a_j is the area of the landscape over which $r = r_j$ and q is the number of discrete values of r. The area a_j can be written as a fraction of the total area A, or

$$a_j = g(r_j) A \tag{19.13}$$

where $g(x_j)$ is the probability that at any point in the landscape $r = r_j$. Substituting the right side of Eq. 19.13 in Eq. 19.12 and arranging terms yield

$$N = A \sum_{j=1}^{q} n_0 e^{r_j t} g(r_j) \tag{19.14}$$

The summation in Eq. 19.14 defines the expected value of the local model when r is a discrete vandom variable with the probability function $g(r)$, or

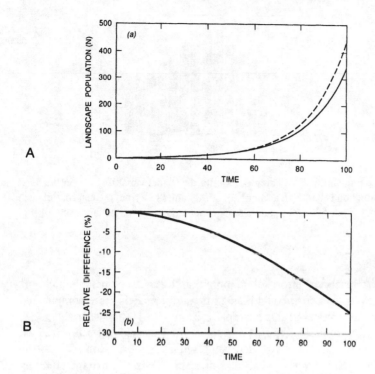

Figure 19.3 (A) Comparison of the landscape population, N, estimated by extrapolation of the local model of exponential population growth using direct extrapolation (broken line) and lumping (solid line). (B) Relative difference between the direct extrapolation and the estimate by lumping. Relative difference, expressed as a percentage, is the difference between the lumped and direct extrapolation estimates as a fraction of the estimate by direct extrapolation. A negative relative difference means that lumping underestimates the result of the direct extrapolation.

$$N = AE[n_0 e^{rt}] \qquad\qquad (19.15)$$

Figure 19.5A compares the landscape population estimated by Monte Carlo evaluation of the expected value of the local model (Eq. 19.15) with the reference simulation obtained by direct extrapolation (Eq. 19.10). The simulations are nearly indistinguishable. Figure 19.5C shows the relative error in the extrapolation by expected value. The difference is quite small (< 1.2%), much smaller than the relative difference in extrapolation by lumping (compare the ordinates of Figs. 19.3B and 19.5C).

The expected value of the local model was calculated with the use of 100 Monte Carlo simulations with Latin Hypercube sampling of r from the frequency distribution in Fig. 19.4 (i.e., $q = 100$ in Eq. 19.14). The direct extrapolation required 2500 model simulations. Thus, a 96% reduction in the number of model runs

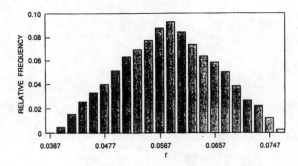

Figure 19.4. The probability distribution describing the spatial variability of r in the local model. The distribution is based on a sample of 2500 values of r from the centroid of each cell of a uniform grid of 50×50 square elements superimposed on the landscape of Fig. 19.2.

introduced a difference, an "error," of no more than 1.2%. A comparable trade-off between model runs and error could be significant for models more complicated than the exponential model of this example.

While direct extrapolation required estimates of r for each grid cell (provided by Eq. 19.8), extrapolation by expected value required information only on the probability distribution for r across the landscape. Thus the extrapolation by expected value could have been applied without knowledge of r as an explicit function of space. Furthermore, the assumptions of smaller-scale grid cell homogeneity required by the direct extrapolation (i.e., the grid-cell mode, Section 19.4.1.2) are absent from the extrapolation by expected value.

19.4.1.5 Extrapolation by Explicit Integration

Extrapolation by explicit integration predicts that

$$N = \int_{y_1}^{y_2} \int_{x_1}^{x_2} n_0 e^{r(x,y)t} \, dx \, dy \tag{19.16}$$

where $r(s,y)$ is the per capita growth rate as a function of the spatial coordinates x and y (i.e., Eq. 19.8), and x_1, x_2, y_1, and y_2 define the spatial limits of the landscape. Evaluating the double integral in closed form yields

$$N = \frac{c^2 n_0}{abt^2} \left(e^{r(x_1,y_2)t} - e^{r(x_2,y_2)t} - e^{r(x_1,y_1)t} - e^{r(x_2,y_1)t} \right) \tag{19.17}$$

The difference between the landscape population estimated by Eq. 19.17 and the reference simulation obtained by direct extrapolation (Eq. 19.10) is negligible (absolute relative difference < 0.014%, Fig. 19.6). Some of the very small dif-

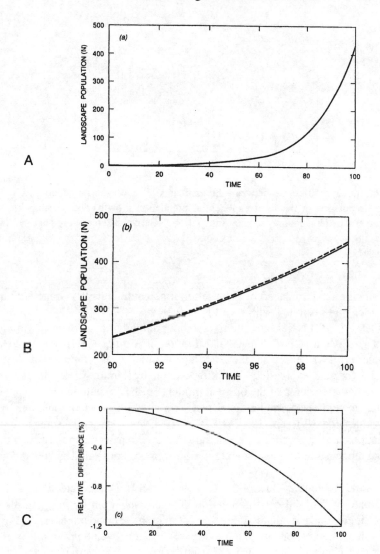

Figure 19.5. Comparison of the landscape population, N, estimated by extrapolation of the local model of exponential population growth using direct extrapolation (broken line) and extrapolation by expected value (solid line): (A) the entire simulation period and (B) a window from $t = 90$ to $t = 100$ that highlights the small difference between the extrapolations. (C) Relative difference between the direct extrapolation and the extrapolation by expected value. Relative difference is as defined in Fig. 19.3.

ference, especially the variation between time steps (Fig. 19.6), is attributable to round-off error. The trend toward increasing difference with time is attributable to error in the direct extrapolation rather than in extrapolation by explicit integration;

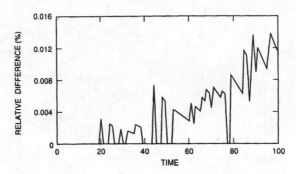

Figure 19.6. Relative difference between the estimates of landscape population, N, obtained by extrapolation of the local exponential population growth model using direct extrapolation and extrapolation by explicit integration. Relative difference is as defined in Fig. 19.3.

there is a small error introduced by representing the continuous landscape with a discrete lattice and assuming within-cell homogeneity.

Note that Eq. 19.17 is a structural reformulation of the model describing population size as a function of space and time (compare Eqs. 19.7b and 19.17). For example, Eq. 19.17 expresses the landscape population as a function of t^2! In contrast, the formula expressing the extrapolation by lumping (Eq. 19.11) is simply a reparameterization of the original model (Eq. 19.7b) with landscape area as a scaling factor. The formulas for direct extrapolation and extrapolation by expected value (Eqs. 19.10 and 19.12–19.15, respectively) do involve some minor transformations, but in practice these methods involve repeated simulations with the original model structure. Functional relationships are not altered as they are in Eq. 19.17.

Furthermore, a single evaluation of Eq. 19.17 yields an accurate, indeed an exact, estimate of the landscape population. The extrapolation by lumping also involves a single model evaluation, but the estimate has considerable error (Fig. 19.3B). Direct extrapolation and extrapolation by expected value provide reasonably accurate estimates of the landscape population, but they require several to many model evaluations. A consideration of accuracy and efficiency suggests extrapolation by explicit integration as the method of choice for this simple combination of landscape and local model.

19.4.2 Model 2: Logistic Growth

19.4.2.1 The Landscape and the Local Model

Consider the landscape of the previous example, but at any point in the landscape, population growth is now described by the differential equation

$$\frac{dn}{dt} = rn\left(1 - \frac{n}{k}\right) \tag{19.18a}$$

and the population at time t is given by

$$n = \frac{k n_0}{n_0 + ((k - n_0)e^{-rt})} \qquad (19.18b)$$

where n, n_0, and r are as in Eq. 19.7b, and k is the local carrying capacity.

The landscape is heterogeneous in both r and k. At any point (x,y) in the landscape, r is given by Eq. 19.8 and k is given by

$$k(x, y) = \frac{p_5 x + p_6}{p_7^{(p_8-y)^2}} \qquad (19.19)$$

(Fig. 19.7). For simplicity, I again assume that the landscape is homogeneous with respect to n_0. As in the previous example, individual points or sites in the land-scape are independent (there is neither immigration nor emigration), and local population dynamics at a site are completely described by Eqs. 19.18a and b. N, the population of the entire landscape at time t, is again the landscape-scale property of interest.

19.4.2.2 Scaling Up by Direct Extrapolation

Imposing the regular grid of the previous example and applying Eq. 19.18b to each cell provide an estimate of the landscape's population by direct extrapolation:

$$N = \sum_{i=1}^{m} a_i \left[\frac{k_i n_0}{n_0 + ((k_i - n_0)e^{-r_it})} \right] \qquad (19.20)$$

where a_i is the area of grid cell i and m is the number of grid cells. Again, I will use this estimate of N to evaluate the other extrapolations.

19.4.2.3 Scaling Up by Lumping

Lumping assumes that

$$N = A \left[\frac{\bar{k} n_0}{n_0 + ((\bar{k} - n_0)e^{-\bar{r}t})} \right] \qquad (19.21)$$

where A is the total landscape area and $\bar{r}$ and $\bar{k}$ are the mean values of r and k for the landscape (the means of Eqs. 19.8 and 19.9, respectively).

Figure 19.8A compares the simulation of N using Eq. 19.21 with the reference simulation obtained by direct extrapolation (Eq. 19.20). The scaling error from lumping in this example is generally small, never exceeding more than about 10% in absolute relative difference (Fig. 19.8B).

19.4.2.4 Extrapolation by Expected Value

Extrapolation by expected value assumes that

$$N = AE\left[\frac{kn_0}{n_0 + ((k_i - n_0)e^{-rt})} \right]$$ (19.22a)

where $E[]$ is the expected value operator, or

$$N = A \int \int \left[\frac{kn_0}{n_0 + ((k - n_0)e^{-rt})} \right] g(r, k)\, dr\, dk$$ (19.22b)

where $g(r,k)$ is the joint probability distribution for the spatial variable r and k. The independent probability distributions for r and k are shown in Figs. 19.4 and 19.9, respectively. Note the skewed distribution of k, which contributes to the error in extrapolation by lumping. A Spearman rank correlation between r and k of 0.1536 (estimated from the sample of 2500 values of r and k from the centroids of the landscape grid) is used to approximate the joint distribution of r and k in the Monte Carlo evaluation of the expected value.

Figure 19.10A compares the landscape population estimated by evaluating the expected value of the local model (Eq. 19.22) with the reference simulation

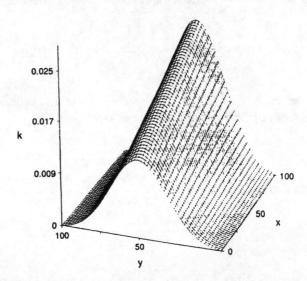

Figure 19.7. The spatial pattern of k, the carrying capacity, in the local model of logistic population growth. The corresponding parameters in Eq. 19.19 (see text) describing r as an explicit function of the normalized spatial coordinates x and y are $p_5 = 0.00125$, $p_6 = 0.0125$, $p_7 = 1.00157$, and $p_8 = 0.0392$.

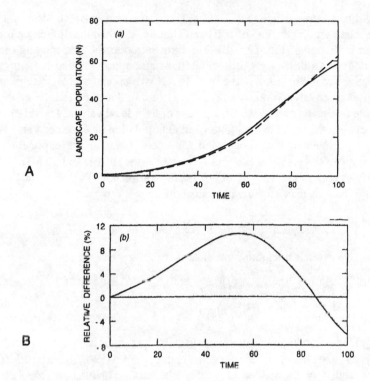

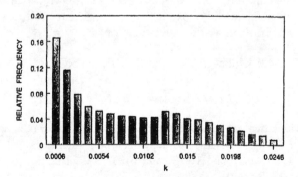

Figure 19.8. (A) Comparison of the landscape population, N, estimated by extrapolation of the local model of logistic population growth using direct extrapolation (broken line) and lumping (solid line). (B) Relative difference between the direct extrapolation and the estimate by lumping. Relative difference is a defined in Fig. 19.3.

Figure 19.9. The probability distribution describing the spatial variability of k in the local model. The distribution is based on a sample of 2500 values of k from the centroid of each cell of a unifrom grid of 50 × 50 square elements superimposed on the landscape of Fig. 19.7.

obtained by direct extrapolation (Eq. 19.20). The differences between these simulations are relatively small (Fig. 19.10B) and slightly less than the difference for extrapolation by lumping (Fig. 19.10C). The error introduced by calculating the expected value with 100 Monte Carlo simulations and approximating the empirical distributions of r (Fig. 19.4) and k (Fig. 19.9) by triangular and logtriangular distributions, respectively, is comparable to the error introduced by estimating the true aggregate landscape r and k with the means of their local values. The relative difference between the direct extrapolation and extrapolation by expected value is larger for this combination of model and landscape than for the exponential model/landscape of the previous example (compare Figs. 19.10B and 19.5B), but the trade-off between model runs and error is still conspicuous (a 96% reduction in model runs for less than 10% relative absolute error).

19.4.2.5 Extrapolation by Explicit Integration

Extrapolation by explicit integration predicts that

$$N = \int_{y_1}^{y_2} \int_{x_1}^{x_2} \frac{k(x,\,y)n_0}{n_0 + ((k(x,\,y) - n_0)e^{-r\,(x,\,y)\,t})} \, dx \, dy \tag{19.23}$$

where $r(x,y)$ and $k(x,y)$ are per capita growth rate and carrying capacity, respectively, as functions of the spatial coordinates x and y (Eqs. 19.8 and 19.9), and x_1, x_2, y_1, and y_2 are the spatial limits of the landscape. I have been unable to explicitly solve the double integral of Eq. 19.23, and a closed-form solution may not exist. Consequently, extrapolation by explicit integration may not be possible for this combination of model and landscape. This example nicely illustrates a major limitation in scaling up by explicit integration: it may be difficult or impossible to find a closed-form indefinite integral for even relatively simple local models and the functions describing landscape heterogeneity. As noted in Section 19.3.2.4, evaluating Eq. 19.23 by numerical quadrature, while possible, is essentially equivalent to direct extrapolation (or Monte Carlo extrapolation by expected value if Monte Carlo integration is used).

19.4.3 Model 3: Competition Between Two Species

19.4.3.1 The Landscape and the Local Model

In the previous two examples, the differential equations modeling the temporal dynamics of the local variable of interest can be explicitly solved with respect to time (Eqs. 19.7a and b and 19.18a and b). This is often not the case for many ecological models.

Consider a landscape with the dimensions of those in the previous examples, but the local and landscape-scale variables of interest are the population sizes of two competing species-populations. At any point in this landscape, the growth of these populations is described by the simultaneous differential equations

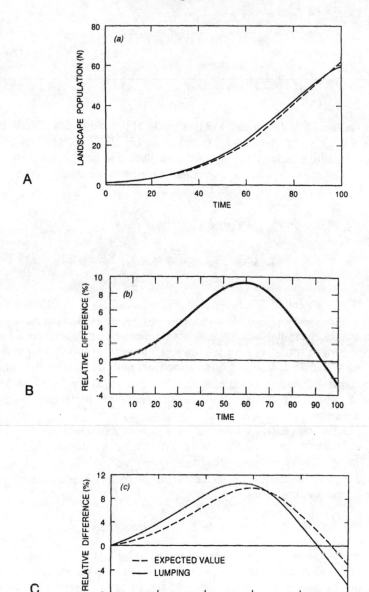

Figure 19.10. (A) Comparison of the landscape population, N, estimated by extrapolation of the local model of logistic population growth using direct extrapolation (broken line) and extrapolation by expected value (solid line). (B) Relative difference between the direct extrapolation and the extrapolation by expected value. (C) Comparison between the relative differences for lumping and extrapolation by expected value. Relative difference is as defined in Fig. 19.3.

$$\frac{dn_1}{dt} = n_1(r_1 - \alpha_{11}n_1 - \alpha_{12}n_2)$$

$$(19.24)$$

$$\frac{dn_2}{dt} = n_2(r_2 - \alpha_{22}\dot{n}_2 - \alpha_{21}n_1)$$

where n_s is the local population size and r_s is the local per capita growth rate for species s ($s = 1, 2$). The parameters α_{ss} (for $ss = 11, 12, 21, 22$) describe, respectively, the competitive influence of species 1 on itself and on species 2 and of species 2 on itself and on species 1. The landscape is heterogeneous with respect to r_s and α_{ss}; at any point (x,y) on the landscape:

$$r_s = r_{s(x,\,y)} = f_r(\mathbf{p}_s, x, y) \qquad \text{(for } s = 1, 2) \qquad (19.25a)$$

$$\alpha_{ss} = \alpha_{ss\,(x,\,y)} = f_a(\mathbf{q}_{ss}, x, y) \qquad \text{(for } ss = 11, 12, 21, 22) \qquad (19.25b)$$

where the $f(\ )$s are explicit functions of the normalized (0 to 100) Cartesian coordinates x and y and a parameter vector ($\mathbf{p}$ or $\mathbf{q}$) for the landscape. As in the previous examples, the initial local populations at time t_0 are homogeneous across the landscape (i.e., n_s at $t_0 = n_{s,0}$ for $s = 1, 2$), and individual points or sites in the landscape are independent. A site experiences neither immigration nor emigration, and local population dynamics at a site are completely described by Eq. 19.24. The populations N_1 and N_2 for the entire landscape at time t are the landscape-scale properties of interest.

Although an explicit general solution of the system of differential equations in Eq. 19.24 does not exist, numerical solutions of the local populations n_1 and n_2 are possible. Without actually evaluating the landscape populations N_1 and N_2, we can consider the impact of this circumstance on the respective application of the different methods for scaling up.

19.4.3.2 Scaling Up by Direct Extrapolation

By imposing the regular grid of the previous examples and solving Eq. 19.24 numerically for each cell, we can derive an estimate of the landscape's populations by direct extrapolation. Denoting the system of Eq. 19.24 as $\dot{n}$ and the numerical solution of this system for n_s at time t as $S[\dot{n}, n_s]$, direct extrapolation estimates that

$$N_s = \sum_{i=1}^{m} a_i\, S[\dot{n}, n_s]_i \qquad \text{(for } s = 1, 2) \qquad (19.26)$$

where i is the grid cell index, m is the number of grid cells, and a_i is the area of grid cell i. Thus, not surprisingly, the absence of an explicit general time-dependent solution for the differential equations describing the local model does not preclude the application of direct extrapolation as a method for scaling up the local model. In practice, however, application with a local model requiring computationally intensive numerical solutions or a landscape with an extremely fine grid and many grid cells might encounter practical limitations of computer time.

19.4.3.3 Scaling Up by Lumping

Extrapolation by lumping assumes that the landscape populations N_1 and N_2 can be estimated by

$$N_s = AS[\dot{n}, \bar{n}_s] \qquad \text{(for } s = 1, 2) \qquad (19.27)$$

where A is the landscape area, and $S[\dot{n}, \bar{n}_s]$ is the numerical solution of the local model (Eq. 19.24) evaluated with the means of the spatially heterogeneous variables:

$$\frac{d\bar{n}_1}{dt} = \bar{n}_1(\bar{r}_1 - \bar{\alpha}_{11}\bar{n}_1 - \bar{\alpha}_{12}\bar{n}_2)$$

$$(19.28)$$

$$\frac{d\bar{n}_2}{dt} = \bar{n}_2(\bar{r}_2 - \bar{\alpha}_{22}\bar{n}_2 - \bar{\alpha}_{21}\bar{n}_1)$$

Lumping can in principle be used to scale up a local model described by differential equations for which there is no general time-dependent solution if a numerical solution is available.

19.4.3.4 Extrapolation by Expected Value

Extrapolation by expected value assumes that

$$N_s = AE[S[\dot{n}, n_s]] \qquad \text{(for } s = 1, 2) \qquad (19.29a)$$

where $E[]$ is the expected value operator, or

$$N_s = A \int \int \int \int \int \int [S[\dot{n}, n_s]](r_1, r_2, \alpha_{11}, \alpha_{22}, \alpha_{12}, \alpha_{21}) \qquad (19.29b)$$

$$dr_1 \, dr_2 \, d\alpha_{11} \, d\alpha_{22} \, d\alpha_{12} \, d\alpha_{21}$$

(for $s = 1, 2$) where $g(r_1, r_2, \alpha_{11}, \alpha_{22}, \alpha_{12}, \alpha_{21})$ is the joint probability distribution function for the spatially distributed variables. Obtaining an explicit formulation for this joint distribution is highly unlikely, but the independent probability distributions and rank correlations between variables could be combined in Latin Hypercube Monte Carlo simulations to estimate the expected value of the numerical solutions (Iman and Conover 1983; Iman and Shortencarrier 1984). Constrained by the rank correlations, values for $r_1, r_2, \alpha_{11}, \alpha_{22}, \alpha_{12},$ and α_{21} would be drawn from their independent distributions and the system of Eq. 19.24 solved numerically with these values. This process would be repeated sufficient times to adequately sample the distribution of spatial variables (≥ 100 iterations), and the means of the resulting collection of numerical solutions for n_1 and n_2 would then be evaluated to approximate the $E[S[\dot{n}, n_s]]$ of Eq. 19.29a. The absence of a general time-dependent solution for the local model does not preclude the potential application of extrapolation by expected value. Indeed, it is in these circumstances that Monte Carlo extrapolation is most valuable.

19.4.3.5 Extrapolation by Explicit Integration

Extrapolation by explicit integration predicts that the landscape populations, N_1 and N_2, are given by

$$\int \int \left\{ \begin{matrix} f_{n_1}(x, y, t) \\ f_{n_2}(x, y, t) \end{matrix} \right\} dx\, dy \tag{19.30}$$

where $f_{n_1}(x,y,t)$ and $f_{n_2}(x,y,t)$ are the general time-dependent solutions of

$$\frac{dn_1(x, y)}{dt} = n_1(x, y)(r_1(x, y) - \alpha_{11}(x, y)n_1(x, y) - \alpha_{12}(x, y)n_2(x, y))$$

$$\tag{19.31}$$

$$\frac{dn_2(x, y)}{dt} = n_2(x, y)(r_2(x, y) - \alpha_{22}(x, y)n_2(x, y) - \alpha_{21}(x, y)n_1(x, y))$$

for n_1 and n_2, respectively, at the point (x,y). In the system of Eq. 19.31, $r_s(x,y)$ (for $s = 1, 2$) and $\alpha_{ss}(x,y)$ (for $ss = 11, 12, 21, 22$) are the spatially heterogeneous variables as explicit functions of the spatial coordinates x and y (Eqs. 19.25a and b), and $n_s(x,y)$ is the local population for species s (for $s = 1,2$) at coordinates (x,y). The absence of general time-dependent solutions for n_1 and n_2 precludes evaluation of Eq. 19.30 and, consequently, the use of extrapolation by explicit integration to estimate the landscape-scale populations from the local model (Eq. 19.24).

As this example has illustrated, the absence of general time-dependent solutions for the differential equations of a local model, as a general rule, only excludes extrapolation by explicit integration as a method for scaling up. Any of the other methods are applicable in principle, but, in practice, considerations of computational requirements, potential error, and available information might influence the selection.

19.4.4 Model 4: Population Growth with Immigration and Emigration

19.4.4.1 The Landscape and the Local Model

In the previous examples there were no flows or interactions among sites in the landscape. In this example, I relax this assumption and consider, in a simple way, how among-site interactions influence scaling up with the methods considered here.

Consider a landscape with the dimensions of those in the previous examples. At any point in this landscape population growth is modeled by

$$\frac{dn}{dt} = rn\left(1 - \frac{n}{k}\right) - bn + c(N - n) \tag{19.32}$$

where n, r, and k are as in Eq. 19.18, b is the local per capita emigration rate, c is the local realized immigration rate, and $N - n$ is the total landscape population N (at time t) less the local population. Local population growth is influenced by all other sites through the immigration term $c(N - n)$. The landscape is heterogeneous with respect to r, k, b, c. At any point (x, y) on the landscape r and k are given by Eqs. 19.8 and 19.19, respectively, and

$$b = b(x, y) = f_b(\mathbf{p}, x, y) \tag{19.33a}$$

$$c = c(x, y) = f_c(\mathbf{p}, x, y) \tag{19.33b}$$

where the $f(\ )$s are explicit functions of the normalized (0 to 100) Cartesian coordinates x and y and a parameter vector $\mathbf{p}$ for the landscape. As in the previous examples, the landscape is homogeneous with respect to the initial local population n_0. N, the population for the entire landscape at time t, is again the landscape-scale property of interest.

Solution of the local model in Eq. 19.32 presumes simultaneous knowledge of both the local and the landscape population at time t (n and N, respectively). But N is the landscape-scale unknown that the scaling up is attempting to estimate from knowledge of the local dynamics, and a differential equation describing changes in N is unknown. N can be eliminated as a term in the local model by consideration of discrete sites and reformulation of the model such that

$$\frac{dn_i}{dt} = r_i n_i\left(1 - \frac{n_i}{k_i}\right) - \sum_{\substack{j=1 \\ j \neq i}}^{z} e_{ij} n_i + \sum_{\substack{j=1 \\ j \neq i}}^{z} h_{ij} n_j \tag{19.34}$$

where z is the number of discrete sites, e_{ij} is the spatially heterogeneous per capita rate of emigration from site i to site j, and h_{ij} is the spatially heterogeneous per capita rate of immigration into site i from site j. This spatially distributed formulation of the model eliminates the circularity of needing N to estimate N by scaling up, but it is replaced by the need to know the local populations at all sites simultaneously. Without arguing the finer points of this model (e.g., whether immigration into site i should include a density dependence on n_i) and ignoring edge effects (e.g., whether the edges of the landscape are closed to emigration and immigration), we can assume knowledge of explicit spatial functions describing the heterogeneity of e_{ij} and h_{ij} and (without actually solving for the landscape population) consider the influence of interactions among sites on the methods for scaling up.

First, either the circularity of knowing N in the model of Eq. 19.32 or the need for simultaneous knowledge of n for all sites in the model Eq. 19.34 can obviously be eliminated if one is willing to assume a time lag in the interaction terms. With this assumption, the local model of Eq. 19.32 can be reformulated so that, in effect, any local site is independent of the combined sites contributing to N. For example, the model could be reformulated as

$$\frac{dn}{dt} = rn\left(1 - \frac{n}{k}\right) - bn + c(N_{t-\varepsilon} - n) \tag{19.35}$$

where ε is some time lag. Given N_0, an initial estimate of N at $t = 0$, this local model can be solved for n at $t = \varepsilon$ and the solution scaled up by some applicable method (direct extrapolation or Monte Carlo extrapolation by expected value are likely candidates) to estimate N at $t = \varepsilon$. This new extrapolated estimate of N can, in turn, be used to solve the local model for n at $t + \varepsilon$ (or $t = 2\varepsilon$). This process is repeated for the duration of the simulation. Similarly, the local model could be reformulated as a finite-difference model so that

$$n(t + \Delta t) = n(t) + rn(t)\left(1 - \frac{n(t)}{k}\right) - bn(t) + c(N(t) - n(t)) \tag{19.36}$$

and $n(t + \Delta t)$ scaled up to estimate $N(t + \Delta t)$.

Similar reformulations could also be applied to the alternative, spatially distributed model of Eq. 19.34 so that, in effect, the sites are independent at time t. Immigration and emigration among sites at time t is a function of the population of each site at some slightly previous time $t - \varepsilon$. For example, immigrants to site i in one year might be dispersing juveniles (or propagules) from site j of the previous year.

"Roving quadrat" models (e.g., Smith and Urban 1988) and transect models with paired interactions (e.g., Shugart et al. 1988a,b) are variations of this approach. These models sweep the entire spatial extent of the model system at each time step, simulating interactions between or among sites at each spatial step of the sweep. The temporal dynamics of interactions within the sweep are assumed to be homogeneous or insignificant. Spatial interactions within the sweep are assumed to be much faster than the temporal grain or time step of the model and, in effect, occur simultaneously. Individual-based models (Huston et al. 1988), with or without an explicit spatial dimension, involve a similar assumption. These models sweep through all the individuals under consideration at each time step; interactions within that sweep are assumed to occur at rates much faster than the time step and are, in effect, simultaneous.

Temporal reformulations and assumptions like these might be sufficient for many or most applications involving local models and landscapes with among-site interactions, and they are commonly employed. However, it is worth considering further how simultaneous interactions might affect the application of each of the general methods for scaling up in those circumstances when the assumptions of a time lag in the interactions or simultaneous interactions within a spatial sweep are inappropriate or the potential error in such assumptions is suspect.

19.4.4.2 Scaling Up by Direct Extrapolation

The discrete-site, spatially-distributed model of Eq. 19.34 is ideally suited to scaling up by direct extrapolation. Indeed, the reformulation from the local model of Eq. 19.32 is a scale translation. Using the regular grid of homogeneous cells from previous examples

$$\frac{dn_i}{dt} = a_i \left[= r_i n_i \left(1 - \frac{n_i}{k_i} \right) - \sum_{\substack{j=1 \\ j\neq i}}^{m} e_{ij} n_i + \sum_{\substack{j=1 \\ j\neq i}}^{m} h_{ij} n_j \right] \qquad (19.37)$$

where a_i is the area of grid cell i, and assuming a numerical solution of Eq. 19.37, $S[dn_i/dt, n_i]$, for the population n_i of grid cell i, direct extrapolation predicts that

$$N = \sum_{i=1}^{m} S \left[\frac{dn_i}{dt}, n_i \right] \qquad (19.38)$$

Solution of dn_i/dt for n_i requires simulataneous knowledge of n for all other cells. Thus an obvious approach is to describe the landsape as a system of m simultaneous differential equations, with each equation in the system modeling the dynamics of an individual cell:

$$\frac{dn_1}{dt} = a_1 \left[r_1 n_1 \left(1 - \frac{n_1}{k_1} \right) - \sum_{j=2}^{m} h_{1j} n_1 + \sum_{j=2}^{m} e_{1j} n_j \right]$$

$$\frac{dn_2}{dt} = a_2 \left[r_2 n_2 \left(1 - \frac{n_2}{k_2} \right) - \sum_{\substack{j=1 \\ j\neq2}}^{m} h_{2j} n_2 + \sum_{\substack{j=1 \\ j\neq2}}^{m} e_{2j} n_j \right]$$

$$\vdots$$

$$\frac{dn_m}{dt} = a_m \left[r_m n_m \left(1 - \frac{n_m}{k_m} \right) - \sum_{\substack{j=1 \\ j\neq m}}^{m} h_{mj} n_m + \sum_{\substack{j=1 \\ j\neq m}}^{m} e_{mj} n_j \right] \qquad (19.39a)$$

or

$$\dot{n} = a \left[rn \left(1 - \frac{n}{k} \right) - Hn + En \right] \qquad (19.39b)$$

Direct extrapolation of this system assumes that the landscape population is given by the summation of the solution vector n, or

$$N = \sum_{i=1}^{m} S[\hat{n}, n_i] \qquad (19.40)$$

This approach is commonly used in modeling the dynamics of populations in patchy environments where there is interaction among patches (see Levin 1976a,b).

Numerical solution of a system of 2500 simultaneous differential equations may be limited by computation time and available memory, especially if the equation is very complex, but the approach is straightforward. Aggregating like cells or patches to reduce patch number can reduce the number of differential equations to tractable levels. The discrete patch model with its ordinary differential equations can also be replaced by a continuous model with fewer partial differential equations. Solution of the partial differential equations has its own analytical and computational constraints, but the approach is transparent and existing tools (e.g., finite element methods) can be brought to bear.

Furthermore, advanced computer architectures (e.g., vector processors and hypercubes or other parallel processors) can be used to at least partially overcome computational constraints on the solution of large systems of both partial and ordinary differential equations (Band and Wood 1988; Casey and Jameson 1988).

19.4.4.3 Scaling Up by Lumping

Application of lumping to the spatially distributed model of Eq. 19.34 is inappropriate. The model is not a local or point model that needs scaling up. As noted above, the model already represents a distributed, heterogeneous area.

However, the original local model of Eq. 19.32 can be scaled up by lumping. In this example, lumping assumes that the landscape population N is given by the product of the landscape area A and the time-dependent solution of

$$\frac{d\bar{n}}{dt} = \bar{r}\bar{n}\left(1 - \frac{\bar{n}}{\bar{k}}\right) - \bar{b}\bar{n} + \bar{c}\,(N - \bar{n}) \qquad (19.41a)$$

This application is confounded by the circularity of needing the landscape N to estimate N by scaling up the average local population n. The circularity can be avoided with the assumption that $N = An$ so that

$$\frac{d\bar{n}}{dt} = \bar{r}\bar{n}\left(1 - \frac{\bar{n}}{\bar{k}}\right) - \bar{b}\bar{n} + \bar{c}\,(A\bar{n} - \bar{n}) \qquad (19.41b)$$

Thus lumping is not in principle excluded by the concept of simultaneous interactions among sites, particularly if the interactions are as represented in the local model of this example. However, the appropriateness of lumping in these circumstances is dependent upon the specific formulation of the model. The potential for error in lumping should also be carefully considered.

19.4.4.4 Extrapolation by Expected Value

Extrapolation by expected value assumes that the landscape population N is equal to the product of the landscape area and the expected value of the time-dependent solution of the small-scale model. As we have seen in this example, those solutions are complicated by either the circularity of needing N to predict N in the local model of Eq. 19.32 or the need for simultaneous simulation of local n for all sites in the alternative model of Eq. 19.34.

Neither of the fixes found for either direct extrapolation or lumping is truly applicable here. If the combination of local model and landscape can be described by a system of linked equations and the system can be solved (direct extrapolation, Section 19.4.4.2), there is no need to estimate the expected value of the local model. Similarly, an approximation of $N = A\bar{n}$ or $n = AE[n]$ like that used in the extrapolation by lumping (Section 19.4.4.3) simply replaces the circularity of knowing N at time t with that of knowing $E[n]$ at time t. Thus extrapolation by expected value cannot be used to estimate the landscape population in this example. The presence of significant flows or interactions among sites violates a fundamental assumption of the approach.

19.4.4.5 Extrapolation by Explicit Integration

The apparent absence of an explicit indefinite spatial integral for the logistic term of the local model in this example (Eq. 19.32, Section 19.4.2.5) effectively precludes extrapolation by explicit integration for this example. Moreover, even if the explicit integral for the logistic term could be found, application of this method would still require either an expression of N (or some approximation of N) as an explicit function of time and space to extrapolate Eq. 19.32 or an infinite nesting of an explicit space-time function for n to extrapolate Eq. 19.34. The first is not available, and the second makes an explicit formulation virtually impossible, even to a reasonable approximation. Thus, extrapolation by explicit integration is unlikely to be generally applicable to scaling up models in landscapes with simultaneous interdependent site dynamics.

19.5 Conclusions

Meeting the challenge of the scaling problem in landscape ecology requires (1) a conceptual or theoretical framework for addressing multiple scales and (2) quantitative methods for implementing that framework. Both exist. Here I have used a basic principle from hierarchy theory, manipulation of grain and extent to link levels in a nested hierarchy, to access four quantitative methods for translating models across spatial scales. The mathematics and computer implementation of these methods are commonplace and accessible. They can be, and have been, readily applied to the spatial extrapolation of local, small-scale models. Other conceptual frameworks exist, and the methods I have presented here in no way exhaust all possible or existing methods for translating models across scales. Scaling up models to the landscape is limited not so much by the existence of

quantitative methods as by the application of existing methods to specific landscape questions.

From the simple demonstrations presented here, a review of existing applications, and my experience with scaling up site-specific ecosystem models to regional scales (King et al. 1987, 1990b), I can recommend a general strategy for
scaling up from local small-scale models to landscape-scale predictions. First,
lumping can be used to obtain a rough first approximation for the landscape-scale.
Then, assuming independence among sites, extrapolation by expected value using
Monte Carlo simulation can be used to refine this first approximation. Interpretation of the lumped approximation must allow for the potential error arising from
nonlinearities in the local model. The refined estimate from the Monte Carlo
extrapolation can be used to assess this scaling error.

This combination of approaches is applicable to a wide range of small-scale
models and landscape phenomena. It is particularly useful when the small-scale
model is large, complicated, or computationally demanding and the landscape is
large and fine-grained. The procedure is easily implemented, requires a minimum
amount of information on landscape heterogeneity that can be accurately estimated with reasonable effort, and is computationally efficient. If the landscape-
scale estimate from the extrapolation by expected value can be tested and is found
inaccurate, or is otherwise questionable, alternative methods should be considered,
especially if careful consideration of the potential errors (e.g., estimates of landscape heterogeneity) fails to resolve the problem.

If the assumption of independent sites is suspect or known to be invalid, model
reformulations or alternative methods may be required. The assumption of a
time-lag in the interactions might be employed, resulting in effectively independent sites. If that assumption is undesirable, the interacting sites might be modeled
by a system of linked ordinary differential equations, or the model could be
reformulated as partial differential equations. The literature on modeling population dynamics in patchy environments is especially rich in its considerations of
strong interactions among sites and should be more aggressively utilized as a pool
of methods for landscape applications.

Similarly, if scale-dependent phenomena are indicated, alternative methods
incorporating these considerations should be employed. I believe that hierarchy
theory will continue to provide guidelines for identifying, modifying, and implementing such methods. Understanding how processes change with scale and
developing general methods for dealing with scale-dependent interactions, processes, and phenomena are exciting challenges in the search for quantitative
methods for translating models across scales in the landscape.

19.6 Summary

Four general and simple quantitative methods for scaling up models by manipulations of model grain and extent have been identified. These methods are most
applicable when one is scaling up over sites without strong influences among sites
and across areal extents that do not involve scale-dependent phenomena. The
methods are (1) lumping—using the means of spatially distributed variables as

input to the small-scale model; (2) direct extrapolation—running the small-scale model for every discrete element or patch in the landscape; (3) extrapolation by expected value—calculating the expected value of the small-scale model; and (4) extrapolation by explicit integration—evaluating the closed-form solution of the indefinite integral of the small-scale model with respect to space.

Demonstrations of these methods using simple models and landscapes allow some conclusions and recommendations about their appropriate application. Selection from this set of general methods is largely determined by the complexity of the local model, the nature of the information available on landscape heterogeneity, and the potential error inherent in their respective assumptions. A combination of lumping as a first approximation and extrapolation by expected value using Monte Carlo simulation to refine this approximation is recommended as a general and useful strategy for predicting the landscape-scale expression of phenomena described by local, small-scale models.

Acknowledgments

I am grateful to Robert V. O'Neill, Alan R. Johnson, and Dean L. Urban for the many frank and open discussions that both stimulated and refined many of the ideas I have presented here. Robert H. Gardner provided the original computer code (PRISM) used in the Monte Carlo extrapolations; his assistance is much appreciated. I also thank Monica G. Turner, Robert H. Gardner, and an anonymous reviewer for their comments and suggestions on this manuscript. The work was sponsored jointly by the Ecosystems Studies Program, National Science Foundation under Interagency Agreement BSR 8417923 and the Carbon Dioxide Research Program, Atmospheric and Climate Research Division, Office of Health and Environmental Research, U.S. Department of Energy, under contract DE–AC05–84OR21400 with Martin Marietta Energy Systems, Inc. Publication No. 3546, Environmental Sciences Division, Oak Ridge National Laboratory.

References

Allen, T.F.H., O'Neill, R.V., and Hoekstra, T.W. 1984. Interlevel relations in ecological research and management: some working principles from hierarchy theory. USDA *Forest Service General Technical Report RM-110*. Rocky Mountain Forest and Range Experiment Station, Fort Collins, Colo.

Allen, T.F.H., and Starr, T.B. 1982. *Hierarchy: Perspectives for Ecological Complexity*. Chicago: University of Chicago Press.

Band, L.E., and Wood, E.F. 1988. Strategies for large-scale, distributed hydrological simulation. *Applied Mathematics and Computation* 27:33–27.

Bartell, S.M., Gardner, R.H., and O'Neill, R.V. 1989. Nutrient patterns across space and time as a consequence of topography and vegetation type. Abstract. *Bulletin of the Ecological Society of America* 70:57.

Bolla, M., and Kutas, T. 1984. Submodels for the nutrient loading estimation on River Zala. *Ecological Modelling* 26:115–43.

Burke, I.C., Schimel, D.S., Yonker, C.M., Parton, W.J., and Joyce, L.A. 1990. Regional modeling of grassland biogeochemistry using GIS. *Landscape Ecology*, 4:45–54.

Cale, W.G., Jr., and Odell, P.L. 1979. Concerning aggregation in ecosystem modeling. In *Theoretical Systems Ecology*, ed. E. Halfon, pp. 55–77. New York: Academic Press.

Cale, W.G., Jr., and Odell, P.L. 1980. Behavior of aggregate state variables in ecosystem models. *Mathematical Biosciences* 49:121–37.

Cale, W.G., Jr., O'Neill, R.V., and Gardner, R.H. 1983. Aggregation error in nonlinear ecological models. *Journal of Theoretical Biology* 100:539–50.

Casey, R.M., and Jameson, D.A. 1988. Parallel and vector processing in landscape dynamics. *Applied Mathematics and Computation* 27:3–22.

Coffin, D.P., and Lauenroth, W.K. 1988. Disturbances and gap dynamics in a semiarid grassland: a landscape-level approach. Abstract. *Bulletin of the Ecological Society of America* 69:101.

Coughlan, J.C., and Running, S.W. 1988. An aggregation algorithm for the efficient application of ecosystem models within a geographic information system. Abstract. *Bulletin of the Ecological Society of America* 69:107.

Curry, J.W. Jr., Shapiro, L.G., and Vanderlip, R.L. 1977. A population simulation model for field crops. Proceedings of the 8th Annual Pittsburg Conference on Modeling and Simulation, April 21–22, 1977.

Dale, V.H., and Gardner, R.H. 1987. Assessing regional impacts of growth declines using a forest succession model. *Journal of Environmental Management* 24:83–93.

DeAngelis, D.L., Post, W.M., and Travis, C.C. 1986. *Positive Feedbacks in Natural Systems.* Berlin: Springer-Verlag.

Duncan, W.G., Loomis, R.S., Williams, W.A., and Hanau, R. 1967. A model for simulating photosynthesis in plant communities. *Hilgardia* 38:181–205.

Forman, R.T.T., and Godron, M. 1986. *Landscape Ecology.* New York: John Wiley and Sons.

Gardner, R.H., Cale, W.G., and O'Neill, R.V. 1982. Robust analysis of aggregation error. *Ecology* 63:1771–79.

Gardner, R.H., O'Neill, R.V., and Carney, J.H. 1981. Spatial patterning and error propogation in a stream ecosystem model. In *Proceedings, Summer Computer Simulation Conference,* pp. 391–95. La Jolla, Calif.: Simulation Councils.

Gates, W. Lawrence. 1985. The use of general circulation models in the analysis of the ecosystem impacts of climatic change. *Climatic Change* 7:267–84.

Grossman, W.D., and Schaller, J. 1986. Geographical maps on forest die-off, driven by dynamic models. *Ecological Modelling* 31:341–53.

Hirata, H., and Ulanowicz, R.E. 1986. Large-scale ecosystem perspectives on ecological modelling and analysis. *Ecological Modelling* 31:79–104.

Huston, M., DeAngelis, D., and Post, W. 1988. New computer models unify ecological theory. *BioScience* 38:682–91.

Huston, M., and Smith, T. 1987. Plant succession: life history and competition. *American Naturalist* 130:168–98.

Iman, R.L., and Conover, W.J. 1983. *A Modern Approach to Statistics.* New York: John Wiley and Sons.

Iman, R.L., and Shortencarier, M.J. 1984. A FORTRAN 77 program and user's guide for the generation of Latin Hypercube and random samples for use with computer models. NUREG/CR–3624, SAND83–2365. National Technical Information Service, Springfield, Va.

Iwasa, Y., Andreasen, V., and Levin, S. 1987. Aggregation in model ecosystems: I. perfect aggregation. *Ecological Modelling* 37:287–302.

Iwasa, Y., Levin, S., and Andreasen, V. 1989. Aggregation in model ecosystems: II. approximate aggregation. *IMA Journal of Mathematics Applied in Medicine and Biology* 6:1–23.

Jarvis, P.G., and McNaughton, K.G. 1986. Stomatal control of transpiration: scaling up from leaf to region. *Advances in Ecological Research* 15:1–49.

Jeffers, J.N.R. 1988. Statistical and mathematical approaches to issues of scale in ecology. In *Scale and Global Changes: Spatial and Temporal Variability in Biospheric and*

Geospheric Process, SCOPE 35, eds. T.R. Rosswall, R.G. Woodmansee, and P.G. Risser, pp. 47–56. Chichester, England: J. Wiley and Sons.

King, A.W., DeAngelis, D.L., and Post, W.M. 1987. The seasonal exchange of carbon dioxide between the atmosphere and the terrestrial biosphere: extrapolation from site specific models to regional models. ORNL/TM–10570. Oak Ridge National Laboratory, Oak Ridge, Tenn.

King, A.W., Emanuel, W.R., and O'Neill, R.V. 1990a. Linking mechanistic models of tree physiology with models of forest dynamics: problems of temporal scale. In *Process Modeling of Forest Growth Responses to Environmental Stress*, eds. R.K. Dixon, R.S. Meldahl, G.A. Ruark, and W.G. Warren, pp. 241–48. Portland, Oreg.: Timber Press.

King, A.W., O'Neill, R.V., and DeAngelis, D.L. 1990b. Using ecosystem models to predict regional CO_2 change between the atmosphere and the terrestrial biosphere. *Global Biogeochemical Cycles*, in press.

Leduc, S.K., and Holt, D.A. 1987. The scale problem: modeling plant yield over time and space. In *Plant Growth Modeling for Resource Management*. Vol 1, Current Models and Methods, eds. K. Wisiol and J. Hesketh. pp. 125–37. Boca Raton, Fla.: CRC Press.

Levin, S.A. 1974. Dispersion and population interactions. *American Naturalist* 108:207–28.

Levin, S.A. 1976a. Population dynamic models in heterogeneous environments. *Annual Review of Ecology and Systematics* 7:287–310.

Levin, S.A. 1976b. Spatial patterning and the structure of ecological communities. In *Some Mathematical Questions in Biology*, Vol. 7, *Lectures on Mathematics in the Life Sciences*, vol. 8, ed. S.A. Levin, Providence, R.I.: American Mathematical Society.

Levin, S.A. 1989. Models in ecotoxicology: methodological aspects. In *Ecotoxicology: Problems and Approaches*, eds. S.A. Levin, M.A. Harwell, J.R. Kelly, and K.D. Kimball, pp. 213–30. New York: Springer-Verlag.

Luckyanov, N.K. 1984. Linear aggregation and separability of models in ecology. *Ecological Modelling* 21:1–12.

Luckyanov, N.K., Svirezhev, Yu.M., and Voronkova, O.V. 1983. Aggregation of variables in simulation models of water ecosystems. *Ecological Modelling* 18:235–40.

Luxmoore, R.J., Tharp, M.L., and West, D.C. 1990. Simulating the physiological basis of treering responses to environmental changes. In *Process Modeling of Forest Growth Responses to Environmental Stress*, eds. R.K. Dixon, R.S. Meldahl, G.A. Ruark, and W.G. Warren, pp. 393–401. Portland, Oreg.: Timber Press.

Meentemeyer, V., and Box, E.O. 1987. Scale effects in landscape studies. In *Landscape Heterogeneity and Disturbance*, ed. M.G. Turner, pp. 15–34. New York: Springer-Verlag.

O'Neill, R.V. 1979a. Natural variability as a source of error in model predictions. In *Systems Analysis of Ecosystems*, eds. G.S. Innis and R.V. O'Neill, pp. 23–32. Fairland, Md.: International Cooperative Publishing House.

O'Neill, R.V. 1979b. Transmutations across hierarchical levels. In *Systems Analysis of Ecosystems*, eds. S.G. Innis and R.V. O'Neill, pp. 58–78. Fairland, Md.: International Cooperative Publishing House.

O'Neill, R.V. 1988a. Hierarchy theory and global change. In *Scale and Global Changes: Spatial and Temporal Variability in Biospheric and Geospheric Process*, SCOPE 35, eds. T.R. Rosswall, R.G. Woodmansee, and P.G. Risser, pp. 29–45. Chichester, England: J. Wiley and Sons.

O'Neill, R.V. 1988b. Perspectives in hierarchy and scale. In *Perspectives in Ecological Theory*, eds. J. Roughgarden, R.M. May, and S.A. Levin. Princeton, N.J.: Princeton University Press.

O'Neill, R.V., DeAngelis, D.L., Waide, J.B.; and Allen, T.F.H. 1986. *A Heirachical Concept of Ecosystems*. Princeton, N.J.: Princeton University Press.

O'Neill, R.V., Elwood, J.W., and Hildebrand, S.G. 1979. Theoretical implications of spatial heterogeneity in stream ecosystems. In *Systems Analysis of Ecosystems,* eds. G.S. Innis and R. V. O'Neill, pp. 79–101. Fairfield, Md.: International Cooperative Publishing House.

O'Neill, R.V., Gardner, R.H., and Carney, J.H. 1982. Parameter constraints in a stream ecosystem model: incorporation of a priori information in Monte Carlo error analysis. *Ecological Modelling* 16:51–65.

O'Neill, R.V., and Rust, B. 1979. Aggregation error in ecological models. *Ecological Modelling* 7:91–105.

Pastor, J., and Post, W.M. 1986. Influence of climate, soil moisture, and succession on forest carbon and nitrogen cycles. *Biogeochemistry* 2:3–27.

Pastor, J., and Post, W.M. 1988. Response of northern forests to CO_2-induced climatic change: dependence on soil water and nitrogen availabilities. *Nature* 334:55–58.

Patten, B.C. 1982. Environs: relativistic elementary particles for ecology. *American Naturalist* 119:179–219.

Prentice, I.C., Webb, R.S., Ter-Mikhaelian, M.T., Solomon, A.M., Smith, T.M., Pitovranov, S.E., Nikolov, N.T., Minin, A.A., Leemans, R., Lavorel, S., Korzukhin, M.D., Hrabovszky, J.P., Helmisaari, H.O., Harrison, S.P., Emanuel, W.R., and Bonan, G.B. 1989. Developing a global vegetation dynamics model: results of an IIASA summer workshop. RR–89–7. International Institute for Applied Systems Analysis, Laxenburg, Austria.

Rastetter, E.B., King, A.W., Cosby, B.J., Hornberger, G.M., O'Neill, R.V., and Hobbie, J.E. Predicting ecosystem response to environmental change: aggregating fine-scale knowledge to anticipate large-scale responses. Manuscript.

Risser, P.G. 1986. Report of a workshop on the spatial and temporal variability of biospheric and geospheric processes: research needed to determine interactions with global environmental change, Oct. 28–Nov. 1, 1985. St. Petersburg, Fla. Paris: ICSU Press.

Risser, P.G. 1987. Landscape ecology: state of the art. in *Landscape Heterogeneity and Disturbance,* ed. M.G. Turner, pp. 3–14. New York: Springer-Verlag.

Rosen, R. 1977. Observation and biological systems. *Bulletin of Mathematical Biology* 39:663–78.

Rosswall, T., Woodmansee. R.G., and Risser, P.G., eds. 1988. *Scale and Global Changes: Spatial and Temporal Variability in Biospheric and Geospheric Process.* SCOPE 35. Chichester, England: J. Wiley and Sons.

Running, S.W., Nemani, R.R., Peterson, D.L., Band, L.E., Potts, D.F., Pierce, L.L., and Spanner, M.A. 1989. Mapping regional forest evapotranspiration and photosynthesis by coupling satellite data with ecosystem simulation. *Ecology* 70:1090–1101.

Sawyer, A.J., and Haynes, D.L. 1985. Simulating the spatiotemporal dynamics of the cereal leaf beetle in a regional crop system. *Ecological Modelling* 38:83–104.

Shugart, H.H. 1984. *A Theory of Forest Dynamics: The Ecological Implications of Forest Succession Models.* New York: Springer-Verlag.

Shugart, H.H., Antonovsky, M.Ya., Jarvis, P.G., and Sandford, A.P. 1986. CO_2, Climatic Change and Forest Ecosystems. In *The Greenhouse Effect, Climatic Change, and Ecosystems,* SCOPE 29, eds. B. Bolin, B.R. Döös, J. Jager, and R. A. Warrick, pp. 475–521. Chichester, England: J. Wiley and Sons.

Shugart, H.H., Bonan, G.B., and Rastetter, E.B. 1988. Niche Theory and community organization. *Canadian Journal of Botany* 66:2634–39.

Shugart, H.H., Goldstein, R.A., O'Neill, R.V., and Mankin, J.B. 1974. TEEM: A Terrestrial Ecosystem Energy Model for forests. *Oecologia Plantarum* 9:231–64.

Shugart, H.H., Michaels, P.J., Smith, T.M., Weinstein, D.A., and Rastetter, E.B. 1988. Simulation models of forest succession. In *Scale and Global Changes: Spatial and Temporal Variability in Biospheric and Geospheric Process,* SCOPE 35, eds. T.R. Rosswall, R.G. Woodmansee, and P.G. Risser, pp. 125–51. Chichester, England: J. Wiley and Sons.

Shugart, H.H., and Urban, D.L. 1988. Scale, synthesis, and ecosystem dynamics. In *Concepts of Ecosystem Ecology*, eds. L.R. Pomeroy and J.J. Alberts, pp. 279–89. New York: Springer-Verlag.

Shugart, H.H. Jr., and West, D.C. 1977. Development of an Appalachian deciduous forest succession model and its application to assessment of the impact of the chestnut blight. *Journal of Environmental Management* 5:161–79.

Smith, T., and Huston, M. 1989. A theory of the spatial and temporal dynamics of plant communities. *Vegetatio* 83:49–69.

Smith, T.M., and Urban, D.L. 1988. Scale and resolution of forest structural pattern. *Vegetatio* 74:143–50.

Urban, D.L., O'Neill, R.V., and Shugart, H.H. Jr. 1987. Landscape ecology. *Bioscience* 37:119–27.

Vlad, M.O. 1988. A new nonlinear model for the growth of age-structured populations living in patchy environments. *Ecological Modelling* 43:251–69.

Vollenvieder, R.A. 1970. Models for calculating integral photosynthesis and some implications regarding structural properties of the community metabolism of aquatic ecosystems. In *Prediction and Measurement of Photosynthetic Productivity*, pp. 455–72. Wageningen, The Netherlands: Pudoc.

Webster, J.R. 1979. Hierarchical organization of ecosystems. In *Theoretical Systems Ecology*, ed. E. Halfon, pp. 119–31. New York: Academic Press.

Welsh, A.H.; Townsend Peterson, A.; and Altmann, S.A. 1988. The fallacy of averages. *American Naturalist* 132:277–88.

Wiens, J.A. 1976. Population responses to patchy environments. *Annual Review of Ecology and Systematics* 7:81–120.

Wilkie, D.S., and Finn, J.T. 1988. A spatial model of land use and forest regeneration in the Ituri forest of northeastern Zaire. *Ecological Modelling* 41:307–23.

Woodmansee, R.G. 1988. Ecosystem processes and global change. In *Scale and Global Changes: Spatial and Temporal Variability in Biospheric and Geospheric Process*, SCOPE 35, eds. T.R. Rosswall, R.G. Woodmansee, and P.G. Risser, pp. 11–27. Chichester, England: J. Wiley and Sons.

Ziegler, B.P. 1979. *Theory of Modeling and Simulation*. New York: Wiley.

20. Future Directions in Quantitative Landscape Ecology[1]

Robert H. Gardner and Monica G. Turner

20.1 Introduction

Landscape studies have always posed difficult quantitative problems for ecologists. Although many different variables may be measured, repeated observations in time and space are not often available at the landscape scale. This situation results in what Allen and Starr (1982) refer to as the dilemma of middle number systems, which lack a clear means of identifying cause and effect relationships. While these issues are not new to ecology (Hurlbert 1984), the traditional solution is the controlled experiment, which may be difficult or impossible at the landscape scale. When one is embarking on a landscape-level study, therefore, it is prudent to examine the broad spectrum of analysis and simulation methods available and choose those which best address the problem at hand and have the most quantitative rigor.

20.2 Quantitative Analyses and Models in Landscape Ecology

This book presents a diversity of tools and techniques from a variety of fields, offering new methods and approaches to the quantitative analysis problem at the landscape scale. This comes at an appropriate juncture in ecological research

[1]This research was funded by the Ecological Research Division, Office of Health and Environmental Research, U.S. Department of Energy, under contract DE–AC05–84OR21400 with Martin Marietta Energy Systems, Inc. Publication No. 3542, Environmental Sciences Division, Oak Ridge National Laboratory.

because the availability of key techniques affects the kinds of questions that are asked. This feedback is important—the advent of new techniques helps stimulate new avenues of research, which, in turn, generate new questions. However, a note of caution is appropriate regarding new approaches. It is important to recognize that the inappropriate application of a technique can result in incorrect answers to important questions. Thus, as with any analysis, the method must be suited to the question being addressed.

Technological improvements have greatly broadened the scope of questions that can be addressed in landscape studies. The revolutions in computer and remote sensing technologies have reduced computational expenses and provided new sources of spatially extensive data. This is fortuitous because many current environmental issues demand information and analysis at broad spatial scales. For example, projecting the spread of a disturbance or the ecological effects of climatic change at the regional scale requires landscape-level data and sophisticated analysis methods. Current methods presented in this book now allow rapid answers to many questions (e.g., what are the land cover patterns in a region? how has the landscape changed through time?) and provide a foundation for more complex investigations concerning the nature of landscape change and potential interaction among the multiple factors that contribute to these changes. As demonstrated in this book, theory and models are now available to predict the relationship between landscape pattern and ecological processes, and many specific hypotheses can be tested against available data.

Landscape processes occur in a spatially heterogeneous context. The multispectral information provided by remote sensing (Quattrochi and Pelletier, Chapter 3; Musick and Grover, Chapter 4; and Luvall and Holbo, Chapter 6) has altered the way that we view and quantify this heterogeneity. The possibility of taking repeated measures at broad scales, combined with methods that are able to handle extensive data sets, has only begun to influence the type and scale of ecological problems that can be tackled. The necessary remote sensing techniques to measure specific processes, such as evapotranspiration, across landscapes (Luvall and Holbo, Chapter 6) will allow fine-grained physiological responses to be projected through time and space.

A host of statistical methods for the analysis of broad-scale data are now available (S.J. Turner et al., Chapter 2). These techniques, combined with an understanding of the key components relating pattern and process, will allow us to determine the appropriate spatial and temporal scales for study. In addition, new methods based on photogrammetric techniques of textural analysis by Musick and Grover (Chapter 4) provide an especially nice method for analyzing edges in continuous gradient data.

Geographic Information Systems (GISs) have become an invaluable tool for the analysis of landscape data. The development of GISs for microcomputers has reduced costs and increased the availability of methods to store, retrieve, and analyze spatial data. The evolution of these methods has resulted in a suite of computer-based tools for problem solving and decision making (Coulson et al., Chapter 7), adding needed power and speed to the processes of data analysis and hypothesis testing. Because the GIS allows a massive amount of information to be manipulated and key landscape attributes to be viewed easily, species-specific questions can be addressed at landscape scales (Dunn et al., Chapter 8).

The quantitative difficulties posed by the problem of assessing the ecological effects of landscape heterogeneity on biogeochemical cycling mandates a simple approach to empirical studies. Shaver et al. (Chapter 5) found that the effect of spatial heterogeneity in element cycling at the landscape level is best studied in simple systems, such as the arctic. The result of these empirical studies is a conceptual model for analyzing and quantifying the importance of element transport between landscape "units," and the approach could be applied to other landscapes.

Methods for spatial analysis of empirical data are also important as tools for the development and testing of landscape models. Although models differ greatly in their representation of space (links and nodes, grids, geometric shapes, etc.) and the complexity of feedbacks between components, all spatial models show that space and time are intertwined and cannot be reduced to two independent components (Sklar and Costanza, Chapter 10). Models of animal population dynamics (Merriam et al., Chapter 16) and herbivory (Hyman et al., Chapter 18) nicely demonstrate that landscape pattern and habitat connectivity affect the spatial and temporal scales at which the problem can be considered.

Gardner and O'Neill (Chapter 11) make the argument that the most effective model for understanding cause and effect relationships at the landscape scale is the simplest model. Several theoretical approaches are suggested, including percolation theory, epidemiology theory, and diffusion-reaction theory. Fahrig (Chapter 17) carries this argument further in the development of spatial population models, giving several examples of a general model for simulating population dynamics in a patchy landscape.

The use of landscape models for testing interactions of biotic and abiotic factors at landscape scales is illustrated with vegetation-soil-water interactions on a barrier island (Rastetter, Chapter 14). A profile through the dune allows a two-dimensional image of a three-dimensional problem to be constructed. Results show that the abiotic processes of dune formation are an integral and dynamic part of succession, demonstrating spatially interactive elements that form the landscape. Therefore, it is necessary to measure and incorporate large-scale processes as dynamic parts of the system, not as extrinsic controlling factors.

Economic factors are often the dominant forces affecting landscape change (Parks, Chapter 12). Conversely, the pattern of resources in a landscape often dominates economic development. However, ecologists often ignore the links between economic theory and ecological studies. Parks's review of economic approaches and examples of their use is helpful for choosing a model for particular questions linking ecology and economics. It is clear that much remains to be done in this area; the interactions among economic factors and landscape patterns should be a fruitful new area for study.

The temporal and spatial scales necessary for understanding disturbance-landscape interactions make models a useful tool for analysis (M.G. Turner and Dale,

Chapter 13). Nearly all landscapes are disturbed in some way, but new models and analysis methods are again needed at the broad scales. Turner and Dale provide a summary of approaches to simulating landscape disturbances, including the advantages and limitations of each approach. Elucidating the relationship between landscape heterogeneity and disturbance remains a challenging research area.

Methods derived from fractal geometry have useful landscape applications (Milne, Chapter 9). The importance of fractals in determining scale-related effects is particularly germane to many landscape studies. The effect of fractal patterns on diffusive movement and the existence of critical landscape phenomena can be treated in a general fashion by fractal analysis. Although simple combinations of variables in spatial models can produce complex results that can be difficult to analyze, Milne shows that equally simple methods of analysis can "decompose" these complex patterns into simpler forms.

Bartell and Brenkert (Chapter 15) provide an interesting example of the interface between models and empirical data that is well suited to the quantitative testing of predicted effects. Using a finite difference model of nutrient movement in the vegetation and soil of a watershed, the chapter explores the implications of heterogeneities in vegetation type and topography on the accumulation of nutrients in a forested watershed. Sensitivity analyses demonstrate the time dependence of model processes, with short-term effects due to the internal cycling of nutrients and long-term effect dominated by changes in topography. Because aggregation in space distorts watershed topography, the spatial and temporal scales adopted for model simulations can be critical.

The issues and problems posed by the various spatial and temporal scales of ecological analyses pervade this book. Therefore, it is appropriate that the final chapter discusses in depth the quantitative issues associated with scale, with particular emphasis on translating information across scales (King, Chapter 19). The challenge of developing a rigorous conceptual framework for addressing multiple scales and the quantitative methods for implementing them is a serious and timely challenge. Success in this area will allow us to: (1) identify where critical fine-grain detail is required, (2) develop reliable simulation tools, (3) recommend new information and/or measurements, (4) and extrapolate this information in a rigorous and reliable fashion. Chapter 19 provides new impetus (and theory) to expand the theoretical foundation of landscape ecology as our tools and data base also expand.

20.3 Future Directions

New methods are usually developed to address a particular set of questions. The successes and failures in answering these questions often stimulates the development of the next generation of techniques. A review of the chapters in this book indicates several areas where the development of new techniques would be immediately useful.

The difficulty in quantifying the effects of landscape heterogeneity stems largely from our inability to distinguish, a priori, important relationships from merely interesting ones. The diversity of species, the variety of their responses to changes in resources, and the complex mechanisms needed to explain these interactions demonstrate that all landscape studies must limit the scope and scale of their measurements. But how can this be accomplished without sacrificing our ability to make reliable (and interesting) predictions? New techniques—similar to sensitivity and uncertainty analysis—that allow the importance of different variables and mechanisms to be evaluated before additional measurements are taken would be extremely useful for establishing realistic limits for landscape studies. Although sensitivity methods have been successfully applied in a variety of ecological studies (see Gardner et al., 1990, for a review and Chapter 15 for an example), these methods have yet to be generally and rigorously applied to spatial systems.

Theoretical studies of spatial effects in physical and chemical systems offer several new concepts that may be useful for landscape studies. For instance, the "backbone" is defined in percolation theory as the critical set of sites through which there is flow of material or energy (Stauffer 1985). The concept of a backbone is of interest to landscape studies because it suggests that disturbance of specific sites may have landscape-scale consequences. For instance, removal of sites adjacent to the backbone will not affect the movement of organisms, but the removal of a single site from the backbone itself can effectively disconnect the cluster and thus disrupt both pattern and process at the landscape scale.

Another statistic that might be adapted from percolation theory is the "correlation" or "connectivity length." The correlation length is estimated by taking the landscape average of the distances between similar sites that belong to the same cluster. Theoretical studies show that sudden increases in the correlation length occur near the critical thresholds in landscape connectivity (Stauffer 1985). Because the abundance of many species can be related to the number, size, and connectivity of suitable habitat sites, the correlation length should provide a useful index describing landscape-scale changes in abundance and diversity of organisms. For instance, sudden reductions in the correlation length as a result of change in landscape heterogeneity (i.e., disturbance) should indicate equally sudden changes in species abundances.

Although much work has been performed in cellular automata (see Casti 1989 for a recent review), these theoretical methods have not been applied to the problem of spatial heterogeneity at landscape scales. Recent simulations of cellular automata that allow the spatial and temporal scales of effects to be varied (Gerhardt et al. 1990) indicate real potential for estimating the broad-scale effects of changes in fine-grained details.

If one could make a single quantitative wish, one might request the development of a technique (or techniques) that would establish, a priori, what the relevant spatial and temporal scales of measurement and prediction should be. Continued interest in this problem is evident (see Dale et al. 1989), and progress is being made in a variety of disciplines (e.g., Chapter 9). Hierarchy theory (Allen and Starr, 1982; O'Neill et al. 1986) predicts that complex systems, including landscapes (Urban et al. 1987) should develop structures that are hierarchically organized—that is, patterns that show significant shifts with changes in scale. Although this effect has been observed in several empirical studies (Anderson 1971; Krummel et al., 1987; O'Neill et al., submitted), it remains a challenge to establish the spatial and temporal scales at which specific mechanisms (e.g., variability in soils, topography, climate, economics, species life-history attributes, etc.) will have the most influence on our ability to measure and predict.

Spatial models must be tested with spatial data. We expect to see rapid progress in the integration of models with GISs, which will allow predictions to be compared with landscape data. This development is both exciting and quantitatively challenging. If the power of the GIS is used to develop a "best fit" between data and predictions, then the spatial display of prediction errors will not be useful. However, if data are divided into separate sets for model development, calibration, and testing, then the spatial attributes of model errors will be extremely interesting. An analysis of these errors, similar to the analysis of regression residuals, should establish confidence limits for predictions and indicate where new measurements will most improve our understanding and prediction of spatial phenomena.

The general relationships between landscape indices, ecological processes, and scale need more study to provide understanding of both the factors that create pattern and the ecological effects of changing patterns on processes. Quantitative measures of landscape heterogeneity can provide appropriate metrics for monitoring regional ecological changes. These applications are of particular importance because changes in landscape patterns (e.g., in response to global change) can be measured with remote-sensing technology, and an understanding of the pattern-process relationship will allow functional changes to be inferred.

Future research should be oriented toward testing hypotheses in actual landscapes. Methods for characterizing landscape structure and predicting changes are now available, but the landscape questions require creative solutions to experimental design. Theoretical and empirical work should progress jointly, ideally through an iterative sequence of model and field experiments. Natural experiments, such as disturbances that occur over large areas or regional development, also provide opportunities for hypothesis testing. Of paramount importance is the development and testing of a general body of theory relating pattern and process at a variety of spatial and temporal scales (Turner 1989).

20.4 Guidelines for Obtaining Robust Results

Studies of spatial systems show that even simple interactions can display a bewildering variety of behaviors dependent upon the spatial arrangement of the component parts. Because landscapes are spatially heterogeneous, it is necessary to understand not only the mechanisms of interaction among the individual components but also how these components are spatially arrayed. In the past this has been difficult because of insufficient data, limitations of equipment, and inadequacies of theory. These objections are now being met to produce a revolution in the way we think about and deal with the great diversity of landscape issues.

We believe that landscape studies will continue to develop and apply new methods for quantifying spatial heterogeneity, comparing patterns and processes in different landscapes, and predicting the broad-scale effects of changes in pattern and process. Because landscape ecology is complex, the development and use of new and untried quantitative methods can lead to spurious results. Therefore, we encourage the continued development of methods that: (1) result in the identification of key landscape components, (2) quantify the spatial and temporal scales over which reliable predictions can .be made, (3) use simulation methods to indicate where critical information and measurements are required, and (4) verify predictions by comparison with existing data.

20.5 Summary

This book presents a diversity of tools and techniques from a variety of fields, offering new methods and approaches to the quantitative analysis problem at the landscape scale. In this chapter, we present an overview of these approaches along with caveats for their application. Future directions in both quantitative methods development and landscape studies are identified. Research needs that are of particular importance include the application and testing of available theory and the identification of appropriate scales for measurement and prediction.

References

Allen, T.F.H., and Starr, T.B. 1982. *Hierarchy*. Chicago: University of Chicago Press.

Anderson, D.J. 1971. Spatial patterns in some Australian Dryland plant communities. In G.P. Patil and W.E. Waters, eds. Statistical Ecology, Volume I. Spatial Patterns and statistical distributions. pp. 271–286. University Park, PA: Pennsylvania State University Press.

Casti, J.L. 1989. *Alternate Realities: Mathematical Models of Nature and Man*. New York: John Wiley and Sons.

Dale, V.H.; Gardner, R.H.; and Turner, M.G. 1989. Predicting across scales: comments of the guest editors of Landscape Ecology. *Landscape Ecology* 3:147–51.

Gardner, R.H.; Dale, V.H.; and O'Neill, R.V. 1990. Error propagation and uncertainty in process modeling. In *Process Modeling of Forest Growth Responses to Environmental Stress*, eds. R.K. Dixon, R.S. Meldahl, G.A. Ruark, and W.G. Warren. Portland, Oregon: Timber Press.

Gerhardt, M.; Schuster, H.; and Tyson, J.J. 1990. A cellular automaton model of excitable media including curvature and dispersion. *Science* 247:1563–66.

Hurlbert, S.H. 1984. Pseudoreplication and the design of ecological field experiments. *Ecological Monographs* 54:187–211.

Krummel, J.R.; Gardner, R.H.; Sugihara, G.; O'Neill, R.V.; and Coleman, P.R. 1987. Landscape patterns in a disturbed environment. *Oikos* 48:321–24.

O'Neill, R.V.; DeAngelis, D.L.; Waide, J.B.; and Allen, T.F.H. 1986. A hierarchical concept of ecosystems. Princeton, NJ: Princeton University Press.

O'Neill, R.V.; Turner, S.J.; Cullinan, V.I.; Coffin, D.P.; Cook, T.; Conley, W.; Brunt, J.; Thomas, J.M.; Conley, M.R.; and Gosz, J. Multiple landscape scales: an intersite comparison. Landscale Ecology (submitted).

Turner, M.G. 1989. Landscape ecology: the effect of pattern on process. *Annual Review of Ecology and Systematics* 20:171–97.

Urban, D.L.; O'Neill, R.V.; and Shugart, H.H. 1987. Landscape ecology. *BioScience* 37:119–27.

Index

Ecological Studies

Volumes published since 1995